World Symposium on Mechanical-Materials Engineering & Science (WMMES 2021)

IOP Conference Series: Materials Science and Engineering
Volume 1190

Prague, Czech Republic
9 – 11 September 2021

Editors:

Lubos Kucera
Levent Cenk Kumruoglu

ISBN: 978-1-7138-4650-5
ISSN: 1757-8981

Printed from e-media with permission by:

Curran Associates, Inc.
57 Morehouse Lane
Red Hook, NY 12571

Some format issues inherent in the e-media version may also appear in this print version.

Institute of Physics
Dirac House, Temple Back
Bristol BS1 6BE UK

Phone: 44 1 17 929 7481
Fax: 44 1 17 920 0979

techtracking@iop.org

Additional copies of this publication are available from:

Curran Associates, Inc.
57 Morehouse Lane
Red Hook, NY 12571 USA
Phone: 845-758-0400
Fax: 845-758-2633
Email: curran@proceedings.com
Web: www.proceedings.com

TABLE OF CONTENTS

IOP Conf. Series: Materials Science and Engineering 1190 (2021) 011001 doi:10.1088/1757-899X/1190/1/011001

World Symposium on Mechanical-Materials Engineering & Science - WMMES 2021

Lubos Kucera [1], Levent Cenk Kumruoglu [2]

[1] University of Zilina, Zilina, Slovakia
[2] Iskenderun Technical University, Iskenderun, Turkey

editor@wmmes.org

Preface

This volume of "IOP Conference Series: Material Science and Engineering" contains the papers presented in the "World Symposium on Mechanical-Materials Engineering & Science" in Prague (Czech Republic) during 9-11 September 2021. The aim of the WMMES 2021 to provide a forum for discussion of the latest findings and technologies in different fields of Mechanical Engineering, Materials Engineering and other related disciplines, to give opportunities for future collaborations, to be a platform for sharing knowledge and experiences in the fields of Mechanical Engineering, Materials Engineering and other related disciplines, and to lead in providing a forum for early career researchers for presentation of their work and discussion of their ideas with experts in different fields of Mechanical Engineering and Materials Engineering.

However, the planned period of the 1st WMMES was 2020, it was postponed to 9-11 September 2021 because of the Covid-19 pandemic, and held in Prague (Czech Republic) by appearing in person of the participants, not on-line format.

All the missions of World Symposium on Mechanical-Materials Engineering & Science – WMMES 2021 were completed, and 35 peer reviewed articles were published. We would like to thank all reviewers for their great effort to achieve the high quality of published articles.

We would like to show our deep gratitude to the honored Scientific Committee given to be list below and Institutional Scientific Partners of WMMES. The Scientific Committee and Institutional Scientific Partners of WMMES was completed by paying strict attention, and all members were selected from well-known, very much appreciated, productive and representatives of the different countries.

We thank to the sponsors of WMMES 2021 (as main sponsors LAMA Energy Group, LAMA Gas and Oil, Cihelna and to be co-sponsor Prague City Tourism) for their very much appreciated supports during the organization of " World Symposium on Mechanical-Materials Engineering & Science – WMMES 2021". Of course, we would like to express my sincere gratitude to all participants different countries all over the world for their interests and contributions in the 1st WMMES 2021. Another thanks goes to the team of WMMES for their very much appreciated high workload during year.

We hope to see the participants of WMMES 2021 once again in 2nd WMMES 2022 that will be held in Prague (Czech Republic).

IOP Conf. Series: Materials Science and Engineering 1190 (2021) 011001 doi:10.1088/1757-899X/1190/1/011001

Committee Members

Prof.Dr. Dimitrios G. AGGELIS
VUB (Vrije Universiteit Brussel), BELGIUM

Prof.Dr. Hashem AKBARI
Concordia University, CANADA

Prof.Dr. Dumitru BALEANU
Institute of Space Sciences, ROMANIA - Cankaya University, TURKEY

Prof.Dr. Antonio CAGGIANO
Technische Universität Darmstadt, GERMANY

Prof.Dr. Daolun CHEN
Ryerson University, CANADA

Prof.Dr. Franco CONCLI
Free University of Bozen, ITALY

Prof.Dr. Francesco DELL'ISOLA
Sapienza Università di Roma, ITALY

Prof.Dr. Luís Pinto FERREIRA
School of Engineering, Polytechnic of Porto, PORTUGAL

Prof.Dr. Mark J. JACKSON
Kansas State University, USA

Prof.Dr. Jaroslav KOVACIK
Slovak Academy of Sciences, SLOVAKIA

Prof.Dr. Duc Truong PHAM
University of Birmingham, UK

Prof.Dr. Ramesh SINGH
University of Malaya, MALAYSIA

Prof.Dr. Ted STATHOPOULOS
Concordia University, CANADA

Prof.Dr. Faris TARLOCHAN
Qatar University, QATAR

Prof.Dr. Songye ZHU
The Hong Kong Polytechnic University, HONG KONG

Dr. Krzysztof Kamil ZUR
Bialystok University of Technology, POLAND

(*listed in alphabetical order of surname*)

IOP Conf. Series: Materials Science and Engineering 1190 (2021) 011002 doi:10.1088/1757-899X/1190/1/011002

Peer review declaration

All papers published in this volume of IOP Conference Series: Materials Science and Engineering have been peer reviewed through processes administered by the Editors. Reviews were conducted by expert referees to the professional and scientific standards expected of a proceedings journal published by IOP Publishing.

- **Type of peer review: Double-blind. All articles (62) had not been sent to the reviewers when they were submitted, and the review process had been completed in 2 rounds. In the 1st round, the articles were evaluated by means of their originality and suitability. These selected 57 articles were taken under consideration and sent to the reviewers.**
  Each of the selected 57 articles was first sent to at least 2 reviewers, and their comments were then sent back to the author for the requested revisions. After the completion of the revisions, the revised paper was sent back to the reviewers for the last check. By considering the reviewers' comments, decision was done.

- **Conference submission management system: On-line submission system of WMESS was used for the submission of the articles. However, the review process was managed by sending the articles to the reviewers by e-mail accounts of the editorial office (editor@wmmes.org). The chairmen (Prof.Dr. Isik Yilmaz – isik.yilmaz@gmail.com, Prof.Dr. Marian Marschalko – marian.marschalko@gmail.com, Prof.Dr. Marian Drusa – md525d@gmai.com) were responsible for the review process.**

- **Number of submissions received: 62**

- **Number of submissions sent for review: 57**

- **Number of submissions accepted: 41**

- **Acceptance Rate (Number of Submissions Accepted / Number of Submissions Received X 100): 66%**

- **Average number of reviews per paper: 2**

- **Total number of reviewers involved: 46**

- **Any additional info on review process: No**

- **Contact person for queries: Isik Yilmaz. isik.yilmaz@gmail.com**

IOP Conf. Series: Materials Science and Engineering 1190 (2021) 012001 doi:10.1088/1757-899X/1190/1/012001

Attributes relevant for sustainable additive manufacturing – material driven approach

Daniela Mladenovska[1], Ana M. Lazarevska[2], Magdalena Krstanoski[3]

[1]Faculty of Technical Sciences, Mother Teresa University, Mirce Acev 4, Skopje, Republic of Macedonia

[2]Faculty of Mechanical Engineering, Ss Cyril and Methodius University, Karpos II, Skopje, Republic of Macedonia

[3]Everglades University, 6001 Lake Osprey Dr, Sarasota, FL 34240, U S A

daniela.mladenovska@unt.edu.mk

Abstract. Additive manufacturing (AM) covers variety of applications in industrial manufacturing, healthcare industry, as well as 3D home printing desktop applications. It has significant contribution in the manufacturing technology development, while simultaneously creates an impact on the economy of scale. The production cost of AM is almost linear, regardless of the size of the production batch. Furthermore, an increasing parts complexity does not necessarily means an increase of the production costs. It offers new degrees of freedom in the product design, enables reengineering and redesign, and facilitates resource optimization (energy and materials) and waste reduction. At the same time, it promotes reducing the number of assembly operations while contributing to more energy efficient and more cost-effective production. As one of the mostly spread AM technologies, which is quite limited in terms of material selection but has numerous advantages, e.g. during crisis that imply supply chain disruptions, the focus of this paper is on the 3D printing features. In particular, with the assistance of Multi Attribute Analysis, this study elaborates and identifies relevant attributes describing the 3D printing process and its features through the perspective of materials susceptible to 3D printing. The holistic approach presented in this paper introduces a new perspective for decision makers in the field of 3D printing, because in addition to the traditional material selection process, introducing attributes relating only to cost and manufacturability, it includes attributes describing the extended sustainability paradigm. The approach presented herein, categorizes attributes describing the 3D printing process in five main groups, i.e. economic, social and environmental (corresponding to the sustainability paradigm), amended by the technical and legal aspects.

1. Introduction

The contemporary manufacturing industry is facing dramatic changes especially with the recent fast developments in the digital technologies. One of the most significant moments of this change is the invention of advanced manufacturing technologies, such as Additive manufacturing (AM), an important driver in terms of small-scale production enabling improvements both in cost and resource efficiency. [1] Namely, the basic principle of AM technology is that a model, initially generated using a three-dimensional Computer Aided Design (3D CAD), can be produced directly without process planning, which usually requires complex and expensive tools and machines. AM technology significantly simplifies the process of producing complex 3D objects directly from a CAD model (digital model), without the need of traditional tools specialized for separate operations (casting, forging, etc.). Following the 3D model features, the product is manufactured directly on the designated building platform, by adding layer over layer of a selected material. [2]

The variety of AM technologies, are classified by a corresponding variety of criteria. One of the classification approaches is according to the used baseline technology (lasers, printer technology, extrusion technology, etc.). [3, 4] Another approach based on layer manufacturing (LM) process and type of used material [5] is presented on figure 1. [2] Although 3D printing falls under a broader set of AM technologies, among the wider public and even professionals, not completely justified, AM is often referred to as 3D printing.

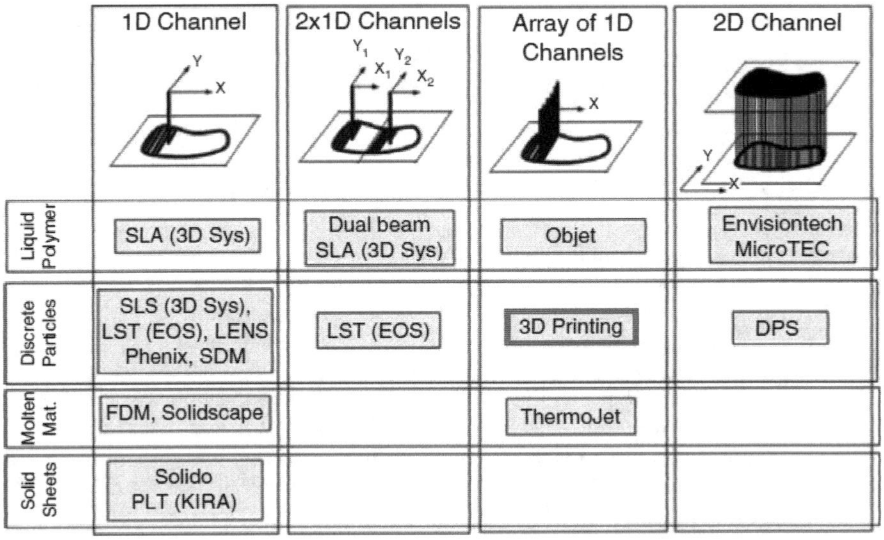

Figure 1. Additive Manufacturing (AM) classification based on Layered Manufacturing (LM) processes and used materials [2][1]

Further development of AM, as well as, other advanced manufacturing technologies will significantly affect the supply and value chains. In terms of reconfiguration of the value chains, shorter and simpler supply chains, more localized production, innovative distribution models and new forms of collaboration, 3D printing significantly influences the structure of material supply chains. Thus, 3D printing affects not only the economy of scale, but also the economy of scope as it reduces the need to hold large inventory. [6, 7]

Since in the context of Industry 4.0, AM technology emerges as one of the key concepts of the next decade, the corresponding aspects of sustainability are very important. Apart from the economic aspects, to date, investigations by researchers into the sustainability implications of AM have either been done at a broad level [7, 8] or have been highly focused on the issue of material and energy consumption. [9, 10] AM technology originally used polymeric materials, waxes and paper laminates. The introduction of composites, metals, and ceramics, was a significant step forward. [2] Materials have been carefully selected to closely suit the operating parameters of different processes and to provide better quality and accuracy of the final parts. In addition, the enriched set of available materials, have resulted in the processes being tuned to produce goods that withstand higher temperatures, smaller feature sizes, and improved productivity. On the other side, constant enriching

[1] DPS – Discrete Particles Scheme; FDM - Fused Deposition Modeling; SLA - Stereolithography; SLS - Selective Laser Sintering; SDM- Shape Deposition Manufacturing; PLT - Paper Lamination Technology

of the set of available materials for 3D printing increases the complexity of the material selection process. AM offers several new degrees of freedom in the product design and production processes. The additive nature of this technology (adding layer over layer), means that products with complex inner structures could be produced (printed) at no additional cost. [11] This enables designing and manufacturing more complex structures (enriching the design spectrum, material saving and weight reduction, reducing assembly efforts in part production, etc.). [12] Simplified assemblies with less material diversity, improve the opportunities for recycling.

The CoViD-19 pandemic brings out the importance of localized, small scale (and on demand) production of variety of products (final or intermediate, such as protection shields or their parts, valves for respirators or their parts, etc.) which have a tremendous impact on the health system support. Moreover, the shift toward a decentralized manufacturing implies that the environmental impact of goods transport is reduced, while supporting and empowering local communities. In addition, AM allows conversion of waste and by-products into new products. [1]

This paper focuses on the specifics of the 3D printing, widely used, not only by professionals but by numerous of enthusiasts in the field of digital technologies. 3D printing, requires relatively simple equipment (home 3D printers), and it is predominantly based on the knowledge and experience of the skilled enthusiasts. Exploring the topic of 3D printing through the attributes of the materials used, a more comprehensive understanding of 3D printing sustainability is provided via multicriteria decision analysis (MCDA).

2. Methodology: Multi criteria decision analysis

Decision-making is the study of identifying and choosing alternatives to find the best solution based on different attributes/criteria, while considering decision-makers' preferences. Materials selection process is a multifold problem, which optimally should result in finding the best option comprising material properties and design features with fulfilment of the products life cycle sustainability requirements. [13] The basic procedures for addressing conflicting and intrinsically not comparable aspect/requirements, are covered by Saaty [14], Goicoechea and Duckstein [15], etc., while their solution is offered via multi-criteria decision analysis (MCDA). This major class of methods is further divided into Multi-objective decision-making (MODM) and Multi-attribute decision-making (MADM). [16] MODM involves multiple conflicting objectives, while MADM deals with small number of alternatives and their assessment against a set of attributes, often hard to quantify. [17] These methods are favorable when decisions are to be delivered in challenging and complex environment and derive solid results. It should be emphasized that different methods and the correspondent results are not necessarily comparable. The first step in performing MCDA is problem definition, performed by setting the main goal, identifying attributes and sub-attributes, as well as considering relevant constrains and uncertainties. Criteria relevant for the decision are defined based on the decision-makers preferences (in this case the manufacturer/designer as in 3D printing those two roles are interwoven). Furthermore, the alternatives are defined, representing possible options for reaching the main goal. Once the model hierarchy is created, the next step is to determine the weights of the criteria against which the alternatives should be assessed and ranked. The criteria weights present their relevant importance versus each group of attributes and sub attributes. After ranking the alternatives, the set of the top - ranked, shall be accepted as an optimal solution addressing the main goal of the predefined problem.

The herein presented work focuses on defining the problem hierarchy and identifying the attributes, sub-attributes and alternatives. Thus, material driven identification of relevant attributes for sustainable 3D printing practically implies evaluating material alternatives versus the attributes describing the extended sustainability paradigm (technical, economic, social, environmental and legal attributes).

3. Selection of material for 3D printing via the extended sustainability paradigm

Sustainable manufacturing, in general, denotes manufacturing products by means of processes with minimal negative environmental impact, remained resource efficiency (energy, water, materials input), employees, communities and consumers safety, while contributing towards improved societal benefits. From the perspective of material usage, 3D printing (as well as AM), in general, is more efficient. Namely, based on layer-by-layer fabrication, material waste in 3D is significantly reduced. [18]. In addition, the properties of the supporting material utilized to make increasingly possible complex geometries, allows recycling into raw material, after the 3D printing of the desired product is finished. Meanwhile, many CNC machined (and in general in subtractive/traditional production) parts are produced from block, which is always significantly bigger than the produced part, leading to up to 19:1 ratios of material waste versus the material embedded in the final part. [19, 20] Besides the fact that 3D printing is inherently less wasteful than the traditional subtractive methods, it has a potential to decouple social and economic value creation from the environmental impact of business activities. [2, 19]

Although the materials used for this technology are not always greener than those used in traditional manufacturing, its sustainability advantages are obvious, regardless of the used material. [10] The material analysis throughout the product Life Cycle Assessment (LCA), significantly affect the sustainability of the overall 3D printing process. Moreover, the fact that the majority of the used materials are based on thermoplastics (extrusion process) [21], originating either from fossil fuel (mostly) or bio-based feedstock (in rare cases), the material analysis versus the herein utilized extended sustainability paradigm gains in complexity. This complexity drives the increased development of a whole range of new materials suitable for 3D printing. Since the main characteristics of interest in terms of manufacturability are the melting temperatures, melting viscosity and coagulation time, the two dominant thermoplastics are acrylonitrile butadiene styrene (ABS) as a fossil-based plastic and polylactid acid (PLA) as a plastic. In relation to 3D printing, combined with (bio) additives to create special properties, a whole range of bio plastics is under development. [22]

Material selection suitable for sustainable 3D printing process, commences with identification of the relevant material attributes and sub-attributes, which correspond to the extended sustainability paradigm. The vision for sustainability encapsulated in the Brundtland Report [23], was ground breaking, but recently numerous authors address the need to extend it on a broader context, having in consideration public policies and legislation, corporate decision-making, and other causes of sustainability threats. [24] Namely, 3D is closely interlinked with legal aspects of this process, focused on intellectual properties, patent law, standardization and consumer protection. [12] On the other side, according Ashby & Johnson [25] classical theory, materials attributes (based on designer perspective) are categorized as basic, technical, aesthetical and environmental. Thus, the material selection should be carefully performed considering not only the availability, costs, (material costs, machining costs, man hour costs, etc.), manufacturability, as well as taking into account the environmental and social impact. The decision maker (in this case, both in the role of designer and manufacturer) should be equally responsible and aware for the product quality, price and its environmental impact. The manufacturing constrains, customer requirements and design guidelines should be balanced in a way that minimizes overall (direct and indirect/external) costs and maximizes product quality and social benefits. [13] The technical aspects present a set of features affecting product quality in the terms of accuracy, surface finish, product lifetime, etc., but in the same time simultaneously encompass certain limitations of product size and shape. Thus, carefully performed material selection is very important precondition for optimal design and manufacturing process, and for obtaining accepted and durable final product with minimal environmental impact. Merging all aforementioned aspects results in the s.c. extended sustainability paradigm, which combines the classical definition of sustainability with the classical theory of Ashby & Johnson for classifying material attributes [25], while upgraded by adding legal aspects, as it is shown on figure 2. Thus, the hierarchy structure (model) of the analysed problem, encompasses attributes and sub-attributes arranged in two hierarchy levels, having in mind the numerous and often conflicting criteria specified by the stakeholders. The first hierarchy level lists the five classes of material attributes (technical, economical, environmental, social and legal), in line with the above proposed extended sustainability paradigm. [12, 25, 26]

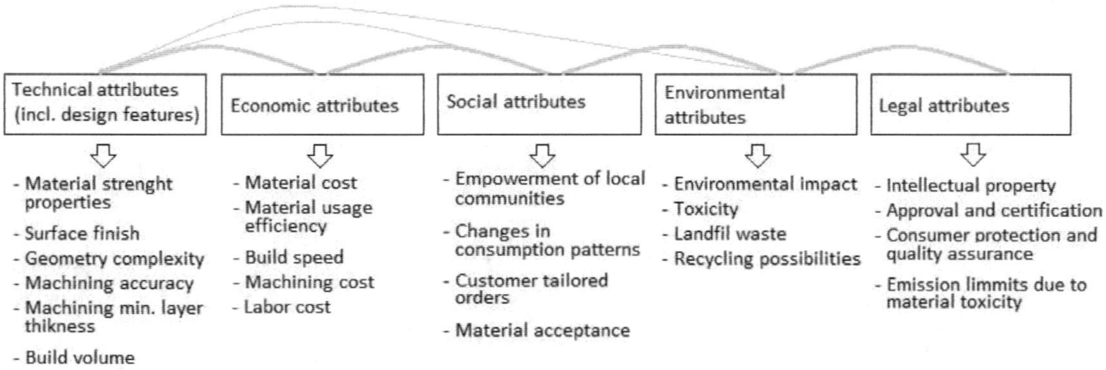

Figure 2. Problem hierarchy in line with the extended sustainability paradigm consisting of five categories of material attributes[2]

3.1 Identification of attributes and sub-attributes

Technical attributes and sub-attributes. Among the identified five main group of attributes, this group gains in complexity because the material properties, machining features and design possibilities are strongly interwoven. E.g. manufacturability depends on the material type, but is also related to the geometry complexity as well as the equipment limitations, whereby all these factors vary significantly for different materials. Further, the design process is a synthesis of numerous factors based on the material properties and manufacturing possibilities, aimed towards obtaining a product suitable for the life-cycle objectives. Moreover, different possibilities exist contributing for an extended product life (based on the type of used materials), by using technical approaches such as repair, remanufacture and refurbishment. On the second level, the following technical sub-attributes are identified:

- *Material strength properties.* (e.g. Thermal expansion Coefficient (K-1), Young Modulus (MPa), Material Yield Strenght (MPa), Hardening factor (-), Anisotropic Strength Coefficients (-). These parameters define material behavior during the manufacturing process, also influence the final product quality and durability)
- *Surface finish.* (as related with the materials microstructure, anisotropy, fatigue cracks appearance and growth, defects, etc. It also depends from the manufacturing accuracy. Product manufacturing that does not require post processing is a significant driver towards sustainable 3D printing.)
- *Geometry complexity.* (Geometry limitations are present during 3D printing, and should be considered combined with the material properties. This sub-attribute includes recommended values in terms of wall thickness to assure accuracy (mm), minimum and maximum feature size (mm), minimal holes size (mm), machining allowance (mm) as well as the aspect ratio – the maximum ratio between the print height and width.) [27]
- *Machining accuracy.* (3D scanning techniques are usually used for machining accuracy process comparison. One of the most used method to determine the deviation between the products from different materials is the Root Mean Square (RMS)(μm).
- *Machining minimum layer thickness.* (Machining or printing minimum layer thickness (mm) is related with the materials strength properties)

[2] Yellow lines denote some of the correlations between different groups of attributes

- *Build volume.* (Limits the building platform space, and affects the product shape, size and cooling properties, depending of the usage of different material)

Economic attributes and sub-attributes. These attributes are predominantly related with material purchase and labor costs, but they also affect material efficiency rates due to the build speed (required time for production). Latter sub-attribute is again strongly correlated with the material properties (temperature distribution along the build product, cooling time and prevention of defects), thus the optimal manufacturing parameters should be adjusted based on the specific material features. Furthermore, by using the material properties as advantage, cost reductions could be obtained if the product shape is modified to fully exploit the 3D efficiency potential.

Social attributes and sub-attributes. All technologies are developed and adopted within the complex network of existing infrastructure, technologies, behaviors, norms and attitudes of its constituent parts. Employment and the distribution of labor, health and safety, ethics, quality of life, community empowerment, creativity and self-expression are just some of the aspects [1] Material properties and sub-attributes related with the surface finish influence the material (product) acceptance and the mainstream in consumption patterns. On demand and customer tailored manufacturing is a serious advantage when used materials are easy affordable and available.

Environmental attributes and sub-attributes. The environmental impact of the selected material can significantly vary depending on origin of the materials and chemical composition. Concerns such as toxic vapors related to some materials. They affect the occupational health of the workers, but could also have an influence in terms of product acceptability, despite their formal alignment with the legal requirements. Landfill waste and recycling possibilities are correlated sub-attributes, since the materials have different recycling potential. In addition, the available recycling infrastructure and know-how for recycling processes of each material type are very important, having in consideration that some of the materials are compostable. While in most regulations mechanical recycling is the preferred end-of-life solution, energy recovery via incineration is preferred over land filling. [28]

Legal attributes and sub-attributes. The decentralized market for production enabled by 3D printing unfortunately facilitates a decentralized market for piracy. In traditional manufacturing, the copying of a design can be readily traced to a source since an infringer requires an infrastructure for manufacturing. Intellectual property and patent law issues relating to the traditional manufacturing are more-or-less common for all types of materials used. Moreover, in terms of used material features, the aspects like the toxicity, consumer protection and quality assurance are in particular challenging for the legal framework.

3.2 Identification of alternatives (materials)

A set of materials for 3D printing is listed (as alternatives) in order to be assessed versus the above noted group of attributes. The list of materials is presented on figure 3.

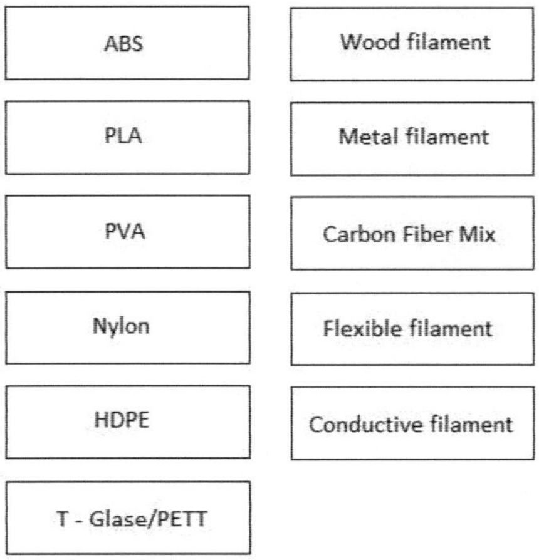

Figure 3. List of materials (alternatives) for 3D printing

The proposed list of materials is based on the focus of this paper i.e. 3D printing for customized manufacturing in small batches, where the manufacturer as well is the designer. The identified materials are easy affordable and available compared to some other more advanced options. At the same time, the intention was to extend the list with as much as possible material types that are fitting the purpose. The sustainability of the 3D printing materials is a challenge since predominately they originate from petroleum products (ABS, PVA, Nylon, HDPE, T-Glass/PETT, etc.) PLA is a biodegradable eco friendly material whose application is mainly concentrated in the food industry. The second column in the figure 3, lists the filaments materials that combine the properties of two of more materials in order to exploit their advantages (composites). Typically, wood filaments contain 70% PLA and 30% wooden fibres, but it all depends on the filament manufacturer. Metal filaments, also have PLA as a base, and contain very fine metal powder such as copper, bronze, brass, and stainless steel. The percentage of metal powder infused in each filament can vary depending on the manufacturer. The presence of the metal powder makes the filament much heavier than standard plastics, but provides much better strength properties. Conductive filament is a new emerging material type that is great for small electronic projects or small electronic circuits. "Conductive" means that electricity can flow through it. This filament often has PLA as a base material, but other materials (ABS, for example) are also used. Carbon fiber filaments are also composite materials that could increase the strength of the printed product. By means of combining the features of thermoplastics and elastomers, flexible filaments are obtained. They are soft and flexible, recyclable, resistant to chemicals and resistant to low and high temperatures. [29] Seen through the extended sustainability point of view, all these materials have their advantages and disadvantages in certain application depending on their cost, manufacturability, strength properties, environmental and social impact as well as legal constrains. Thus, the selection of the most suitable material responsive to all these requirements becomes more and more difficult task for decision makers.

4. Results and discussions

There are number of available and affordable materials for 3D printing applications. Their usage for manufacturing specific product could be fostered or inhibited depending on a complex set of attributes and sub-attributes, relevant for the sustainable material selection to be used for 3D printing. As aforementioned, the *technical attributes* are the most complex, but it must be emphasized as they have a strong correlation and influence over all other classes of attributes. Moreover, there is interdependence among sub-attributes, not only within their paternal class, but as well between the other four comprising the extended sustainability paradigm. Elaboration of the most important and notable interdependences are described below:

- *Technical attributes.* Manufacturability of certain material is in correlation with the equipment features and process parameters, but it also affects the manufacturing costs, energy consumption, waste generation, recyclability, etc. Holding a database of digital designs allows products to be manufactured on demand. Optimization of process parameters in order to control temperature distribution and total deformation of the produced good (especially of complex parts) is also significantly affected by the selected material properties. Design optimization itself enables products improvements resulting in mass reduction. The potential for mass reduction of the products depends on the materials behavior in terms of geometry complexity sub-attributes.

- *Economic attributes.* Economic attributes are defined by material costs (directly) as well as by productivity and resource efficiency (indirectly). They affect not only the direct cost (i.e. material purchase), but they also promote cost reduction by means of mass reduction (via design optimization), manufacturing time (related to the material manufacturability and required operational parameters), post processing and finishing operation requirements, as well as the product end of life costs. Moreover, the availability and easy access to the required (selected) material is very important.

- *Social attributes.* The *social dimension* of 3D printing is very important in terms of local communities' empowerment. Namely, the potential for local availability of the selected material as well as the potential for its local production, could affect the decision for material selection, having in perspective the benefits for the community. In addition, the democratization of production, the combination of information and communication technology (ICT), widely available CAD software and 3D printers is changing consumption pattern. It creates an environment where the communities' cultural specifics and habits have key importance in materials acceptance. However, the consumers' awareness both in terms of material quality and environmental impact must be taken in consideration.

- *Environmental attributes.* E.g. potential material toxicity needs to be analyzed carefully, since it could offset the benefit from materials savings during the manufacturing process. Moreover, the energy usage related with specific material has a significant impact on the processing itself. The material recycling potential as well as locally available recycling infrastructure are also significant aspect of the environmental attributes. Thus, the full environmental performance of the entire product life cycle must be evaluated from a system's perspective.

- *Legal attributes.* *Legal aspects* of the problem are very complex since it involves new technology that requires specific legal and regulatory environment. Protection of consumers and quality assurance are important when it comes to new materials applications. There are still many open questions in this area, reaching from issues of intellectual property to complicated procedures for approval and certification. To enable a successful 3D printing surroundings, these regulatory challenges need to be solved in order to set the right boundary conditions. Since a number of 3D printing materials are toxic and cause toxic emissions during processing, occupational health and safety issues requires a serious review.

The sustainability in the process of 3D printing material selection could be improved not only by means of attributes and sub-attributes identification and analysis, but also by means of research and development of new more sustainable materials. The wider penetration of biodegradable plastics (PLA) and other eco-friendly filaments, will contribute for a more effective shift towards sustainable 3D printing.

5. Conclusions

There are numerous criteria related with the 3D printing material selection in accordance with the extended sustainability paradigm. Some of the criteria are not directly measurable and cannot be quantified. This is an additional burden for the decision makers. Moreover, the greater diversity of available materials brings additional complexity in the decision making analysis. The herein proposed concept identifies relevant attributes and sub-attributes that describe materials. It must be emphasized that these attributes and sub-attributes are closely interlinked both in terms of different classes and into the same class. The most complex is the technical class of attributes since it covers material mechanical properties, machining limitations, design advantages and disadvantages as related with the selected materials. They not only affect time and costs of the manufacturing, but also influence the environmental impact, changes in consumers behaviour and consumption patterns, requiring various levels of legal support. From a material selection perspective, all the above noted attributes have different importance. Due to the complexity of the analyzed problem, and the identified correlations, future work should be focused on quantifying the correlation and addressing this complexity. This extends to a correspondent selection of materials that optimally suit the predefined decision-maker (in this case in the role of both designer and manufacturer) criteria.

Abreviations and acronyms

AM	Additive Manufacturing
LM	Layer Manufacturing
CNC	Computer Numerical Control
ABS	Acrylonitrile Butadiene Styrene
PLA	Polylactid Acid
MCDM	Multi Criteria Decision Making
MCDA	Multi Criteria Decision Analysis
MADM	Multi-Attribute Decision Making
MODM	Multi-objective decision-making
PVA	Polyvinil Alcohol
HDPE	High Density Polyethilene
RMS	Root Mean Square
ICT	Information and Communication Technology

References

[1] S. Ford, and M. Despeisse, “Additive manufacturing and sustainability: an exploratory study of the advantages and challanges,” *Journal for Cleaner Production*, vol. 137, pp. 1573–1587, 2016.
[2] I. Gibson l D. W. Rosen, and B. Stucker, "Additive Manufacturing Technologies: Rapid Prototyping to Direct Digital Manufacturing," *Springer - Verlag*, New York, 2015
[3] J. P. Kruth, M. C. Leu, and T. Nakagawa, “Progress in additive manufacturing nad rapid prototyping,” *Ann CIRP*, vol. 47(2), pp. 525-540, 1998.
[4] M. Burns, “Automated fabrication: improving productivity in manufacturing,” *Prentice Hall*, Engelwood Cliffs, New Jersey, 1993.
[5] C. K. Chua, and K. F. Leong, “Rapid prototyping: principles and applications in manufacturing,” *Wiley*, New York, 1998.
[6] M. Gebler, A. J. M. Shoot Uiterkamp, and C. Visser, “A global sustainability perspective on 3D printing technologies, ” *Energy Policy*, vol. 74(C), pp. 158-167, 2014.
[7] P. Brody and V. Pureswaran, "The new software-designed supply chain", *IBM Institute for Business Value*, New York, pp. 1-18, 2013.
[8] C. Kohtala, “Addressing sustainability in research on distributed production: an integrated literature review,” *Journal of Cleaner Production*, vol. 106, pp. 654-668, 2015.
[9] M. Baumers, C. Tuck, D.L. Bourell, R. Sreenivasan, and R. Hague, “Sustainability of additive

manufacturing: measuring the energy consumption on the laser sintering process," *Proc. of the IMechE*, Vol. 225, *Part B: J. Eng. Manuf.*, pp. 2228-2239, 2011.
[10] J. Faludi, C. Bayley, S. Bhogal, and M. Iribarne, "Comparing environmental impacts of additive manufacturing vs traditional machining via life-cycle assessment," *Rapid prototyp.,* vol. 21(1), pp. 14-33, 2015.
[11] M. Eisenhut, and B. Langerfeld, "Additive manufacturing: A game changer for the manufacturing industry?," *Roland Berger Strategy Consultants Holding GmbH*, Munich, 2013.
[12] L. Bethold, V. Fischer, A. Hainzlmaier. D. Hugenroth, Lj. Ivanova, K, Kroth, B. Römer, E. Sikorska and V. Sitzmann, "3D printing: A qualitative assessment of applications, recent trends and the technology's future potential," *Studien zum deutschen innovations system, Expertenkommision Forschung und innovation (EFI),* Berlin, 2015.
[13] M. Ashby, H. Shercliff, and D. Cebon, "Materials: Engineering, Science, Processing and Design," *Butterworth-Heinemann*, London, 2002.
[14] T. L. Saaty, "The Analytic Hierarchy process: planning, priority setting, recource allocation," *Mc Graww Hill*, New York, USA, 1980.
[15] A. D. Goicoechea, and L. Duckstein, "Introduction to Multiobjective Analysis with Engineering and Bussiness applications," *John Wiley*, New York, USA, 1982.
[16] J. J. Wang, Y.Y. Jing, C. F. Zhang, et al., "Review on multi-criteria decision analysis aid in sustainable energy decision-making," *Renew Sust Energ Rev.,* vol. 13, pp. 2263–2278, 2009.
[17] C. L. Hwang, and K. Yoon, "Multiple attribute decision making: methods and applications," *Springer*, Berlin, 1981.
[18] M. Mani, K. W. Lyons and S. K. Gupta, "Sustainability Characterization for Additive Manufacturing," *Journal of Research of the National Institute of Standards and Technology*, vol. 119, pp. 419 - 428, 2014.
[19] S. Veronneau, G. Torrington and J. P. Hlavka, "3D Printing: Downstream Production Transforming the Supply Chain", *RAND Corporation*, Santa Monica, CA, 2017.
[20] Z. Liu, Q. Jiang, Y. Zang, T. Li and H. C. Zhang, "Sustainability of 3D printing: A critical review and recommendations," *Proceedings of the ASME 2016 International Manufacturing Science and Engineering Conference MSEC2016*, Blacksburg, Virginia, USA, pp. 1-8, 2016.
[21] L. Chaunier, S. Guessasma, S. Belhabib, G. Della Valle, Denis Lourdin and E. Leroy, "Material extrusion of plant biopolymers: Opportunities & challenges for 3D printing," *Additive Manufacturing, Elsevier,* vol. 21, pp. 220 - 233, 2018.
[22] A. van Wijk, and I. van Wijk, "3D printing with biomaterials towards a sustainable and circular economy," *IOS Press, imprint Delft University Press*. (available at: http://ulib.iupui.edu/static/pdfs/3DPrintingBiomaterials.pdf)
[23] G. H. Brundtland, "Our Common Future: Report of the World Commission on Environment and Development," Geneva, 1987.
[24] P. Johnston, M. Everard, D. Santillo and K. H. Robèrt, "Reclaiming the Definition of Sustainability," *Environmental Science and Pollution Research*, vol. 14(1), pp. 60-66, 2007.
[25] M. Ashby and K. Johnson,"Materials and Design: The art and science of material selection in product design," *Butterworth – Heinemann*, Oxford, 2002.
[26] N. H. Afgan, "Sustainability Concept for Energy, Water and Environment Systems, ", *Springer*, 2008.
[27] B. Redwood, F. Schöffer and B. Garret, "The 3D Printing Handbook: Technologies, design and applications," *3D Hubs B. V,* Amsterdam, 2017.
[28] G. Eriksson and G. Finnveden, "Plastic waste as a fuel - CO_2 neutral or not?," *Energy Environ Sci,* vol. 2, pp. 907-914, 2009.
[29] All3DP, The World's #1 3D Printing Magazine for Beginners and Professionals [Online] 2021 [Accessed on 08.06.2021] Available at: https://all3dp.com/.

IOP Conf. Series: Materials Science and Engineering 1190 (2021) 012002 doi:10.1088/1757-899X/1190/1/012002

Heat Transfer in Double Annular due to Natural Convection

Jan Pařez[1], Pavel Rohan[2], Tomáš Vampola[1]

[1]Department of Mechanics, Biomechanics and Mechatronics, Faculty of Mechanical Engineering, Center of Advanced Aerospace Technology, Czech Technical University in Prague. Technicka 4, Praha 6, Czech Republic
[2]Department of Manufacturing Technology, Faculty of Mechanical Engineering, Czech Technical University in Prague. Technicka 4, Praha 6, Czech Republic

Jan.Parez@fs.cvut.cz

Abstract. Presented paper is focused on 2D numerical simulation of turboprop engine cooling in the section of a gas turbine. The paper describes the issue of heat transfer by natural cooling of a turboprop engine in the turbine region in order to determine the time-dependent temperature field in the engine. The analysis is represented on a simplified double annular geometry, where the inner tube represents the engine rotor, the middle tube represents the flow path and the outer ring replaces the engine case. Simultaneous heat transfer by natural convection (Boussinesq model) and radiation (model S2S) is solved on this geometry. Numerical data were compared with data that was gained from measuring on an experimental stand. The surface temperatures of the tubes were measured on the experiment stand during the cooling of the engine model. The temperatures on the walls of the flow path were based on real temperatures of the running engine as initial conditions together with the temperature of the outer tube. The results will be further applied to the study of temperature-dependent deformation of aircraft engine parts, which have a significant effect on the safety and trouble-free operation of the aircraft engine.

1. Introduction

After shut off the turbine engine, the asymmetrical distribution of the temperature field on the components of the turbine engine occurs due to natural convection. This unequal distribution of the temperature field has many adverse effects that can lead to irreversible engine damage or critical engine failure. A certain time after shutting off the engine, there is a temperature difference between the upper and lower side of the shaft, and also other parts of the engine construction. This difference is changed with time and his maximal is just at one specific time. After this time, the temperature difference begins to stabilize until the engine is cooled down completely.

Due to the temperature difference caused by natural convection of hot engine parts, the thermal expansion of the parts occurs. This expansion is more noticeable with the length of the part. The solution on the rotor shaft assembly is significant. Based on the temperature difference, the shaft bends and causes i.e. thermal deflection called "Thermal Bow". Due to this bending, the balance distribution of the rotor parts is changed and this unbalance has many of side effects. These effects include, for example, significant vibrations of the rotor parts, stresses on the connecting surfaces and last but not least bearing loadings. In extreme cases, unintentional contact may occur during the subsequent starting of the engine due to the reduction of the gaps between the rotor and stator parts, especially in the area of the labyrinth seals or the area of the blades.

The state of the art has long been known in the field of energy, where the development of natural convection is significant due to huge rotating mass. This mass accumulated heat energy, which is released by natural convection after engine shut off, but sufficient cooling is prevented by insulating the entire steam turbine and a large temperature gradient. After restarting the turbine, the thermal deflection may not be sufficiently eliminated and the entire system may be destroyed. In the field of aviation, this phenomenon has not been significant for a long time, due to sufficient clearances and limited temperatures. With increasing demands on engine efficiency, designers are forced to choose constructions with smaller gaps between the stator and rotor parts, temperatures are rising and, finally, there are field of military aircraft and application, where in some cases sufficient cooling of the engine after landing is impossible.

2. Analytical Model

The presented research describes the situation of engine shut off and subsequent cooling by natural convection. The solution is simplified to a 2D analysis in the section of a gas turbine, where the initial and boundary conditions defined below are assumed. The presented assembly consists of 3 tubes corresponding to the dimensions and thicknesses on the aircraft engine. Further the middle section, the hot gas flow path section of the turbine and the inner section of the shaft. The initial conditions are then prescribed for the outer tube, where is tube cooled by an external air flow. The temperatures on the middle turbine tube and shaft tube are prescribed.

If the aircraft fly at cruising altitude (CRUISE), the installed engine is cooled by a forced air flow flowing in the longitudinal direction. Due to the high speed of the air and its low temperature (corresponding for example to the temperature of the standard atmosphere at cruising altitude), cooling is usually sufficient in this flight mode. The worst case occurs when the aircraft is in standby mode or taxiing on the ground at low speed. Even in this case, however, the heat transfer from the surface of the engine is sufficient to cool the engine and its lower parts due to forced convection. In this case, the air temperature can reach much higher values than the corresponding temperature at cruising altitude, e.g. the temperature corresponding to an extremely hot day can be up to 50 [°C]. The analysis of the cooling of the outer tube has been previously studied and measured on a real turbine engine and is presented in previous research [1]. The results of previous paper are used for comparison and validation with the current study. The initial state is assumed to be shutting of the running engine and the associated heating of the middle and shaft tubes from hot gas streams to an isothermal field of 600 [°C]. The outer tube is heated to an isothermal field at a temperature of 150 [°C], this condition approximately corresponds to the external cooling of the engine due to forced convection of air during flight. Figure 1 shows a 2D diagram of the monitored area including dimensions, temperatures and emissivity values for heat radiation.

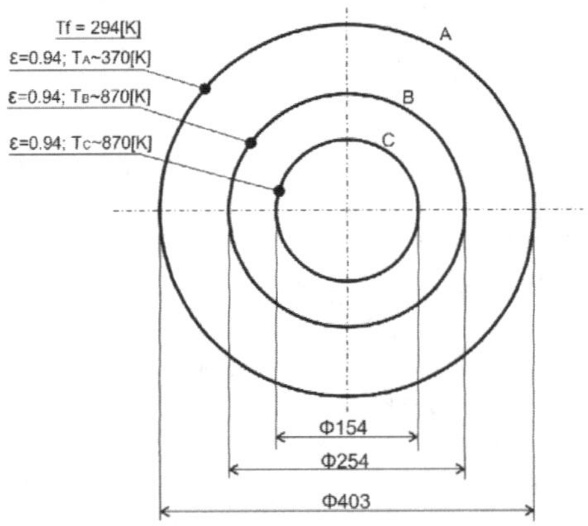

Figure 1. Scheme of 2D turbine section

IOP Conf. Series: Materials Science and Engineering 1190 (2021) 012002 doi:10.1088/1757-899X/1190/1/012002

All areas of the gas are stationary in the initial area of the solution and a temperature of 25 [°C] is prescribed for the middle part, then a temperature of 150 [°C] for the area of the flow path and the shaft space. This simplification is chosen with regard to the subsequent experimental validation of the simulated data. When the heat source is removed, the areas are blown through and then the measuring of natural convection is recorded, the initial condition is based on this assumption.

3. CFD Analysis

Numerical analysis of heat transfer by natural convection and heat radiation is compared and validated with experimental data measured on a measuring stand. The stand model is based on the real dimensions of the turbine section of the turboprop engine and it was presented in the previous section. Material properties and emissivity parameters are considered.

The CFD software ANSYS 2019R2 was used to calculate the heat transfer. The problem is solved in 2D with RANS (Reynolds Averaged Navier-Stokes) transport equations supplemented by the equilibrium energy balance. It used the Boussinesq model and precise density model to express the temperature dependence of thermophysical properties. The turbulent flow is modeled using the $k - \omega$ SST (Shear Stress Transport) turbulence model. It used the S2S (surface-to-surface) radiation model to predict the radiative heat transfer.

The meshing was chosen with the assumption of double annular profiles and smoothing "Inflation" layers around the walls in the area of the boundary layers. The meshing scheme is shown in Figure 2.

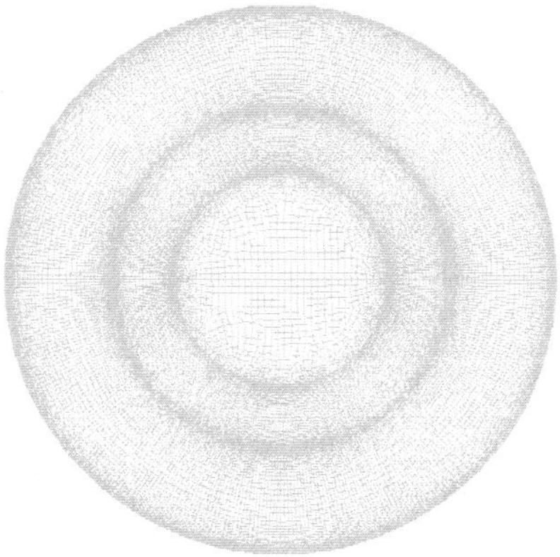

Figure 2. Scheme of 2D turbine section

The research of heat convection in the annulus was presented, for example, in articles [2-4]. The heat transfer due to the natural convection in the section between two coaxial cylinders was studied by Kuehn and Goldstein [5, 6].

Kuehn and Goldstein experimentally study heat transfer during natural convection between two horizontal cylinders using an optical method (Mach–Zhender interferometer). Authors state that most of the papers are about study for a limited range of Rayleigh's number $10^1 < R_{aL} < 10^6$, where Rayleigh's number is based on radius difference, i.e. $R_{aL} = g\beta L^3 \Delta T/\mu a$, where $L = R_{outer} - R_{inner}$ and ΔT is the temperature difference between cylinders.

Due to the huge temperature differences between the individual tubes, especially between the outer tube and the middle tube after the start of the measurement and before the temperatures stabilized, a large Rayleigh number is assumed. The study of high Rayleigh numbers in natural convection in annuloid was presented by Fant [7] and Charter-Moitabi. [8]

Natural convection is the driving force in the system and it is caused by a change in air density. These changes in the density of the air in the gravitational field cause the fluid to flow and move. The density differences are mostly due to temperatures differences. In natural convection is areas, where temperature is higher than means the density is smaller and the fluid moves up. The tube is thus unevenly cooled and reheated at the same time. In the case of the double annulus component, this phenomenon is even more pronounced due to the dependences of the partial annulus and the reheating of the outer parts by the inner ones.

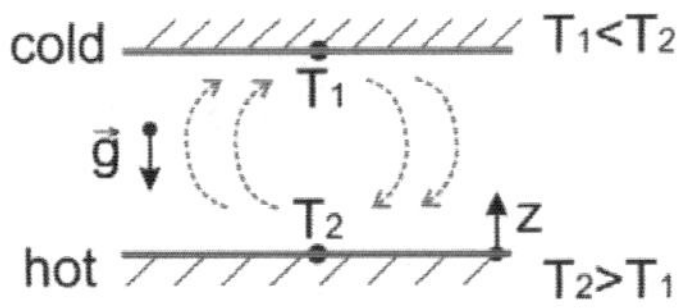

Figure 3. Fluid flow due to heat transfer

Equation 1 and Equation 2 show the dependencies of the changes shown in Figure 3.

$$T_2 > T_1 \rightarrow \rho_2 > \rho_1 \tag{1}$$

$$\frac{dT}{dz} < 0 \rightarrow \frac{d\rho}{dz} > 0 \tag{2}$$

Boussinesq approximation assume the density depends on temperature and its approximate by the following Equation 3.

$$\rho \approx \rho_0 + \frac{\partial \rho}{\partial T}|_{T_0}(T - T_0) + \cdots, \tag{3}$$

where ρ_0 is initial density and $\frac{\partial \rho}{\partial T}$ is time depend density gradient. This Equation 3 is may be present with thermal expansion coefficient β to Equation 4.

$$\rho = \rho_0 + \beta\rho_0(T - T_0) \tag{4}$$

In some case is this approach not valid because of high temperature differences. Precise approach of calculation exact dependency density on temperature for this case is used. For transient simulation with $\rho = \rho(T, p \ldots)$ and high temperature differences is used incompressible ideal gas model correspond to temperature ideal gas model.

$$\rho v = RnT \tag{5}$$

Analysis of NS equation using dimensionless variables in gravitation field is gained dimensionless velocity profile

$$\vec{u} = f(\vec{x}^*, Gr, Pr), \tag{6}$$

where $\vec{x}^*$ is coordinate, Gr is Grasshof number and Pr is Prantl number.

$$Gr = \frac{g\beta\Delta TL^3}{\mu^2}, \tag{7}$$

where g is gravitation constant, L represent characteristics length and μ kinematics viscosity.

$$Pr = \frac{\mu}{a} \tag{8}$$

The previous equation 8, where υ is kinematics viscosity and a is thermal conductivity. Multiple of Grasshof number and Prandtl number is Rayleigh's number gained. This number to heat transfer and calculation natural convection is used. Nusselt number is then calculated with Rayleigh's number and another geometry depend parameters.

$$R_{aL} = Gr * Pr = \frac{\beta g \Delta T L^3}{\mu a} \quad (9)$$

ANSYS software allows the selection of many types of equations and models described and presented in [9] [10]. The solver can choose the required turbulence model and approach to the solution of natural convection and heat transfer by radiation. The used turbulence model determines the "expected" flow regime in the solved geometry mentioned in the previous chapters. In the case of problems with natural convection, the Rayleigh number determines the flow mode [7]. The value of the Rayleigh number may not be accurate, because the temperature-dependent parameters in the case of natural convection cannot be determined for eat time step. The outer diameter of the middle tube of the annulus, shown in Figure 1, and the properties of air at an average temperature of 25 [°C] were used as a characteristic dimension. The calculated value of the Rayleigh number is of the order of $7 * 10^7$ to $9 * 10^8$ and this value is between the turbulent mode with the transition after initial natural convection to the laminar flow. From the point of computational it is a model of turbulence.

A comparison of different approaches to turbulence modeling has been made in academic work [11]. Based on numerical experiments, the Transition SST (Shear Stress Transport) model was recommended for turbulence modeling. The author also dealt with the selection of a suitable approach to modeling natural convection. He compared the model based on the Boussinesq linearization of the ascending force and the model of the incompressible ideal gas. When comparing the results of both models, he states that both models provide practically the same results in a given task. Finally, the author compared two models of radiant heat transfer - model S2S (surface to surface) and P1.

The results show that the P1 model overtakes about 12% of the radiant heat flux above the S2S model. However, when comparing heat fluxes on both surfaces, the P1 model shows better results. Using the S2S model, the task could not be brought closer, so the total heat fluxes on both sides of the gondola were not the same. This imbalance was caused by the radiation component of the heat flux and it was approx. 6% of the total heat flow. On the other hand, the P1 model managed to calculate the task with zero resulting heat flow imbalance. For these reasons, the author recommends using the radiation model P1.

4. Experimental Stand

4.1. Stand Design

The numerical analysis is compared and validated with experimental data measured on a measuring stand. The construction of the stand is based on the real dimensions of the turbine section of the turboprop engine and corresponds to the model given in the analytical model. Stainless steel W.Nr. was chosen as the pipe material. 1.4301 (AISI304) with defined parameters considered in CFD analysis. The solution of the 2D problem on the measuring stand was realized by thermal insulation of the measured area from the surrounding contact surfaces and lateral transition surfaces of the intermediate tube. Ceramic Fiber Insulation was chosen for the insulation.

A Dawell Inverter Heater shown in the Figure 6 was used for heating the tubes Heating of the inner and middle tubes to the isothermal heat field was performed by resistive removable "blankets" with which the tube was wrapped. After heating and stabilizing the temperature field, these heating elements were removed and the measuring space was closed with an insulated cover. Temperatures were measured in a perpendicular section of six equally spaced K-type thermocouples on each tube showed on Figure 4. The results were recorded in the Almemo measuring center. Numerical and experimental data are further compared and evaluated at the end of this paper. (Figure 5)

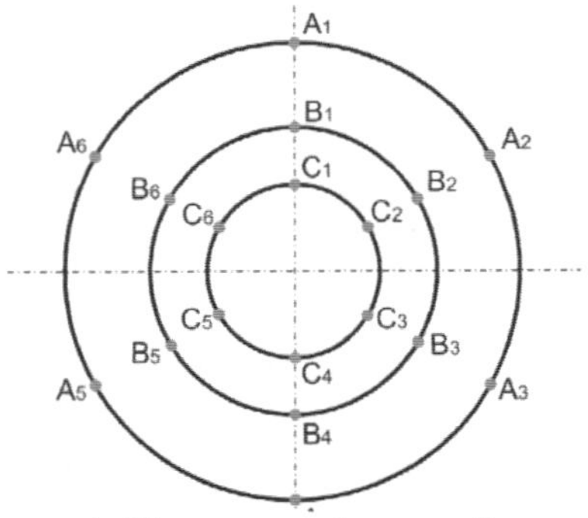

Figure 4. Thermocouples equally spaced on tubes

Figure 5. Experimental stand

4.2. Thermal Camera

The course of the measurement was recorded with a FLIR Thermal Camera to determine heat fluxes and areas with heat loss and transition. Due to the dependence of the emissivity parameter ε on the temperature, the angle of reflection and the variable state of the surface on the assembly, it was not possible to precisely set the emissivity value in each time step. The Thermal Camera was controlled by the temperature at a specific point, where this temperature was detected using a placed K-type thermocouple. Thermal analysis with FLIR Thermal Camera is shown in Figure 6.

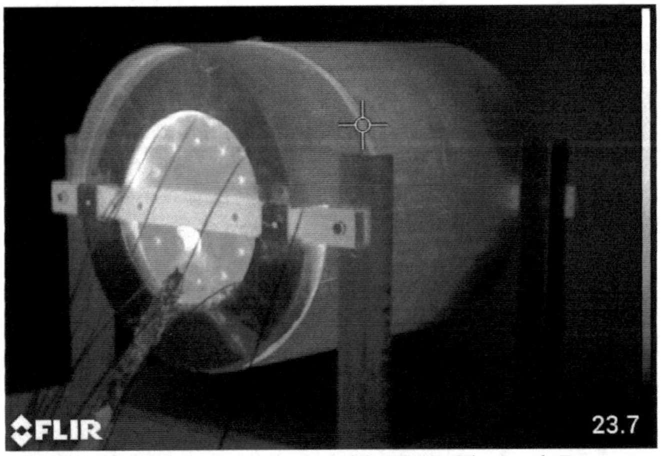

Figure 6. Thermal analysis with FLIR Thermal Camera

5. Results

A CFD model of turboprop engine cooling was described, which was further compared with the results from experimental measurements. Time step 1 [sec] was selected for the numerical calculation with the number of iterations 50 in each time step. the measurement was performed in the range of 0 to 60 minutes, however, the most significant phenomenon takes in the range between 0-45 minute, then the courses are decreases exponentially until engine is completely cooled. The temperature and velocity profiles for the selected time steps are shown in Figure 7 to Figure 10. Next, the points corresponding to the location of the thermocouples in Figure 4 have been defined, in which graphs of temperature versus time are shown for each tube in Figure 11, Figure 12 and Figure 13. The temperature versus time for experimental measurements is shown and compared in the same graph.

IOP Conf. Series: Materials Science and Engineering 1190 (2021) 012002 doi:10.1088/1757-899X/1190/1/012002

The graph shown on Figure 11 to Figure 13 compares the corresponding numerical and experimentally measured values. It can be evaluated that under the same initial and boundary conditions, the numerical values of the temperature are higher and at the same time the decrease of temperatures with time is lower than the experimental values. This is due to several influences. The main reason is the spread of heat transfer in a real 3D assembly. It is not structurally possible to ideally insulate all connecting members and pipe lengths.

The heat dissipation in the assembly was observed by a FLIR Thermal Camera shown in Figure 6. Furthermore, the difference between the numerical and experimental data is due to the cooling surface and the manipulation when removing the heating elements. Last but not least, this is due to non-compliance with all temperature-dependent factors such as emissivity or heat transfer coefficient. These parameters would have to be studied in detail for each specific case and temperature. Finally, the numerical analysis showed good agreement with the measured values from the experimental stand and at the same time the external temperature field on the engine case with the characteristic corresponds to the previously measured temperature field on a real turboprop engine. Difference between measured and numerical values of the engine temperature field is about 20[°C] after 30 minutes. From this result it is possible to evaluate the possibility to use numerical analysis to obtain the temperature field of the engine after switching off. The temperature field can be used directly for strength analysis and shaft deflection analysis.

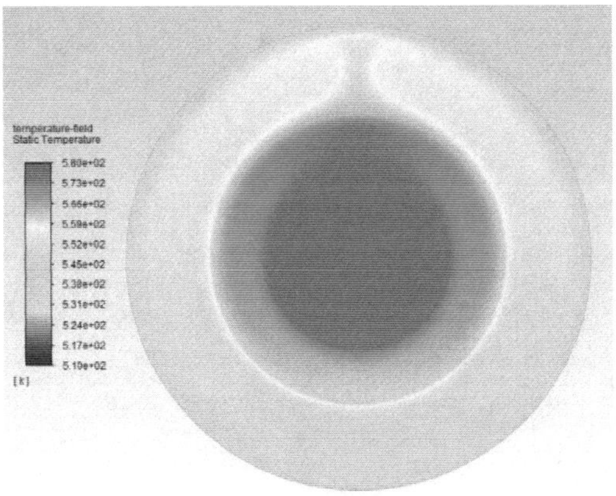

Figure 7. Temperature field t=15min

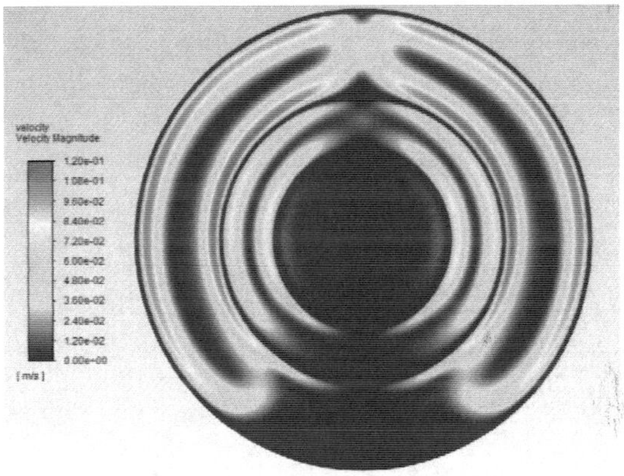

Figure 8. Velocity field t=20min

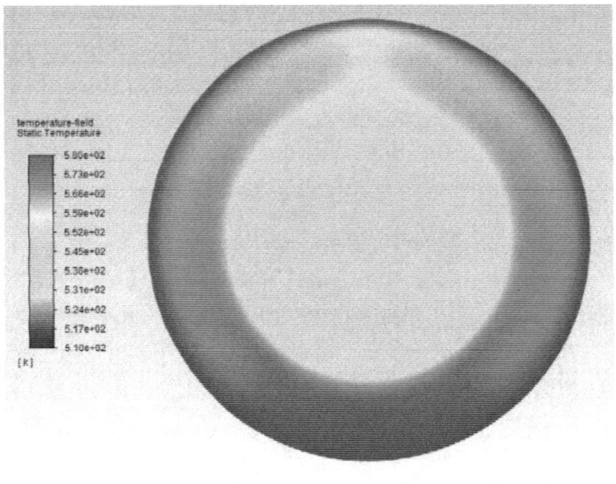

Figure 9. Temperature field t=30min

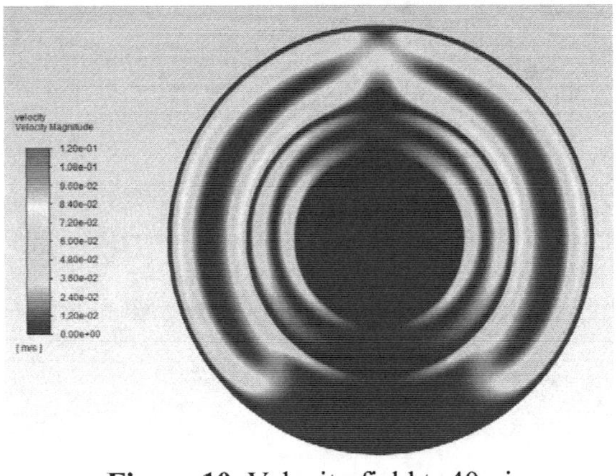

Figure 10. Velocity field t=40min

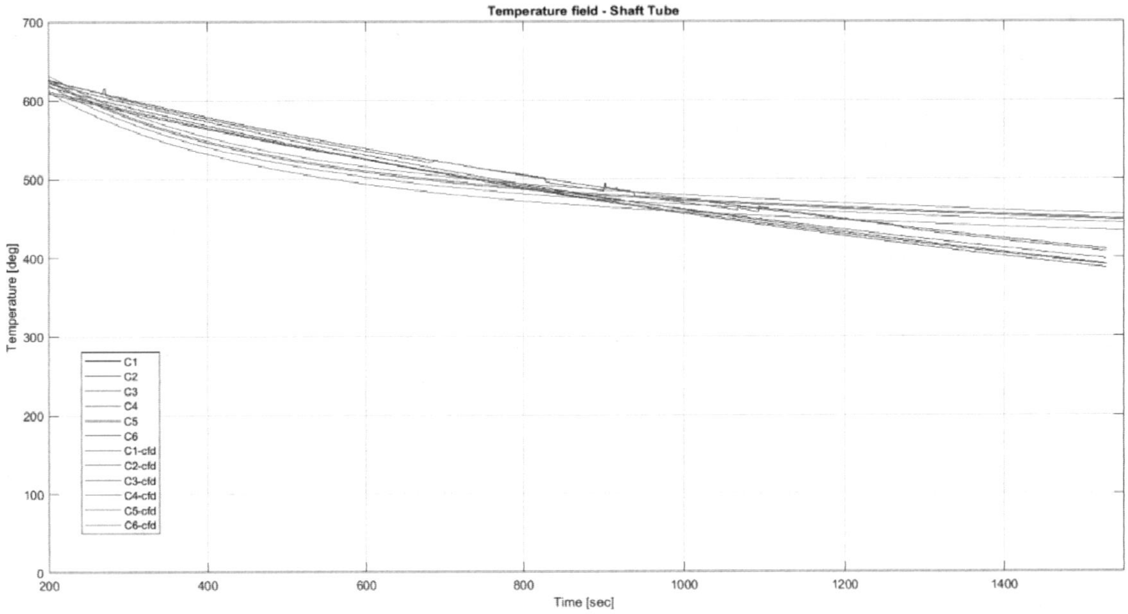

Figure 11. Comparison of numerical and experimental temperatures on shaft tube

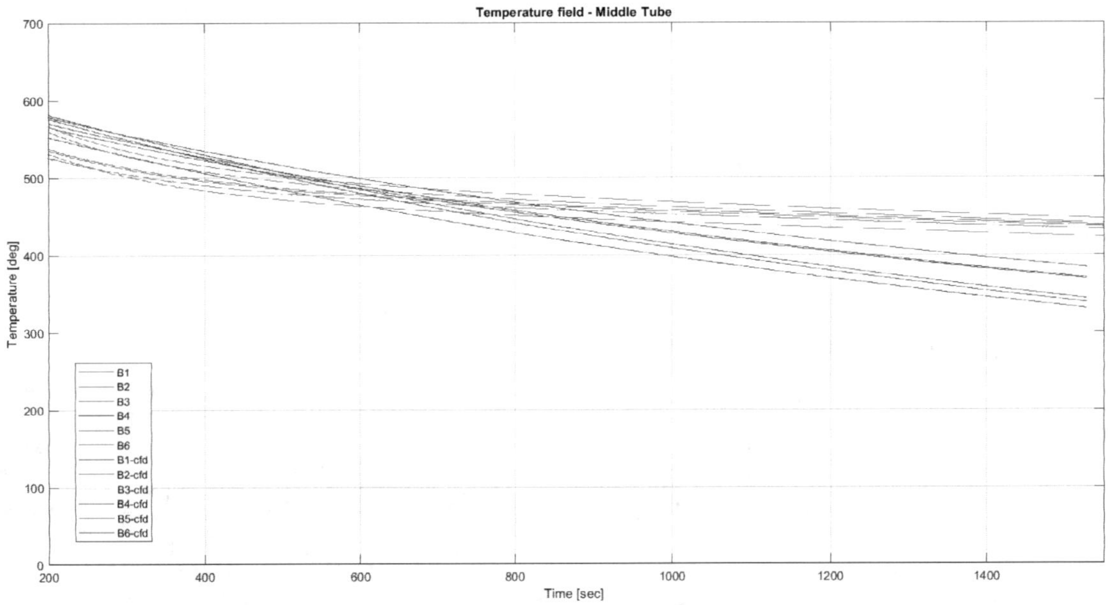

Figure 12. Comparison of numerical and experimental temperatures on middle tube

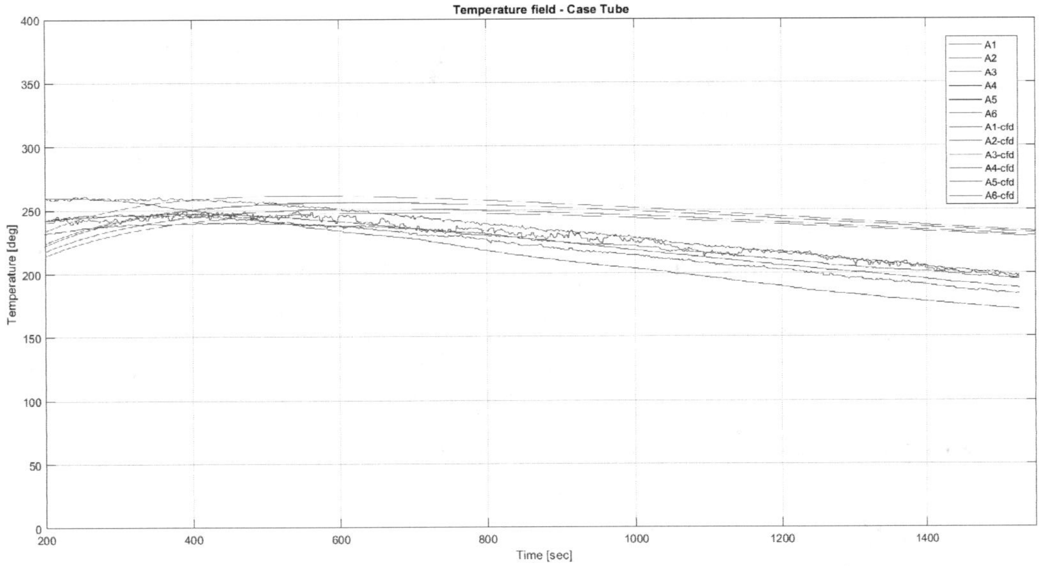

Figure 13. Comparison of numerical and experimental temperatures on case tube

6. Conclusions

The problem of cooling the turboprop engine after shut off in the area of the gas turbine was studied, assuming simplification to 2D problems. The cooling of the engine due to natural convection and the effect of heat radiation in a double annulus consisting of three concentric tubes of different diameters and thicknesses were solved. The analysis was performed in two steps. First, an analysis of natural convection with the effect of heat radiation was performed in the Ansys CFD program. The geometry corresponding to the actual dimensions was designed, and the material and initial conditions corresponding to the actual values were defined. The results of the numerical simulation also show that the effect of radiation is significant and must be considered.

The numerical simulation was compared with experimental data from measurements on a measuring stand. The construction of the experimental stand is similar to the conditions in the numerical simulation. The geometries and materials were chosen similar to the numerical simulation. The conversion of the 3D measuring stand into a 2D task was performed using ceramic fibre insulation and partitions separating the interior space. Contact surfaces and sidewalls were insulated to achieve heat sharing only in the radial direction. Heating was provided by a Dawell Inverter Heater. Temperatures were recorded with 6 K-type thermocouples equally spaced on each tube. The obtained experimental data were compared with data from numerical analysis.

The results of the numerical simulation of the temperature field are higher than the measured application temperatures. The reasons for these differences were evaluated. An adverse effect in terms of experimental and calculated temperature differences is probably non-compliance with the conditions of pure natural convection during the measurement (especially at the beginning of the measurement). Furthermore, it was not possible to maintain the heat transfer by conduction at the connecting screws and the outer shell of the assembly. However, the results show the possibility of numerical simulation of cooling a turboprop engine, determination of its temperature field for the purposes of other temperature-dependent phenomena under real environmental conditions. The results can significantly affect the service life of individual engine parts and its closer understanding of the issue of temperature-dependent processes.

Acknowledgment
Authors acknowledge support from the ESIF, EU Operational Programme Research, Development and Education, and from the Centre of Advanced Aerospace Technology (CZ.02.1.01/0.0/0.0/16_019/0000826), Faculty of Mechanical Engineering, Czech Technical University in Prague.

References

[1] S. Solnař, L. Popelka, M. Dostál, Cooling turboprop engine during shutdown. 13th International Conference on Experimental Fluid Mechanics (EFM 2018): EPJ Web of Conferences Volume 213, 2018. ISBN 9781510889668.
[2] E. Abu-Nada, Z. Masoud, and A. Hijazi, Natural convection heat transfer enhancement in horizontalconcentric annuli using nanofluids [online]. International Communications in Heat and Mass Transfer 35, pp. 657-665. Jordan: elsevier, February 2008. Available at: https://doi.org/10.1016/j.icheatmasstransfer.2007.11.004.
[3] D. F. Francis, T. M. Itamura, W. S. Webb, L. D. James, CFD Calculation of Internal Natural Convection in the Annulus between Horizontal Concentric Cylinders [online]. Sandia National Laboratories Albuquerque, New Mexico and Livermore, California: Sandia National Laboratories, October 2002 [Accessed 2021-5-23]. Available at: https://core.ac.uk/download/pdf/208903821.pdf
[4] K. Ranganathan, Study of natural convection in horizontal annuli [online]. Int. J. Heat Mass Transfer Vol. 30 No. 6, pp. 1137-1148. Clemson, U.S.A.: Peergamon Press, 1988. Available at: https://doi.org/10.1016/0017-9310(88)90056-7.
[5] T. H. Kuehn, R. J. Goldstein, An experimental and theoretical study of natural convection in the annulus between horizontal concentric cylinders, .Z. Fluid Mech. 74, pp. 695-719, 1976.
[6] T. H. Kuehn, R. J. Goldstein, Correlating equations for natural convection heat transfer between horizontal circular cylinders, Int. Heat Mass Transfer 19, pp. 1127-1134, 1976.
[7] D. B. Fant, The numerical and analytical study flow instability due to natural convection between narrow horizontal isothermal cylindrical annuli at high Rayleigh numbers [online]. Iowa: Iowa State University Capstones, 1987 [Accessed 2021-5-10]. Available: https://lib.dr.iastate.edu/cgi/viewcontent.cgi?referer=https://www.google.com/&httpsredir=1&article=12676&context=rtd.
[8] M. C. Charter-Mojtabi, A. Mojtabi, J. P. Caltagirone, Numerical solution of a flow due to natural convection in horizontal cylindrical annulus, Jounal of Heat Transfer 101, pp. 171-173 (1979)
[9] ANSYS, Inc. Natural Convection and Buoyancy-Driven Flows - ANSYS Fluent. Section 13.2.5, 2006a.
[10] ANSYS, Inc. Surface-to-Surface (S2S) Radiation Model Theory - ANSYS Fluent. Section 13.3.7, 2006c.
[11] M. Hodek, Bachelor's thesis, Czech Technical University in Prague (2018) Available: https://dspace.cvut.cz/handle/10467/79943.

Increasing the Reliability of FMEA Evaluation by Modifying Rating Scales and Applying Pairwise Comparison Method

Edina Kulcsár[1], István Gábor Gyurika[1], Tamás Csiszér[2]

[1]Department of Mechanics, University of Pannonia, H-8200 Veszprém, Hungary
[2]Sándor Rejtő Faculty of Light Industry and Environmental Protection Engineering, Óbuda University, H-1034, Budapest, Hungary

kulcsar.edina@mk.uni-pannon.hu

Abstract. Failure Mode and Effect Analysis (FMEA) is a frequently applied risk assessment methodology in different fields of industry. It provides the opportunity for deep analysis of risky events, connecting to their potential causes and effects. Due to its robust method, risk can be identified and managed but, because of its complexity, the assessment can be time-consuming, and the answers can be inconsistent. In the authors' experience, the scale used to quantify the priority number can influence these outcomes. To identify and measure the significance level of this connection, a survey was organized with experts of digital manufacturing, who were asked to rank the probability of risk occurrence in planning a production process based on digital manufacturing principles. Participants had to use three assessment methods: 1) traditional 'linguistic' scale; 2) 'ratio scale; 3) pairwise comparison. In the case of the linguistic scale, probabilities were defined by their descriptions and an integer between number 1 and 7. These numbers did not equal the probability percentage. In contrast, the ranks of the ratio scale equalled the ranges of probability percentage. In the pairwise comparison, risks were compared by the frequencies of their occurrences. The assessment results were compared by time-need, difficulty, priority list, and consistency. Time and difficulty were ranked by the survey participants on a 4-level scale. The priority list was created by the summarized rank values. In the consistency assessment, we looked at how the amended scale affects the assessment uncertainty. In addition, pairwise comparisons were used to examine the consistency of each expert's evaluation and filter out those that were not accurate enough.

1. Introduction

Failure Mode and Effects Analysis (FMEA) is a widely used risk assessment method. Its purpose is to identify and quantify the risks associated with the manufacture or use of a product to properly address them [1]. The FMEA can be considered as a "measurement system" for risk assessment, the elements and characteristics of which significantly influence the outcome of the study. As in any measurement system, we distinguish between the object to be measured (risk), the measuring device (FMEA), the measurement method (evaluation scales), the person performing the measurement (evaluator), and the environment (historical data, assumptions, etc.). For the measuring system to give adequate results, the following must be ensured. The sensitivity can determine the value to be measured with adequate sensitivity. Accuracy ensured that the measured value shows the actual value. Repeatability means that successive measurements on the same batch must give the same result if all the other variables in the measurement system are identical. Reproducibility is that successive measurements on the same batch should give the same result if another of the other variables in the measurement system, usually the person performing the measurement, changes.

Compliance with expectations is affected by all variables in the measurement system. Our research aims to examine how the expectations can be interpreted in the case of FMEA, and in the case of the method as a measurement system, which variable affects the extent to which it meets the criteria related to the outcome of the risk assessment. In this dissertation, we examine the effect of the scale used to assess risk factors on FMEA reproducibility (R&R).

During the Fourth Industrial Revolution, customer requirements for product manufacturing changed significantly. The most important element of the change is that cyclical, in many cases extreme, changes in customer requirements can be observed in the entire design and production planning process of the product. Companies producing products need to respond with sufficient flexibility to the frequent changes in customer demand generated during the product manufacturing process to remain successful market players. These new property requirements provided the basis for the creation of digital manufacturing science, one of the main pillars of which is design and manufacturing design based on a digital manufacturing system. The basic idea of a digital manufacturing system is that companies and manufacturing associations working together to produce products to meet customer needs are independent of each other in time and space, and that information flows between members of a manufacturing association in a digital environment: needs, conceptual designs, material requirements, design and manufacturing design in a virtual environment, and simulation. Given the importance of this topic area, the risk assessment focuses on the risks of design based on a digital manufacturing system [2].

In this context, after a brief literature review, we describe the theoretical background of the FMEA and R&R methodology and then present the impact of the traditionally used scale and a possible alternative scale and a pairwise comparison method on each outcome of the risk assessment by presenting a case study.

2. Background

2.1 Subjectivity of the method

We hypothesize that an improperly designed scale reduces the objectivity of evaluation, which is supported by various researches [3]. The combination of the FMEA method with Monte Carlo simulation does not ensure the elimination of subjectivity either [4]. It was performed severity assessment by varying the amount of error-specific input information. The results of the study showed that participants' understanding was influenced by the availability of information, and therefore the risk classification became inconsistent [5]. The appropriate scale was selected for the use of the AHP (Analytic Hierarchy Process) methodology, and the consistency test that can be performed with it have been proposed [6]. Other research has used the R&R method to measure FMEA consistency. In one study, two different but equally trained teams assessed the risks, producing different results [7].

2.2 Failure Mode and Effects Analysis

The FMEA is a risk assessment tool that provides a structured, transparent method for collecting and evaluating potential errors, identifying measures to reduce the likelihood of error, and documenting the entire process [8]. The experts assess the risks according to three aspects, which are the severity of the effect of the error (Severity -S) and the frequency (Occurrence - O) and detectability (Detection - D) of the error. For each aspect, potential errors are rated with a predetermined scale [9]. These scales typically range from 1 to 5 or 1 to 10, with the highest value representing the worst case. The product of the three scale values (S, O, D) gives the RPN (Risk Priority Number), which indicates the priority of the risk: the higher the value of the RPN, the riskier the event under study. Measures should normally be set for risks with a product of more than 125 for the 10-point scale with the highest RPN.

The new AIAG manual replaces the RPN with the Action Priority (AP) value to prioritize risks, according to which risks can be divided into three groups based on the values of the factors: low, medium, and high priority. In the latter case, it is mandatory to specify the appropriate measures [10].

2.3 Repeatability and Reproducibility examination
R&R is a procedure used to jointly evaluate the repeatability and reproducibility of measurements. [11]. The adequacy of the measurement system is assessed by fluctuating the results of tests performed several times while keeping its elements constant (repeatability) or changing (reproducibility). In both cases, the goal is to consistently produce values that can be considered identical based on expectations while ensuring the robustness of the measurement system [12].

2.4 Pairwise comparison based FMEA
The application of pairwise comparison in the ranking of problems is already possible when using the FMEA methodology. The aim of the method is to reduce the uncertainty and subjectivity of the evaluation by pairwise comparison. The pairwise comparison is based on a customized version of the basic scale of the AHP method. The consistency of the result of the pairwise comparison can also be evaluated by the method. The FMEA team performs a comparison of n (n –1) / 2 in a pairwise comparison matrix. To avoid human fatigue and increased inaccuracy, it is advisable to ensure that the number of assessments is kept to a minimum [13].

3. Case study
The study created an FMEA team of 6 people who are experts in design based on a digital manufacturing system. For the assessment, they identified 17 risks, which are included in table 1. This ensures that the risks are clear to everyone. Three scales were developed for evaluation. A scale of ratios that indicates the probability of occurrence of each risk in percentage terms. A linguistic scale with no numerical values or a scale for pairwise comparison, which is a modified version of the AHP methodology scale. All three scales range from 1 to 7, with 0 as the same occurrence value for pairwise comparison. During the risk assessment, participants had to estimate the probability of the risks occurring. The scales are shown in table 2. To estimate the incidence of risks, three questionnaires were prepared according to the three scales, which were completed several weeks apart, thus ensuring that team members did not remember their previous answers. During the evaluation, participants were also asked how long it took to perform the evaluation with the method and how difficult it was found to use it. The risk assessment was performed separately by the participants so that we could also measure the accuracy relative to themselves.

Table 1. Identified risks for the evaluation and their identification numbers

Identification of the risks	Identified Risks
1	Inadequate in-depth assessment of customer needs due to lack of precision. Possible Consequence: The customer needs assessment process is therefore not fast enough. The customer may need to be asked several times during the product design process for clarifications.
2	When developing your own product, the market survey is not as detailed as is necessary for proper planning.
3	When developing your own product, the wrong target audience is selected for the market survey.
4	Improperly defining your own competencies (at the development company) required for a product manufacturing process before searching for and selecting members of a virtual manufacturing association.
5	In the case of a product production process, the conceptual design required for the distribution of one's own competencies is not of sufficient depth (the conceptual design does not take sufficient account of the machine capacity within the plant).
6	Inadequate definition of competencies when creating alternative competency teams within a virtual manufacturing alliance. Explanation: Alternative competency teams are assembled within the virtual manufacturing alliance, selected from professionals in the federal companies according to the specific competency.
7	Selection of inappropriate professionals for the creation of alternative competency teams in the distribution of competencies within a virtual manufacturing alliance.
8	Inadequate consideration of customer needs for collaborative design.
9	Stricter manufacturing requirements are set by designers in construction plans than the customer expects. The consequence may be that both manufacturing and production may be more expensive.
10	Improper use of designs in collaborative manufacturing design.
11	In collaborative manufacturing design, you do not have the necessary expertise to properly apply the designs in structural design. The consequence may be that the wrong product to meet customer needs is produced.
12	In collaborative production planning, product production parameters that do not correspond to the production environment are determined due to a lack of expertise.
13	In collaborative production planning, product production parameters that do not correspond to the production environment are determined due to excessive management pressure. (e.g. to increase productivity).
14	Selection of an inappropriate quality control method to verify the conformity of the product during production and final inspection.
15	Non-compliant product parameters are checked during production and final inspection to verify the conformity of the product.
16	Incomplete assessment of risks related to product non-conformity.
17	Lack of adequate spare parts during product support.

Table 2. Scales used for risk assessment

Occurrence scale (percent)		Occurrence scale (linguistic)		Occurrence (pairwise comparison)	
				0	Equally frequent
1	O=0%	**1**	The probability of occurrence is trifling	**1**	Slightly more frequent
2	0%<O<=20%	**2**	The probability of occurrence is low	**2**	Moderately more frequent
3	20%<O<=40%	**3**	The probability of occurrence is minor	**3**	More frequent
4	40%<O<=60%	**4**	The probability of occurrence is medium	**4**	Strongly more frequent
5	60%<O<=80%	**5**	The probability of occurrence is common	**5**	Very strongly more frequent
6	80%<O<100%	**6**	The probability of occurrence is very common	**6**	Extremely more frequent
7	O=100%	**7**	The probability of occurrence is high	7	Extremely more (+) frequent

4. Results and discussions

The assessment was performed by all 6 group members on three different scales for all 17 risks. The results are shown in table 3, where P is the result of the answers on the percentage scale, L is the result of the answers on the linguistic scale, and PW is the result of the pairwise comparison method. In the table, the value of the standard deviation is marked in the table with S, and the value of the mean for the answers to the same risks is calculated for all respondents. Identical answers were marked in green for a filler using a different scale or method.

Table 3. Result of the risk assessment with the three scales, P as the percentage scale, L as the linguistic scale and PW as the pairwise comparison method. The green columns mark the same occurrence result on different scale.

	P	L	PW	P	L	PW	P	L	PW	P	L	PW	P	L	PW	P	L	PW	P	L	PW	P	L	PW	P	L	PW
	1	1	1	2	2	2	3	3	3	4	4	4	5	5	5	6	6	6	7	7	7	8	8	8	9	9	9
1.	6	3	7	6	3	8	4	3	7	6	3	7	4	4	6	3	4	6	5	4	6	5	3	7	3	5	3
2.	4	4	6	5	5	6	5	4	6	3	4	6	4	5	4	4	3	6	1	2	7	5	5	6	4	3	5
3.	4	4	4	5	5	6	3	3	6	4	4	8	4	5	6	5	3	8	5	3	7	4	5	6	6	6	5
4.	5	6	4	6	6	5	3	3	7	5	3	7	4	3	5	2	2	9	7	3	9	3	5	3	2	4	3
5.	2	5	5	4	4	6	5	6	6	2	3	8	3	2	7	6	4	7	6	2	7	3	3	6	7	5	2
6.	6	4	3	3	4	5	5	3	5	4	6	7	5	3	9	3	4	7	2	4	7	5	4	5	6	6	6
S	1.52	1.03	1.47	1.17	1.05	1.09	0.98	1.21	0.75	1.41	1.17	0.75	0.63	1.21	1.72	1.47	0.81	1.16	2.34	0.89	0.98	0.98	0.98	1.38	1.97	1.17	1.55
X	5	4	5	5	5	6	4	4	6	4	4	7	4	4	6	4	3	7	4	3	7	4	4	6	5	5	4

	P	L	PW	P	L	PW	P	L	PW	P	L	PW	P	L	PW	P	L	PW	P	L	PW	P	L	PW
	10	10	10	11	11	11	12	12	12	13	13	13	14	14	14	15	15	15	16	16	16	17	17	17
1.	4	4	6	4	4	6	5	4	6	5	5	3	4	4	4	5	4	6	4	4	6	6	5	4
2.	3	3	5	3	3	6	4	6	6	4	2	6	4	6	7	1	5	6	1	3	7	2	5	7
3.	4	4	5	4	2	5	5	3	6	6	3	7	4	2	6	5	3	6	5	2	6	5	3	3
4.	3	2	5	5	3	8	4	3	6	5	6	3	2	4	8	2	5	8	2	3	8	3	4	4
5.	4	3	5	2	2	7	1	4	8	4	5	2	5	3	8	3	2	8	5	5	6	5	4	2
6.	4	5	6	2	4	7	3	5	6	5	4	6	3	4	6	4	4	7	5	5	6	4	5	5
S	0.52	1.05	0.52	1.21	0.89	1.05	1.51	1.17	0.82	0.75	1.47	2.07	1.03	1.33	1.52	1.63	1.17	0.98	1.75	1.2	0.83	1.47	0.81	1.72
X	4	4	5	3	3	7	4	4	6	5	4	5	4	4	7	3	4	7	4	4	7	4	4	4

The results can be examined from several perspectives based on what has changed in our evaluation system. One such case is when we examine how different the results of different respondents on the same scale are, thus examining the robustness of the scale. The standard deviation values show that different respondents did not give very different answers to the same risks using the same scale or method. 1-2 scale value deviations are the most typical. This can also come from the different opinions of the team members, which can be resolved by consensus in a group discussion. Figure 1 shows the distribution of the occurrence values for each respondent in case of the three scale.

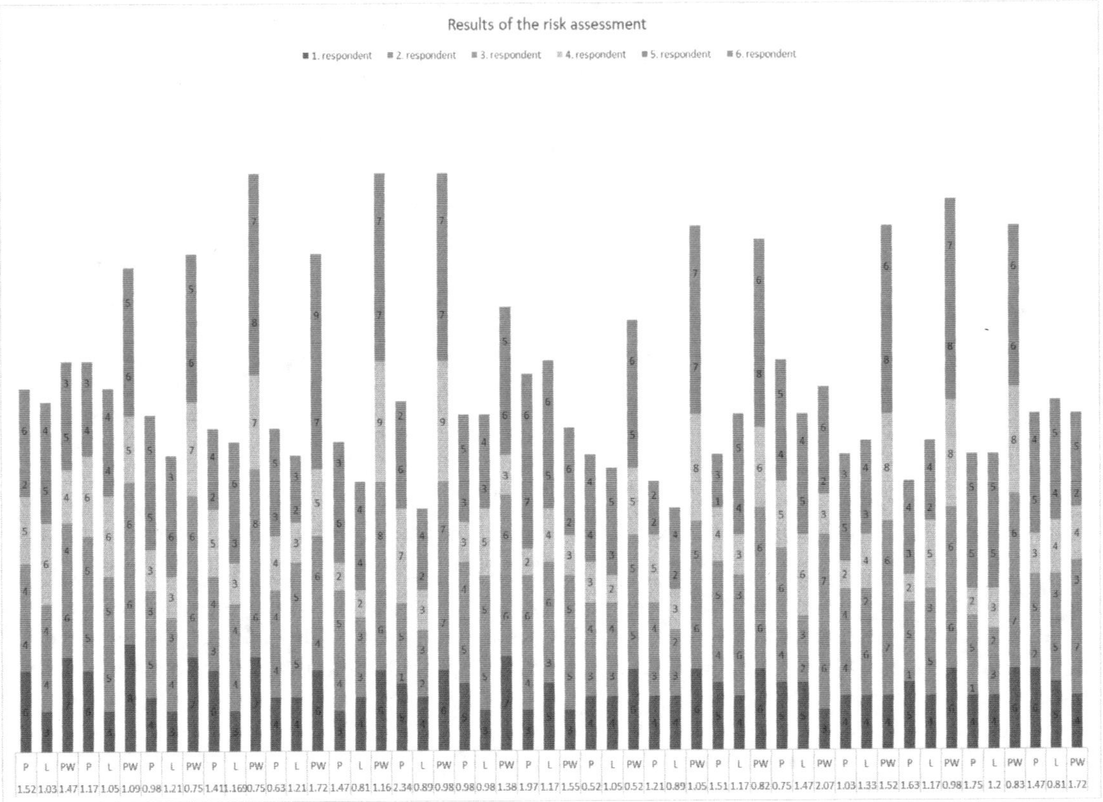

Figure 1. Distribution of occurrence values for the three scale and the standard deviation.

The other variable in our system is the scale, which also changed for each participant. The distribution of the standard deviation given on each scale shows which scale was used to scatter the data the most. The results can be seen in figure 2. Based on the sum of the values of the standard deviation, the standard deviation of the values given on the linguistic scale is the smallest, and on the percentage scale the largest. In addition, it is worth examining the average and median values of the standard deviations, the results of which are shown in table 4. Average and median values show the lowest value when using the linguistic scale and the pairwise comparison method.

Table 4. Average and median values of the standard deviations.

	P	L	PW
Average	1.31	1.095	1.198
Median	1.41	1.169	1.09

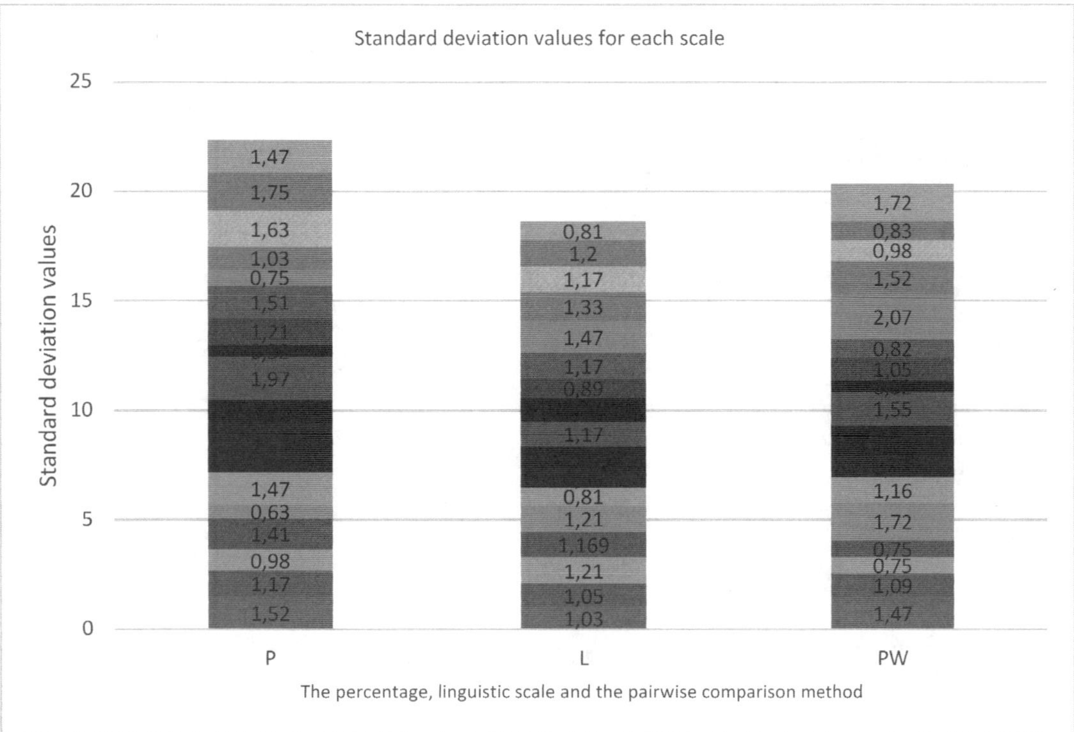

Figure 2. Distribution of the standard deviation of the occurrence values in case of the three scale.

From these results, it can be concluded that the design of the scales influences the final result of the evaluation even if the same content is displayed in different ways. This can already be deduced from figure 1, where the responses of different scales are often very different from each other.

Opinions on the scales and method are given in table 5. The notations for the scales and method are the same as those used in table 3. Based on the opinions, no significant difference in difficulty was observed when using each scale. One participant assessed that evaluation by pairwise comparison is very simple. In terms of lead time, each of its applications was considered long or very long. The difficulty of the percentage and language scales was different for the evaluators, with most responses coming from difficult or easy.

Table 5. Respondent's opinion about the scales and the pairwise comparison method.

	1.			2.			3.			4.			5.			6.		
	P	L	PW	P	L	PW	P	L	PW	P	L	PW	P	L	PW	P	L	PW
How easy was it to apply the scale in the risk assessment?																		
Very hard			X			X									X			X
Hard		X		X						X		X	X	X			X	
Simple	X				X		X	X								X		
It's very simple									X		X							
How long did the risk assessment take to use each method?																		
For a very short time (less than 10 minutes)					X		X									X	X	
For a short time (between 10 and 20 minutes)	X									X	X		X	X				
For a long time (between 20 and 30 lawsuits)		X		X		X						X						X
For a very long time (more than 30 minutes)			X						X						X			

The level of inconsistency in the pairwise comparison result was also evaluated. The methodology initially considers RE (A) <0.1 to be acceptable, but this can only be considered for a few factors. For 17 factors, it is difficult to avoid fatigue during evaluation [13]. The results of the evaluation are presented in table 6.

Table 6. Inconsistency of the measurements in the case of the six respondents.

Inconsistency of the risk assessment		
1.	0.02753	3%
2.	0.796235	80%
3.	0.708084	71%
4.	0.398368	40%
5.	0.446742	45%
6.	0.538912	54%

For one team member, the level of inconsistency became very low at only 3%. The value of outstanding inconsistencies can be experienced in the case of the 2nd and 3rd fillers; here, the consistency and accuracy of the evaluation can be questioned.

5. Conclusions

The risks of the design process based on a digital manufacturing system were assessed on three different scales, where there was no difference in meaning between the scales. The evaluation was based on the change of 2 factors, whether the respondent or the scale used for the evaluation changed.

When changing the evaluator, the largest deviation was 2 scale values in terms of standard deviation. If we look at the difference of the standard deviations when changing the scales, it can be said that the sum of the standard deviations became the largest on the percentage scale. Similarly, the average and median values were the highest on the percentage scale, while they were the lowest on the linguistic scale and in case of the pairwise comparison.

In conclusion, changes in scales and assessment methods affect the uncertainty of the assessment, with the same value given in few cases for the probability of occurrence. Regarding the standard deviation, average and median values, the language scale and the pairwise comparison seem to be a more accurate evaluation. The consistency of the pairwise comparison can also be measured, which was very high in two cases for the six evaluators and very low in one case. The most difficult and time-consuming method proved to be the pairwise comparison, but based on the evaluations, the evaluation on the other 2 scales was not so easy, but it took less time. This also follows from the fact that in a pairwise comparison, all pairs have to be compared, which means much more judgment.

Acknowledgement (s)
We would like to thank Dr Tibor Szalay, Dr Ádám Jacsó, Miklós Boleraczki, Dr Géza Haidegger and Dr Gábor Erdős for their participation in the evaluation and for their professional knowledge.

References
[1] L. Petrescu, M.C. Petrescu, E. Cazacu, V. Ionita. Risk Priority Number vs Action Priority in Electrical Systems. In 2021 12th International Symposium on Advanced Topics in Electrical Engineering (ATEE) (pp. 1-6). IEEE, 2021.
[2] Z. Zhou, S.S. Xie, and D. Chen. Fundamentals of digital manufacturing science. Springer Science & Business Media, 2011.
[3] M. Banghart, et al. "Subjectivity in Failure Mode Effects Analysis (FMEA) Severity Classification within a Reliability Centered Maintenance (RCM) Context." International Journal of Aviation, Aeronautics, and Aerospace 5.1, 2018.
[4] S.J. Rhee, and I. Kosuke. "Life cost-based FMEA incorporating data uncertainty." ASME 2002 International Design Engineering Technical Conferences and Computers and Information in Engineering Conference. American Society of Mechanical Engineers Digital Collection, 2002.
[5] M. Banghart, et al. "Subjectivity in Failure Mode Effects Analysis (FMEA) Severity Classification within a Reliability Centered Maintenance (RCM) Context." International Journal of Aviation, Aeronautics, and Aerospace 5.1, 2018.
[6] D.C. Aguiar, H. J. C. Souza, and V. A. Salomon. "AHP Application to Evaluate Scoring Criteria for Failure Mode and Effect Analysis (FMEA)." International Journal of Analytic Hierarchy Process 2, 3-13, 2010.
[7] M. T. Oldenhof, et al. "Consistency of FMEA used in the validation of analytical procedures." Journal of pharmaceutical and biomedical analysis 54.3, 592-595, 2011.
[8] K. Johnson, "It's fun to work with an FMEA." Quality Progress 35.1,152, 2002.
[9] R.J. Mikulak, R. McDermott, and M. Beauregard. The basics of FMEA. Productivity Press, 2008.
[10] Stamatis, D. H. Risk Management Using Failure Mode and Effect Analysis (FMEA). ASQ Quality Press, 2019.
[11] H. Urdhwareshe, "Validating measurement system using measurement systems analysis (MSA)." Quality World, 41-47, 2006.
[12] Ch. Dubois, and P. Courtois. "Choice of parameters for an R&R study and its impact on the measurement capability." 19th International Congress of Metrology (CIM2019). EDP Sciences, 2019.
[13] E. Kulcsár, T. Csiszér, J. Abonyi. Pairwise comparison based failure mode and effects analysis (FMEA). MethodsX, 7, 101007, 2020.

Wear behaviour of austempered, ductile iron microalloyed with boron under different contact load by dry sliding wear conditions

Garegin Zakharov[1], Nugzar Khidasheli[1,2], Zurab Aslamazashvili [1], Giorgi Gordeziani[2], Mikheil Chikhradze[1,2], Salome Gvazava[2], Irakli Maisuradze[1]

[1]F. Tavadze Metallurgy and Materials Science Institute, 8b E. Mindeli str., Tbilisi 0186, Georgia
[2]Georgian Technical University, 77 Kostava srt., Tbilisi 0175, Georgia

m.chikhradze@gtu.ge

Abstract. In the preformed work were studied the characteristics of Ni, Mo and Cu – free austempered ductile iron microalloyed by Boron under different contact load. For microalloying liquid cast iron with boron, the ferroboron by synthesized SHS - innovation technology was used. The upper and lower bainitic austempered ductile irons were studied with different content of metastable retained austenite. The experimental samples were tested with a fixed linear sliding speed of 2 m / s at a different contact load values, which were changed in the range of 25, 50 and 100 Newtons. It was defined that the value of the contact load has a significant effect on the tribotechnical characteristics of bainitic cast irons in conditions of dry sliding friction. Particularry, they determine the structural changes character and the degree of surface layers degradation of friction. High strength cast irons with the lower bainite structure are more resistant to adhesive wear. According to the obtained data, the presence of up to 80% metastable retained austenite in the metal matrix structure of the of experimental cast irons makes it possible to stabilize the friction coefficient under the investigated contact loads. Thus, to ensure maximum wear resistance, the isothermal quenching of high-strength cast irons must be carried out in the temperature range 220-280 ^{0}C with a minimum treatment time.

1. Introduction

It is known that the tribotechnical properties of constructional materials are largely determined by their phase composition and microstructure, also by singularities of their production technology. The main task of increasing reliability and durability of parts under friction and wear conditions is the microstructure formation of surface layer, which makes it possible to prevent or slow down the formation processes of deformation defects in materials and afterward increase their strength and tribotechnical properties [1-5].

During developing wear-resistant materials, it is necessary to take into account not only the mechanics of contact interaction, but also the microstructure transformation processes of surface layers under various contact loads. This fully applies to high-strength bainitic cast irons, in which it's possible to regulate the relation of structural components, differing by level of physical and mechanical parameters, which have various effects on the formation the functional properties of bainitic cast irons. This refers to the amount of upper and lower bainite, martensite and retained austenite, the stability of which can be varied within wide range.

The task of structure optimization is more and more complicated when microalloying a melt of bainitic cast irons, for example, with boron micro additive [6-9].

In this case, the interphase boundaries are strengthened by dispersed inclusions of borides, carbides and boron nitrides, the hardness of metal matrix increases about 5-7 HRC units, also the plastic deformation resistance is improved. At the same time, boron microadditives accelerate the carbon diffusion and contribute increasing in the amount of retained austenite, which can cause decreasing wear resistance of these alloys [10]. Thus, boron has a multi-vector effect on the formation of bainitic cast irons exploitation characteristics and therefore it's necessary to carry out additional researches in order to use it effectively.

In the presented work, the wear out processes have been studied of high-strength cast irons microalloyed by boron and with various types of bainitic matrix under dry sliding friction, depending on the values of contact load, which is known [11-12], as significantly effects on the features of interaction and destruction of friction surfaces in the process of frictional loading. During researches, the effectiveness of method of formation the metastable retained austenite structure was tried out in bainitic cast irons, which can be subjected to phase transformations under contact load influence.

2. Experimental procedures

2.1 Microalloyed by Boron austempered ductile iron elaboration
The cast iron was produced in an induction furnace by melting raw materials consisting from 3.62% C, 2.00% Si, 0.35% Mn, 0.03% S, 0.06% P and about 10% steel scrap. For receiving high strength cast iron the modification procedures were carried out with metallic Mg vapour and 0.5 % Fe-Si ligature. For microalloying liquid iron with boron, the 10% ferroboron was used, which was synthesized by SHS - an innovative technology [13]. That was ensuring high (92% B) assimilation and equable distribution of Boron. The ductile iron was poured at 1400 °C into Y block ingots (30 mm). The Chemical composition of melted ductile iron is shown in the Table 1.

Table 1. Chemical composition of experimented ductile irons (wt, %)

Element	**C**	**Si**	**Mn**	**S**	**P**	**Mg**	**B**	**Fe**
Weight (%)	3.47	2.10	0.27	0.003	0.035	0.037	0.03	bal

2.2 Heat treatment
The samples were austenitized at 900 ^{0}C in a thermal furnace for 60 min, then soaked in a melted metal bath at 280 ^{0}C and 400 ^{0}C temperature under 10- 40 sec time range for receiving the metastable austenite. The samples were austenitized at 900 ^{0}C in a thermal furnace for 60 min, then soaked in a melted metal bath at 280 ^{0}C and 400◦C temperature under 10- 40 sec time range for receiving the different amount of metastable austenite with lower and upper bainite.

For microstructural and phase composition studies was used the metallographic, scanning electron microscopy and X-ray diffraction analysis methods. Optical microscopy gave us the qualitative picture of received phases. Quantitative study was carried out by X-Ray diffraction technique. X-ray diffractometer (model ДРОН-4) with Cu Kα radiation (λ=1.54 Å) was used as the source.

Hardness was measured for all the heat treatment conditions before and after wear test. Hardness values were taken on a Rockwell hardness testing machine in Rc scale for a load of 150kg.

2.3. Block-on-Ring Wear Testing

Dry sliding wear testing procedures were performed using the СМЦ-2 tribometer (block-on-ring): model, capable of maintaining a constant unidirectional, sliding velocity between the block and ring. Wear tests were carried out at 25, 50 and 100 N contact loads, under 2.3 m/s sliding speed. The test specimens were subjected to a cyclic load with a total sliding distance of 6.280 m. The sample mass loss calculations and working surfaces topographic analyses was performed after each 1256-meter continuous sliding. The quenched carbon steel (0.9% carbon content) disk with 62 HRC hardness was used as a counter-body (Figure 1).

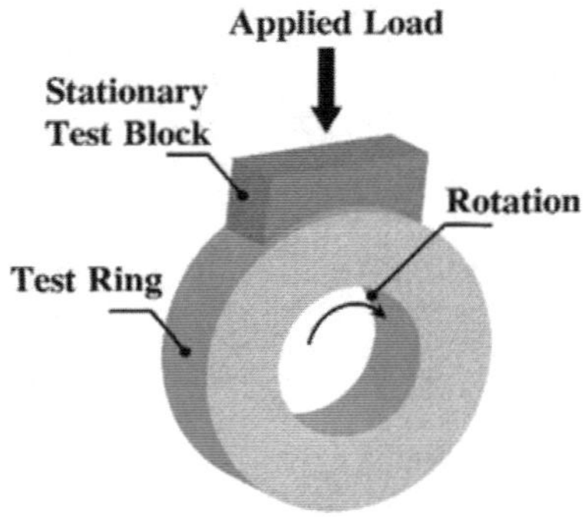

Figure 1. Schematic diagram of the block-on-ring friction and wear test

During each test, the evolution of the friction coefficient was recorded. The worn out surfaces of specimens were studied by Energy Dispersive X-ray (EDX) microanalysis using scanning electron microscope. Friction-induced vibration caused by dry sliding were defined using vibrometer El-Calc during wear tests.

3. Results and discussion

In the presented work, 4 groups of samples were studied, which differed in the type of metal matrix and the amount of metastable retained austenite. The formation of various amounts of metastable retained austenite in the structure was achieved by interrupting the bainitic transformation at different time intervals. As a result of researches, it was determined that, under contact load of 25 N, isothermally quenched at 280 °C and 400 °C cast irons have practically same coefficient of friction (Figure 2 a, b).

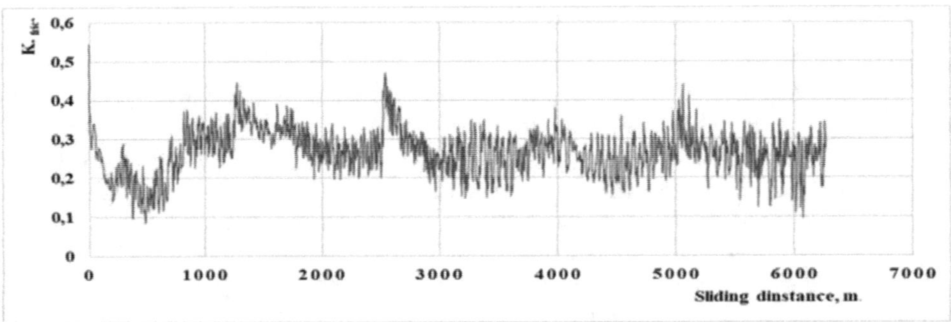

a

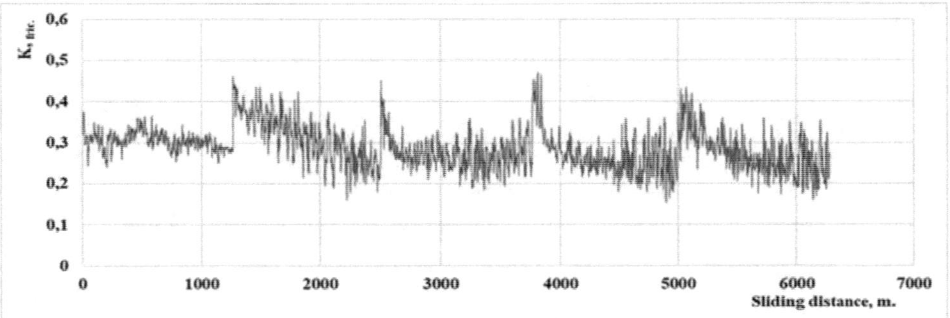

b

Figure 2. The friction coefficient evolution of ADI under sliding distance a –Isothermal quenched at 400^{0}C; b –Isothermal quenched at 280^{0}C;

The cast irons with a content of ≈ 75% metastable retained austenite are characterized by most stable values of this parameter during tribological testings. In the cast irons with a lower amount (≈ 60%) of retained austenite, a cyclic fluctuation of friction coefficient is observed in the range 0.22 ... 0.45. During sliding, the periodic decreases of the friction coefficient reflects the self-hardening processes of the material surface layers and allows us to conclude, that under load, the partial γ — α' (martensitic) transformation is going on. Further mechanical-thermal action contributes the decomposition of surface martensite with the formation of high dispersed carbide particles, which causes relaxation of maximum internal stresses and ensures high wear resistance of bainitic cast iron. During the process of contact interaction under a load of 25 N, the temperature of investigated samples reached to 150, 200° C in the friction zone. With an increasing the contact load to 50 N, the temperature was risen to 350 ... 400 °C, and under load of 100 N, the surface volumes of cast iron were heated to 700 ... 730° C.

Analysis of friction coefficient values change kinetics and the wear surface topography of experimental samples allows us to conclude, that the destruction of cast-irons surface layers, microalloyed with boron has a fatigue nature during the contact interaction process under a load of 25 N on the friction pair. The surface of the samples indicates a low wear intensity, are not observed the traces of selective transfer individual phases and overcovering of graphite inclusions, which effectively perform as a solid lubricant for a long time.

At higher contact loads (50 N and 100 N), the intensity of frictional interaction and power of formed heat flow increases. This contributes to an increase in the temperature of the sample's active volumes and the passing of plastic hardening processes of the surface layers. At the same time the range of oscillations and values of friction coefficient is increasing (Figure 3).

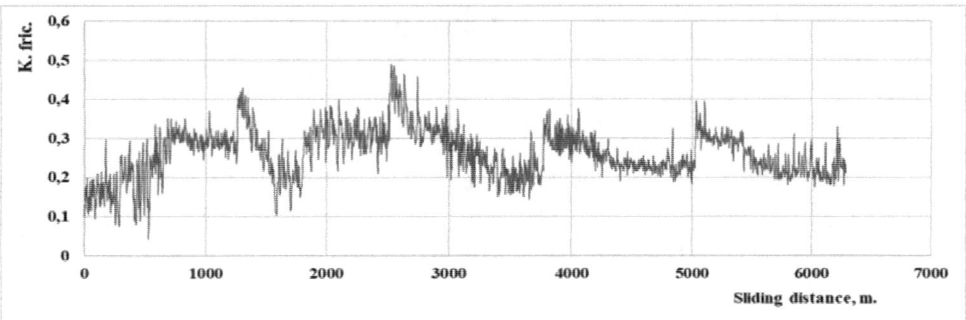

a

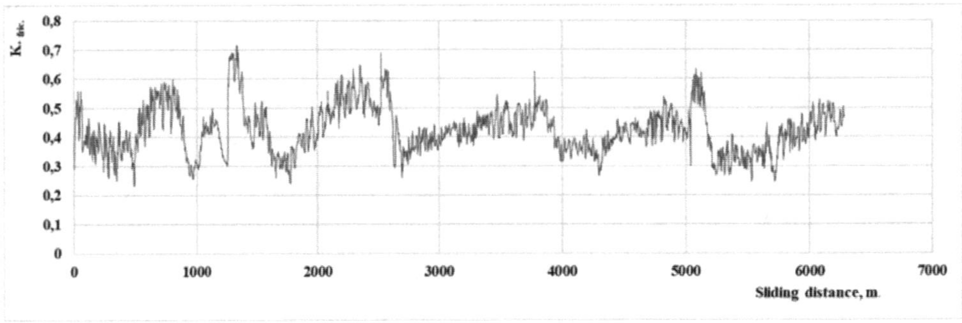

b

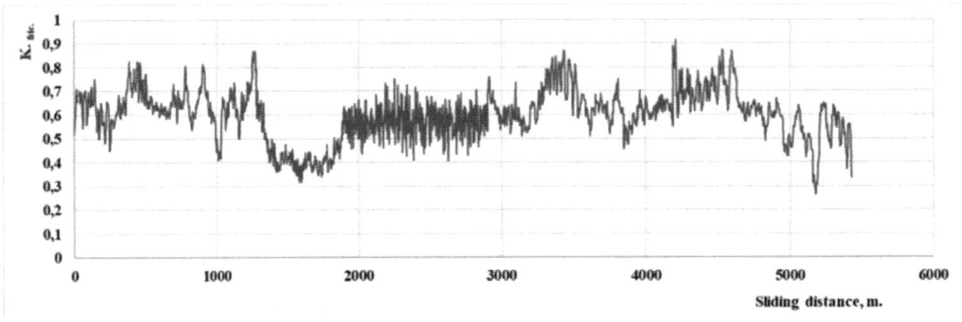

c

Figure 3. The friction coefficient of ductile iron isothermal quenched at 400°C (60% retained austenite) depending under different contact load: a- 25N; b-50N; c-100N;

High temperature and plastic deformation accelerate diffusion processes and coagulation the structural constituents, increase the rate of damage accumulation, softening and destruction the working layers of high-strength cast iron. According to the data obtained, isothermally quenched at 280 °C high strength cast iron are differed by more stable values of friction coefficient under the studied conditions of dynamic loading. This allow to conclude that the structure of lower bainite more efficient provides relaxation of internal stresses and the deformation processes keep homogeneity. The advantages of bainitic cast irons with a lower bainite structure are demonstrated to a greater extent under 100 N dynamic loads. The wear resistance of these alloys is 1.21 -1.6 times higher than cast irons with corresponding indicators, quenched at 400° C (Figure 4).

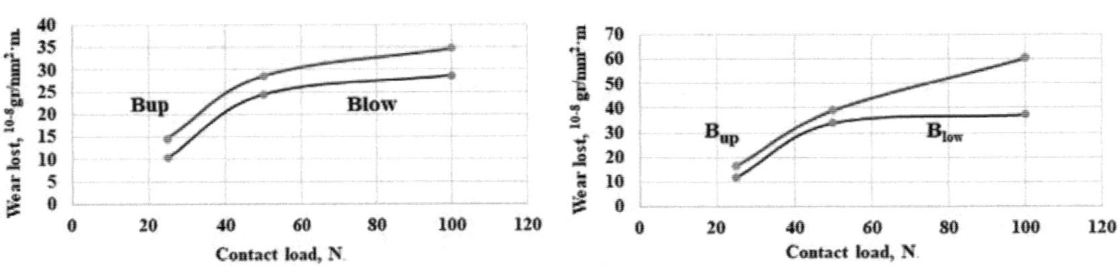

Figure 4. Total wear volume of austempered ductile iron with 60% austenite content as the function under different loads

It should be noted that cast irons with a more content of metastable austenite at high contact loads are differed by stability of tribological properties and a low wear rate (Figure 5).

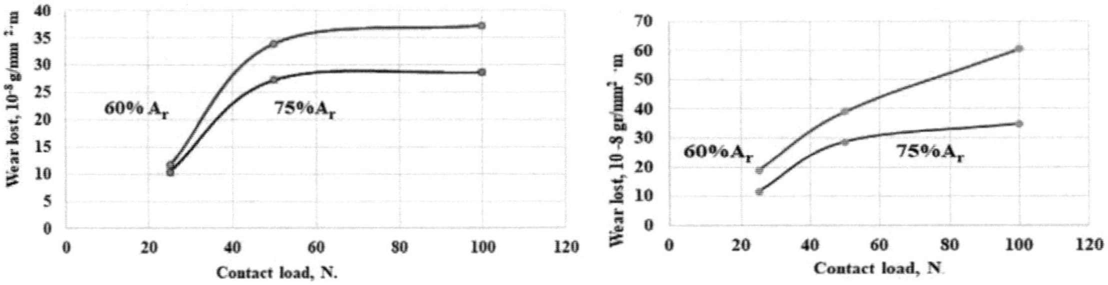

Figure 5. Total wear volume of austempered ductile iron with 75% austenite content as the function under different loads

Energy dispersive spectroscopy (EDS) analysis of friction surfaces showed that the structure of the upper bainite is oxidized more intensity under a contact load of 100 N and approximately 52% more oxygen (O) is defined in the surface layers of the experimental samples (Figure 6).

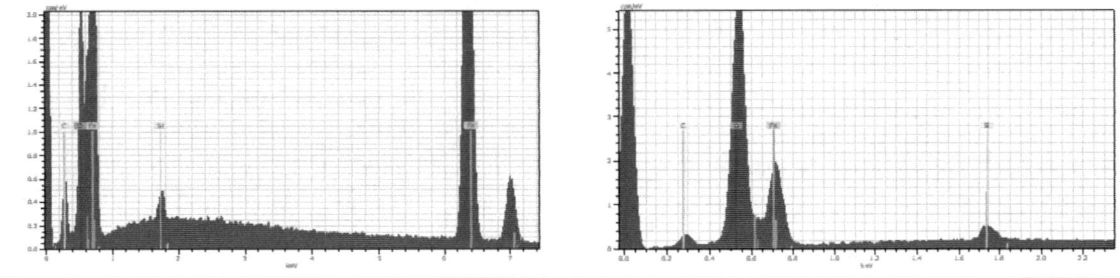

Spectrum: Point

Element	AN	Series	norm. C [wt.%]	Atom. C [at.%]
Iron	26	K-series	83.20	60.16
Oxygen	8	K-series	13.67	28.27
Carbon	6	K-series	2.28	10.57
Silicon	14	K-series	0.85	1.00
		Total:	100.00	100.00

Spectrum: Point

Element	AN	Series	norm. C [wt.%]	Atom. C [at.%]
Iron	26	K-series	62.78	35.28
Oxygen	8	K-series	33.67	58.55
Carbon	6	K-series	2.29	4.92
Silicon	14	K-series	1.26	1.25
		Total:	100.00	100.00

Figure 6. EDS spectrum of austempered ductile iron, isothermal quenched at 280°C (a) and 400 °C (b) with 60% retained austenite content

The datas of vibration-frequency studies show that during the testing of cast iron samples with a lower bainitic structure, the more low amplitude of frictional vibrations are defined and a noise level is 7 - 9 dB lower than in samples with an upper bainitic structure in the process of dynamic loading (Figure7).

So, we can conclude that bainitic cast irons microalloyed by boron have high wear resistance under rational structuring and investigated conditions of frictional loading.

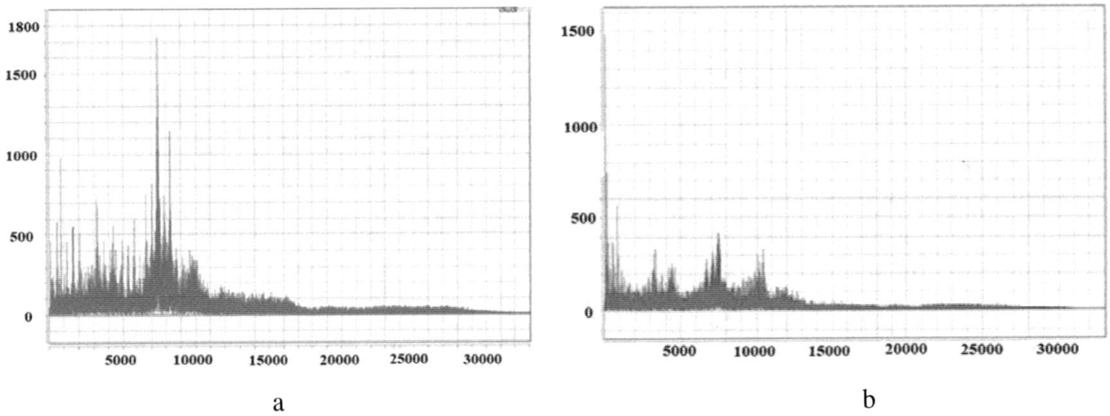

Figure 7. Typical time-domain frequency spectrum induced by friction vibrations of ductile iron under dry sliding: a -isothermal quenched at 400ºC; b - isothermal quenched at 280ºC

This is explained by the fact that dispersed borides, carbides, and boron nitrides located along the grain boundaries strengthening the metal matrix and increase its resistance to plastic deformation. In addition, the formation in the structure optimal amounts of metastable austenite provides cyclic strengthening of surface layers under contact loads influence and slows down the process of internal stresses accumulation, which in accordance with the data of works [14, 15]. This also ensures the stability of tribological characteristics during long-term frictional interaction.

4. Conclusions

- It was defined that the value of the contact load has a significant effect on the tribotechnical characteristics of bainitic cast irons in conditions of dry sliding friction.
- High strength cast irons with the lower bainite structure are more resistant to adhesive wear.
- According to the obtained data, the presence of up to 80% metastable retained austenite in the metal matrix structure of the of experimental cast irons makes it possible to stabilize the friction coefficient under the investigated contact loads.
- Thus, to ensure maximum wear resistance, the isothermal quenching of high-strength cast irons must be carried out in the temperature range 220-280^{0}C with a minimum treatment time.

Acknowledgment(s)

This work was supported by Shota Rustaveli Georgian National Science Foundation of Georgia (SRGNSFG) under GENIE project [grant number № CARYS-19-802].

References

[1] A.V. Kolubaev, E.A. Kolubaev, O.V. Sizova. Structure, deformation, and fracture of hard coatings during sliding friction Russian Physics Journal. T. 62. № 8. C. 1363-1397. 2019.

[2] B. Wang, F. Qiu, G.C. Barber, Y. Pan, W. Cui, and R. Wang. *Microstructure, wear behavior and surface hardening of austempered ductile iron. Journal of Materials Research and Technology*, 9(5), 9838–9855, 2020.

[3] J. K. Kaleicheva, and V. Mishev. *Wear Resistance of Austempered Ductile Iron with Nanosized Additives. IOP Conference Series: Materials Science and Engineering*, 295, 012034, 2018.

[4] T. Sarkar, and G. Sutradhar. Investigation on mechanical properties and wear behavior of Cu-alloyed austempered gray cast iron (AGI). Sādhanā, 43(10), 2018.

[5] B. Wang, X. Han, G.C. Barber, and Y. Pan. *Wear Behavior of Austempered and Quenched and Tempered Gray Cast Irons under Similar Hardness. Metals*, 9(12), 1329, 2019.

[6] A. Bedolla-Jacuinde, F.V. Guerra, M. Rainforth, I. Mejía, and C. Maldonado. *Sliding wear behavior of austempered ductile iron microalloyed with boron. Wear*, 330-331, 23–31, 2015.

[7] P. Sellamuthu, D. Samuel, D. Dinakaran, V. Premkumar, Z. Li, and S. Seetharaman. *Austempered Ductile Iron (ADI): Influence of Austempering Temperature on Microstructure*, Mechanical and Wear Properties and Energy Consumption. Metals, 8(1), 53, 2018.

[8] A. Bedolla-Jacuinde, F.V. Guerra, M. Rainforth, I. Mejía, and C. Maldonado. *Sliding wear behavior of austempered ductile iron microalloyed with boron. Wear*, 330-331, 23–31, 2015.

[9] R. Kumar, R.K. Dwivedi, and S. Ahmed. *Influence of Multiphase High Silicon Steel (Retained Austenite-RA, Ferrite-F, Bainite-B and Pearlite-P) and Carbon Content of RA-Cγ on Rolling/Sliding Wear. Silicon*, 2020.

[10] A. Bedolla-Jacuinde, F.V. Guerra, M. Rainforth, I. Mejía, and C. Maldonado. Sliding wear behavior of austempered ductile iron microalloyed with boron. Wear, 330-331, 23–31, 2015.

[11] S. Yazdani, and M.A. Rahimi. *Wear Behavior of an Austempered Ductile Iron Containing Mo-Ni-Cu. Materials Science Forum*, 475-479, 199–202, 2005.

[12] B. Li, and Z. Rui. *Friction and wear behaviours of YG8 sliding against austempered ductile iron under dry, chilled air and minimal quantity lubrication conditions. Advances in Mechanical Engineering*, 11(5), 168781401984706, 2019.

[13] G. Oniashvili, G. Tavadze, G. Zakharov, and Z. Aslamazashvili. Self-propagating High-temperature Synthesis of Ferroalloys. LAP Lambert Academic Publishing. 2018.

[14] J.S. Sokolan. Effect of external friction on deformational aging of steel taking into consideration its thermal treatment. Problems of Tribology, № 4,з.73-80, 2014.

[15] A. P. Cheiliakh. Ekonomnolegirovannye metastabil'nye splavy i uprochniaiushchie tekhnologii. [Economically alloyed metastable alloys and the strengthening technologies]. Mariupol': PGTU, 483 p., 2009.

Challenges in developing of 3D nonlinear viscoelastic models

Liva Pupure[1], Leonids Pakrastins[1], Janis Varna[1,2]

[1]Institute of Structural Engineering and Reconstruction, Riga Technical University, Riga, Latvia
[2]Division of Materials Sciences, Lulea University of Technology, Lulea, Sweden

Liva.Pupure@rtu.lv

Abstract. Due to development of different material systems and structural elements, there is growing need for more complex material models, since linear elasticity or even more complicated linear viscoelasticity cannot capture the nuances of behaviour of these materials. This study describes the process and difficulties of development of 3D nonlinear viscoelastic model. Study proposes two models with different level of complexity and methodology for experimental parameter identification. The limitations of both of these models have been discussed.

1. Introduction

The advancement in materials, requires more and more complicated material models. For example, Bio-based materials offer an environmentally friendlier alternative to conventional composite materials [1]. However, it has been demonstrated that behaviour of bio-based composites and their constituents cannot be characterized by linear elasticity or even more complicated linear viscoelasticity (VE) [2-6]. There has been a great effort made to use full potential of different materials, thus significantly reducing weight, volume etc. of different components. This would require materials to function in more demanding environments e.g. elevated temperature and humidity, used at partially cured stages etc. Often these factors significantly contribute to nonlinear behaviour of materials. Thus, it is crucial to develop material models, that could capture all these nuances.

The most widely used nonlinear VE models have been developed by Schapery [7]. With advancement of structures, the complexity of elements also increases. VE analysis of structures using finite element modeling requires VE models in 3D formulation accounting for changes in material state and environmental conditions that can be easily implemented in computational codes This paper is focused on the problems and difficulties of developing a 3D model and experimental parameter identification. Two models of different complexity have been proposed. The difficulties of experimental setup for parameter identification have been discussed.

2. Material models

Schapery developed stress formulation for nonlinear VE model based on thermodynamics [7-8]. This model was further modified to account for viscoplasticity (VP) and damage [9] and has been successfully applied to characterize different materials under load controlled tests [10-11]. In uniaxial loading it follows:

$$\varepsilon = d(\sigma_{max})\left(\varepsilon_{el} + g_1 \int_0^t \Delta S(\psi - \psi')\frac{d(g_2\sigma)}{d\tau}d\tau + \varepsilon_{VP}(\sigma, t)\right) \quad (1)$$

where $d(\sigma_{max})$ represents effect of damage, $\varepsilon_{VP}(\sigma, t)$ is VP strain component and the rest is VE, where ε_{el} is initial elastic response and $g_1 \int_0^t \Delta S(\psi - \psi')\frac{d(g_2\sigma)}{d\tau}d\tau$ is time dependent part. In order to obtain all parameters a large number of time-consuming tests must be performed – tests introducing damage and measuring stiffness degradation, multiple-step creep tests at different stress levels for VP and VE.

2.1. Damage

In different material systems, such as composites, microcracks, cracks, delamination etc. can appear during service life. This can cause degradation of elastic modulus. This degradation can accumulate as a result of previous loading history to higher stresses/strains. In cases when the damage state is very complex, as it is in short fibre composites, a more pragmatic approach is to find the modulus reduction as a function of the applied strain (or stress) in tensile test.

Microdamage parameter $d(\sigma_{max})$ in [9] has been expressed as ratio between modulus in undamaged state E_0 and modulus in the damaged $E(\sigma_{max})$ state that usually corresponds to the highest stress σ_{max} experienced during the previous service life:

$$d(\sigma_{max}) = \frac{E_0}{E(\sigma_{max})} \quad (2)$$

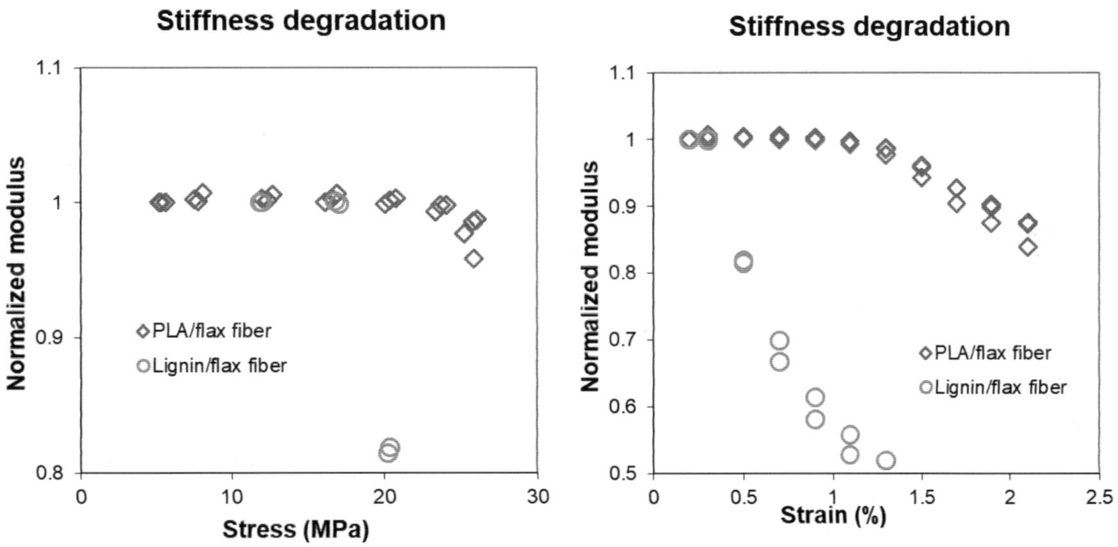

Figure 1. Normalizes modulus E/E_0 (Elastic modulus at damaged state versus modulus at undamaged state) as a function of stress (left) and strain (right) for Flax fiber/lignin and flax fiber/polylactic acid (PLA) composite. Data from [10,12,13]

The tensile loading ramp for damage parameter identification in details has been described in [12,13]. The proposed tests consist of a sequence of four main steps:

1) new damage introduction by loading to certain (high) strain or stress level and unloading to "almost zero" stress;

2) recovery period at "almost zero" stress for a time at least 5 times longer than the length of the previous step for decay of all VE effects;

3) the elastic modulus determination applying high strain rate and low- load level loading-unloading;

4) VE-strain recovery after modulus measurement for the time that is at least 5 times longer than the time interval for modulus measurement.

After the last step, the sequence is repeated with higher strain level than in the first step. The data of modulus obtained from these tests then is plotted and fitting curve for damage is obtained. An example of elastic modulus reduction is presented in Figure 1. VP strain may also develop during the first steps of the loading sequence thus resulting in situations where the same damage might be corresponding to different strain levels. Therefore, it is more suitable to analyze damage as a function of applied stress not strain.

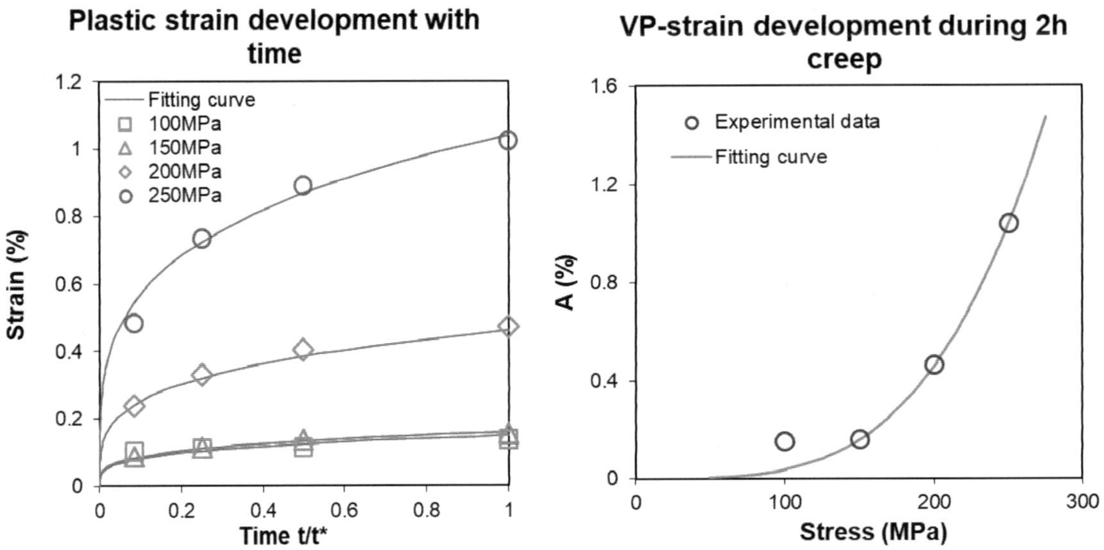

Figure 2. VP strain development in regenerated cellulose fibers with respect to time (left) and stress (right). Data from [15]

2.2. Viscoplasticity

Term $\varepsilon_{VP}(\sigma, t)$ represents irreversible VP strains that develop at high stresses and increase with time of loading. In [14] a model for VP strains was suggested:

$$\varepsilon_{VP}(\sigma, t) = C_{VP}\left\{\int_0^{\frac{t}{t^*}} \left(\frac{\sigma(\tau)}{\sigma*}\right)^M d\tau\right\}^\alpha \tag{3}$$

C_{VP}, M and α are material constants, that are dependent on different environmental conditions, temperature, moisture, degree of cure etc., t/t^* is normalized time where t^* is an arbitrary chosen characteristic time constant and σ^* is an arbitrary chosen reference constant stress value. The validity of this particular model has to be verified experimentally. This VP model was successfully used in [9-11,14] to characterize VP strain development.

In case of constant applied stress $\sigma = \sigma_0$, the VP strain dependency on the creep time t becomes very simple:

$$\varepsilon_{VP}(\sigma, t) = C_{VP}\left(\frac{\sigma_0}{\sigma*}\right)^{M\alpha}\left(\frac{t}{t*}\right)^\alpha \tag{4}$$

IOP Conf. Series: Materials Science and Engineering 1190 (2021) 012005 doi:10.1088/1757-899X/1190/1/012005

Thus, if the model (3) is applicable, the time dependence of VP strains in constant stress (creep) test should follow a power function with respect to time:

$$\varepsilon_{VP}(\sigma, t) = \mathrm{A}\left(\frac{t}{t*}\right)^{\alpha} \tag{5}$$

Where A has power law dependence on the applied stress level in the creep test:

$$A = C_{VP}\left(\frac{\sigma_0}{\sigma*}\right)^{M\alpha} \tag{6}$$

All parameters in the VP law can be determined from sequence of creep loading and VE strain recovery performed at different stress levels. After each recovery period, the remaining strain is considered as irreversible. Both equations (5)-(6) in ln-ln axis become linear, thus fitting experimental data in these axes are the easiest way to determine constants in VP law (see figure 2). However, this methodology requires specimen with no previous loading history – at each stress level different specimen must be used. This increases complexity of data analysis for materials with high scatter – the trends of VP can be lost in variations between specimens. Thus another, more complicated methodology has been developed for such cases, where the described tests at different loads are performed on the same specimen. Details of both methodologies can be found in [15].

2.3. Nonlinear viscoelasticity

In material model (1) for VE reduced time is used:

$$\Psi = \int_0^t \frac{dt\prime}{a_\sigma} \quad \text{and} \quad \Psi' = \int_0^\tau \frac{dt\prime}{a_\sigma} \tag{7}$$

In (1), the initial elastic response ε_{el}, generally speaking, may be nonlinear function, g_1 and g_2 are nonlinearity functions and a_σ is the shift factor. Some materials behave or have a region of linear VE, where $g_1 = g_2 = a_\sigma = 1$. Schapery showed that the linear VE creep compliance can be written in form of Prony series [8]:

$$\Delta S(\Psi) = \sum_i C_i\left(1 - exp\left(-\frac{\Psi}{\tau_i}\right)\right) \tag{8}$$

where C_i are constants and τ_i are retardation times. When damage and VP has been identified, they can be removed from the total strain to extract pure VE response. Afterwards the VE during creep and recovery can be characterized by equations:

$$\varepsilon_{creep} = \varepsilon_0 + g_1 g_2 \sigma \sum_i C_i\left(1 - exp\left(-\frac{t}{a_\sigma \tau_i}\right)\right) \tag{9}$$

$$\varepsilon_{rec} = g_2 \sigma \sum_i C_i\left(1 - exp\left(-\frac{t_1}{a_\sigma \tau_i}\right)\right) exp\left(-\frac{t-t_1}{\tau_i}\right) \tag{10}$$

where t_1 is the length of the creep test. First the coefficients in Prony series are identified – this procedure is much simpler, if linear VE region can be found. Afterwards, one can move to higher stress regions and find nonlinearity functions. Figure 3. shows the pure VE strains in creep tests for bio-based resin "EpoBioX" and VE creep compliance curves for pure high-density polyethylene (HDPE). It can be seen that with increasing stress level the compliance increases, thus indicating that material is nonlinearly VE, even at relatively low stress levels. In Figure 4. Examples of the VE nonlinearity functions for resin "EpoBioX" and Lignin/flax fiber composite are presented.

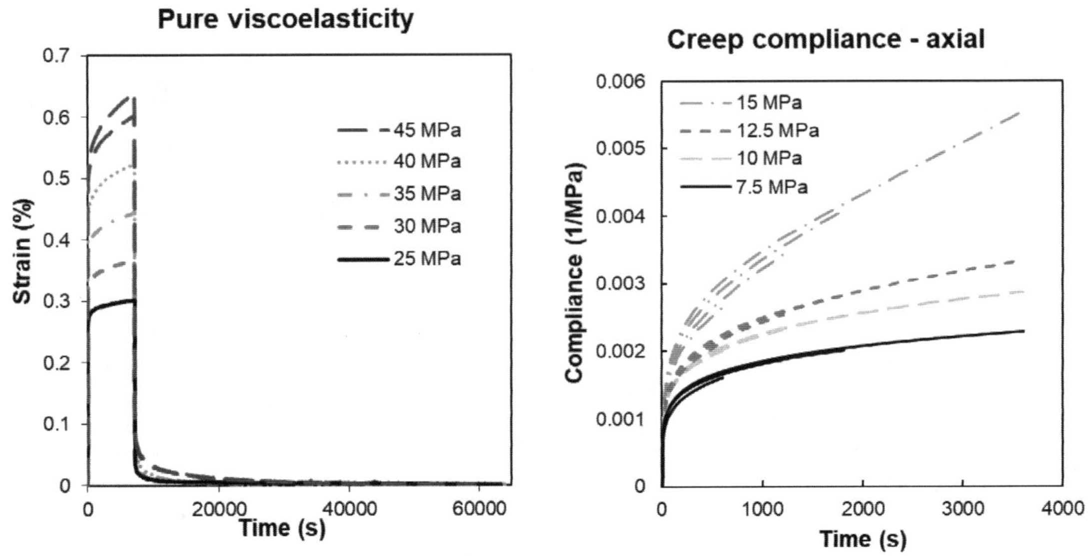

Figure 3. Pure VE strain in creep test for "EpoBiox" (left) and creep compliances for HDPE (right). Data from [16,17]

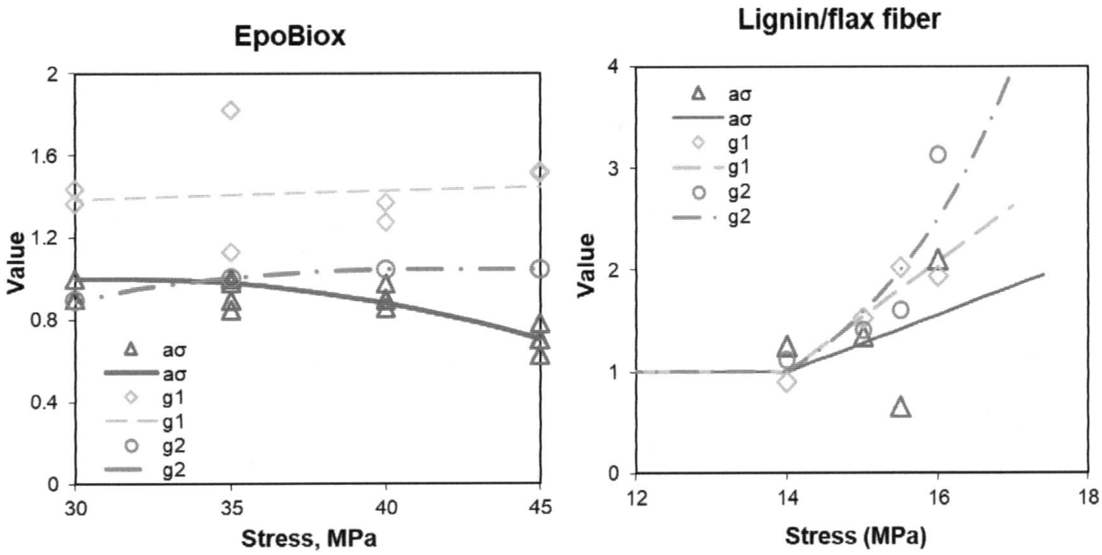

Figure 4. VE nonlinearity functions for bio-based resin "EpoBioX" (left) and Lignin/flax fibers (right). The dots represent experimentally obtained values and lines the fitting function. Data from [12, 16]

2.4. Strain formulation

Although stress-formulation of the material model can be successfully applied to characterize nonlinear material behavior, there is growing need to simulate strain-controlled test. This is because most of the numerical structural analysis, analytical micromechanics models (rule of mixture, concentric cylinder assembly model etc.), classical laminate theory require model, where stresses are expressed as a function of strain, time etc. Moreover, many of experiments are performed in strain-controlled mode.

Similarly, to stress-formulation, Schapery [18,19] has developed nonlinear VE model where VE strain is used as an independent variable using expansion of Helmholzs' free energy. In one-dimensional case, the constitutive law is in form:

$$\sigma = E_r \varepsilon + h_1(\varepsilon) \int_0^{\psi_\varepsilon} \Delta E(\psi_\varepsilon - \psi_\varepsilon') \frac{d(h_2 \varepsilon)}{d\psi_\varepsilon'} d\psi_\varepsilon' \quad (11)$$

In (11) E_r is equilibrium modulus that, generally speaking, may be nonlinear function of strain. The "reduced time" ψ_ε is introduced in (11) as:

$$\psi_\varepsilon = \int_0^t \frac{dt'}{a_\varepsilon} \text{ and consequently } \psi_\varepsilon' = \int_0^\tau \frac{dt'}{a_\varepsilon} \quad (12)$$

The transient part of the VE response is characterized by $\Delta E(\psi_\varepsilon)$ which does not depend on stress and have a form of Prony series:

$$\Delta E(\psi_\varepsilon) = \sum_i E_i exp\left(-\frac{\psi_\varepsilon}{\tau_i}\right) \quad (13)$$

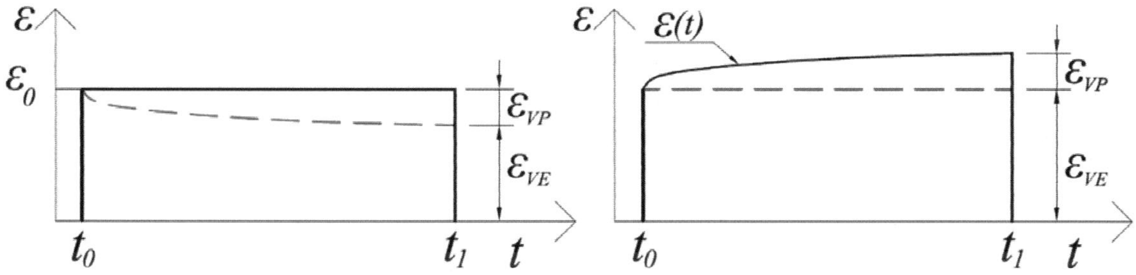

Figure 5. Schematic drawing of VP and VE strain development during relaxation test with applied constant strain (left) and increasing applied strain to account for VP strain development (right). Red line represents pure VE strain.

In (13) E_i are constants and λ_i are relaxation times. Nonlinearity functions h_1 and h_2 and the shift factor a_ε are strain invariant dependent. If it is possible to find region where E_r is a linear function of the applied strain and $h_1 = h_2 = a_\varepsilon = 1$, equation (11) is reduced to stress-strain relationship for linear VE material. As demonstrated in [18,19], in case of nonlinear VE both forms of the material model are not compatible and cannot be directly inverted to each-other. Moreover, in order to obtain parameters and functions for the model in strain formulation, relaxation tests where VE strain is kept constant need to be performed. However most of materials have also VP strain component, thus to perform such experiments becomes very complicated. If the strain is kept constant, during time, the VP strain develops, thus in reality, the VE strain is not constant but decreases. This would require the strain to be increased, to compensate for the VP strain. The visual description of this problem is presented in Figure 5. and discussed in more details in [20]. These issues can be resolved by using an inverted incremental form of the stress-formulation derived in [21]. This model is used to simulate pure VE relaxation tests using parameters that were obtained from creep tests, avoiding the problems with experimental relaxation test needed for parameter identification for strain formulation. Then these simulated relaxation tests are used to obtain parameters for strain formulation. This methodology was successfully used in [22] to simulate strain controlled tensile tests. The agreement between experiments and simulations was great if the characterization in creep was at stress levels close to rupture. However, in cases when it was not possible to characterize material behavior at high stresses due to creep rupture, the agreement between simulations and experiment was good only in the stress region where creep characterization was performed. At higher stresses, the method of extrapolation has significant effect on final curves. It was demonstrated that further adjustments of nonlinearity functions can be done with help of simple tensile tests. In order to use models described in previous sections, a large number of time consuming tests and fittings needs to be performed. In case of more complex material systems such as composite materials, due to large number of available matrix and reinforcement (or filler) combinations, each time one component or morphology changes, the full characterization needs to be performed again. Thus, it would be more convenient to use models, where only the constituents have to be fully characterized in nonlinear VE. Such approach has been described in [16].

2.5. 3D nonlinear viscoelastic material model with parameters dependent on material state and temperature
In this Section the VE related Internal State Variables (ISV) have notation $\boldsymbol{\xi_m}$. All matrixes, vectors (matrix-column) that have constant elements have upper index 0. Everything without "0" may be temperature and strain dependent. Indexes i, j, l 1,...,6 are used for stress and strain and related functions. Indexes m, n, k are used for ISV related functions. Free thermal expansion and curing shrinkage strains are not included in this derivation. This means that "strain" is actually "the applied strain minus the free expansion strain". Following the scheme suggested by Schapery [8] the stress – strain relationship follows from:

$$\sigma_i = \frac{\partial H}{\partial \varepsilon_i}, \qquad H = H^e\left(\alpha, T, \varepsilon_j\right) - \sum_m A_m(\alpha, T, \hat{\varepsilon})\xi_m + \frac{1}{2}\sum_{mn} B_{mn}(\alpha, T, \hat{\varepsilon})\xi_m\xi_n \tag{14}$$

where H is the Helmholtz free energy presented in form of expansion with respect to VE internal state variables (ISV) ξ_m, $m = 1, M$. Parameter α represents the effect on VE of the material state (degree of cure, aging etc.), T is temperature and $\hat{\varepsilon}$ strain invariant dependent function. We assume that in the initial state $\sigma(0) = \varepsilon(0) = \xi_m(0) = 0$. B_{mn} is symmetric matrix with dependence on $\alpha, T, \hat{\varepsilon}$ as follows:

$$B_{mn} = B_{mn}^0 a_2(\alpha, T, \hat{\varepsilon}) \tag{15}$$

Using (14) and neglecting quadratic terms with respect to ISV's:

$$\sigma_i = \frac{\partial H^e}{\partial \varepsilon_i} - \sum_m \frac{\partial A_m}{\partial \varepsilon_i}\xi_m \tag{16}$$

We assume that the evolution rate of ISV is linear with respect to conjugated thermodynamic forces

$$\frac{d\xi_m}{dt} = \frac{1}{a_1(\alpha,T,\hat{\varepsilon})}\sum_n c_{mn}^0 f_n, \text{ where } f_n = -\frac{\partial H}{\partial \xi_n} = A_n(\alpha, T, \hat{\varepsilon}) - \sum_k B_{kn}^0 a_2(\alpha, T, \hat{\varepsilon})\xi_k \tag{17}$$

In (17) c_{mn}^0 is a positive definitive, symmetric matrix, $a_1 > 0$, can be assumed equal to 1 in the reference state. Using notations [ ] for square-matrix and { } for a column and introducing reduced time ψ as:

$$d\psi = \frac{a_2(\alpha,T,\hat{\varepsilon},)}{a_1(\alpha,T,\hat{\varepsilon})}dt, \qquad \psi(t) = \int_0^t \frac{a_2(\alpha,T,\hat{\varepsilon})}{a_1(\alpha,T,\hat{\varepsilon})}d\zeta \tag{18}$$

from (17) follows:

$$[c^0]^{-1}\frac{d}{d\psi}\{\xi\} + [B^0]\{\xi\} = \frac{1}{a_2}\{A\} \tag{19}$$

Then this system of linear differential equations with constant coefficients is rearranged by introducing new set of ISV (as linear combination of the previous ISVs) so that for the new ISVs the constant matrices have diagonal form and the system (6) diverges to independent equations. Using for these new ISV and for the diagonal elements of these matrices the same notation as for the old ones, ξ_n, c_m^0 and B_m^0:

$$c_m^0 \frac{d\xi_m}{d\psi} + B_m^0 \xi_m = \frac{A_m}{a_2} \qquad m=1,2,3.... \tag{20}$$

Solution (20) is as follows:

$$\xi_m = \frac{A_m}{a_2 B_m^0} - \frac{1}{B_m^0} \int_0^\psi exp\left(-\frac{\psi-\psi'}{\tau_m}\right) \frac{d}{d\psi'}\left[\frac{A_m}{a_2}\right] d\psi', \text{ where } \tau_m = \frac{c_m^0}{B_m^0} \tag{21}$$

At this point, no assumptions have been made regarding A_m. Two models will be presented now with different complexity regarding A_m. The aim is to compare the resulting stress-strain equations, to see the differences and to discuss their potential regarding applicability and complexity.

First, following suggestion by Brouwer [23] we assume (referred as **Model A**) that for all strain components the dependence of A_m on T, α and $\hat{\varepsilon}$ is the same:

$$A_m = h(\alpha, T, \hat{\varepsilon}) \sum_j D_{mj}^0 \, \varepsilon_j \tag{22}$$

$$\frac{\partial A_m}{\partial \varepsilon_i} = h(\alpha, T, \hat{\varepsilon}) D_{mi}^0 + \frac{\partial h}{\partial \varepsilon_i} \sum_j D_{mj}^0 \, \varepsilon_j \tag{23}$$

Equation (22) is substituted in (21) for ξ_m. After that, the result and (23) are used in (16) obtaining the following stress-strain relationship:

$$\sigma_i = \Gamma_i^e(\alpha, T, \hat{\varepsilon}) + h \cdot \sum_j S_{ij} + \frac{\partial h}{\partial \varepsilon_i} \cdot \sum_j \varepsilon_j \sum_l S_{il} \tag{24}$$

$$S_{ij} = \int_0^\psi \Delta E_{ij}(\psi - \psi') \frac{d}{d\psi'}\left(\frac{h}{a_2} \varepsilon_j\right) d\psi' \tag{25}$$

$$\Delta E_{ij}(\psi - \psi') = \sum_m E_{ij}^{0(m)} exp\left(-\frac{\psi-\psi'}{\tau_m}\right) \tag{26}$$

Obtaining these expressions, new unknown constants and functions were introduced instead of the used above, for example $E_{ij}^{0(m)} = D_{mi}^0 \cdot D_{mj}^0 / B_m^0$. This model, in addition to Prony coefficients and the equilibrium response Γ_i^e, contains three unknown functions dependent on strain invariants, T and α: $h(\alpha, T, \hat{\varepsilon})$, $\frac{h(\alpha,T,\hat{\varepsilon})}{a_2(\alpha,T,\hat{\varepsilon})}$ and $\frac{a_1(\alpha,T,\hat{\varepsilon})}{a_2(\alpha,T,\hat{\varepsilon})}$ (in definition of ψ), the same number of functions as in the 1D model. However, the material class, where the assumption (22) is applicable, is *a priori* not known. Thus, the model's applicability has to be verified in experiments.

The second model (**Model B**) has more "flexibility" in the sense that each strain component may have a different material state, environment and loading dependent function:

$$A_m = \sum_j D_{mj}^0 f_j(\alpha, T, \hat{\varepsilon}) \varepsilon_j \tag{27}$$

$$\frac{\partial A_m}{\partial \varepsilon_i} = D_{mi}^0 f_i + \sum_j D_{mj}^0 \frac{\partial f_j(\alpha,T,\hat{\varepsilon})}{\partial \varepsilon_i} \varepsilon_j \tag{28}$$

The stress-strain relationship is as follows:

$$\sigma_i = \Gamma_i^e(\alpha, T, \hat{\varepsilon}) + f_i \cdot \sum_j J_{ij} + \sum_j \frac{\partial f_j}{\partial \varepsilon_i} \cdot \varepsilon_j \sum_l J_{jl} \tag{29}$$

$$J_{ij} = \int_0^\psi \Delta E_{ij}(\psi - \psi') \frac{d}{d\psi'}\left(\frac{f_j}{a_2} \varepsilon_j\right) d\psi' \tag{30}$$

The eight unknown functions $f_i, i = 1,2,..,6$, a_2 and a_2 have to be found experimentally. Hence. **Model B** requires larger amount of more complex tests for model identification than **Model A**. For example, one test has to be bi-axial creep test, hence significantly increasing the difficulty of experimental parameter identification. On the other hand, this inconvenience may be justified for anisotropic materials. For example, for composites the stress nonlinearity effect can be very different in fiber direction than in shear loading. One solution for experimental parameter identification could be, that some of the parameters and functions are adjusted until simulations fit the experimental data. This, of course, would require several iterations, where parameters are changed. It is not clear that this identification procedure has to lead to unique solution. Unfortunately, it is possible that for certain loading ramps different combinations of parameters could lead to the same or very similar result.

3. Conclusions

The need for more complex 3D nonlinear model that in addition to viscoelasticity also accounts for viscoplasticity and damage has been demonstrated. The 1D model that includes all of the above-mentioned phenomena and that has been used successfully to characterize complex material behaviour has been used to illustrate the inelastic effects. Stress and strain formulations of the nonlinear viscoelastic model, developed by Schapery, have been discussed. The pros and cons of each approach has been considered.

Two 3D nonlinear viscoelasticity models have been introduced. Model A is simpler, due to fever experimentally determined functions needed as input. The experimental procedure for parameter identification for this model is also relatively simple – the same number of tests as for 1D models' case. However, it is unclear if the assumptions used developing this model would be valid for anisotropic materials. This model could be more suitable for isotropic materials. Model B is more complex with more functions needed for input as well as the experimental identification of parameters is more sophisticated. However, this model could be more accurate in predicting behavior of complex material systems.

Acknowledgment(s)

This work has been supported by the European Regional Development Fund within the Activity 1.1.1.2 "Post-doctoral Research Aid" of the Specific Aid Objective 1.1.1 "To increase the research and innovative capacity of scientific institutions of Latvia and the ability to attract external financing, investing in human resources and infrastructure" of the Operational Programme "Growth and Employment" (No.1.1.1.2/VIAA/4/20/646).

References

[1] S. V. Joshi, L. T. Drzal, A. K. Mohanty, S. Arora. "Are natural fiber composites environmentally superior to glass fiber reinforced composites?" *Composites Part A: Applied Science and Manufacturing*, vol. 35, pp. 371-376, 2004.

[2] E. Marklund, J. Varna and L. Wallström. "Nonlinear Viscoelasticity and Viscoplasticity of Flax/Polypropylene Composites," *Journal of Engineering Materials and Technology*, vol. 128, pp. 527-536, 2006.

[3] J. Varna, E. Sparnins, R. Joffe, K. Nattinen and J. Lampinen. "Time dependent behavior of flax/starch composites," *Mechanics of Time-Dependent Materials*, vol. 16, pp. 47-70, 2012.

[4] K. M. Almgren, E. K. Gamstedt and J. Varna. "Contribution of wood fibres to the moisture-induced dimensional instability of composite plates," *Polymer Composites*, vol. 31, pp. 762-771, 2010.

[5] L.-O. Nordin and J. Varna. "Nonlinear viscoelastic behavior of paper fiber composites," *Composite Science and Technology*, vol. 65, pp. 1609-1625, 2005.

[6] C. Martins, V. Pinto, R. M. Guedes and A. T. Marques "Creep and stress relaxation behaviour of PLA-PCL fibres – A linear modelling approach" *1st International Conference on Structural Integrity, Procedia Engineering,* vol. 114 pp. 768-775, 2015.

[7] Y. C. Lou and R. A. Schapery. "Viscoelastic Characterization of a Nonlinear Fiber-reinforced Plastic," *Journal of Composite Materials*, vol. 5, pp. 208-234, 1971.

[8] R. A. Schapery. "Nonlinear Viscoelastic and Viscoplastic Constitutive Equations Based on Thermodynamics". *Mechanics of Time-Dependent Materials*, vol. 1, pp. 209-240, 1997.

[9] E. Marklund, J. Eitzenberger and J. Varna. "Nonlinear viscoelastic viscoplastic material model including stiffness degradation for hemp/lignin composites," *Composites Science and Technology*, vol. 68, pp. 2156-2162, 2008.

[10] J. Varna, L. Rozite, R. Joffe and A. Pupurs. "Nonlinear behavior of PLA based flax composites" *Plastics, Rubber and Composites*, vol. 41, pp. 49-60, 2012.

[11] K. Giannadakis, P. Mannberg, R. Joffe and J. Varna. "The source of inelastic behavior of Glass Fibre /Vinilester non-crimp fabric [±45]s laminates," *Journal of Reinforced Plastics and Composites,*vol. 30, pp. 1015-1028, 2011.

[12] L. Rozite, J. Varna, R. Joffe and A. Pupurs. "An analysis of nonlinear behaviour of lignin based flax composites," *Mechanics of Composite Materials*, vol. 49, pp 139-153, 2013.

[13] L. Rozite, J. Varna, R. Joffe and A. Pupurs. "Nonlinear behavior of PLA and lignin-based flax composites subjected to tensile loading," *Journal of thermoplastic composite materials*, vol. 26, pp. 476-496, 2011.

[14] L. J. Zapas and J. M. Crissman. "Creep and Recovery Behavior of Ultra-high Molecular Weight Polyethylene in the Region of Small Uniaxial Deformations," *Polymer*, vol. 25, pp. 57-62, 1984.

[15] L. Pupure, J. Varna and R. Joffe. "On viscoplasticity characterization of natural fibers with high variability," *Advanced composites letters*, vol. 24, pp. 125-129, 2015.

[16] L. Pupure, J. Varna and R. Joffe. "Methodology for macro-modeling of bio-based composites with inelastic constituents," *Composite Science and Tehcnology*, vol. 163, pp. 41-48, 2018.

[17] Z. Al-Maqdasi, L. Pupure, G. Gong, N. Emani and R. Joffe. "Time-dependent properties of graphene nanoplatelets reinforced high-density polyethylene", Journal of Applied Polymer Science, vol. 138, 2021.

[18] R. A. Schapery. "Further development of a thermodynamic constitutive theory: stress formulation," *Purdue University report* No. 69-2, 1969.

[19] R. A. Schapery. "On the Characterization of nonlinear viscoelastic materials," *Journal of Polymer Engineering and Science*, vol. 9, pp. 295-310, 1969.

[20] L. Pupure, R. Joffe, J. Varna and B. Nystrom. "Development of constitutive model for composites exhibiting time dependent properties," *IOP Conference series: Materials Science and Engineering*, vol. 48, 2013.

[21] J. Varna, L. Pupure and R. Joffe. "Incremental forms of Schapery's model: convergence and inversion to simulate strain controlled ramps," *Mechanics of Time-Dependent Materials*, vol. 20, pp. 353-552, 2016.

[22] L. Pupure, J. Varna and R. Joffe. "Natural fiber composites: challenges simulating inelastic response in strain controlled tensile tests," *Journal of Composite Materials*, vol. 50, pp. 575-587, 2016.

[23] R. Brouwer. "Nonlinear viscoelastic characterization of transversely isotropic fibrous composites under biaxial loading", *PhD thesis Free univ of Brussels*, 1986.

IOP Conf. Series: Materials Science and Engineering 1190 (2021) 012006 doi:10.1088/1757-899X/1190/1/012006

Advanced calculation of the deflection of worm shafts with FEM

Johannes Gründer[1], Alexander Monz[1], Philipp Norgauer[2]

[1]Technische Hochschule Nürnberg Georg Simon Ohm, Keßlerplatz 12, 90489 Nürnberg, Germany

[2]Bauer AG, BAUER-Straße 1, 86529 Schrobenhausen, Germany

johannes.gruender@th-nuernberg.de

Abstract. Worm gears are gear units with a shaft angle of mostly 90°. They are used in a variety of industrial applications due to large gear ratios and high transmittable torques. The high sliding ratio in the tooth contact requires a hard/soft material combination that is insensitive to scuffing. Worm shafts usually are made of case-hardened steel and worm wheels of a copper-tin-bronze alloy. Since damage predominantly occurs at the wheel (pitting, wear, tooth fracture, etc.), a wide range of calculation methods for load capacity and service life were developed. Regarding the worm shaft, only the deflection is examined, because high displacements of the shaft shifts the contact pattern, which leads to transmission errors and may lead to increased wear. Common calculation methods, e.g. according to DIN 3996 and ISO/TS 14521 provide a good approximation of the deflection. However, simplifications are made in favour of the manageability of the calculation method. Geometric modifications on the worm such as reduced tooth thickness or different lead angles cannot be taken into account with common methods. In order to close this gap, a new approach was presented by Norgauer in 2021, however, significantly overestimating the stiffening effect of the gearing in some cases by neglecting the helical winding of the worm teeth. For efficiency-optimal design of minimally thin worm shafts, approaches that go beyond the deflection are also missing. Investigations by the authors on worm gears using the finite element method (FEM) show a deviation of the bending line of worm shafts from the common calculation methods. The FEM calculations simulate the tooth meshing under load by a driving torque on the shaft, whereas the standard calculations simplify the load distribution on the worm shaft as a radial force introduced at a point. In addition to the magnitude of the radial forces, the axial tooth force component was identified as an influence factor on the bending line since it causes a displacement of the maximum bending location. Especially for increasing diameter factors the displacement of the bearings under load must be taken into account since they may be in the same range as the deflection. Furthermore, the helical winding of the teeth around the shaft leads to an average stiffening effect of the gearing. The cross-section depends on the gear meshing position and causes a periodic fluctuation of the area moment of inertia and a wobbling motion of the shaft. The changing load distribution on several teeth causes a periodic change of the lever arms and an additional dependence of the deflection from the mesh position. An analytical calculation method for the deflection of worm shafts is proposed. It aims at the precise calculation of the spatial expression of the bending line, extends the commonly used simplified model and takes the newly identified factors into account. The bending lines in both radial directions are considered separately. The values of the new introduced factors are derived from the FEM results. Finally, the calculated bending lines are compared with the FEM results and verified.

1. Introduction

Worm gear units enable high torque and speed ratios in a small installation space with an axis cross angle of usually Σ = 90 °. They are helical gears and, in addition to the rolling component in the tooth contact, also have a high design-related sliding component. Due to the sliding, they are preferably designed in a hard-soft material pairing that is insensitive to scuffing, with the worm usually made of case-hardened steel and the wheel of copper-tin bronze as shown in Figure 1. Due to the pairing of the hard steel with the relatively soft bronze material, worm gears usually fail in operation due to excessive damage to the worm wheel.

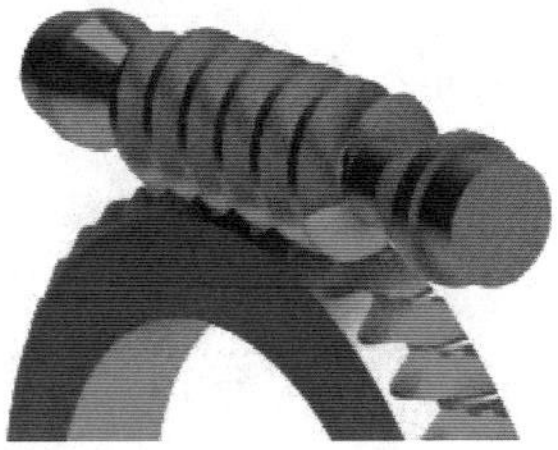

Figure 1. Worm shaft and worm wheel.

Therefore, previous research on the wear of worm gears in Germany has concentrated almost exclusively on the investigation of the wheel in terms of geometry and materials. Worm shafts, on the other hand, are designed exclusively against deflection, since this may cause a displacement of the contact pattern and consequently to increased local wear due to unfavorable load transmission. The exact course of the bending line remains unknown, since only the maximum value is calculated by the simplifications made in common calculation methods. In some cases, the geometry of worm shafts is modified in favor of other advantages (e.g. duplex worms, reduction of the tooth thickness to increase the durability of the worm wheel teeth). This cannot be considered in common calculation methods.

2. State of the art

Calculation methods for determining the deflection of the worm shaft are proposed in DIN 3996 [1], ISO/TS 14521 [2], AGMA 6022 [3] as well as the British Standards BS 721 [4]. Norgauer [5] compared the proposed calculation methods and showed the differences between them. These methods are all based on the simplification of the worm geometry as a round profile which is radially loaded by a force and calculated according to Bernoulli's principle. The radial load is calculated in [1] and [2] according to Niemann/Winter [6] and considers the friction in the tooth contact. The deflection of the worm shaft is approximated according to [1] for symmetrical arranged bearings by equation (1):

$$\delta_m = 2 \cdot 10^{-6} \cdot l_1{}^3 \cdot F_{tm2} \cdot \frac{\sqrt{tan^2\left(\gamma_m + arctan(\mu_{zm})\right) + tan^2\alpha_0/cos^2\gamma_m}}{\left(1{,}1 \cdot d_{f1}\right)^4} \quad (1)$$

Both DIN 3996 and the ISO/TS 1451 use the root diameter extended by 10% as substitute diameter to consider the stiffening effect of the worm teeth. The factor is based on experiments carried out by Lutz [7], who loaded various worm shafts on a pulser with a radial force, measured the deflection and derived the factor from the results. AGMA 6022 calculates the deflection with the root diameter and therefore always calculates higher values than [1] and [2]. The calculation is on the safe side.

Norgauer [5] developed an extended analytical approach for calculating the moment of inertia for standard and geometric modified worm teeth. The tooth form of the worm is re-calculated from the machining tool simplified as trapezoid in the face section and added to the moment of inertia of the root diameter. Both the rotation of the trapezoids due to the position as well as the distance of the centroids of the tooth areas to the common center according to Steiner's theorem are considered. Based on the calculations, he derives an equivalent diameter to consider the stiffening effect of the toothing. While this approach allows for the consideration of geometric modifications, it overestimates the stiffening effect of the gearing in some cases because of the neglection of the helical winding of the worm teeth over the length of the gearing.

The finite element method offers the possibility to analyze the bending of the worm shaft with increased accuracy, but also increased computational effort. Norgauer [5] proved the suitability of the finite element method for calculating the deflection of worm shafts but did not simulate the complex load transmission in the tooth contact; therefore those results highly agree with the common standard calculation methods [1], [2], [3] and the experiments according to Lutz [7]. However, the full potential of the finite element method is not exploited due to the simplified load on the worm. Hermes [8] successfully calculate the contact behavior and load transfer between the flanks with FEM, Reißmann [9] investigated the tooth root stress on worm wheels by simulating the loaded tooth contact. However, the subject of their investigations was not the deflection of the worm shafts.

3. Advanced calculation of the worm shaft deflection with FEM

The FEM modelling aims at the realistic calculation of the worm shafts bending line by simulating the load distribution in the tooth contact as well as the deformation of the worm teeth, the wheel and the displacement of the bearings. The load transmission in the contact is evaluated for reference.

3.1. Scope of investigation, FEM- modelling and calculation method

The investigated geometries shown in Figure 2 according to DIN 3975 [10] where modelled with CAD. The axis distance is a = 100mm, gear ratio i = 20,5, axial modulus m_x = 4mm, normal pressure angle $\alpha_0 = 20°$ and number of threads $z_1 = 2$. One geometry of each diameter factor is modelled with normal and one with reduced tooth thickness on the worm. The main gear data is listed in Table 1:

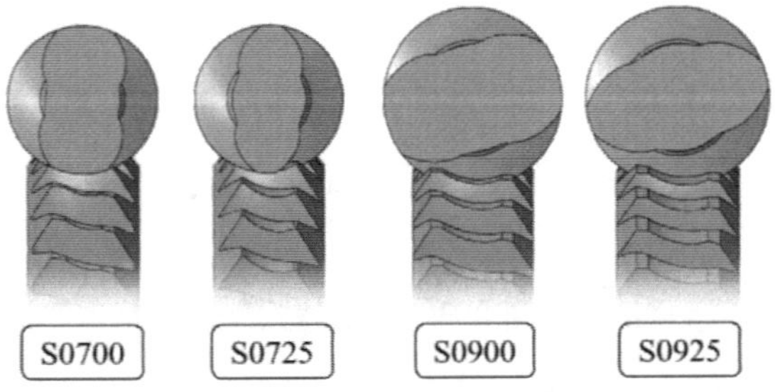

Figure 2. CAD models of the investigated gear geometries (ZI-flank)

Table 1. Main gear data

		S0700	S0725	S0900	S0925
q	[-]	7,00		9,00	
γ_m	[°]	15,95		12,53	
x_s	[-]	0,00	-0,25	0,00	-0,25
s_{mx}	[mm]	6,28	4,28	6,28	4,28
A_{St}	[mm^2]	629,70	508,00	1.028,80	871,00
b_1	[mm]	75,00		64,90	
d_{f1}	[mm]	18,00		26,00	
d_{m1}	[mm]	28,00		36,00	
d_{a1}	[mm]	36,00		44,00	
b_2	[mm]	26,70		29,40	
d_{f2}	[mm]	162,00		154,00	
d_{m2}	[mm]	172,00		164,00	
d_{a2}	[mm]	180,00		172,00	
d_{e2}	[mm]	184,00		176,00	

The load is introduced into the FEM model by applying the drive torque T_1 on the worm shaft. The different values of the drive torque cause a normal force F_n on the tooth flank of 10,000 N for each geometry. By converting the equations for normal and tangential force according to DIN 3996 [1] and Niemann/Winter [6], T_1 is calculated as follows:

$$T_1 = \frac{d_{m1}}{2000} \cdot F_n \cdot \cos\alpha_n \cdot (\sin\gamma_m + \mu_z \cdot \cos\gamma_m) \quad (2)$$

The friction coefficient μ_z between the tooth surfaces is calculated according to DIN 3996 [1] and assumed to be constant in the flank contact of the FEM model. For realistic modelling, a fixed/loose bearing arrangement with a double-row angular contact ball bearing (A) at $z = l_1$ as fixed and a single-row deep groove ball bearing (B) at $z = 0$ as loose support is used with symmetrical arrangement of the bearings and a spacing of $l_1 = 135$mm. To determine the stiffness of each roller bearing, the reaction forces of the bearings were calculated in a first step using FEM and assuming undeformable bearings. In a second step, the bearing stiffness of each bearing were calculated according to ISO/TS 16281 [11] using the reaction forces from the first step. The gear housing stiffness is neglected in the calculation. The schematic of the model is shown in Figure 3, the introduced load, friction coefficient and bearing stiffnesses are summarized in Table 2:

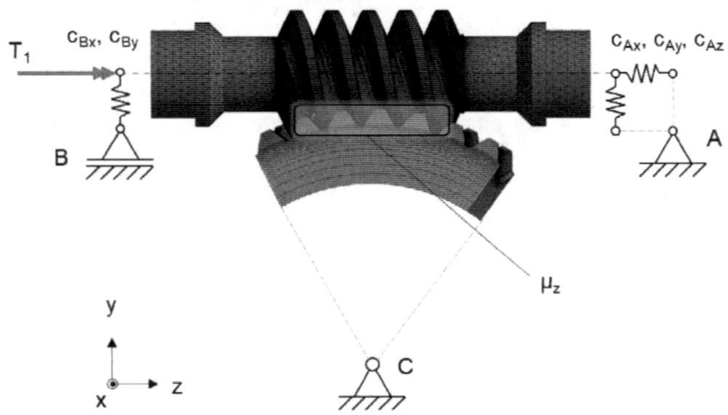

Figure 3. Schematic of the FEM model

Table 2. Load and boundary conditions

Größe	unit	S07	S09
T_1	[Nm]	49,4	54,1
μ_z	[-]	0,1048	0,1057
cAx	[Nmm^{-1}]	174.467	178.546
cAy	[Nmm^{-1}]	174.466	178.551
cAz	[Nmm^{-1}]	272.811	272.981
cBx	[Nmm^{-1}]	69.912	73.504
cBy	[Nmm^{-1}]	69.385	73.024

The components are simulated as deformable bodies made of the commonly used reference materials 16MnCr5 (worm) and 12SnCu-C-GZ (wheel) according to DIN 3996. At the worm wheel, a certain amount of plasticization of the material is expected under load due to the small yielding strength, Young's modulus and Poisson Ratio of bronze. To consider this effect, a bilinear isotropic hardening model as shown in Figure 4 is used. It allows to approximate the plastic deformation of the material above the yield strength by expanding the stress-strain-curve above the yield point with the tangential modulus E_T. The material data used in the simulations is summarised in Table 3:

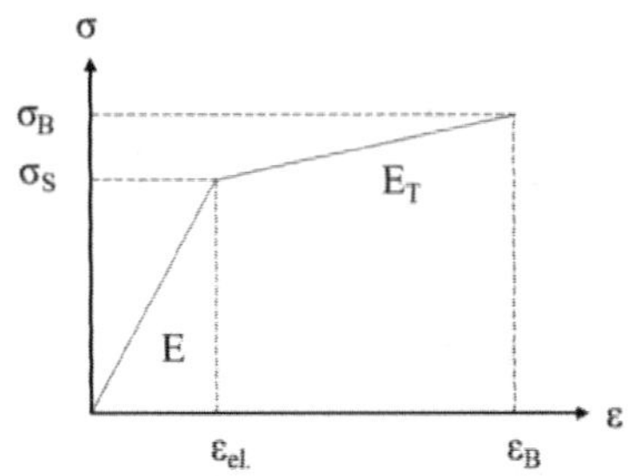

Figure 4. Bilinear isotropic hardening

Table 3. Material properties according to [12] and [13]

	unit	16MnCr5	12CuSn-C-GZ
E	[Nmm^{-2}]	206.000,00	88.300,00
E_T	[Nmm^{-2}]	1.000,00	500,00
ν	[-]	0,30	0,35
σ_B	[Nmm^{-2}]	1.000,00	280,00
σ_S	[Nmm^{-2}]	695,00	150,00

3.2. Results of the FEM-calculations

The result data from the FEM calculations for all analysed engagement positions are evaluated and further processed in post-processing tools. The comparison with established calculation programs such as SNETRA [7] provides very good agreements. The contact lines between the flanks are determined for each contact position by calculating a regression line from the contact elements with pressure $p > 0$ MPa and projected into the xy-plane. Secondly, the load distribution over the contact lines is calculated from the pressure transmitted by the surface contact elements and the associated element area, depending on the mesh size. In the model, modifications of the tooth thickness leads to a changed load transmission in the contact for both worm geometries as shown in Table 4:

Table 4. Load transfer between worm and wheel flanks on the contact lines in [N/mm]

	xs = 0,00	xs = -0,25
q = 7	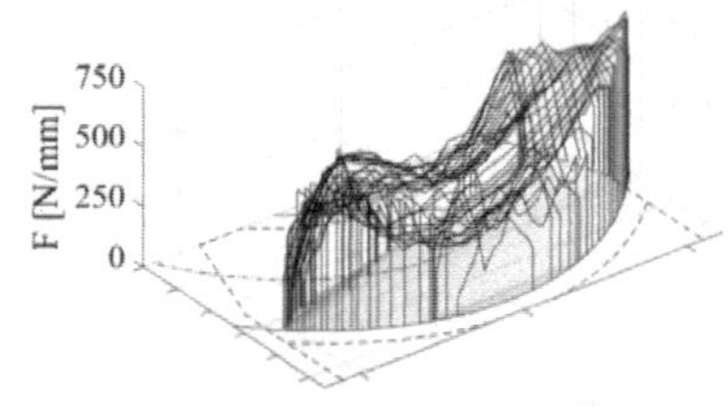	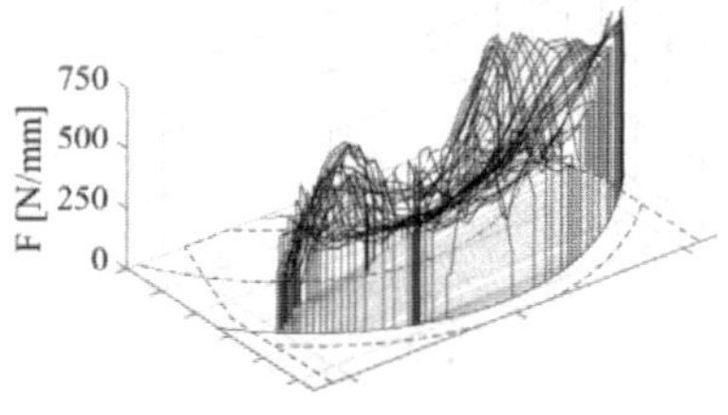
q = 9	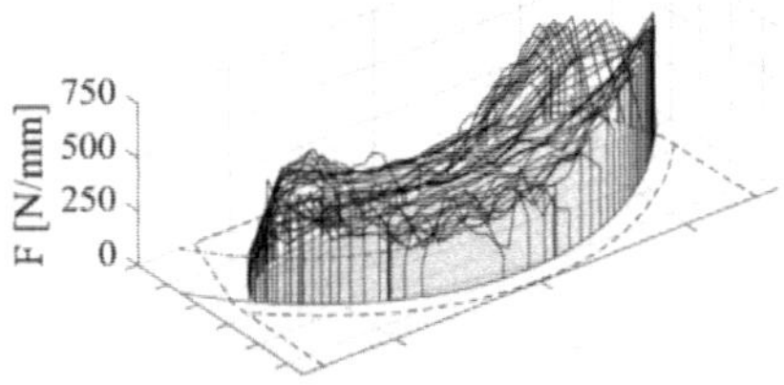	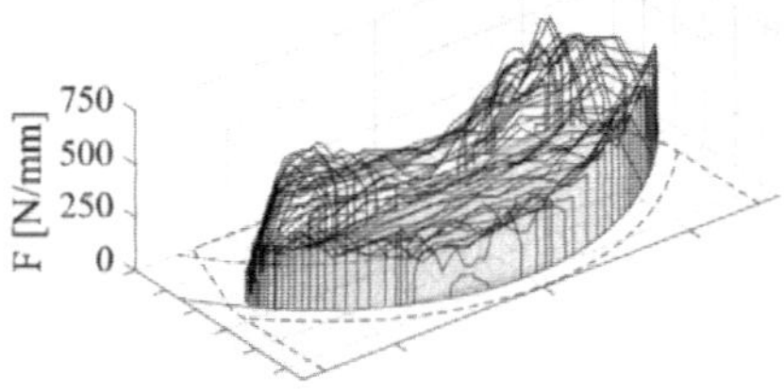

The tooth modification causes locally increased line loads in the run-out area and a shift in direction of the tip diameter which, according to the model, would have a negative effect on the wear of the flanks and thus the service life of the gearing. The changes may also have a minor effect on the deflection of the worm due to the shifted load transmission. The smaller gear geometries of the worms with diameter factor q = 7 show increased line loads and stronger gradients than the ones with q = 9.

The bending lines of the worm shafts are evaluated as the directional displacement in x and y of the mesh nodes on the worm axis. The displacement in axial direction z is not considered. In Figure 5, the bending lines of the worm shafts for different engagement positions in the various planes and in three dimensions are summarized. In direction of the tangential force (xz-plane), the bending line forms symmetrically with the maximum almost exactly between the bearing points while in the yz-plane, it is significantly shifted in direction of the axial force in z-direction caused by the lever arm to the x-axis. The displacement of the bearings is more relevant for higher diameter factors as the maximum bending value decreases. The xy-plane shows the spatially curved course of the bending line as a result due to different bearing displacements which is very similar for all simulated geometries. Due to the changing load transmission, the change of the force reactions on the bearings and the variable stiffness of the helically-shaped toothing, all simulations show a dependency of the worm angle φ_1. To quantify those influences on the bending lines, the maximum deflection values and their corresponding locations on the z-axis as well as the amplitudes between the highest and the lowest maximum value are calculated in the xz- and the yz-plane and summarised in Table 5. The maximum three-dimensional value is defined as the maximum of all values δ_{xy} calculated from δ_x and δ_y for every z coordinate by $\delta_{xy} = \sqrt{\delta_x(z)^2 + \delta_y(z)^2}$. The location of the maximum in z is not necessarily the same as for the deflection values in x- and y-direction. All results show a good correlation to the deflection results calculated with DIN 3996. For higher diameter factors, the deviation increases due to the consideration of the bearing stiffness in FEM.

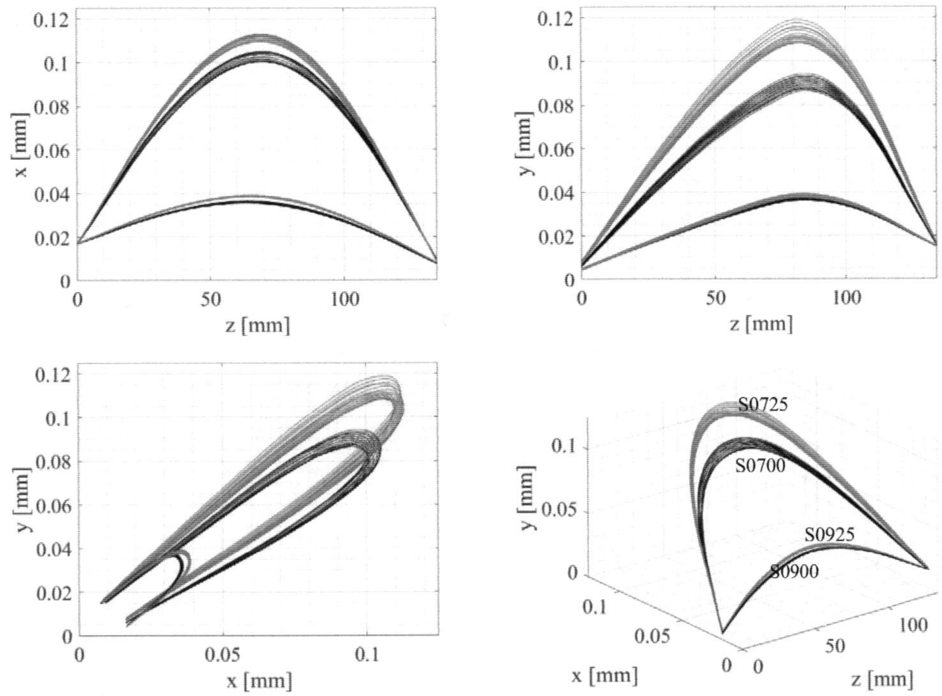

Figure 5. Bending lines in the xz-, yz- and the xy-plane and in 3 dimensions as calculated with FEM

All simulations show a shifting of the maximum value of the bending lines in direction of the axial force as well as a dependence of the course on the angle φ_1. For smaller diameter factors, considering the modification of the tooth thickness is more important while for higher diameter factors the displacement of the bearings is more relevant. The maximum deflection values calculated according to DIN 3996 [1] are the same for standard and modified geometries which is highly inaccurate.

Table 5. FEM-Results compared to DIN3996

		S0700	S0725	S0900	S0925
$\delta_{y,max} \mid z_{y,max}$	[mm]	0,0941 | 85,0	0,1190 | 82,5	0,0375 | 85,0	0,0394 | 84,0
$\delta_{x,max} \mid z_{x,max}$	[mm]	0,1050 | 70,0	0,1131 | 69,0	0,0366 | 64,0	0,0391 | 65,0
$\delta_{xy,max} \mid z_{xy,max}$	[mm]	0,1379 | 77,0	0,1609 | 77,0	0,0507 | 77,0	0,0538 | 76,0
δ_m	[mm]	0,1479	0,1479	0,0331	0,0331
$\delta_{xy,max} / \delta_m$	[-]	0,9324	1,09455	1,5317	1,6253
$\delta_{Ay,max}$	[mm]	0,0077	0,0107	0,0014	0,0020
$\delta_{Ax,max}$	[mm]	0,0047	0,0036	0,0011	0,0008
$\delta_{Ayx,max}$	[mm]	0,0077	0,0082	0,0013	0,0016

4. Extended analytic approach for calculating the deflection of the worm shaft

In order to calculate the bending line for different worm geometries with higher accuracy and avoiding the extensive effort of FEM, an analytical approach is presented in the following. The influencing factors identified from the FEM results as shown in 3.2. as well as the relevant dimensions are shown in Figure 6 for two rotation angles φ_1. The bending line is defined in the xz- and the yz-plane and is calculated by equation (3):

$$w_{total} = \sqrt{w_{xz}(z,\varphi_1)^2 + w_{yz}(z,\varphi_1)^2} \tag{3}$$

IOP Conf. Series: Materials Science and Engineering 1190 (2021) 012006 doi:10.1088/1757-899X/1190/1/012006

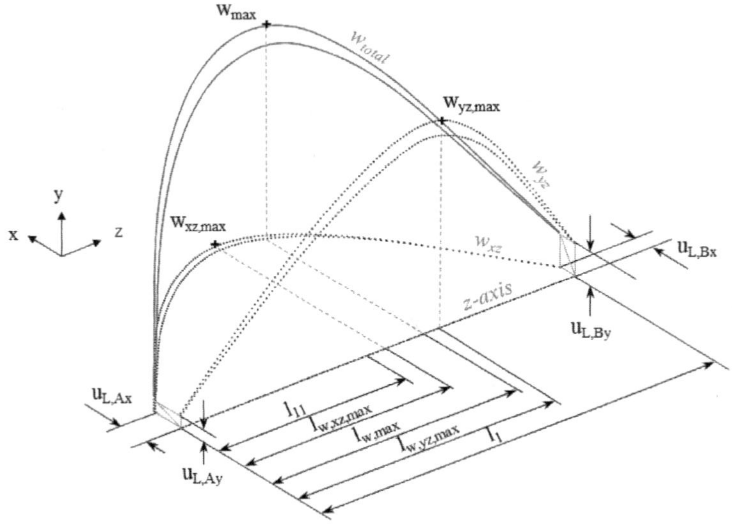

Figure 6. Dimensions and influences on the course of the bending line

4.1. Influences on the course of the bending line

Due to the changing load transmission in the contact between the flanks, the resulting **focus of the force transmission** changes periodically depending on the engagement position. This may cause a change of the lever arms of the toothing forces, the bending moments as well as the bending lines in different engagement positions. The coordinates of the load transmission centre as shown in Figure 7 were derived from the FEM-results. The offset of the graphs on the left side compared to the ones on the right is due to the different starting angle of the worm in the simulation. For $\varphi_1 = 112{,}5°$ the FEM-calculation for model S0725 did not converge, hence this value is missing in the graphs.

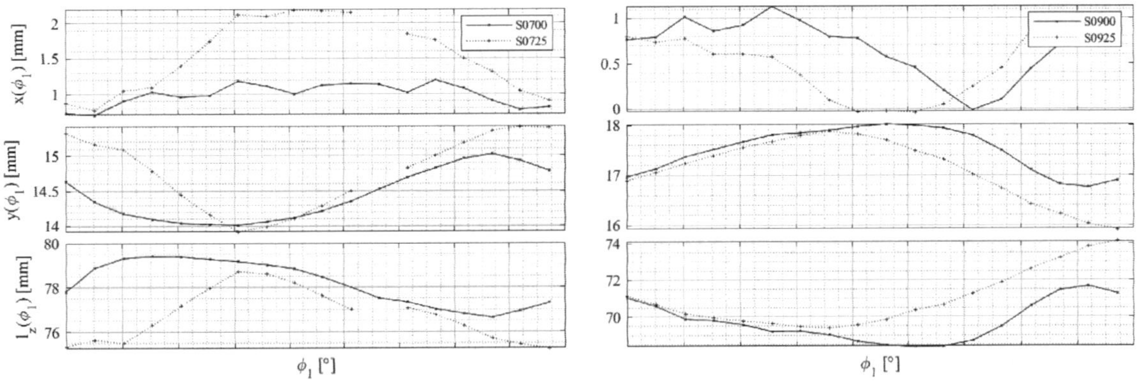

Figure 7. Coordinates of the force introduction points as a function of the rotation angle, derived from FEM

The **axial tooth force** causes a tilting of the worm shaft and a displacement of the maximum deflection in working direction of the axial force in the yz-plane as shown in 3.2. due to the eccentric load transmission relative to the worm axis with an average of $d_{m1}/2$. Although it is not significant in the xz-plane, the influence of the axial force is taken into account for both directions in the presented approach. Roller bearings exhibit compliance under load due to their geometric design and the elasticity of the materials. Depending on the **stiffnesses of the bearings**, the displacement may be in a similar range like the deflection of the shaft. Since the displacement of the bearings of the worm shaft additionally reduces the actual bending, it should be considered. For the calculation, the bearings supporting the worm wheel are considered as non-deformable due to their relatively larger size compared to the bearings on the worm shaft. The displacements u of the bearings in x- and y-direction due to the reaction forces are summarised in matrix U_L and calculated by equation 4:

$$U_L = F_L \cdot c_L = \begin{Bmatrix} F_{Ax} \cdot c_{L,Ax} & F_{Bx} \cdot c_{L,Bx} \\ F_{Ay} \cdot c_{L,Ay} & F_{By} \cdot c_{L,By} \end{Bmatrix} \tag{4}$$

According to Norgauer [5], the surface moment of inertia of the worm teeth in the face section can be approximated by a calculation from the tool shape. All dimensions of the tooth form in the face section, which is approximated as a trapezoid, are calculated from the geometric diameters of the worm, the tooth thickness, the axial modulus, the number of teeth as well as the pressure angle. The dimensions of the trapezoid as shown in Figure 8 (*) are calculated according to Norgauer [5]. The moment of inertia of the cross-section of the worm is calculated by summing the moment of inertia of the root diameter, the rotated tooth areas depending on the number of teeth and the Steiner's theorem. With an extreme value analysis, Norgauer calculates the worst case of the rotated cross-section of the worm with the smallest moment of inertia representing the whole worm shaft. This consideration still overestimates the stiffening effect since the teeth are wound around the base body of the shaft in a helical shape. The resulting rotation of the helical toothing along the worm axis corresponds to a rotation as a function of the axis coordinate z according to Figure 8:

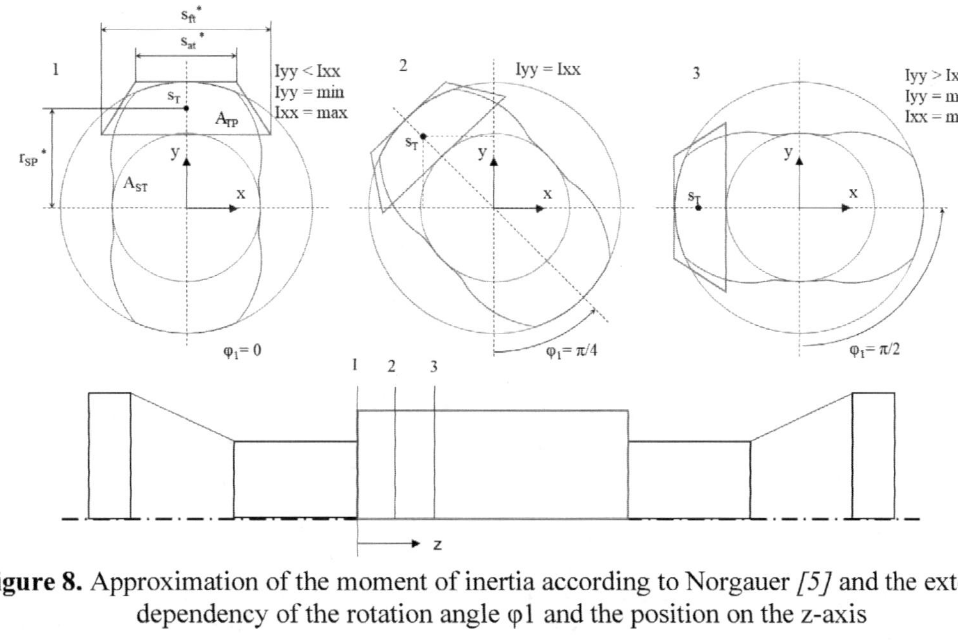

Figure 8. Approximation of the moment of inertia according to Norgauer *[5]* and the extended dependency of the rotation angle φ1 and the position on the z-axis

The moment of inertia of the worm shaft is calculated according to [14] and considering the dependence of the rotation angle φ_1 of the worm shaft:

$$I = \frac{\pi}{64} \cdot d_{ers}(\varphi_1)^4 \tag{5}$$

The used substitutional diameter d_{ers} is described by a sinus function according to equation (6):

$$d_{ers} = d_{f1} \cdot k_{stiff} = d_{f1} \cdot \{a_I \cdot \sin[b_I \cdot (\varphi_1 - c_I)] + d_I\} \tag{6}$$

The amplitude a_I of the oscillation of the moment of inertia of the worm shaft is calculated by equation (7) for the investigated geometries. It is based on the empiric data calculated with FEM and considers the gearing length compared to the worm shaft length:

$$a_I = \frac{\min(I)}{\max(I)} \cdot \frac{b_1}{l_1} \tag{7}$$

The period length b_I =2 is constant because the moments of inertia are equal in size every half rotation. The phase shift factor c_I describes the offset between the moments of inertia around the x- and y-axis. Derived from the FEM-results, in y-direction it is $c_I = 0$ and in x-direction $c_I = \pi/4$. The factor d_I corresponds to the mean moment of inertia of the shaft geometry related to the toothing length b_1. For the investigated geometries, d_I is calculated by equation (8):

$$d_I = \sqrt[4]{\left[{d_{f1}}^4 \cdot \frac{\pi}{64} + mean(I_m) \cdot \frac{b_1}{l_1} \cdot \frac{d_{f1}}{d_{a1}} \cdot \frac{\gamma_m}{\alpha_0}\right] \cdot \frac{64}{\pi}} \cdot {d_{f1}}^{-1} \tag{8}$$

Equations (7) and (8) due meet the characteristics of the investigated geometries evaluated from the FEM analysis but need to be further validated by a more comprehensive study. The equations are likely to be modified in the future to accurately calculate a larger variety of different geometries. The moment of inertia is calculated by summing up every value for every cross-section of the worm shaft divided by the amount n_z of calculated cross-sections:

$$I(z) = \sum_{z=0}^{b1}\left(\sum_{i=1}^{z_1}\left\{\frac{I_{xx0}+I_{yy0}}{2} + \frac{I_{xx0}-I_{yy0}}{2} \cdot cos(2 \cdot \varphi_{geom}) + \left[sin(\varphi_{geom}) \cdot r_{SP}\right]^2 \cdot A_{TP}\right\}\right) \cdot {n_z}^{-1} \tag{9}$$

The geometric rotation angle is described as a function of the z-coordinate:

$$\varphi_{geom} = \frac{i-1}{z_1} \cdot 2\pi + \frac{z}{m_x \cdot z_1 \cdot \pi} \cdot 2\pi \tag{10}$$

To calculate the mean stiffening effect of the toothing, the mean moment of inertia of the worm shaft is further calculated for every rotation angle φ_1 divided by the amount $n_{\varphi 1}$ of calculated rotation angles:

$$I_m = \frac{\sum_{\varphi_1=0}^{2\pi} I(z)}{n_{\varphi_1}}, \quad with\ \varphi_{geom} = \frac{i-1}{z_1} \cdot 2\pi + \frac{z}{m_x \cdot z_1 \cdot \pi} \cdot 2\pi + \varphi_1 \tag{11}$$

In **Table 6** the values of the factors a and d calculated from the FEM-results are compared to the ones calculated with the analytic approach. Overall, they show a good agreement for the investigated gear geometries.

Table 6. Accuracy of the analytical approach vs FEM-results

		S0700	S0725	S0900	S0925
d_{FEM}	[-]	1,2639	1,2174	1,0954	1,0795
a_{FEM}	[-]	0,0207	0,0193	0,0092	0,0099
d_{analyt}	[-]	1,2618	1,2149	1,1329	1,1074
a_{analyt}	[-]	0,0181	0,0190	0,0091	0,0094
d_{dev}	[%]	-0,17	-0,20	3,42	2,58
a_{dev}	[%]	-12,53	-1,37	-1,50	-5,16

4.2. Extended mechanical calculation model

The identified factors are considered in the mechanical model shown in Figure 9 with the tooth force components of the worm calculated according to DIN 3996 [1]. Assuming small deformations, the calculation of the bending line in the xz- and the yz-plane is done by applying the superposition principle. The position of the force introduction is variable in all three directions and described by the unknown functions $x(\varphi_1)$, $y(\varphi_1)$ and $l(\varphi_1)$. The dependence on the angle φ_1 is caused by the changing load transmission. To calculate the internal reactions, the beam is divided into two sections 1 and 2 at the load introduction point:

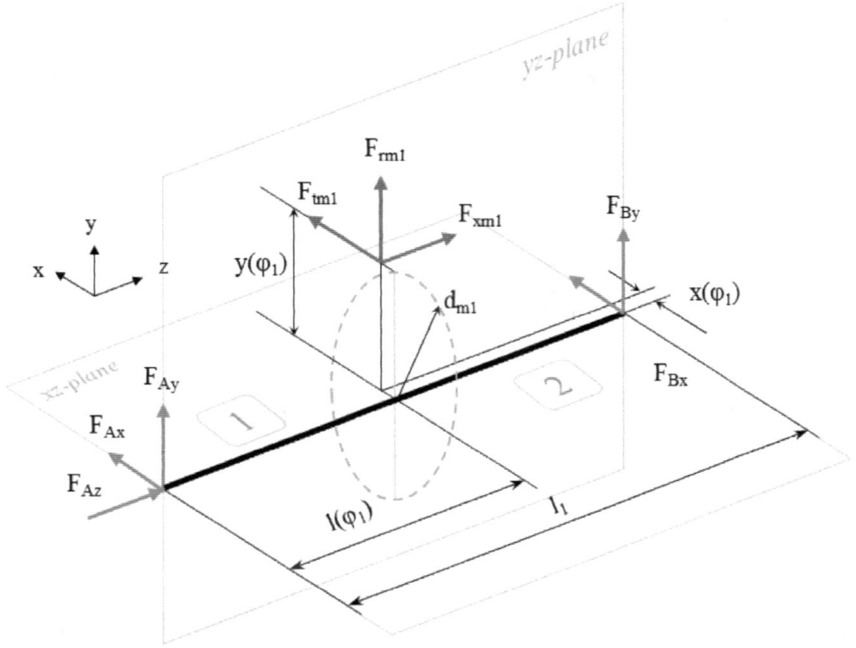

Figure 9. Mechanical model of the worm shaft with load and reaction forces

The reaction forces F of the bearings are calculated by equilibrium of forces and moments in the xz and yz-planes:

$$F_L = \begin{Bmatrix} F_{Ax} & F_{Bx} \\ F_{Ay} & F_{By} \\ F_{Az} & F_{Bz} \end{Bmatrix} = \begin{Bmatrix} \frac{(F_{tm1}\cdot l(\varphi_1) - F_{xm1}\cdot x_{SP}(\varphi_1))}{l_1} - F_{tm1} & \frac{(F_{xm1}\cdot x_{SP}(\varphi_1) - F_{tm1}\cdot l(\varphi_1))}{l_1} \\ \frac{(F_{rm1}\cdot l(\varphi_1) - F_{xm1}\cdot y_{SP}(\varphi_1))}{l_1} - F_{rm1} & \frac{(F_{xm1}\cdot y_{SP}(\varphi_1) - F_{rm1}\cdot l(\varphi_1))}{l_1} \\ -F_{xm1} & 0 \end{Bmatrix} \tag{12}$$

For calculating the moment load inside the material, the beam is divided into two sections at the force initiation point. The moments M in the beam are calculated by:

$$M = \begin{Bmatrix} M_{xz,1} & M_{xz,2} \\ M_{yz,1} & M_{yz,2} \end{Bmatrix} = \begin{Bmatrix} F_{Ax}\cdot z & F_{Ax}\cdot z + F_{tm1}\cdot (z - l(\varphi_1)) + F_{xm1}\cdot x_{SP}(\varphi_1) \\ F_{Ay}\cdot z & F_{Ay}\cdot z + F_{rm1}\cdot (z - l(\varphi_1)) + F_{xm1}\cdot y_{SP}(\varphi_1) \end{Bmatrix} \tag{13}$$

Considering the dependence of the moment of inertia I from the rotational angle φ_1 of the worm shaft as shown in 4.1. the differential equation of the second derivation of the bending line is solved according to [14]:

$$w'' = -\frac{M}{E \cdot I(\varphi_1)} \tag{14}$$

By double integrating (14) for each section and plane, the bending line is described by four equations as summarised in the matrix (15):

$$W_{ges} = \begin{Bmatrix} w_{xz,1} & w_{xz,2} \\ w_{yz,1} & w_{yz,2} \end{Bmatrix} \cdot E^{-1} \cdot I(\varphi_1)^{-1} \tag{15}$$

With the equations for each section in each plane as shown in equations (16) to (19):

$$w_{xz,1} = \left(-\frac{1}{6} \cdot F_{Ax} \cdot z^3 + c_{I,xz,1} \cdot z^2 + c_{I,xz,2}\right) \tag{16}$$

$$w_{yz,1} = \left(-\frac{1}{6} \cdot F_{Ay} \cdot z^3 + c_{I,yz,1} \cdot z^2 + c_{I,yz,2}\right) \tag{17}$$

$$w_{xz,2} = \left(-\frac{1}{6} \cdot (F_{Ax} + F_{tm1}) \cdot z^3 + \frac{1}{2} \cdot \left(F_{tm1} \cdot l(\varphi_1) - F_{xm1} \cdot \frac{x(\varphi_1)}{2}\right) \cdot z^2 + c_{I,xz,3} \cdot z^2 + c_{I,xz,4}\right) \tag{18}$$

$$w_{yz,2} = \left(-\frac{1}{6} \cdot (F_{Ay} + F_{rm1}) \cdot z^3 + \frac{1}{2} \cdot \left(F_{rm1} \cdot l(\varphi_1) - F_{xm1} \cdot \frac{y(\varphi_1)}{2}\right) \cdot z^2 + c_{I,yz,3} \cdot z^2 + c_{I,yz,4}\right) \tag{19}$$

The equations for the bending lines contain eight unknown integration constants c_I. They can be determined by formulating a linear equation system from four boundary conditions for each plane. The boundary conditions are derived from the displacement of the bearing points and the continuity in the transition between the sections of the bending line. They are similar in both planes, for the xz-plane those are formulated as follows: $w_{xz,1}(z=0) = u_{L,Ax}$, $w_{xz,2}(z=l_1) = u_{L,Bx}$, $w_{xz,1}(z=l(\varphi_1)) = w_{xz,2}(z=l(\varphi_1))$ and $w_{xz,1}'(z=l(\varphi_1)) = w_{xz,2}'(z=l(\varphi_1))$. Finally, the resulting bending line is calculated by equation (3).

5. Comparison between the analytic and the FEM results and discussion

The presented analytic approach is used to recalculate the geometries and loads calculated with FEM. In the analytic approach, all identified influencing factors on the bending line are considered. The tooth forces are calculated according to Niemann/Winter the bearing stiffnesses correspond to those from the FEM models. Due to a missing analytical approach for the force application point, it is taken into account on the basis of the evaluation of the FEM results with regard to the center of the load transmission as shown in 4.1. A general approximation as the reference diameter for the force application point may be appropriate, but the intention is to show the potential of the approach.

Table 7 shows the results of the bending lines calculated with FEM (black) and with the analytic approach (red). The calculated results show a good agreement with the ones calculated with FEM but with a slight deviation thus to the complex load transmission between the flanks which cannot be considered analytically. For the diameter factor of 7, the results are on the safe side. The bending lines for the diameter factor of 9 show a good agreement. For all calculations, the displacement of the bearings is smaller in the FEM-calculation than the analytical calculated displacements.

Table 7. Comparison of the bending lines calculated with FEM and analytically

xs = 0,00 | xs = -0,25

q = 7

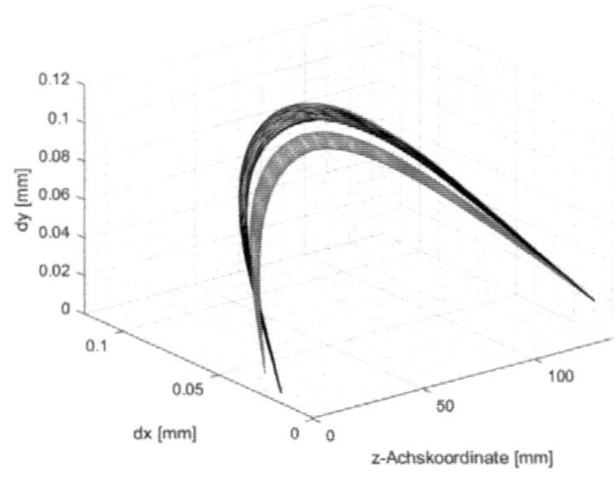

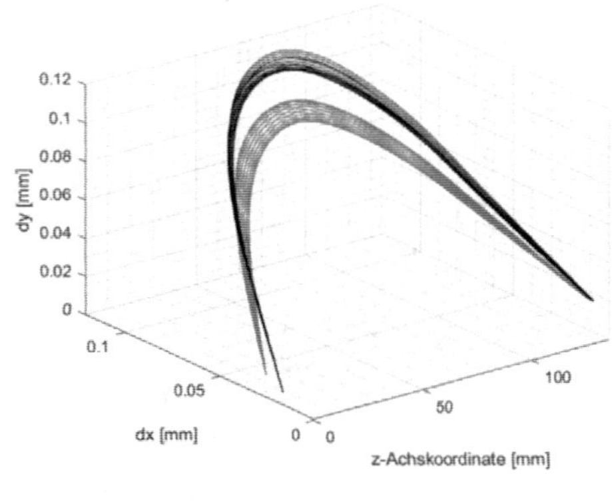

q= 9

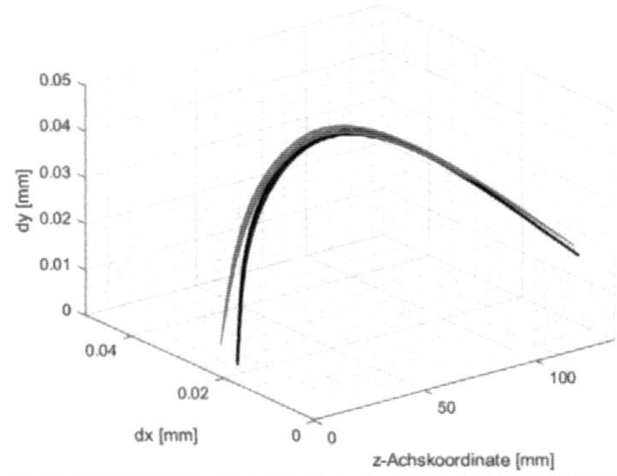

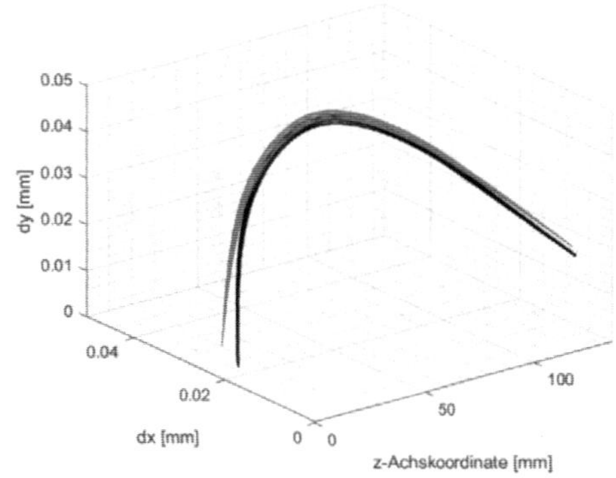

6. Parameters, indices and abbreviations

All parameters, indices and abbreviations used in this article are listed in Chyba! Nenalezen zdroj odkazů..

Table 8. Parameters, indices and abbreviations

Symbol	Unit	Parameter
a	mm	Center distance
b_1	mm	Tooth width of the worm
b_2	mm	Tooth width of the wheel
$c_{L,A/B}$	mm/N	Bearing stiffness in spatial direction
c_I	-	Integration constants
d_f	mm	Root diameter
d_m	mm	Reference diameter
d_a	mm	Tip diameter
d_e	mm	Outer diameter
i	-	Gear ratio
l_1	mm	Distance between bearings A and B
l_{11}	mm	Distance from bearing A to tooth force
m_x	mm	Axial modulus
p	N/mm^2	Contact pressure between surfaces
q	-	Diameter factor
r_{SP}	mm	Radius of center of gravity of tooth area
s_{mx}	mm	Tooth thickness in axial section
$u_{L,A/B}$	mm	Displacement of the bearings
w	mm	Equation of the bending line
w"	-	Second derivation of w
α	°	Pressure angle
γ_m	°	Lead angle
$\delta(z)$	mm	Deflection at coordinate z
δ_A	mm	Amplitude of the deflection
δ_m	mm	Maximum deflection according to DIN3996
δ_x	mm	Deflection in x-direction
δ_y	mm	Deflection in y-direction
δ_{xy}	mm	Deflection in three dimensions
ε	-	Elastic strain
η	-	Efficiency factor
μ_z	-	Mean tooth coefficient of friction
ν	-	Poisson ratio
σ_B	N/mm^2	Tensile strength
σ_S	N/mm^2	Yield strength
φ_1	°	Rotation angle of the worm
Σ	°	Axis cross angle

x	mm	Carthesian coordinate in x-direction
$x(\varphi_1)$	mm	Lever of tooth force in x-direction
x_s	-	Tooth modification factor
y	mm	Carthesian coordinate in y-direction
$y(\varphi_1)$	mm	Lever of tooth force in y-direction
z1	-	Number of threads of the worm
z2	-	Number of teeth of the wheel
z	mm	Carthesian coordinate in z-direction
$z(\varphi_1)$	mm	Lever of tooth force in z-direction
A_{st}	mm^2	Area of worm in transverse section
A_{TP}	mm^2	Area of the worm tooth
E	N/mm^2	Young's Modulus
E_T	N/mm^2	Tangential modulus
$F_{A/B}$	N	Bearing reaction force
F_{xm1}	N	Axial tooth force component
F_{rm1}	N	Radial tooth force component
F_{tm1}	N	Tangential tooth force component
I_{xx}	mm^4	Moment of inertia of x-bending axis
I_{yy}	mm^4	Moment of inertia of y-bending axis
M	Nm	Bending Moment
T	Nm	Torque moment

Index	Explanation
1	Worm *or* section 1 of bending beam
2	Wheel *or* section 2 of bending beam
max	Maximum value
n	Normal direction
x	Axial section of the worm shaft
x, y, z	Associated carthesian direction
xy, xz, yz	Associated plane in carthesian space
A, B	Fixed and loose worm shaft bearing
C	Bearing of the worm wheel
I	Integration constant
L	Bearing
R	Radial direction

Abbreviation	Explanation
AGMA	American Gear Manufacturers Association
CAD	Computer Aided Design
DIN	Deutsches Institut für Normung
FEM	Finite Element Method
ISO	International Organization for Standardization
SNETRA	EDV program for calculating contact patterns of loaded worm gears
ZI	Involute flank form

7. Conclusion

The new developed analytical approach allows a more accurate calculation of the bending line considering the axial force, the load transmission, the bearing displacements as well as the periodically changing moment of inertia without using FEM. By neglecting different input parameters, the level of detail can be chosen by the user. For a general and fast design approach, the method according to DIN 3996 is recommended for gear units with high geometric similarity to the standard reference.

The new calculation method may be used for increased accuracy in the calculation of the load distribution in the tooth contact, for the calculation of geometries deviating from the standard - in particular for efficiency-optimized thin worm shafts, optimized design for bend-critical geometries as well as an approach for the calculation of bending stresses in the worm shaft. Due to the small scope of the FEM-investigation, the results need to be further verified for different geometries and loads. Also, further experimental verification is reasonable.

Acknowledgments

This work has been supported by funds of the Technische Hochschule Nürnberg. Special thanks to Philipp Norgauer for contributing his knowledge on the subject.

References

[1] DIN 3996 - Tragfähigkeitsberechnung von Zylinderschneckengetrieben, 2019 (in German).

[2] ISO/TR 14521:2009(E) - Gears - Calculation of load capacity of wormgears, 2009.

[3] ANSI/AGMA 6022-C93 - Design Manual for Cylindrical Wormgearing, Alexandria, Virginia: American Gear Manufacturers Association, 2014.

[4] B. Standards, BS721 : Part 2 : 1983 : Worm Gearing, Part 2. Metric units, British Standards, 1983.

[5] P. Norgauer, G. Kneiprecht, M. Hein and Prof. Dr. Ing. K. Stahl, "A new approach for the calculation of worm shaft deflection in worm and crossed helical gear drives," *GearSolutions,* 15 February 2021.

[6] G. Niemann and H. Winter, Maschinenelemente Band 3, 2. Auflage, Berlin: Springer-Verlag Berlin Heidelberg GmbH, 1986 (in German).

[7] M. Lutz, Methoden zur rechnerischen Ermittlung und Optimierung von Tragbildern an Schneckengetrieben; Dissertation, München: Technische Universität München, 2000 (in German).

[8] J. Hermes, Tragfähigkeit von Schneckengetrieben bei Anfahrvorgängen sowie Last- und Drehzahlkollektiven; Dissertation, Bochum: Ruhr-Universität Bochum, 2008 (in German).

[9] J. Reißmann, Beitrag zur Entwicklung einer verbesserten Berechnungsmethode für die Zahnfußtragfähigkeit von Zylinderschneckengetrieben; Dissertation, Chemnitz: Technische Universität Chemnitz, 2016 (in German).

[10] DIN 3975-1 - Begriffe und Bestimmungsgrößen für Zylinder-Schneckengetriebe mit sich rechtwinklig kreuzenden Achsen - Teil1: Schnecke und Schneckenrad, 2017 (in German).

[11] ISO/TS 16281 - Rolling bearings - Methods for calculating the modified reference rating life for universally loaded bearings, 2008.

[12] D. Kupferinstitut, CuSn12-C, Werkstoff-Datenblatt, 2005 (in German).

[13] H. Wittel, D. Muhs, D. Jannasch and J. Voßiek, Roloff/Matek Maschinenelemente, Tabellenbuch 21. Auflage, Springer-Vieweg (in German).

[14] D. Gross, W. Hauger, J. Schröder and W. A.Wall, Technische Mechanik 2 - Elastostaik, 13. Auflage, Springer-Verlag GmbH Deutschland, 2017 (in German).

IOP Conf. Series: Materials Science and Engineering 1190 (2021) 012007 doi:10.1088/1757-899X/1190/1/012007

Validation of the Time-Domain Boundary Element Method in Acoustics regarding Flexible Multibody Simulations and Acoustic Measurements

Simon Schneider[1], Bernd Graf[1], Timo Giese[2], Ivaylo Haralampiev[2]

[1]University of Applied Sciences Ulm, 89075 Ulm, Germany
[2]FunctionBay GmbH, 80339 Munich, Germany

simon.schneider@thu.de

Abstract. With flexible multibody simulation, the dynamical behaviour of vehicle components as well as entire systems can be simulated in an efficient way. In addition to structural vibrations, noise emissions play an important role in terms of comfort in the vehicle industry. While the dynamics of the structures can be simulated directly in the time-domain, the sound radiation is usually calculated in the frequency-domain. This process enables an efficient investigation of steady-state operating conditions but not of transient operating conditions. The time-domain Boundary Element Method (TD-BEM) in acoustics allows to determine the sound pressure directly in the time-domain, without complicated transformation from the frequency-domain. Existing work concentrates on stability criteria and general theoretical applicability. The validation on real test cases with measurement results, however, has rarely been focused. On this work, we aim to close this gap between theory and practical applications with the help of real housing structures and test cases. In order to set up an acoustic validation, meaningful measurements need to be combined with validated structural simulations that match the physical world with regard to the sound radiation as closely as possible. Therefore, several test set-ups are used to measure the real dynamic and acoustic behaviour in an anechoic chamber with multiple accelerometers, a laser vibrometer, various microphones and a sound intensity PU probe. Based on these measurements, multibody and BE-models are created, representing the same scenarios with numerical methods. Derived from the simulated surface accelerations, the acoustic radiation is calculated in the time-domain. Comparisons at a simple oval-shaped housing show that the method generates very promising results. The mean deviation between the measured and simulated sound pressure levels (time-domain) is less than 1 dB and the frequency spectra show a high quantitative correlation as well.

1. Introduction

Nowadays, main aspects such as costs, quality and time must be taken into account during the early product development in order to produce a promising product. This is summarized by the iron triangle of product development [1]. According to the Rule of Ten, error detection also plays an enormous role in the early R&D phase [2]. That is the reason why simulation methods are used more and more in mechanical engineering. In the vehicle industry, driving comfort plays a significant role for the customers choosing an automobile. Therefore machine acoustics is an important part of today's engineering disciplines [3].

In the field of mechanical engineering, vibrating structures are responsible for sound radiation. This is why acoustic considerations of components are often coupled with dynamic ones. In recent years, flexible multibody simulation is used to investigate the dynamic behaviour of structures. A practicable approach is to employ flexible bodies in order to consider the dynamic behaviour of entire systems directly in the time-domain. This leads to a reasonable amount of computational resources.

If a numerical acoustic analysis is coupled with a structural vibration analysis, the calculation usually takes place in the frequency-domain. This is an efficient way to examine steady-state but not transient operating conditions for exterior acoustic problems.

The time-domain Boundary Element Method enables determining the sound parameters, like the acoustic pressure p, directly in the time-domain by solving the boundary integral equation problem (1) [4].

$$\epsilon p(\vec{x}_p, t) 4\pi = \int_{\partial B} -\left[\frac{q(\vec{x}, t_{ri})}{r} + \frac{\partial \vec{r}}{\partial \vec{n}} * \left(-\frac{p(\vec{x}, t_{ri})}{r^2} - \frac{1}{cr}\frac{\partial p(\vec{x}, t_{ri})}{\partial t}\right)\right] dS(\vec{x}) \tag{1}$$

The first term of the integral equation (1) describes the monopole source of an acoustic radiation. The second term, which derivative the distance vector $\vec{r}$ with respect to the normal direction $\vec{n}$, describes the dipole source. The dipole component considers the reflection of incoming waves. A practicable approach that seems promising for complex real structures is the solution of the discretized equation (1) using the collocation method [4].

In most publications, this method is examined for stability criteria only or there are shown some theoretical applications. The comparison with real test structures usually plays a minor role or no role at all. Therefore, this paper presents a quantitative comparison between results of the TD-BEM method in acoustics and physical measurements.

The workflow to validate an acoustic calculation is shown in **Figure *1***. In the modal analysis step, an initial comparison between the real test body and the associated FE model is conducted using a numerical (NMA) and experimental (EMA) modal analysis. The aim is the best possible correlation between the physical and the simulation model regarding their modal parameters. The dynamic behaviour of the excited test object is determined by the vibration analysis. This is done on the one hand with measurements from the test bench and on the other hand with the flexible multibody simulation describing the test set up. Similar to the modal analysis, the goal is to ensure the simulation model matches the real measured values as closely as possible. The damping and the boundary conditions such as contacts and excitations play a major role in this process.

Only with a very good validation of the calculated surface vibrations by measurement, it is possible to ensure that the boundary conditions of the sound radiation of the physical and numerical model are equal.

The final step is to use the TD-BEM to compute the acoustic sound pressures directly in the time-domain in order compare these results with the measured signals.

IOP Conf. Series: Materials Science and Engineering 1190 (2021) 012007 doi:10.1088/1757-899X/1190/1/012007

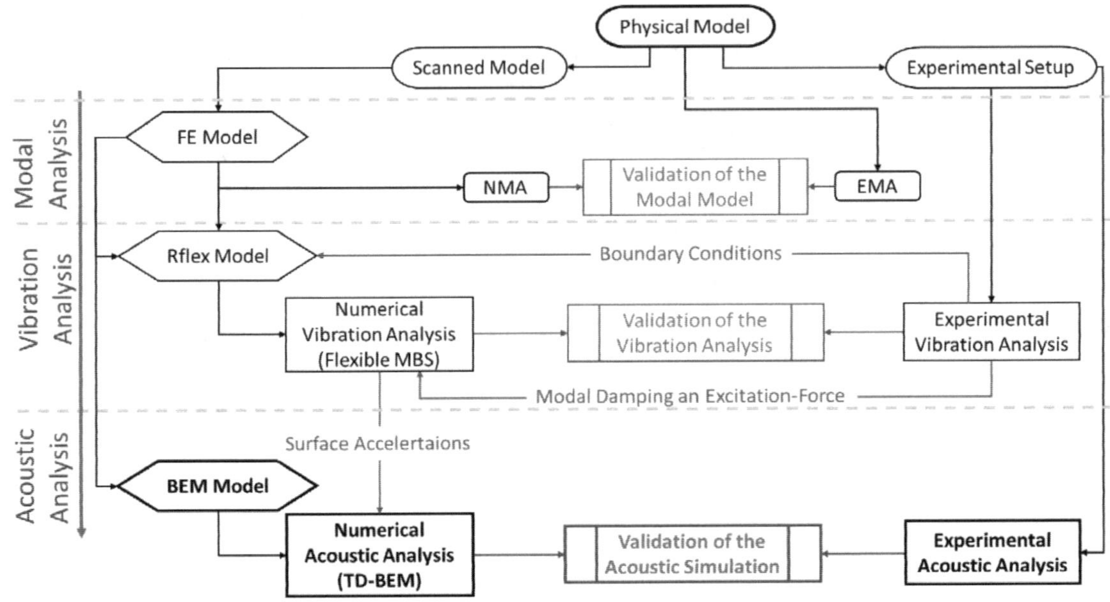

Figure 1. Procedure to validate an acoustic simulation with measurements

2. Test object and experimental setup

A thin-walled oval principle housing (OPG) made of cast aluminium is used as a test body, shown in **Figure** *2*. The simple design holds the complexity of the geometry and the boundary conditions at a manageable level. For these investigations the OPG represents a simplified gearbox housing.

To validate the operational vibration and acoustic analysis, a test bench is set up in a state-of-the-art anechoic chamber. In the anechoic chamber, it is possible to achieve free field conditions in the frequency range under survey and external acoustic disturbances are prevented.

The OPG is placed on a foam with very low stiffness. Below the foam is an aluminium plate, which first eigenfrequency is above the highest OPG natural frequency to be examined. This plate is fixed with rubber elements on a steel frame with sand filled tubes. The foam, the aluminium plate and the steel frame decouple the test object from the environment.

Five uniaxial miniature accelerometers are attached at the housing to measure the dynamic behaviour during operation in normal direction at different structure points. The sensor positions are selected in a way that the structural modes and the fluid modes of the air inside can be recorded well.

During the measurements of the OPG the sound pressure near the surface as well as in the far field is recorded with 12 microphones. In addition, the measurement points of the microphones are chosen in a way that as many modes as possible in the frequency range of interest can be detected.

IOP Conf. Series: Materials Science and Engineering 1190 (2021) 012007 doi:10.1088/1757-899X/1190/1/012007

Figure 2. Experimental test bench

3. Modal analysis

To validate the simulation model at the level of natural structural vibrations, the modal model of the OPG was got by an experimental modal analysis under free-free boundary conditions and the roving hammer method.

In order to minimize possible geometrical deviations of the simulation model, the real test object was digitized using a 3D scanning process. Based on this, the scan data was expanded to an FE mesh and the numerical modal analysis of the digital mock-up was carried out. The correlation between these two modal models is shown in the MAC matrix in **Figure** ***3***.

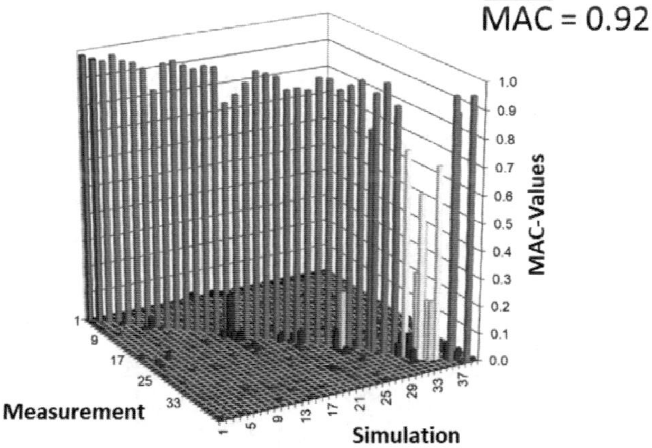

Figure 3. MAC matrix of the scanned and the real housing

The first 39 natural dynamic modes in the frequency range up to 2500 Hz were examined. In this region, the maximum frequency deviation is about 2 % and the mean frequency deviation is reduced to 0 % by optimizing the E-modulus. The correlating mode pairs with an average MAC-Value of 92% [5] are a good basis for further steps of the analysis.

4. Vibration analysis

The aim of the vibration analysis is to validate the dynamic behaviour of the numerical model with the help of measurements from the physical test body. In this step, we evaluate the surface accelerations of the oval principle housing which represent the boundary conditions of the acoustic analysis.

4.1. Experimental transient vibration analysis
During the experimental transient vibration analysis, the housing is excited by an impact hammer, which includes an integrated uniaxial force sensor.

This results in a pure normal oriented force, which doesn't change the boundary conditions of the OPG.

The polyamide tip of the hammer primarily excites the lower frequency range (up to approx. 2000 Hz). The recorded force signal of the measurement is shown in Figure 4.

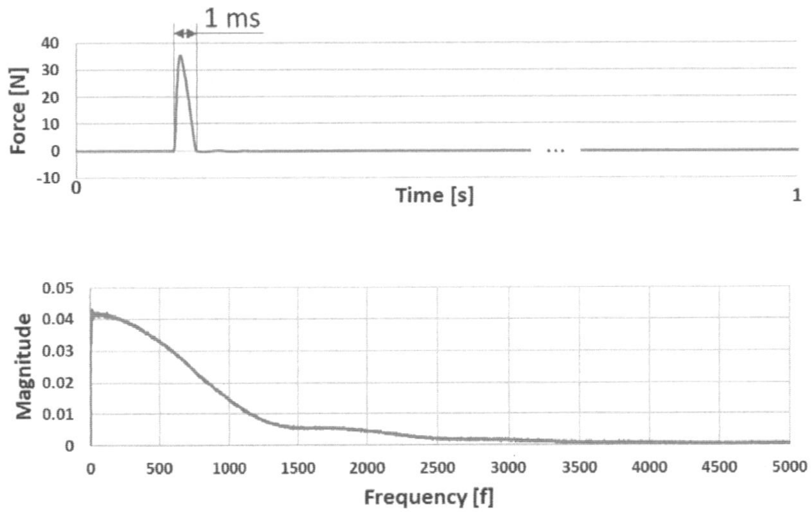

Figure 4: Force-Signal of stimulation

To ensure a strong dynamic answer over the first 39 modes, the excitation point is located in an area where most of the mode shapes show a high amplitude. The modal damping values are extracted for this test set-up by a modal analysis of the frequency response functions of the accelerometers and forces. These modal damping factors are used to update the simulation model (see **Figure *1***) taking into account the test bench boundary conditions.

4.2. Flexible multibody simulation
Based on the FE model [5], a reduced flexible body (RFI) is created for a multibody simulation in the MBS-software RecurDyn. The RFI includes the foam, representing the only boundary condition of the OPG.

To consider the foam, 25 circular node sets were created evenly distributed over the flange, which are connected to CBush elements via RBE3 elements, shown in **Figure 5**. The ends of the CBush elements are fixed to the ground and their stiffness depends on the measured modal results. The sensor masses of the acceleration sensors are also taken into account with mass points on the structure's outer surface.

This modelling technique allows to apply the individually measured damping ratios of the rigid body modes of the RFI.

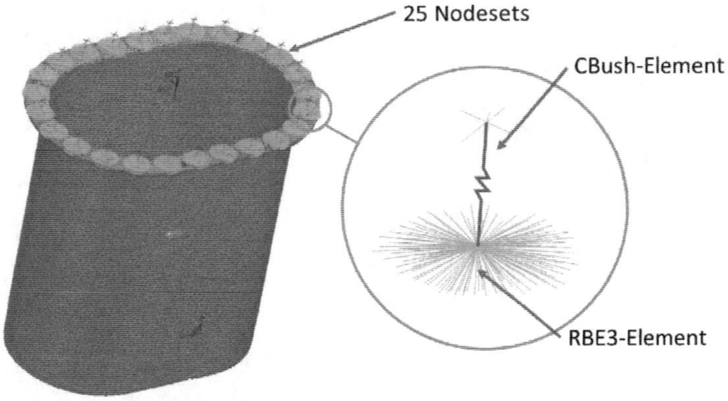

Figure 5. Flexible Body of the OPG

The modal damping values of the measured elastic mode shapes were also assigned directly to the correlating numerical ones. Poorly correlated or unassigned analytical modes got their damping from a linearly extrapolated damping function as shown in **Figure *6***.

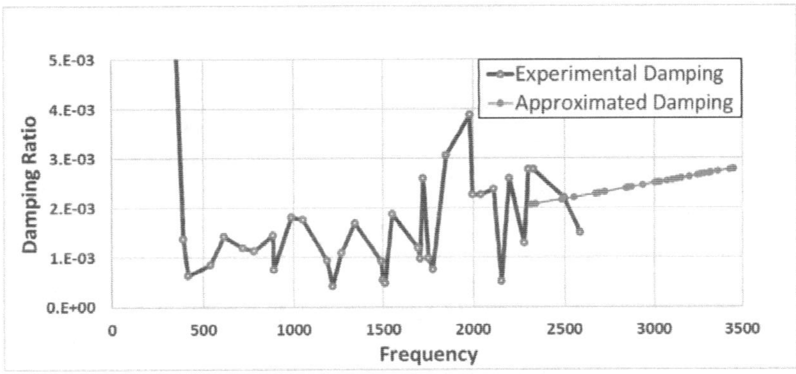

Figure 6. Modal damping of the RFlex body

A spline curve, based on the measured force signal from Figure 4, is used as the analytical force signal of the MBS model.

4.3. Correlation of the transient dynamic behaviour

To evaluate the vibration simulation results, the calculated accelerations at the sensor points are compared with those of the measurement.

In the time-domain (see **Figure** *7*), the experimental and numerical results fit very well over the complete simulation time of 1 s. For a more detailed insight, the observed frequency spectra are also of great interest. **Figure** *7* shows a superb correlation until 2 kHz. Above, the results show slight deviations, but still match the general trend of the amplitudes.

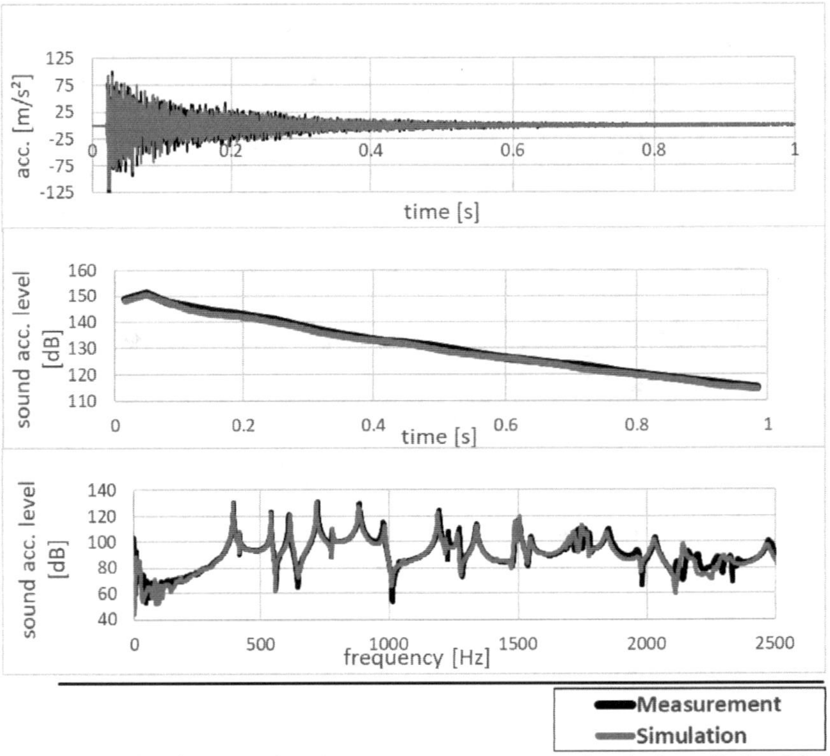

Figure 7. Surface acceleration - measurement vs. simulation - Point 56

The comparison of the simulation regarding the remaining four acceleration sensors and the laser vibrometer are of the same high quality. The absolute mean deviations with regard to the level values in the time range between measurement and simulation are noted in Table 1.

Table 1. Mean deviation of the accelerations between measurement and simulation

	Point 56	*Point 116*	*Point 152*	*Point 184*	*Point 241*	*Point 282*
Type	Acc.	Vel.	Acc.	Acc.	Acc.	Acc.
Direction	X+	X+	Y-	Z-	X-	X-
Mean dev. [dB]	0.6	0.2	0.4	0.1	0.5	0.4

Due to the high level of accordance, the dynamic multibody simulation is considered to be validated successfully. This basis can be used for the following acoustic simulation with respect to the surface vibrations i.e. the boundary conditions of the acoustic radiation.

5. Acoustic analysis

To solve the discretized form of the boundary integral equation (1) Stütz's approach [4] is followed by using the collocation method. The housing surface is meshed with linear two-dimensional elements. Further, their centres are defined as collocation points and the systems of equations at the collocation points are solved with the help of the mesh geometry [4].

5.1. Acoustic BEM-Mesh

The mesh for the numerical acoustic analysis is derived directly from the outer surfaces of the FE body. With the approach of the linear elements it is possible to eliminate the strong singularity part in the discretized form of (1). The choice of the average element edge length is also based on Stütz. He describes the advantages in terms of stability if the mesh fulfil a Helmholtz number (2) smaller than 0.2 and a relationship (3) between the time and space discretization higher than 0.8.

$$He = \frac{l_{el}}{c_{air}/f_{max}} \quad (2)$$

$$\beta = \frac{c_{air} * \Delta t}{l_{el}} \quad (3)$$

The surface, shown in **Figure 8**, is discretized with an average element edge length of l_{el} = 12 mm. The highest frequency to be examined is f_{max} = 2500 Hz, the simulation step size Δt = 1/25600 s and the speed of sound c_{air} = 343 m/s. From that, the following mesh parameters can be derived:

- $He = 0{,}09$
- $\beta = 1{,}1$

5.2. Boundary conditions

The vibrations of a structure's outer surface represent the boundary conditions of the acoustic simulation. These accelerations are required for each BEM element (i.e. collocation point) with respect to the structures normal direction. The element accelerations are responsible for the monopole radiation part in the boundary integral equation (1).

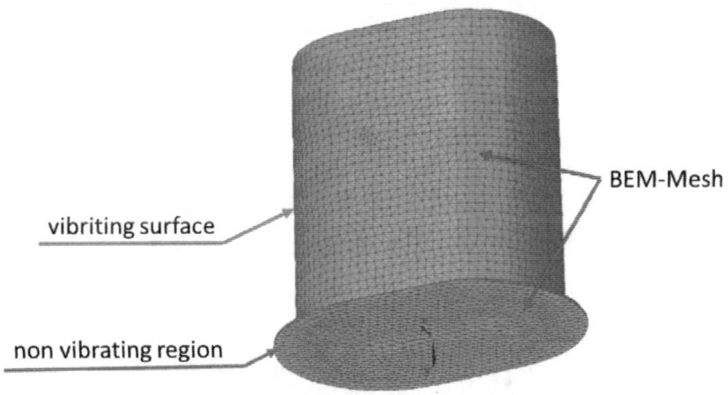

Figure 8. Vibrant regions of the BEM-housing

Since the aluminium plate under the OPG (Chapter 3) is designed to be overcritical, its vibrations are negligible. Thus, the elements of the closed lower boundary surface consider no dynamic behaviour in the numerical acoustic analysis. The accelerations of the outer surface (the green region in **Figure 8**) are defined with the calculated results from Chapter 4.

Finally, the acoustic pressure at the outer surface and at the observation points (i.e. at the collocation points) can be calculated with the computed accelerations (Chapter 4) of the structure.

6. Validation of the time-domain Boundary Element Method

At the validation step, the measured and calculated sound pressures of the microphones are compared and evaluated. The results for microphone 1 are depicted in **Figure 9**. This individual measurement point is located 0.5 meters away from the test object.

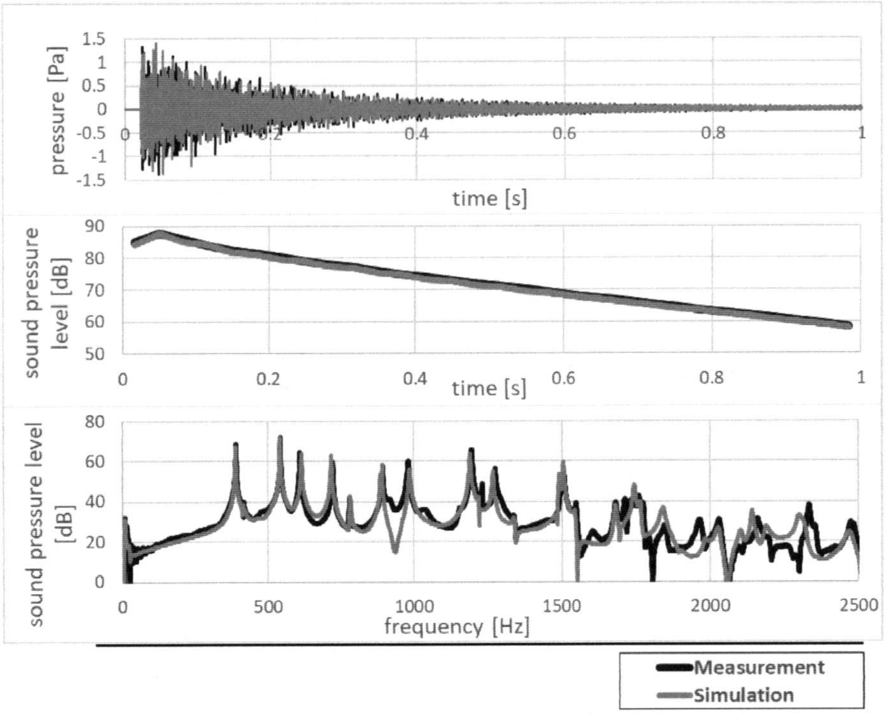

Figure 9. Acoustic pressure - measurement vs. simulation - microphone 1

It is impressive how the signals match in the time-domain as well as in the frequency-domain. The RMS - sound pressure level curves are almost identical over the entire observed time range. For a more detailed insight, the data is transformed into the frequency-domain. Especially the peaks at the natural frequencies are highly similar as a result of the high correlation degree of the time functions. More significant differences only exist in the frequency range above approximately 1800 Hz. This is justified by the weak excitation of the hammer in this region. Hence, the sound pressure levels are completely below 40 dB in this frequency range.

With respect to the mean deviations of the RMS values in the time-domain, the correlations of the remaining 11 microphones are comparatively good, as shown in **Figure *10***.

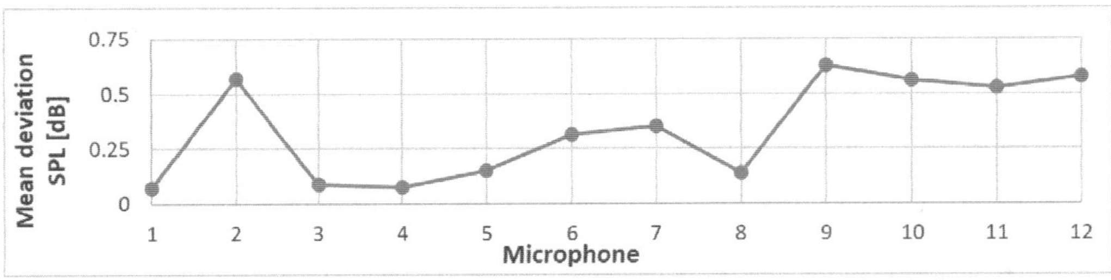

Figure 10. Acoustic pressure deviation - measurement vs. simulation

In a further experimental and numerical investigation, ten of the twelve microphones are positioned at least 1 meter away from the OPG. This separate experiment focuses on the validation of the acoustic results in the far field. Equally good correlations of the signals can be observed. It turns out, however, that in these investigations the random noise of the microphones has a very strong influence on the measured sound pressures. The reason is the low SPL, which declines quadratic to the distance reducing the signal and noise ratio. Therefore, higher excitation levels with a shaker are planned for far field analysis.

Note, that the acoustic simulation is set up with a relatively fine BEM mesh, because this is a first comparison to assess the method in a practical application. This choice is useful in terms of the complexity of the method. For application in development processes, possibilities will be examined in order to reduce the complexity of the mesh.

7. Conclusion

This paper demonstrates the complex process up to the acoustic validation of the TD-BEM calculated results with the help of physical measurements. The numerically calculated sound pressure values of this method are very similar to the measured ones, in the time-domain as well as in the frequency-domain. Within an examining time of one second, the absolute mean error between the calculated and the measured sound pressure level ranges between 0.1 and 0.65 dB. Above all, the curves as well as the peaks of the numerical and the physical sound pressure level in the frequency range are highly similar. With the TD-BEM method in acoustics, both the radiation from structural vibrations and the reflection of incoming waves are considered. Therefore, the acoustic signal at one point is a superposition of all acoustic signals of the structure. Thus, it is assumed that the validation of the acoustic signals at the analysed observation points are representative for the complete structure. Final considerations of the results shows that the TD-BEM method coupled with a validated flexible multibody simulation delivers good results.

Further investigations will show the applicability of the TD-BEM at steady-state and critical operating points such as the influence of a complex ribbed structure geometry. According to Stütz, the numerical eigenfrequencies of the BEM mesh can lead to instability of the method. The next examinations will show the influence of this effect on practical structures and vibrations with regard to the calculated sound radiation.

References

[1] J. L. Y. B. S. F. C.-R. Jose´ Luis Ayuso Munoz, Project Management and Engineering Research, 6330 Cham, Switzerland: Springer Nature Switzerland AG, 2020.

[2] D. Robers, Integrierte Marketing-Kommunikation von Konzernen, Wiesbaden, Germany: Springer-Verlag, 1999.

[3] F. G. Kollmann, Maschinenakustik, Berlin Heidelberg, Germany: Springer-Verlag, 1993.

[4] M. Stütz, Stabilitätsverhalten und Anwendung der transienten Randelementmethode in der Akustik (PhD Thesis), Berlin, Germany: TU Berlin, 2014.

[5] J. Neher, Rechnerische und experimentelle Untersuchungen der Schallabstrahlung bei Fahrzeuggetrieben (PhD Thesis), Illmenau, Germany: TU Ilmenau, 2012.

IOP Conf. Series: Materials Science and Engineering 1190 (2021) 012008 doi:10.1088/1757-899X/1190/1/012008

A Cooling design for an end cell in fuel-cell stack using forced air flow in metal foam: Modelling and Experiment

Ali A. Hmad, Nihad Dukhan
Department of Mechanical Engineering, University of Detroit Mercy, 4001 W. McNichols Rd., Detroit, MI 48221, USA

dukhanni@udmercy.ed

Abstract. The effect of temperature on an end-cell plate in a proton exchange membrane (PEM) fuel cell stack is presented in this paper. Numerical analysis and experimental work investigated the impact of a heat exchanger filled with metal foam on the end-cell plate used for cooling a PEM stack. One side of the edge plate was subjected to constant heat flux (1.55 W/cm2) while the other side was insulated. A metal foam with a pore density of 40 and having a porosity of 60% was analysed. The model was simulated in ANSYS R3-CFD, and the thermal problem was solved using the local thermal non-equilibrium (LTNE) to investigate the heat transfer between the fluid and solid phases. The thermal solution studied the effect of metal foam heat exchangers at different airflow velocities between 0.2 and 1.5 m/s on end-cell plate's temperature. The results showed heat transfer enhancement in plate temperature, proving that the use of the LTNE assumption illustrated the difference between the end-cell plate and fluid flow temperatures. The effect of inlet air velocity on the plate temperature was also examined. An experimental model of a commercial PEM stack was constructed out of aluminium. This model was tested at the same conditions as those employed in the numerical solution. The heat generated by a fuel cell was simulated by a surface heater providing the same heat flux as the actual PEM cell. The numerical and experimental results for the end-cell plate for both at flow direction and different inlet velocities were assessed. There was a good agreement between the numerical and experimental results.

1. Introduction

One of the most critical points in designing fuel cells is appropriate thermal management. Most fuel cells work in a consistently safe range of operating temperatures [1]. The PEMFC operating temperature is between 60-80 °C. Keeping these temperatures in the reliable range and reducing the gradients of cell temperature is the aim of thermal management. The uniform distribution of cell temperature decreases the risk of stack flooding and dehydration of the membrane. The application and size of the PEMFC stacks are responsible when choosing suitable thermal management [2]. Some researchers investigated the thermal management of PEMFC stacks, but only a few studied the use of heat exchangers in PEMFC stack cooling applications. Significant studies have been conducted on the different strategies of thermal management for fuel cell stacks, and water-cooled methods were recently used in the thermal management of PEMFC stacks. This method requires an additional water-cooling loop that uses extra accessories, such as a radiator. So, to avoid increasing expenses, researchers were looking for something as an alternative. The heat exchanger was introduced to the field of the fuel cell stack industry to keep the stack at operating temperatures safe range and to remove the waste heat produced by the stack during operation [3]. The air-cooled systems in fuel cell stacks use air as flow coolant, where the air is forced to flow through the air cooling channels [4].

In some cooling systems, a cooling fan is placed at the bottom of the unit and blows the air across the cooling channels [5]. A few researchers have thermally modeled the PEMFC stack and analytically and numerically studied the thermal management of the fuel cell stacks. Chupin et al. [6] investigated a thermal design of water-cooling for the PEMFC stack. This was followed by Promislow and Wetton [7], who studied the thermal model to investigate temperature distribution in straight flow cooling channels for a fuel cell stack. Yu and Jung [8] numerically established a 2D thermal model of PEMFC; the water-cooled method was used. Their proposed model studied the conduction of heat transfer inside the membrane and the convection of waste heat removed by cooling water. A 2D model was numerically developed by Sasmito et al. [9], wherein the heat transfer by convective forced air was investigated. The study represented the effects of pressure drop in PEMFC stack performance. Previous research has been established by Sinha and Wang [10]. They examined the PEMFC performance when operated at high temperatures, where the applied temperature was constant at the external boundary conditions of the stack. A qualitative study by Ramousse et al. [11] studied the thermal management of the PEMFC and evaluated a thermal model of a fuel cell stack and studied the effects of the distribution of temperature on the stack. Several studies focused on using metal foam to enhance heat transfer in many applications [12]. Some parameters such as pore diameter and porosity play a major effect on the thermohydraulic behavior of metal foams. Due to the thermal properties of metal foam, heat exchangers are extensively used to improve heat transfer [13]. Odabaee et al. [14] numerically and experimentally studied the difference between water and air-cooled heat exchangers for design optimization. In their study, the heat transfer and pressure drop across the heat exchanger in a PEMFC stack were investigated. The results presented the temperature distribution and airflow rate of varied models. Recently, Boyd et al. [15] numerically simulated a 3D model to study the thermal management of a metal foam heat exchanger that was used to remove waste heat from PEMFC. The results showed uniform distribution of the plate temperature at high inlet velocities. This paper presents the effect of high temperature on the edge plate of a PEMFC stack. The model will be simulated in ANSYS R3-CFD, and the thermal equations will be solved by LTNE to investigate the heat transfer between the fluid and solid phases. A metal foam with a pore density of 40 having a porosity of 60% will be used. This paper will study the effects of using metal foam heat exchangers at different air inlet velocities on end-cell plate temperature.

2. Modelling and numerical approach

In this model, the edge cooling plate was machined, where all shoulders between the channels were removed and became one large air-cooling channel. The large air channel was filled with metal foam, as shown in Figure 1(a). The resulting convection heat transfer problem and its boundary conditions are given schematically in Figure 1(b).

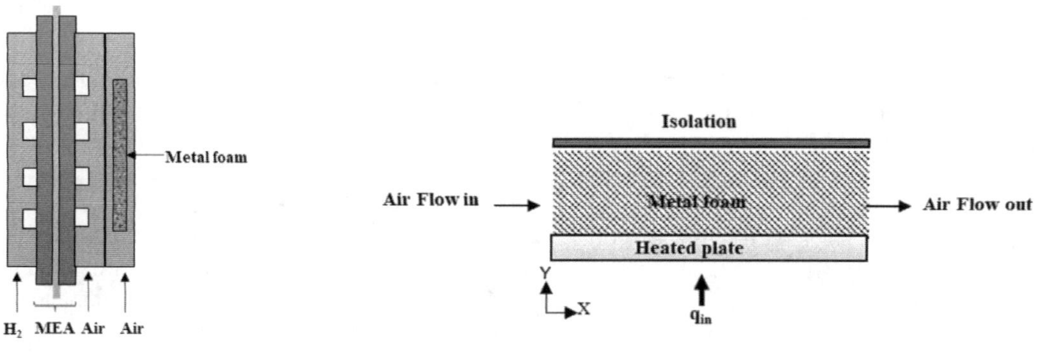

Figure 1. (a) End plate filled with metal foam, (b) Air-cooled end cell: channel filled with metal foam

The morphological properties are the metal foam used are given in Table 1.

Table 1. The sample specifications and metal foam properties.

Metal foam (PPI)	Dimension (mm)	Porosity ε (%)	Heated plate	Heat flux (W/m^2)
40	3.56 x100	60	Aluminium	1.55

The following governing equations were solved using ANSYS- CFD:

Continuity: $$\frac{\partial u}{\partial x}+\frac{\partial v}{\partial y}=0 \quad (1)$$

Momentum: $$\frac{\partial p}{\partial x}+\frac{\mu\partial^2 u}{\varepsilon\partial x^2}-\frac{\mu}{K}u-\frac{\rho f}{\sqrt{K}}u^2=0 \quad (2)$$

where ε and k are the porosity and permeability of the foam, respectively.

Fluid energy equation: $$k_f\left[\frac{\partial^2 T_f}{\partial y^2}+\frac{\partial^2 T_f}{\partial x^2}\right]+h_{sf}\sigma(T_s-T_f)=\rho c_p\left[\frac{\partial(uT_f)}{\partial x}+\frac{\partial(vT_f)}{\partial y}\right] \quad (3)$$

Solid energy equation: $$k_s\left[\frac{\partial^2 T_s}{\partial y^2}+\frac{\partial^2 T_s}{\partial x^2}\right]+h_{sf}\sigma(T_f-T_s)=0 \quad (4)$$

where u and v are the velocity components in the flow direction (x) and perpendicular to the flow direction (y), and p is the pressure. ρ and μ are the density and viscosity of the fluid, respectively. f is the form drag coefficient for the foam. k_f and k_s are the effective thermal conductivities of the fluid and the solid, respectively. ε and K are the porosity and permeability of the foam, respectively. h_{sf} is the interfacial heat transfer coefficient, and σ is the interfacial surface area density (m^2/m^3) inside the foam. T_f and T_s are the fluid and solid temperatures, respectively.

The following boundary conditions were imposed:

At the channel inlet: $$u=u_{in}, v=0, \text{ and } T_f=T_{in} \quad (5)$$

At the heated walls: $$y=0,\ q''=-k_f\frac{\partial T_f}{\partial y}=-k_s\frac{\partial T_s}{\partial y},\ u=v=0 \quad (6)$$

At isolated walls: $$q''=0,\ u=v=0 \quad (7)$$

At the channel outlet: $$p=0 \text{ and } \frac{\partial T_f}{\partial x}=\frac{\partial T_s}{\partial x}=0 \quad (8)$$

3. Experimental work

An identical sample to the commercial stack, EFC-100-03-6-ST, was built precisely in the lab from aluminium to study heat transfer on the end-cell plate and to employ this application on the fuel cell stacks air-cooled system. In this model, all shoulders between air channels in the bipolar plate were removed, which made the air channels one large channel filled with metal foam as a heat exchanger. The stack produces 500 Watts for an active area of 100 cm^2. The stack has three cells, and each cell provides 166.7 watts. The heat flux applied to each cell was 1.55 W/cm^2. Figure 2(a) shows the new plate design of the stack with large, air-cooled channels. The bipolar plate has large channels filled with 40 PPI having a porosity of 60% of aluminium foam, and the air was forced to flow inside the air-cooled channels. Figure 2(b) shows the back side of the plate with a surface heater to apply constant heat flux.

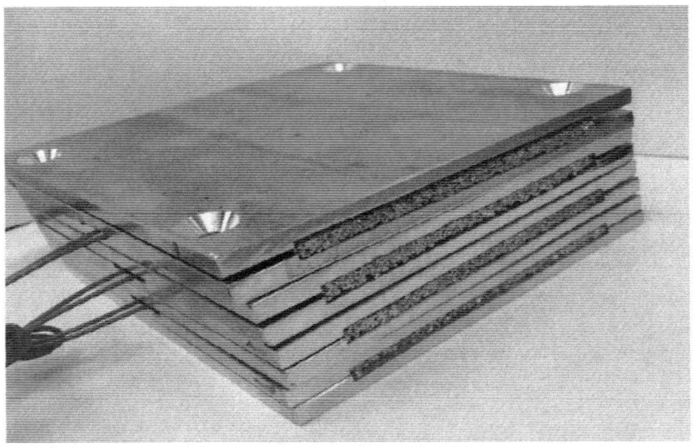

Figure 2. (a) The new plate design of the PEMFC stack with large, air-cooled channels, (b) The installation of the thermofoil heater on the back of the end-cell plate

Super Flow: SF600 was used in this experiment, as shown in Figure 3. The high rates of airflow produced by this machine make it efficient to force the air through the sample's thin channels. Using a flow direction controller, one must simply control the intake and exhaust of the airflow direction. This machine also has a flow controller to maintain the desired air velocity in the end-open tunnel. A square open-end tunnel has a length of 60 cm, and identical cross-section size of 12.50 cm was tightly connected to the machine inlet section. The fabricated test section made of clear Plexiglas has a square cross-section size similar to the air tunnel outlet that was connected to the end of the tunnel by latches. For pressure drop measurements, a heavy-duty pressure transducer made by Honeywell was installed, and a Nanometer was used for inlet temperature and velocity measurements.

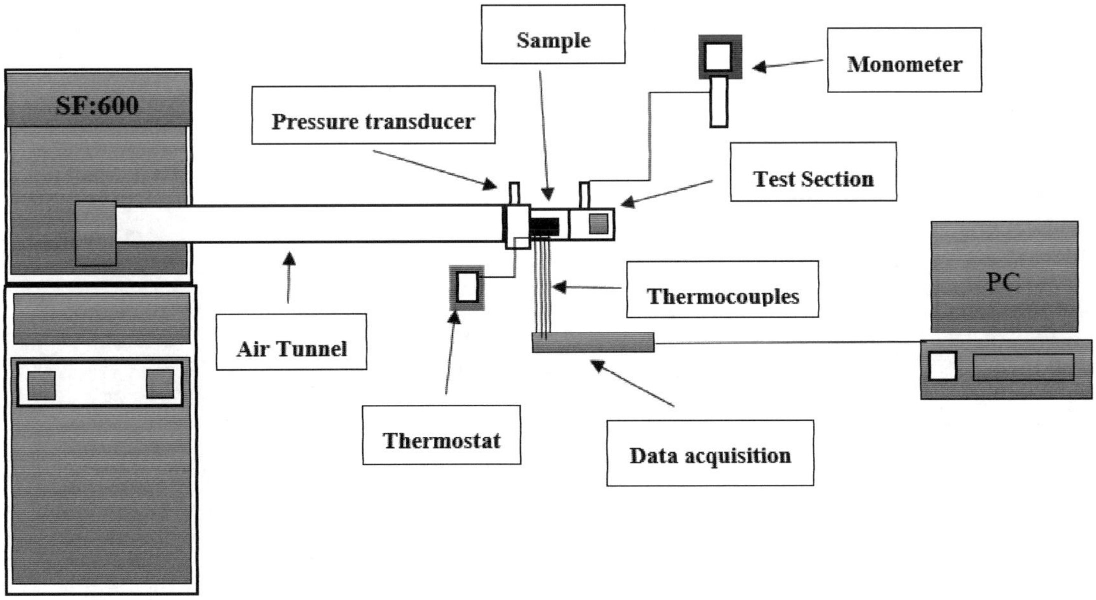

Figure 3. Schematic diagram of the experimental setup

Two holes were drilled to install the speedometer probe and pressure drop device sensor. The sample was inserted into the test section and isolated on all sides by the Plexiglas and sealed by thermal silicon. No air leakages were allowed in this experiment. So, the open-end wind tunnel must tightly connect to the test section by latches. A foam was inserted into the test section entrance to ensure a fully developed flow state. Both the differential pressure transmitter device and twelve thermocouples were connected to the data acquisition. The measurement of the plate's temperature and pressure drop was measured according to the desired inlet velocities. As recommended, the measurements were taken three times and the average was considered.

The root-sum-squares Eqn. (9) was used to calculate the uncertainty in the pressure drop measurement [16]. Fixed and reading errors were declared by the instrument manufacturer. A fixed error of 2% and a reading error of 0.25%were stated by the heavy-duty pressure transducer's manufacturer.

$$\delta p = \pm \sqrt{{e_f}^2 + {e_r}^2} \tag{9}$$

where $\boldsymbol{\delta}_p$ is the pressure uncertainty, e_f is the fixed error, and e_r is the reading error. The inlet velocity was measured by a digital manometer. The uncertainty of 3% as a fixed error and 0.015 m/s as a reading error were reported by the device manufacturer. The pressure uncertainty is 2.06%, and air velocity uncertainty is 3%.

4. Results and discussions

The numerical and experimental results are shown in Figure 4 (a) for end-cell plate temperatures at 2.5 cm near the entrance in the flow direction. Increasing inlet velocity decreased both temperatures and brought them to the safe operating temperature range. The numerical and experimental temperature results in the middle of the plate at different inlet velocities for the end-cell plate. In Figure 4(b), both temperature trends had the same behaviour, decreasing with the increase of the inlet velocity until they reached the lowest temperatures at 1.5 m/s. Shown in Figure 4(c) is the result of the end-cell plate temperature at the end of the plate, which was exactly at 7.5 cm. The result showed a good agreement between the numerical and experimental temperatures measurement. The effect of the inlet velocity was observed, where increasing the velocity decreased the total plate temperatures.

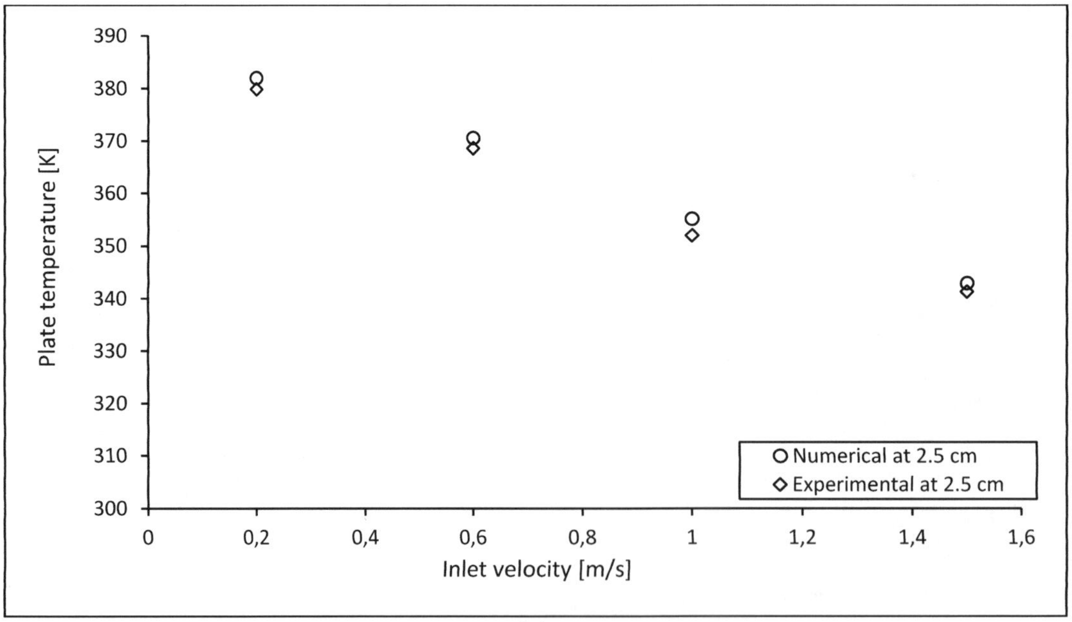

(a)

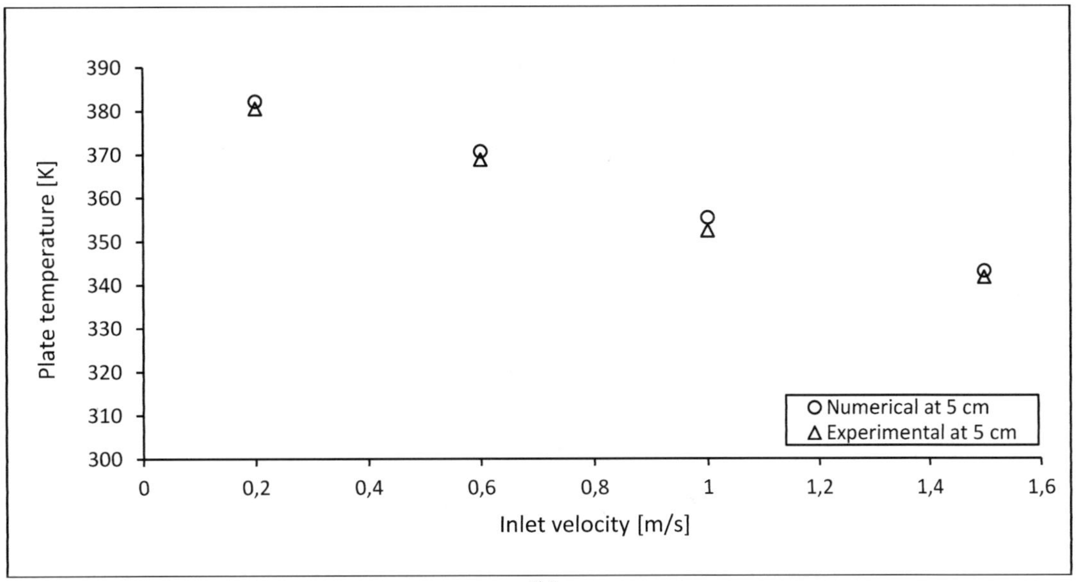

(b)

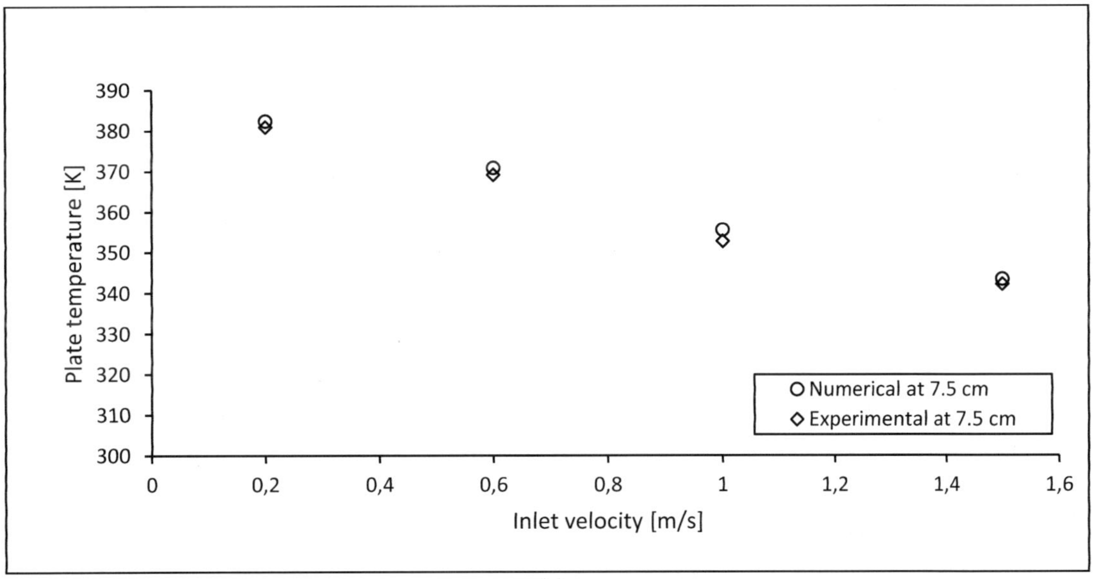

(c)

Figure 4. Numerical and experimental temperatures for end-cell at (a) 2.5 cm, (b) 5 cm, (c) 7.5 cm

The numerical and experimental results for the end-cell plate at 0.2 m/s of inlet velocity at flow direction are shown in Figure 5 (a). This figure shows the effect of inlet airflow velocity on the plate temperature at different locations. At 2.5 cm from the channel entrance, the temperature was about 380 for both temperature readings. The plate temperature slightly increased with the increase of the velocity and that was because of the heat produced from heaters on the plate base. Increasing the inlet velocity from 0.2 to 0.6 m/s decreased the plate temperature by about 10 degrees at 2.5 cm in the flow direction as shown in Figure 5(b). The numerical temperature results were higher than the experimental temperature results. There was a slight increase in the plate temperature, but the total plate temperature decreased when the inlet velocity was increased to 0.6 m/s. Figure 5(c) shows the comparison between the numerical and experimental temperature measurements for different plate locations at the same inlet velocity. There was a big drop in the plate temperature; increasing the inlet velocity to 1 m/s decreased the plate temperatures by about 20 degrees total plate temperature. The temperature measurements for the end-cell plate at 1.5 m/s of inlet velocity in the flow direction are shown in Figure 5(d). Increasing airflow velocity to 1.5 m/s decreased the numerical and experimental temperature measurements. Both temperature measurements had the same behaviour. The plate temperature slightly increased, but generally decreased in the total plate temperature. The numerical and experimental results had a good agreement.

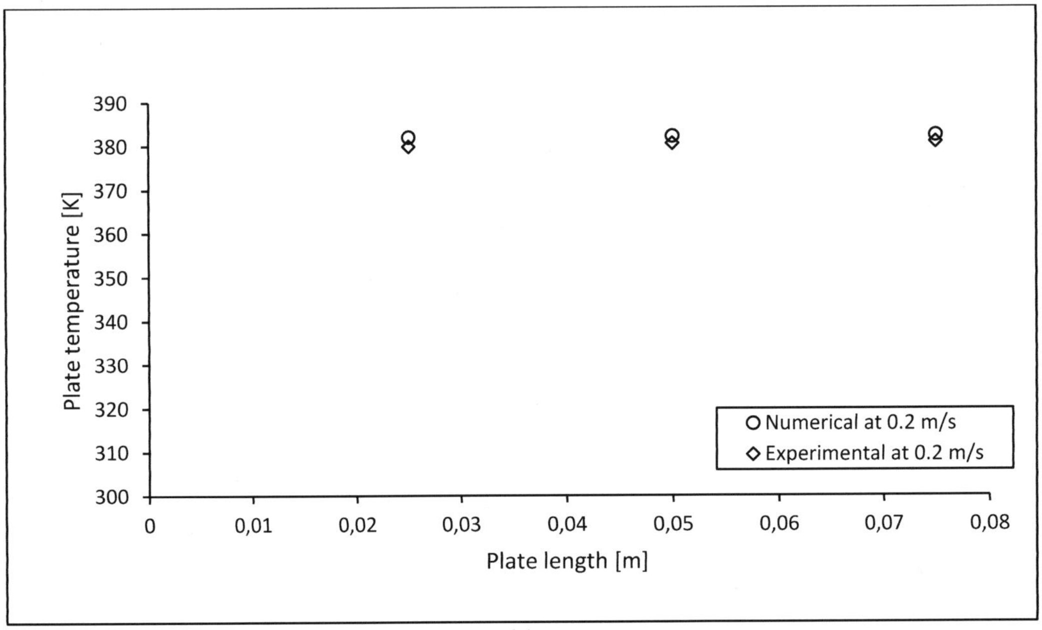

(a)

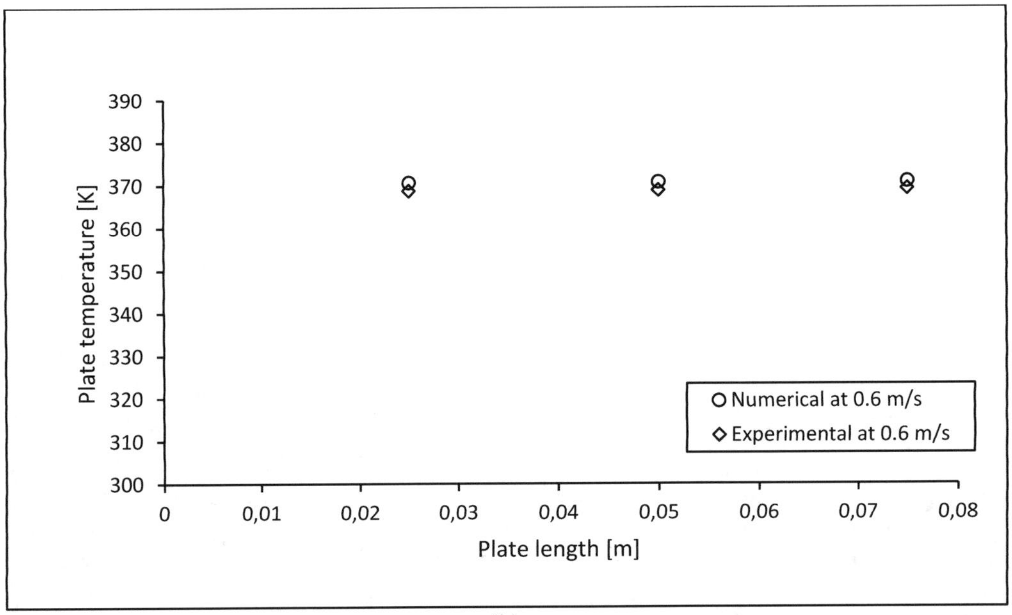

(b)

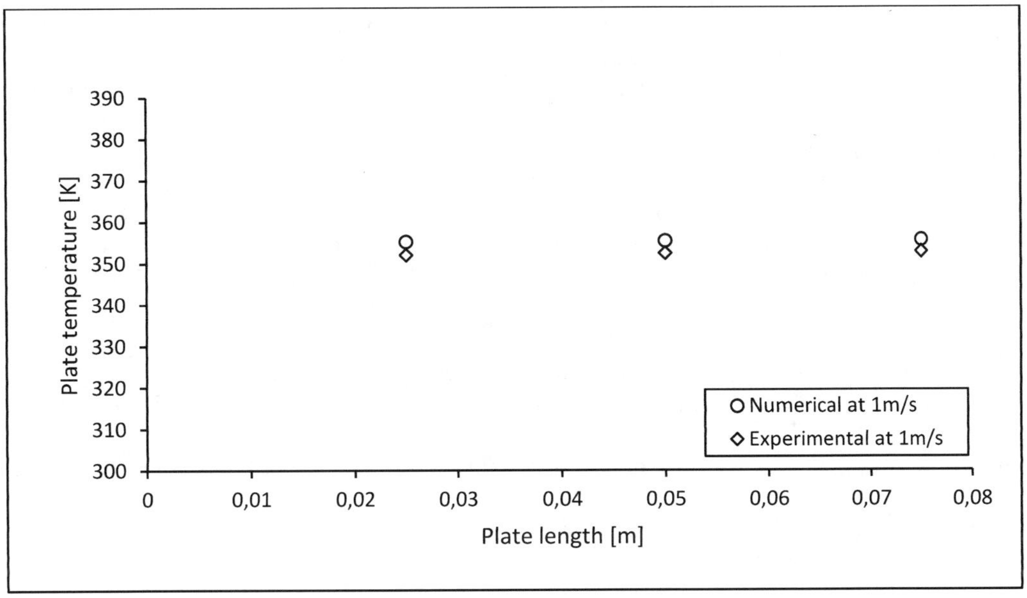

(c)

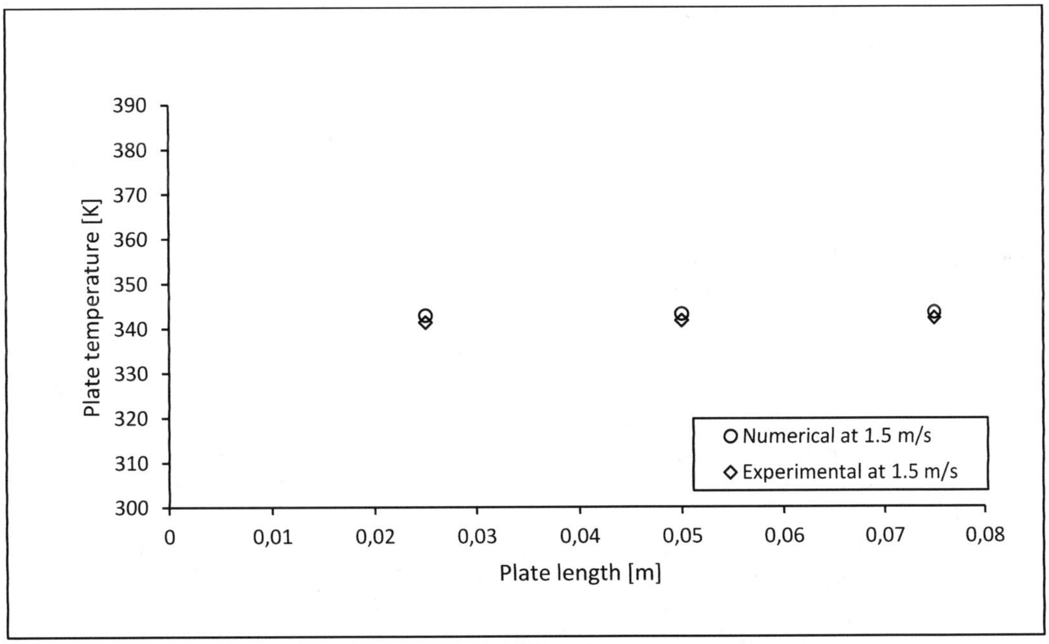

(d)

Figure 5. Numerical and experimental for end cell at (a) 0.2 m/s, (b) 0.6 m/s, (c) 1 m/s, (d)1.5 m/s

5. Conclusions

The impact of maximum temperature on an edge plate filled with metal foam and used for cooling the PEMFC stack was investigated in this paper. The thermal equation of both fluid and solid phases was solved by LTNE. The study was performed for metal foam with 40 PPI and having porosities of 60 % at different airflow field velocities. The results showed the heat transfer enhancement at plate temperature and proved that the use of the LTNE assumption illustrated the difference between the edge plate temperature and fluid flow temperature. Increasing inlet air velocity decreases the plate temperature and there was a good agreement between numerical and experimental results. The difference in temperature between numerical and experimental results for the end-cell plate was investigated. The lowest difference was about 0.3% in 7.5 cm at 1.5 m/s of inlet velocity, while the highest difference was about 0.8 % at 1 m/s.

References

[1] A. Faghri, Z. Guo, Challenges and opportunities of thermal management issues related to fuel cell technology and modeling, International Journal of Heat and Mass Transfer, 48(19-20), 3891–3920, 2005.
[2] J. Larminie, A. Dicks, Fuel Cell Systems Explained, John Wiley, 2002.
[3] S. Cheong, et al., Analysis of water and thermal management with coolant operating conditions for a proton exchange membrane fuel cell. Current Applied Physics, 10(2, Supplement): pp. S22-S25, 2010.
[4] F.C. Chen et al., Analysis of optimal heat transfer in a PEM fuel cell cooling plate, Fuel Cells 3 (4) 2003.
[5] K.P. Adzakpa et al., Transient air cooling thermal modeling of a PEM fuel cell, Journal of Power Sources 179 (1),164–176, 2008.
[6] S. Chupin, T. Colinart, S. Didierjean, Y. Dube´, K. Agbossou, G. Maranzana, et al. Numerical investigation of the impact of gas and cooling flow configurations on current and water distributions in a polymer membrane fuel cell through a pseudo-two-dimensional diphasic model. Journal of Power Sources, 2010.
[7] K. Promislow, and B. Wetton. A simple, mathematical model of thermal coupling in fuel cell stacks. Journal of Power Sources, 2005.
[8] S. Yu, and D. Jung. Thermal management strategy for a proton exchange membrane fuel cell system with a large active cell area. Renewable Energy, 2008.
[9] A.P. Sasmito, K.W. Lum, E. Birgersson, and A.S. Mujumdar. Computational study of forced air-convection in open cathode polymer electrolyte fuel cell stacks. Journal of Power Sources, 2010.
[10] P.K. Sinha, C.Y. Wang, and U. Beuscher. Transport phenomena in elevated temperature PEM fuel cells. Journal of the Electrochemical Society, 2007.
[11] J. Ramousse, K.P. Adzakpa, Y. Dube, K. Agbossou, M. Fournier, and A. Poulin, et al. Local voltage degradations (drying and flooding) analysis through 3D stack thermal modeling. Journal of Fuel Cell Science and Technology 2010;7:041006e10.
[12] M. Alipanah, and X. Li, Numerical studies of lithium-ion battery thermal management systems using phase change materials and metal foams, Int. J. Heat Mass Transf. 102, 1159–1168, 2016.
[13] J. Bock, and A.M. Jacobi, Geometric classification of open-cell metal foams using Xray micro-computed tomography, Mater. Charact. 75, 35–43, 2013.
[14] M. Odabaee, and K. Hooman, Metal foam heat exchangers for heat transfer augmentation from a tube bank, 36, 456–463, 2012.
[15] B. Boyd and K. Hooman, "Air-cooled micro-porous heat exchangers for thermal management of fuel cells," International Communications in Heat and Mass Transfer, vol. 39, no. 3, pp. 363–367, 2012.
[16] N. Dukhan and M. Ali, "Strong wall and transverse size effects on pressure drop of flow through open-cell metal foam," Int. J. Therm. Sci., vol. 57, pp. 85–91, 2012, doi: 10.1016/j.ijthermalsci.2012.02.017.

IOP Conf. Series: Materials Science and Engineering 1190 (2021) 012009 doi:10.1088/1757-899X/1190/1/012009

Experimental calibration methodology for numerical parameters in a DEM software based in granular material static tests

Luis González Calderón[1], Rodrigo Estay Huidobro[1], Sebastián Maggi[2]

[1]Universidad Técnica Federico Santa María, Department of Metallurgical and Materials Engineering – Av. España 1680, Valparaíso, Chile, 2340000
[2]Engineering Simulation and Scientific Software (ESSS) – Alfredo Barros Errázuriz 1954, Santiago, Chile, 7500000

rodrigo.estayh@usm.cl

Abstract. Simulations are one of the most used tools in different investigations and topics, being a powerful tool to replicate a real-life system or to solve real engineering problems without interrupting any process. To accomplish a reliable replication of the real system, it is necessary to make good assumptions in the numerical parameters selection of the simulation and its configuration. The main goal of this research is to propose a calibration methodology of numerical parameters based on static tests simulations in the Rocky-DEM software. A laboratory flow test experiment was developed to measure the repose (a_r) and emptying (a_e) angles of a granular material in a static state. In the simulation experiment, these angles are dependent of numerical parameters, such as the static friction coefficient (m_f), the rolling coefficient (m_r) and the adhesive distance (d_a). To get the best numerical parameters that explained the behavior in the laboratory experiment, an optimization procedure was done. Two materials were used for this research: chickpeas, for the development of the calibration methodology, and sulfide copper ore, to apply this methodology. The numerical parameters for the chickpeas experiment were $m_f = 0.183$, $m_r = 0.105$ and $d_a = 0.938$ [mm], resulting in $a_r = 31.23°$ and $a_e = 28.85°$. These simulated angles have a variation of 2.85% and 3.03% with respect to the values obtained in the laboratory experiment. In the case of the mineral experimentation, a $m_f = 0.892$ and $m_r = 0.269$ were obtained, simulating a $a_r = 30.1°$ and $a_e = 42.31°$. Also, these results have a variation of 2.84% and 3.65% with respect to the values of the angles measured in the laboratory model. As a conclusion of the study, DEM simulation is a powerful tool to predict the behavior of granular materials. However, it is necessary to use calibrated values and settings in the simulations to increase the accuracy of the results.

1. Introduction

The discrete element method (DEM) consists of representing the material discontinuously by a set of discrete particles. The discontinuities are considered as boundary conditions. The contact forces and displacement of the set of particles under a stress distribution are determined through a series of calculations involving their motion.

The DEM simulations are governed by mathematical equations that describe the behavior of the model and therefore its results. Every equation has numerical parameters that control the magnitude of these results [1], which is why these parameters are important to know and calibrate. Also depending on the experiment type that is being replicated with simulations, it will be very important to know which and how many parameters will have a greater effect in the behavior of the model.

The main goal of this research is to develop a calibration methodology to estimate the numerical parameters in a DEM simulation. This methodology was applied in the estimation of the repose and drawdown angles in bulk material performed in a granular flow test [2]. This research is focused for DEM simulations because it is the best method for replicating a bulk material [3]. In this research we use the Rocky-DEM software.

In DEM simulations, the same procedure of the laboratory experiment is replicated. At the end of each simulation the repose and drawdown angles are measured. Finally, these results are compare with those obtained in the granular flow test.

2. Experimental set-up

The experiment chosen was a granular flow test, whose procedure and its respective set-up allowed to estimate the repose and drawdown angles. The repose angle [4] is commonly used to be measured in bulk material, while drawdown angle measures the resulting slope when discharging gravitationally through the center of the base of bulk material [5] (Figure 1).

Figure 1. (a) Repose and (b) drawdown angles

The bulk material chosen was chickpeas due to its shapes, similar to spheres, allowing us to reduce the computational cost in the DEM simulations [6].

Both in the experimental test and DEM simulation, constant environmental values of temperature and humidity were use. Chickpea density is indirectly calculated by volume displaced in a test tube, resulting in a value equal to 1333 kg/m^3, while for sulfide copper ore a value equal to 2600 kg/m^3 with the same measurement method. The average size of chickpea grains was 8.5 millimeters in diameter with 15 kilograms of material used, while for the ore, it has a size distribution (table 1) with a maximum size of 19.1 mm and a total mass of 17.5 kg.

Table 1. Size distribution for sulfide copper ore experiment and DEM simulation.

Size (mm)	Cumulative (%)
19.1	100
12.5	90.05
9.5	87.3
6.35	0.1

Using factorial design [7], the number of measurements is determined with respect to the number of parameters involved in the experiment (equation 1). In this case were: static friction coefficient, rolling resistance and adhesive distance. For each parameter, a lower and a higher level of possible values are considered. For static friction coefficient the values are between 0.1 and 0.7, between 0.1 and 1.0 mm for adhesive distance, and between 0.1 and 0.5 for rolling resistance. As a result of the above, it is necessary to develop 8 experiments and simulations.

$$Numbers\ of\ Experiments = Parameter\ Levels^{Parameter\ Numbers} \quad (1)$$

For those parameters, whose values are not modified in the model, there is the following setting:

- Particle – Particle interaction: Both static and dynamic friction coefficients have a value equal to 0.5, 0.3 value for coefficient of restitution, 6.25 millimeters for adhesive distance, and a force fraction of 0.4.
- Particle – Boundary interaction: Static and dynamic friction coefficient have a value equal to 0.7, a coefficient of restitution of 0.3, an adhesive distance of 6.25 millimeters, and a force fraction of 0.4.
- Rolling resistance equal to 0.28.

Finished the preparations of experiments in laboratory and numerical model, it starts the experimental procedure of the granular flow test. The experimental mounting and the experimental procedure are detailed in Figure 2.

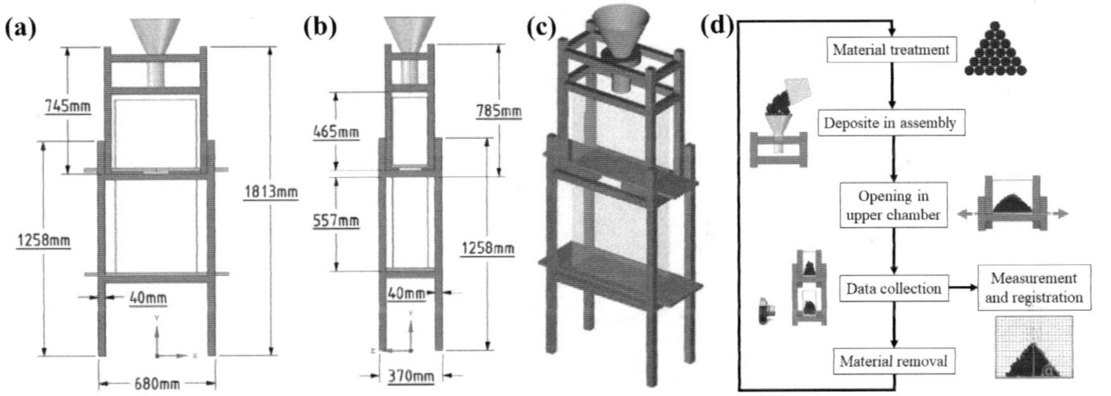

Figure 2. Design of the granular flow test mounting experiment (a) frontal view, (b) lateral view, (c) isometric projection and (d) experimental procedure diagram

The procedure to perform the granular flow test is as follows:

1. **Material treatment:** For maintain the same assumptions of bulk material such as distribution and particle shape, is done a treatment of material before of each measurement. Treatment as weight, screening and homogenization.
2. **Deposite in assembly:** Treated bulk material is deposited on upper part of the mounting, entering to the first chamber.
3. **Opening in upper chamber:** With material at rest, separation of boards is continued at a distance established by the experimenter (20 centimeters for this research). These tables boards correspond to the base of the first chamber. This allows the material to pass into the second chamber.
4. **Data collection:** Once the material comes to rest again, photographs are taken to the upper chamber, and to the lower chamber, where the drawdown and repose angle will be located, respectively. The photographs are taken with corresponding measurements of distance and height of the photographic camera, to reduce as much as possible the influence of perspective

in the angle photographs. To facilitate the image treatment task that will take place later, a green sheet is placed on the chamber background.

5. **Measurement and registration:** Photographs are processed by MATLAB software to measure the angles.
6. **Material removal:** Finally, the experiment is completed by removing the bulk material from the chambers. Then, it is return to point 1.

The experimental procedure must be reproduced into the numerical model, making a copy as exact as possible of the real experiment, changing only in each simulation the values of the numerical parameters of interest.

3. Calibration methodology

The diagrams of the stages of the resulting calibration methodology are shown in Figure 3 and Figure 4.

The first third of the methodology corresponds to the preparation of experimentation in the laboratory and DEM simulations, the choice of bulk material to be calibrated, the corresponding numerical parameters, the response variables and how and where they will be measured. The necessary considerations and simplifications for the simulations are also decided in order to avoid a higher computational processing cost, without compromising the calibration approach and its results.

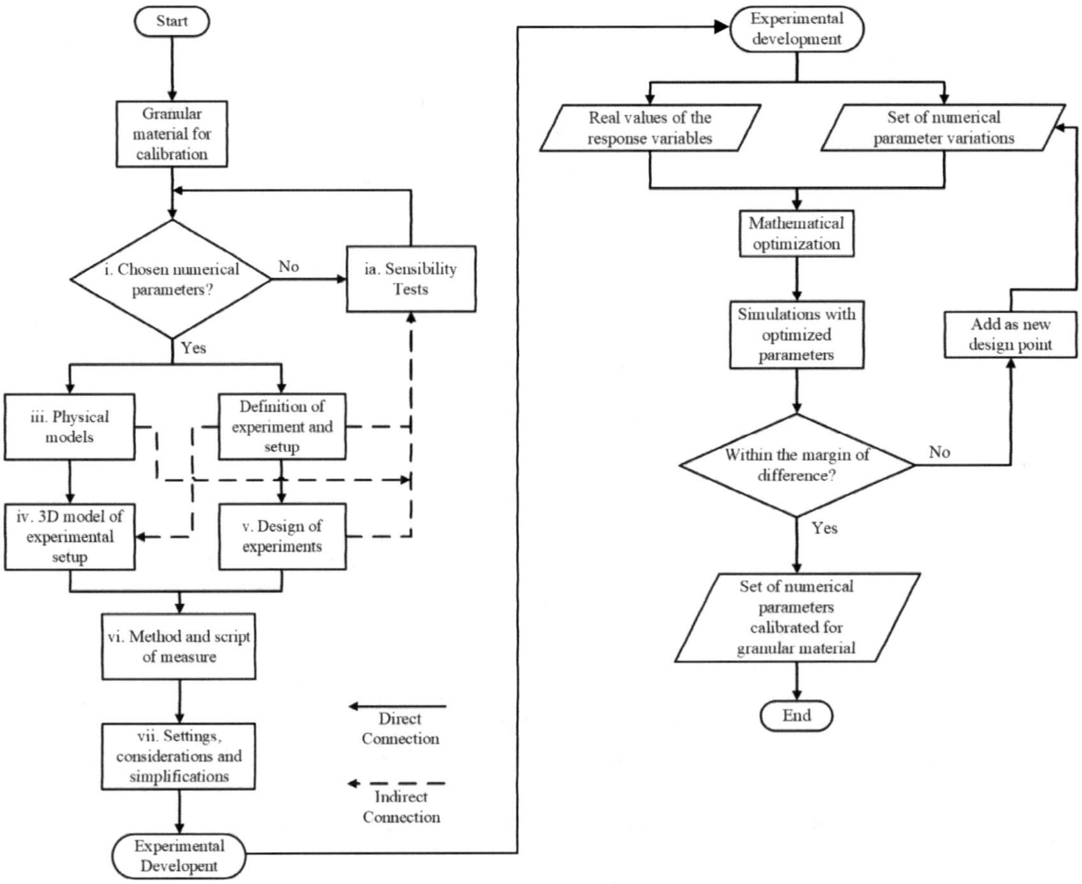

Figure 3. Diagram of the first stage: preparation of base of the experiments

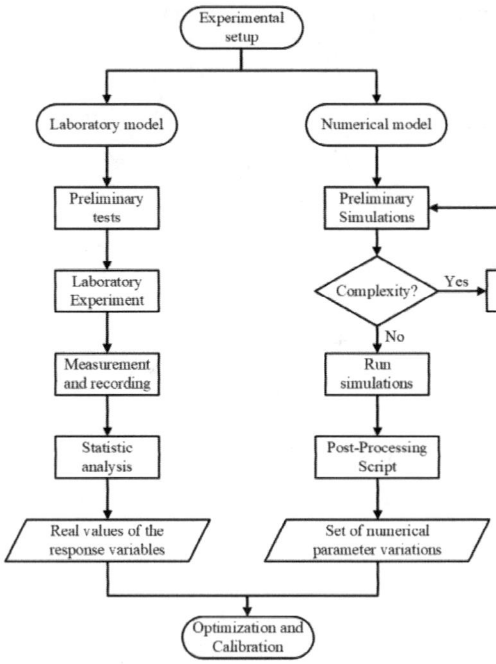

Figure 4. Second stage: development of laboratory experiment and DEM simulations

The intermediate stage corresponds to the experiment execution, simultaneously carrying out the laboratory procedure and the DEM simulations. Preliminary testing is recommended to observe aspects that may not have been considered in the preparation stage.

With the DEM simulations results, a response surface [8] of all possible values of angles is generated regard to numerical parameters that were varying in each simulation (static friction coefficient, rolling resistance and adhesive distance). Ansys DesignXplorer is used through the constant kriging interpolation type to create this surface [9].

An optimization to the response surfaces is carried out, using the MOGA type interpolation algorithm [10], to find the numerical parameters which assure to replicate the values obtained from the angle measurement of chickpeas and the ore in the experimental procedure.

For a higher accuracy in the calibration, it is used as a criterion the difference margin. This margin is associated to the standard deviation of the experimental measurement. Any optimization result that generates a response variable in DEM simulations outside the difference margin, is added to the data set to create a new response surface and be optimized again.

4. Results

4.1. Angle measurement

4.1.1. Experimental model. The experimental angle measurement is done by taking photographs to the repose and drawdown angles of the bulk material. The angle was determined by adding a green sheet to the chamber background and the bulk material. Then the photograph is converted into a binary image [6], that is, in an image with a white area that represent the material and a black area that correspond to the background. Subsequently, binary image is cropped, leaving the right and left side, discarding the upper part due to the sudden change in angle (Figure 5).

IOP Conf. Series: Materials Science and Engineering 1190 (2021) 012009 doi:10.1088/1757-899X/1190/1/012009

Figure 5. Schematic of the cut made to the binary image for the repose angle. In the background the original photograph of the granular material

A set of points are generated at the intersection of both areas (black and white), in which it is possible to apply a linear regression to obtain the slope and subsequently de angle (Figure 6).

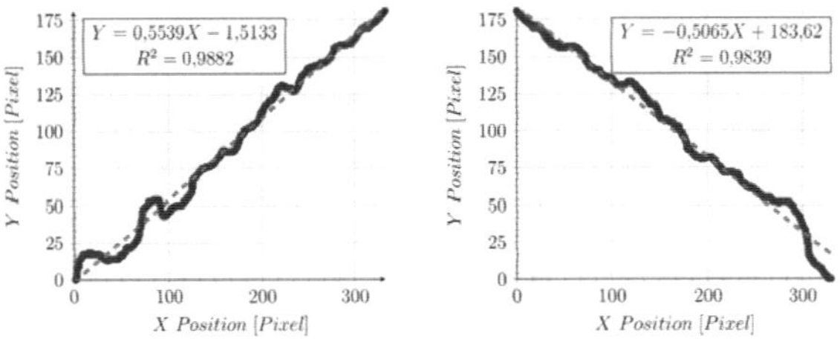

Figure 6. Plots of point set that represent the edge of the bulk material on its left and right sides, respectively, and the trend line that they describe

With angles obtained, the average angles of left and right sides are calculated and recorded, having an average repose and drawdown angle per measurement.

4.1.2. Numerical model. Angle measurement procedure in DEM model consists of performing a post-processing at the end of each simulation. Eulerian statistics are used, a tool present in RockyDEM [11], which allows the chamber volume to be discretized at the width and depth of the chamber, being divided several times with respect to the maximum particle size.

The tool makes it possible to request the particle centroid that reaches the highest position with respect to the height for each division made across the width for a division with respect to the depth of the volume chamber. By taking these points to a plot, the pile silhouette of bulk material is described and where it is possible to determine the left and right angles of this (Figure 7).

This process is repeated for each division regard to the chamber volume depth, from the chamber front to the rear, where left and right side are recorded in each one. Final calculation corresponds to average of previous data for each side and the side with a lower standard deviation is kept as the final result.

4.2. Chickpeas experiment
According to the experimentation and process carried out in laboratory to measure response variables for this bulk material, it was 30.36° ± 1.32° of repose angle and 28.00° ± 2.28° of drawdown angle.

From simulations, there are 8 design points where for values of the minimum levels for the numerical parameters that are 0.1 for static friction coefficient, 0.1 millimeters for adhesive distance and 0.1 for rolling resistance, is reached 29.45°±0.7° in repose angle and 27.89°±1.17° in drawdown angle. While for maximum values of 0.7 of static friction, 1 millimeter for adhesive distance and 0.5 for rolling resistance, is reached 33.88°±1.50° in repose angle and 47.78°±2.01°.

With the design points, the response surface and corresponding optimization are generated, reaching a calibration for chickpeas bulk material with values of numerical parameters 0.18 of static friction coefficient, 0.94 millimeters of adhesive distance and 0.105 of rolling resistance, resulting in DEM simulations response variable of 31.23°±1.34° for repose angle and 28.85°±0.77° for drawdown angle.

4.3. Sulfide copper ore experiment

Due to the complexity of its particle shape and the effect in the computational processing time, it was decided to simulate spheres. In this case, the numerical parameter of the adhesive distance is discarded, since this parameter tries to represent the cohesion /adhesiveness of bulk materials generally in the presence of water, however the ore material was treated with a low moisture content [5].

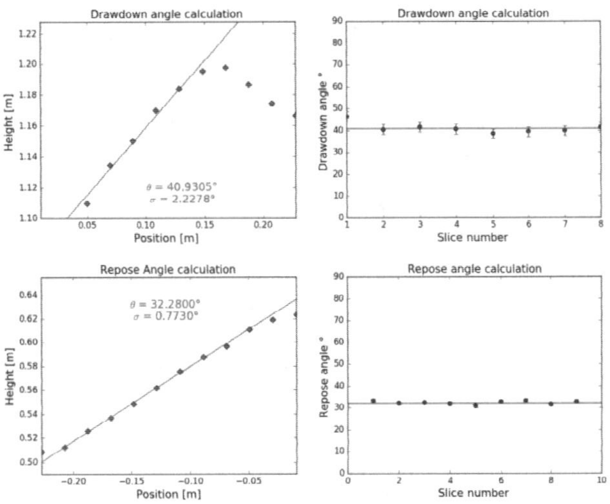

Figure 7. Plot with simulation results, on the left the points that describe the repose and drawdown angle, and on the right side the angle magnitude with respect to the chamber volume depth

Average angle for copper sulfide ore material obtained from laboratory experiment it was 30.98°±1.71° for repose angle and 40.82°±3.21° for drawdown angle. In DEM simulation, the results were 0.892 for static friction coefficient and 0.269 for rolling resistance, generating a result of 30.10°±0.80° of repose angle and 42.31°±1.40° of drawdown angle.

5. Conclusions

The results produced by computational models depend strongly on what input values are used, which is why it is necessary to develop a calibration methodology. In this research it is proposed a methodology with 3 important stages and two types of models: experimental and DEM simulations.

The first stage corresponds to prepare the experimental procedure and the DEM simulations. This stage is essential because it lays the foundation for the experimentation and the research to achieve a good calibration process. In the second step we develop the experiment and the simulations. And in the last stage it is optimize the numerical parameters of the simulation, which best represent the behavior of the measurable variables in the experiment.

Regarding the results of calibration methodology proposed in this research, it is recommended:

- Chickpeas: a value of 0.18 for static friction coefficient, 0.94 millimeters of adhesive distance and 0.105 of rolling resistance.

- Copper sulfide ore: a value of 0.892 for static friction coefficient and 0.269 of rolling resistance.

Some of the important aspects to consider when carrying out the calibration methodology are the simulation time, the number of numerical parameters, the measurable variables and therefore the experiment.

Finally, the validation of the proposed methodology presented in this research should be done. For this, it is suggested to apply the methodology for the same granular material, using two different experiments and/or response variables, calibrating the same numerical parameters. If the optimized values of the parameters are similar, the methodology can be accepted as validated.

References

[1] J. Banks, J. S. C. II, B. L. Nelson, and D. M. Nicol, Discrete-event system simulation, 2001.
[2] T. Donohue. Calibration for material handling equipments. Webinar TUNRA Bulk Solids. Duration 50:05 minutes, 2017.
[3] T. Groger, and A. Katterfeld. Application of the discrete element method in materials handling. part 1: Basics and calibration. Bulk Solids Handling, 2007.
[4] Y. Qiu, Y. Chen, G.G.Z. Zhang, L. Yu, and R.V. Mantri. Developing solid oral dosage forms (second edition). Chapter 10 - Particle, Powder, and Compact Characterization, 271-293, 2017.
[5] F. Cabrejos. Effect of moisture content on the flowability of crushed ores, 2017.
[6] M. Mohammad, T. Mehari, and R.A. Kurt. Calibration and validation of a discrete element model of corn using grain flow simulation in a commercial screw grain auger, 2017.
[7] K. Haerling (Adamson), and S. Prion, Two-by-Two Factorial Design, Clinical Simulation in Nursing, Volume 49, Pages 90-9, 2020.
[8] G.E.P. Box, and N.R. Draper. Empirical model-building and response surfaces. Wiley series in probability and mathematical statistics, 1987.
[9] B. Büttner. Calibration of a dem simulation with ansys optislang. Webinar Dynardo GmbH. Duration 41:11 minutes, 2017.
[10] A. Konak, D.W. Coit, and A.E. Smith. Multi-objective optimization using genetic algorithms: A tutorial. Reliability Engineering and System Safety, 2006.
[11] ESSS. Dem technical manual - rocky 4.2. 3-30, 2018.

Weight Optimization by Using Both Aluminum and Composite Material on the Big Excavator

Luigi Solazzi, Andrea Buffoli, Federico Ceresoli

Department of Mechanical and Industrial Engineering, Via Branze 38, 25123 Brescia Italy

luigi.solazzi@unibs.it

Abstract. The aim of this research is the development of a lightweight large excavator. This is due to the need to reduce the system suitable for its movement and to increase the hourly production of the excavator. This lightweight is done on all the main components by adopting aluminum alloy and composite materials. The study, developed in analytically and in numerically way, considers different loading conditions to which the excavator can be subjected during its use. The new solution was developed in order to have almost the same performance for safety coefficient, displacement and so the stiffness and critical buckling load. The excavator in its minimum weight configuration, obtained by applying composite materials for arms and hydraulic cylinders while aluminum alloy for frame, has an operating weight that goes from the value of 577.6 kN to the value of 425.8 kN with a reduction of over 26%. With this result, three important factors are to be underlined: increase in capacity which depends on the geometric configuration of the arms (about +41.5%); a significant reduction in ground pressure allowing the use of the excavator in previously inaccessible places and finally greater ease in transporting the machine.

1. Introduction

The aim of this work is to go and study an innovative solution for the lightweight of a common excavator. It was decided to address the issue of lightweight as the continuous development of the society that requires a constant increase in the performance of the production and transport systems. However, this increase in performance must also be accompanied by a greater attention to the environment, looking at eco-friendliness and eco-sustainability.

For these reasons, a method to be able to have an increase in performance (translatable in our case into a higher hourly production) trying to maintain a low environmental impact (reduction of consumption and of harmful gas emissions) can be found in the lightweight of the components.

In fact, after the works [1-3] in which the methodologies and construction solutions inherent the development of arms and hydraulic cylinders in composite materials were presented to ensure a significant lightweight of the components, this work focuses on design study of a typical excavator frame using a light aluminum alloy [4] compared to the typical construction steel [5, 6].

For these reasons, we will try to identify the optimal material for the construction of the excavator frame by comparing the use of the typical S355 construction steel with a light aluminum alloy [7] and then going to estimate the optimal geometry for the construction [8, 9], it will therefore be possible to obtain a significantly lighter final structure also thanks to the use of components made of composite material. In addition, we will also study the counterweight and the balance of the machine according to the different construction solutions and the different load conditions.

2. Description of the machine

The excavator used for the following study is the Komatsu PC600 (Figure 1) as it is available for detection of dimensions (Table 1) and for a campaign of experimental tests aimed at determining accelerations during its normal use [10].

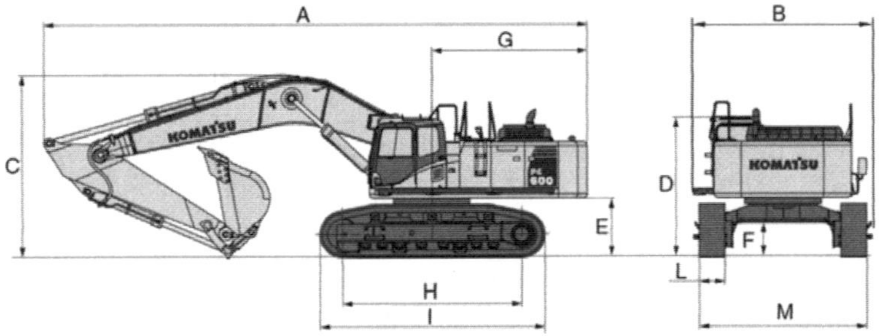

Figure 1. Excavator Komatsu PC600

Table 1. Main dimensions of excavator

Boom: 7660		[mm]
Stick: 3500		
A	Overall length	12810
B	Overall width	4210
C	Overall height (to top of boom)	4300
D	Overall height (to top of cab)	3290
E	Ground clearance	1365
F	Ground clearance	780
G	Tail swing radius	3800
H	Track length on ground	4600
I	Track length	5690
L	Shoe width	600
M	Width of crawler	3900

An excavator can be divided into three main sections, Figure 2:

1. tracked wagon: allows the movement of the machine;
2. arm group: allows the handling of loads is composed of the arms, the bucket and the hydraulic cylinders;
3. upper structure: composed of the frame on which the boom is fixed, the relative cylinders, the counterweight, the cabin and the engine.

Figure 2. Subdivision of the excavator into three subgroups

3. Frame: structural steel and aluminum alloy

The aim is to study the frame of Figure 3 and therefore optimize it by using a light alloy [11, 12], to do this it is necessary to have the steel frame as a starting base so that you can create the solid model of the same. It was assumed that the original frame was made of S355 [13, 14] construction steel.

Below are some images of the model created and the main dimensions of the S355 construction steel frame.

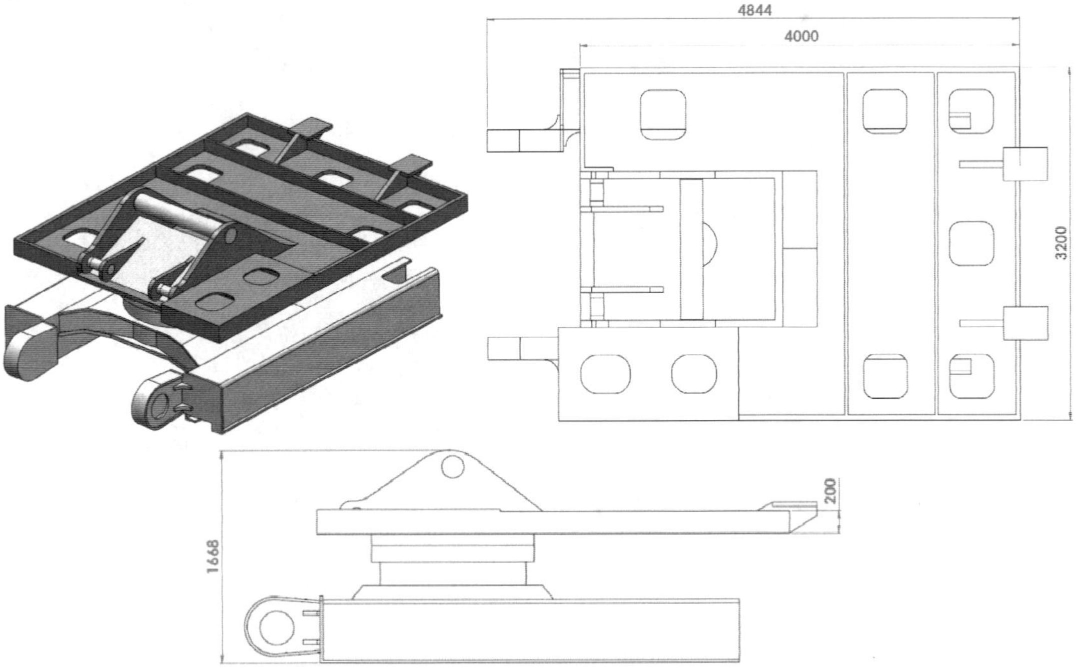

Figure 3. Solid model and main dimensions of the frame and undercarriage in S355 [mm]

For the construction of the light alloy frame, the use of 6060-T5 aluminum alloy (UNI EN 573-3) was assumed, this material presents a good compromise between performance, weldability and costs, the characteristics of the studied materials are shown in Table 2.

Table 2. Main characteristics of the materials

Material	Density ρ [kg/m³]	Young modulus E [MPa]	Poisson ratio ν	σ_{sn} [MPa]	σ_R [MPa]
S 355 (UNI EN 10025-3)	780	210000	0.28	355	510
Al 6060 T5 (UNI EN 573-3)	270	69000	0.33	120	165

In order to create the light alloy frame, it was decided to adopt the same safety factor as the design criterion between the two construction solutions both in terms of stresses and displacements in order to have, for the same mechanical behavior, the same structure, but lighter. To do this and to be able to conduct a comparison between the two solutions, it is necessary to establish the load conditions.

4. Load Conditions

The main load conditions adopted also for the geometric and structural development of the innovative composite arms are shown below, these load conditions have also been studied under the dynamic profile [15, 16] and also used to compare two different geometric solutions [17].

- Load condition 1 (leveling): in this phase the arms are fully extended with the bucket resting on the ground, the movement sees the approach of the bucket to the body of the excavator as can be seen from Figure 4.

(b)

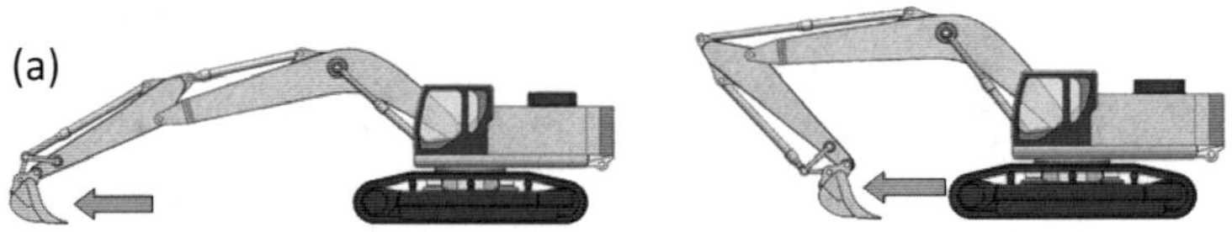

Figure 4. Load condition 1 (leveling): (a) start, (b) end

- Load condition 2 (lifting at minimum distance): during this phase the arms are positioned to be able to carry out a lifting action with the bucket positioned as close as possible to the body of the vehicle in question, as can be seen from the Figure 5.

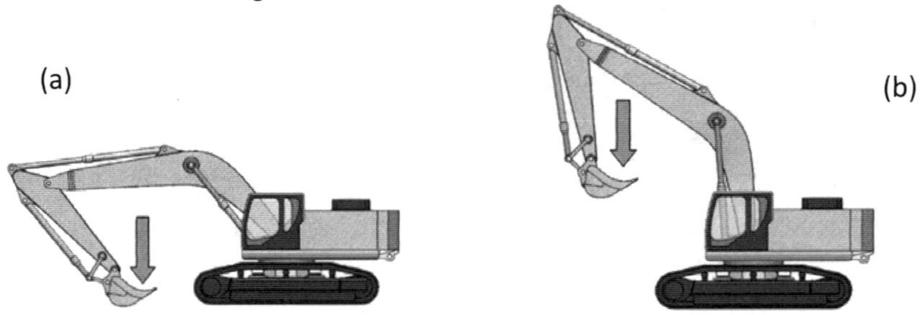

Figure 5. Load condition 2 (lifting at minimum distance): (a) start, (b) end

- Load condition 3 (lifting at maximum distance): with this configuration the machine looks like in leveling condition 1 with the only difference that in this case the arms will not have to carry out a dragging action, but lifting and therefore the bucket it will have to perform a movement in a vertical and not a horizontal direction, as can be seen from Figure 6.

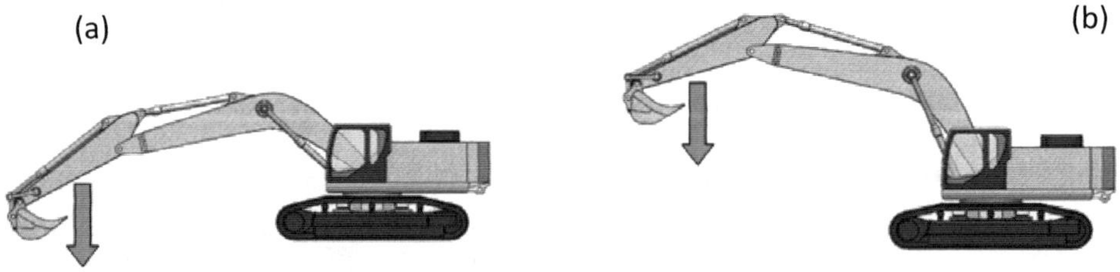

Figure 6. Load condition 3 (lifting at maximum distance): (a) start, (b) end

The magnitude of the applied forces, in correspondence with the bucket, during the simulations of the loading conditions derives from the load diagram of the machine, the values of these forces are shown in Table 3.

Table 3. Forces applied according to load conditions

Load case	F[N]
1	323000
2	243500
3	93500

5. FEM simulation results for S355 frame

Once the constraints and loads [18, 19] relating to each individual load condition were set, finite element numerical analyzes were carried out [20, 21]. The FEM model was created using 4-node brick elements and is made up of approximately 29000 elements and approximately 58000 nodes. The software used

IOP Conf. Series: Materials Science and Engineering 1190 (2021) 012010 doi:10.1088/1757-899X/1190/1/012010

for the finite element analyzes is SolidWorks®. The results obtained from the finite element simulations are shown in Figure 7 as regards the load condition 1, in Figure 8 as regards the load condition 2 and in Figure 9 for regarding load condition 3.

- Load condition 1

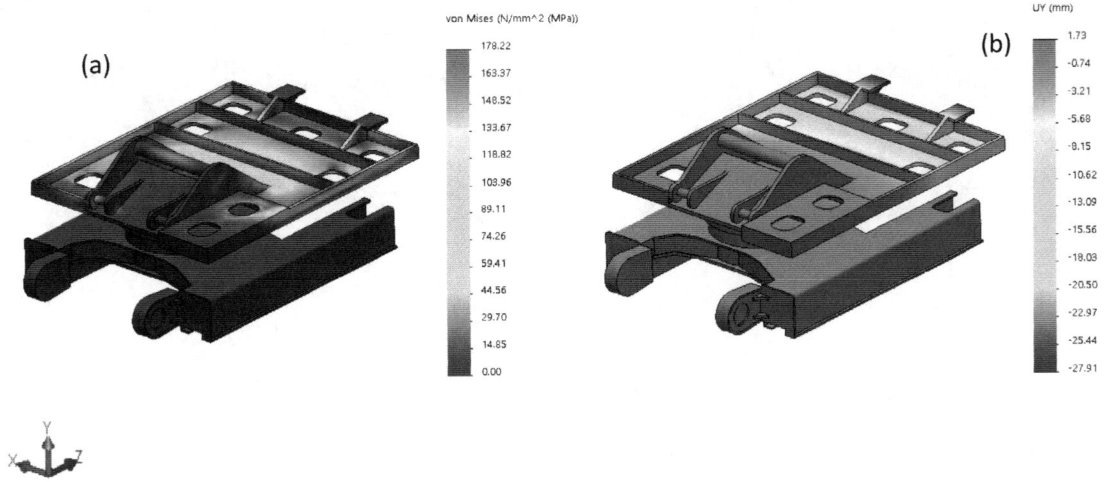

Figure 7. Frame S355 load condition 1: (a) stress, (b) displacement

- Load condition 2

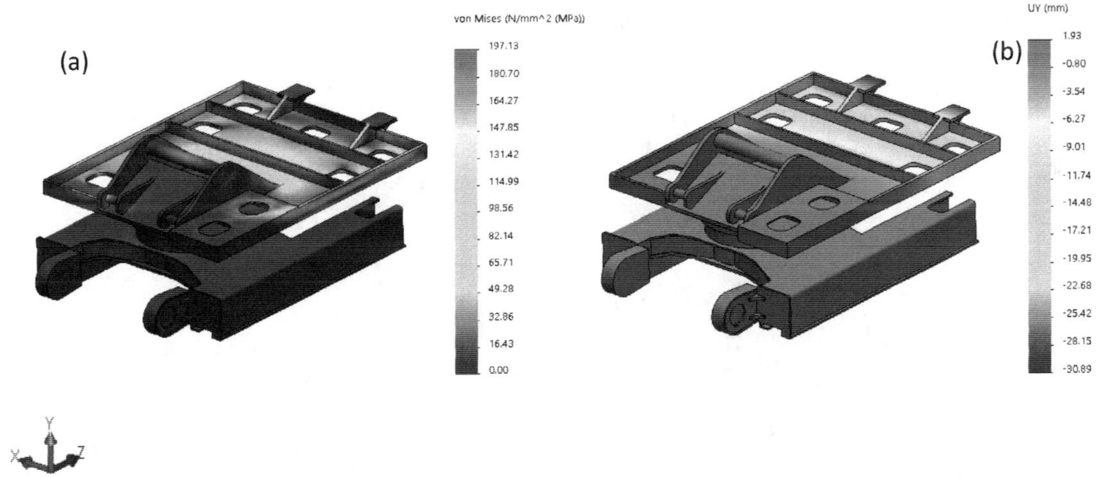

Figure 8. Frame S355 load condition 2: (a) stress, (b) displacement

IOP Conf. Series: Materials Science and Engineering 1190 (2021) 012010 doi:10.1088/1757-899X/1190/1/012010

- Load condition 3

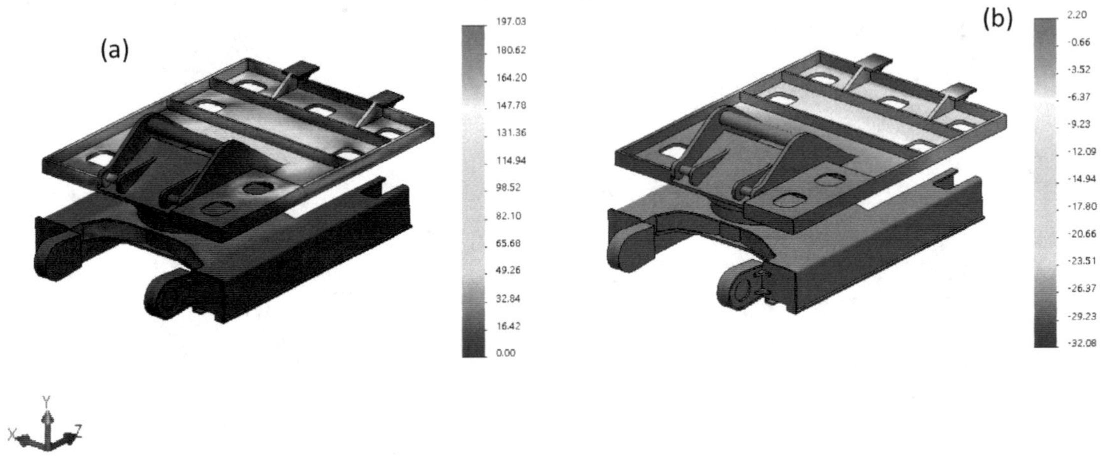

Figure 9. Frame S355 load condition 3: (a) stress, (b) displacement

6. Light alloy frame

In order to be able to create the new frame by adopting an alternative material the first step, is to maintain the same geometry as the S355 steel frame, but using the 6060-T5 so as to identify the critical areas of the component. Once the critical areas were identified, iterations were carried out to improve them through thickness increases and modifications in the basic geometries in order to obtain a frame with the same characteristics and the same behavior as the starting one. In this way what has been obtained can be seen in Figure 10 and in Table 4. At this point it is possible to proceed with the simulation at the finite elements whose results are reported in Figures 11-13.

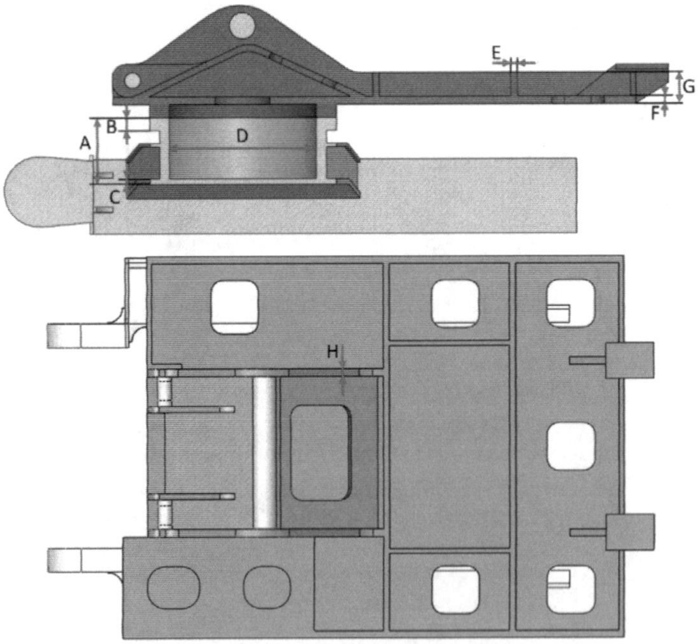

Figure 10. Frame in 6060-T5

Table 4. Variation in size and weight between frame in S355 and frame in 6060-T5

[mm]	S355	6060-T5
A	460	510
B	80	100
C	25	40
D	1250	1150
E	30	50
F	50	60
G	200	240
H	50	65
Weight [kN]	122.9	54.7

- Load condition 1

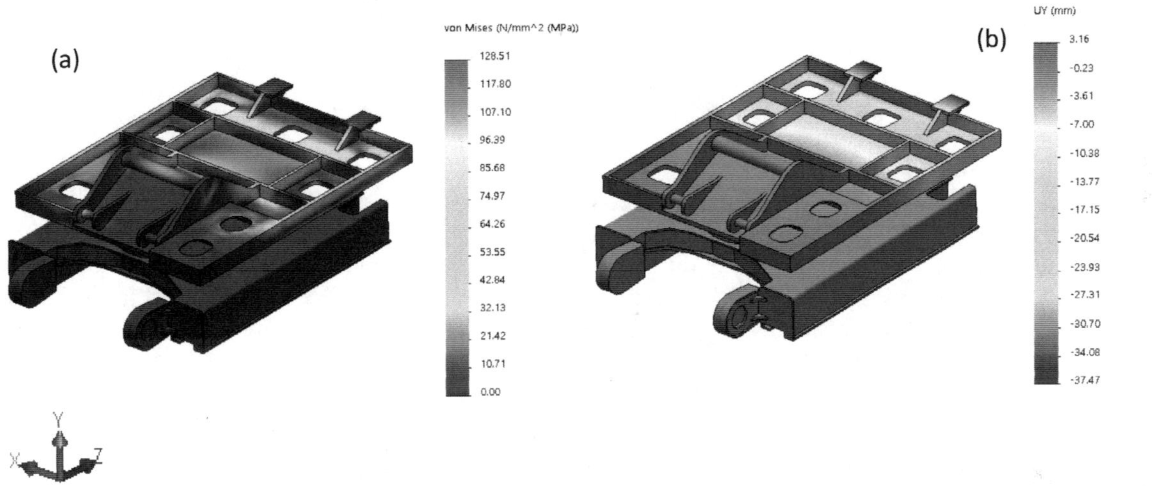

Figure 11. Frame 6060-T5 load condition 1: (a) stress, (b) displacement

- Load condition 2

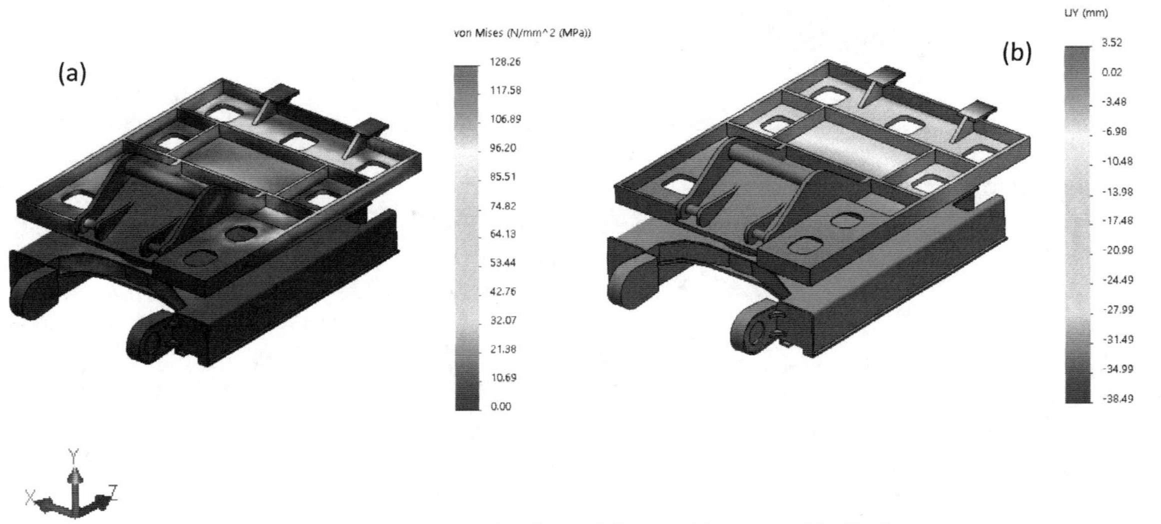

Figure 12. Frame 6060-T5 load condition 2: (a) stress, (b) displacement

- Load condition 3

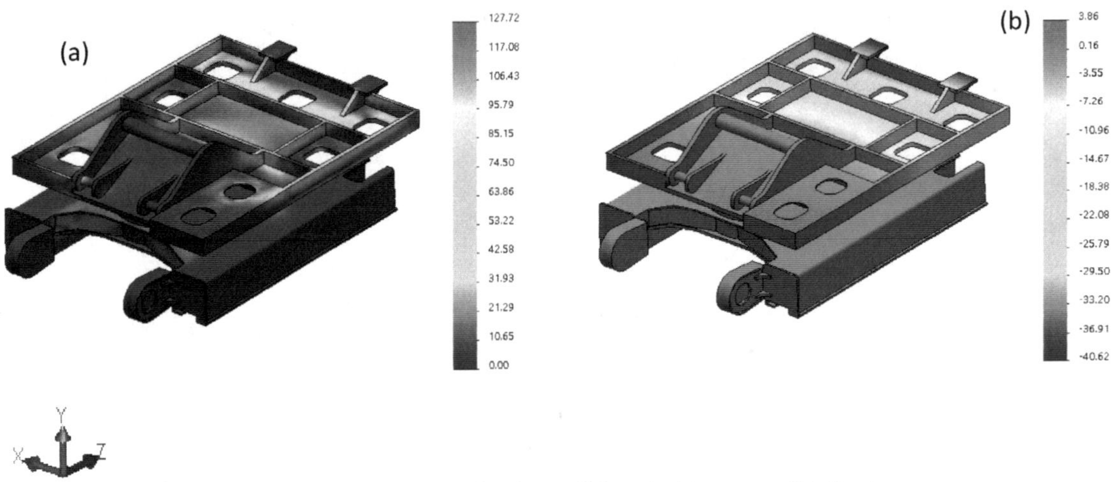

Figure 13. Frame 6060-T5 load condition 1: (a) stress, (b) displacement

7. Evaluation of the counterweight

To balance the excavator in the different load conditions, it is necessary to use a counterweight that must be properly sized [22]. To do this, it is necessary to calculate the excavator's center of gravity and weight by considering the arrangement of the different loads and the position of the arms, so as to be able to calculate the overturning moment of the machine. The front part of the tracks, point A of Figure 14, was chosen as the point around which the excavator can overturn in flat ground conditions.

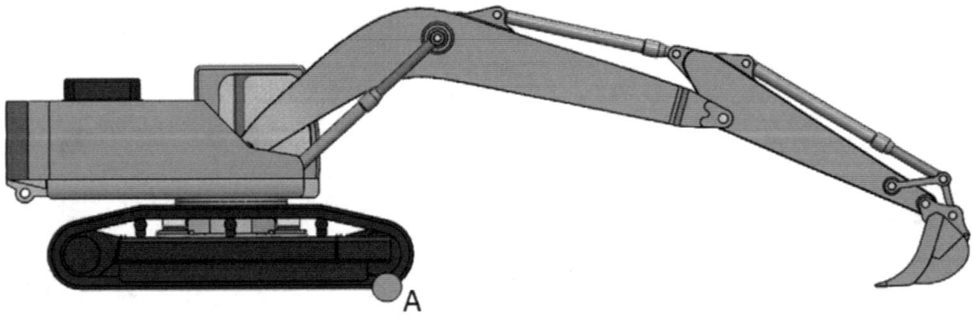

Figure 14. Point of interest for the calculation of the overturning moment (point A)

In order to correctly size the counter-weight it is necessary to use a safety coefficient defined as the ratio between the overturning moment and the stabilizing moment:

$$\eta = \frac{M_s}{M_r}$$

Where the overturning moment M_r is generated by the load present in the bucket, by the weight of the boom and the stick, while the stabilizing moment, which keeps the machine in a safe position, is given by the weight of the load-bearing structure of the excavator and the counterweight (Figure 15). The optimal value is considered equal to 1.1 obtained as the average of the values 1 and 1.2 shown in Table 5.

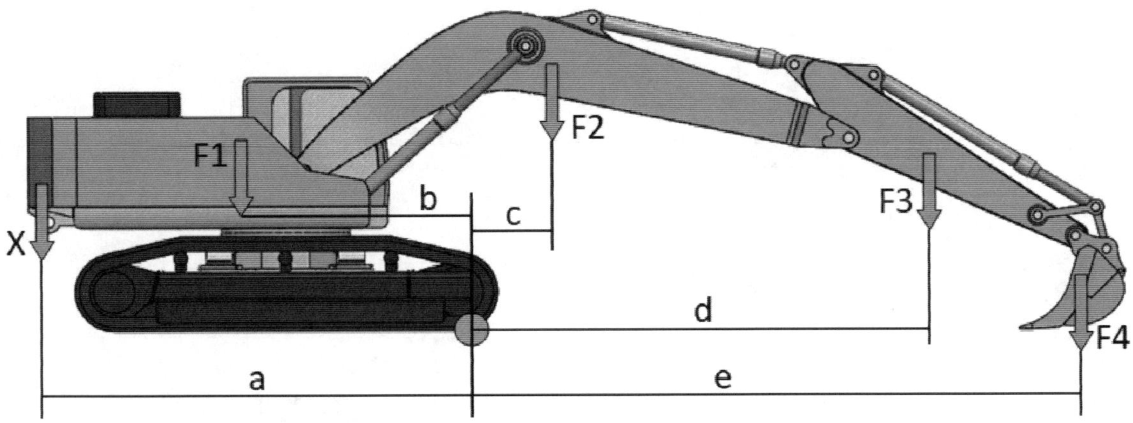

Figure 15. Forces and distances schematic for counterweight calculation

$$\begin{cases} X \cdot a + F1 \cdot b = \eta(F2 \cdot c + F3 \cdot d + F4 \cdot e) \\ X = \dfrac{\eta(F2 \cdot c + F3 \cdot d + F4 \cdot e) - F1 \cdot b}{a} \end{cases}$$

Table 5. Counterweight value as a function of the different construction combinations

Configurations / Weight [kN]	A	B	C	D	E
F1	360,0	266,7	360,0	360,0	266,7
F2	35,3	18,6	18,6	12,4	12,4
F3	14,7	9,0	9,0	5,8	5,8
F4	93,5	93,5	93,5	93,5	93,5
X η=1	94,8	117,0	82,0	75,7	110,6
X η=1.2	140,9	160,4	125,4	117,8	152,8

Configurations:
A) all part in constructional steel S355;
B) all part in aluminum alloy 6060 T5;
C) frame in constructional steel S355, arms in aluminum alloy 6060 T5;
D) frame in constructional steel S355, arms in composite material;
E) frame in aluminum alloy 6060 T5 arms in composite material.

Table 5 as in table 6 shows the variations in terms of weight of the arms (Table 5) and variation of the counterweight value (Table 6) according to the different construction configurations. In fact, these different configurations have been analyzed on the basis of results obtained from previously works [1-2-3-10], in which the arms and the hydraulic cylinders in composite material (carbon fiber 60% and epoxy resin at 40%) have been studied and developed as shown in Figure 16. In particular, the arms in composite material were designed considering an elliptical truncated cone section instead of a classic rectangular hollow section for both steel and aluminum materials. The ends of the arms and the cylinder connections zones are made of aluminum. For the hydraulic cylinders, both the tube and the rod are made of composite material while the connection ends are made of aluminum.

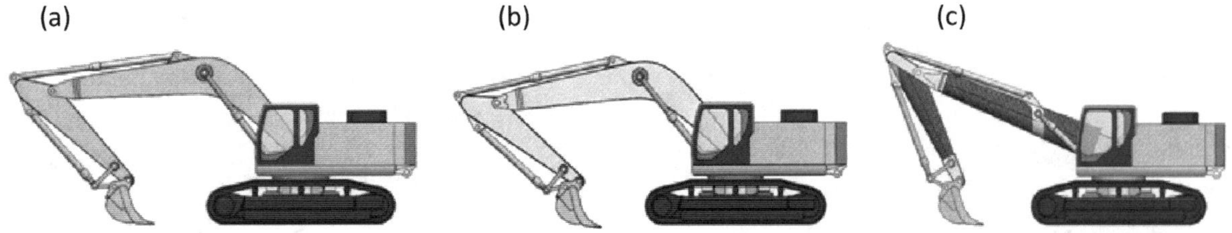

Figure 16. Different configurations of excavator (a) S355; (b) Al6065-T5; (c) Composite material

Table 6. Variation of the counterweight and the operating weight of the machine according to the different construction solutions

Configurations/ Weight[kN]	A	B	C	D	E
Counterweight (F)	117.9	138.7	103.8	96.8	131.8
Operative weight	577.9	483.1	541.4	484.1	425.8
Reduction	-	16.4%	6.3%	16.2%	26.3%

Configurations:
A) all part in constructional steel S355
B) all part in aluminum alloy 6060 T5, hydraulic cylinders in constructional steel S355
C) frame in constructional steel S355, arms in aluminum alloy 6060 T5, hydraulic cylinders in constructional steel S355;
D) frame in constructional steel S355, arms and hydraulic cylinders in composite material;
E) frame in aluminum alloy 6060 T5 arms and hydraulic cylinders in composite material.

8. Maximum slope rating

We want to study the effect of the overturning moment as a function of the inclination of the ground and the configuration of the excavator in static condition (at the start and end of the movement without considering the inertial effects of the movement). The load conditions shown are 2 (lifting at minimum distance) Figure 5 and 3 (lifting at maximum distance) Figure 6, for both loading conditions the initial and final phase of the movement were considered. The determination of the maximum slope that can be overcome is made by determining the intersection between the curve of the overturning moment and the stabilizing moment as a function of the slope angle of the ground. This parameter is of considerable importance since the machine does not always work on a flat surface and therefore it is necessary to estimate the maximum inclination for safety reasons. The inclination of the ground that can vary from 0 ° to 90 ° is represented by the angle α in the graphs of Figure 18, 19 for condition 2 and Figure 21, 22 for condition 3.

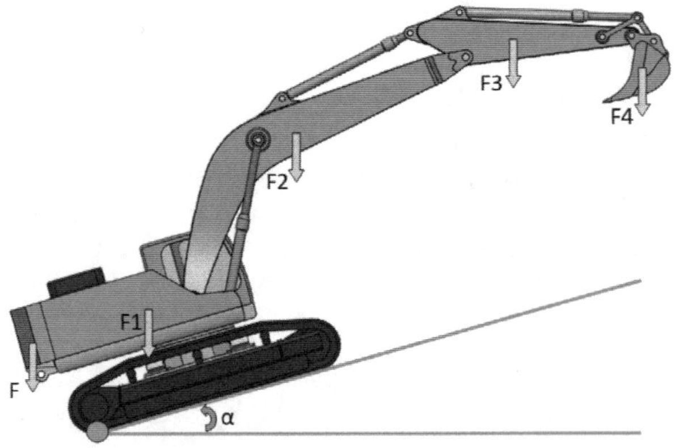

Figure 17. Schematization to calculate the maximum slope to be overcome

- Uphill load condition 2

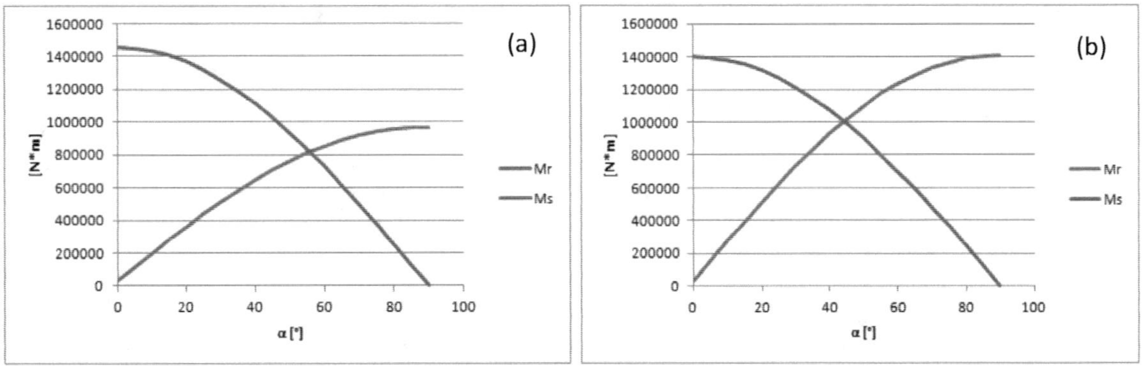

Figure 18. Graphs of the maximum gradient that can be overcome under load condition 2: (a) load condition start, (b) load condition end

- Uphill load condition 3

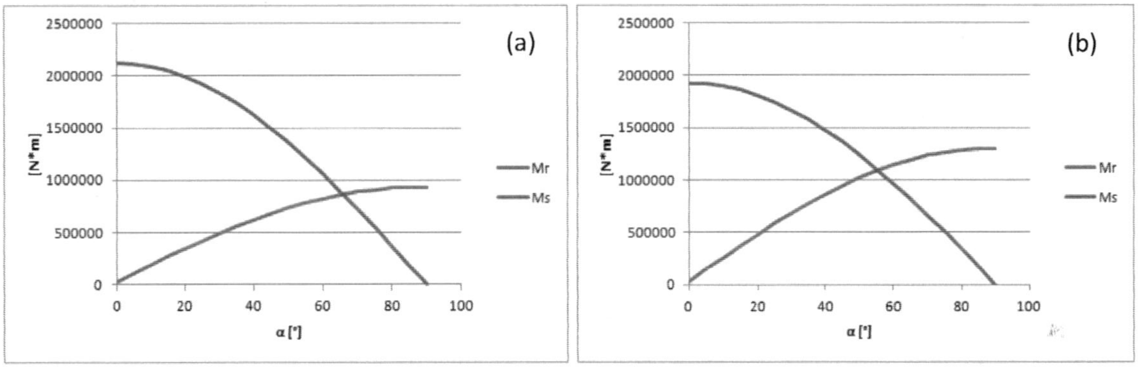

Figure 19. Graphs of the maximum gradient that can be overcome under load condition 3: (a) load condition start, (b) load condition end

The maximum slope before overturning is given by the intersection point of the curves in the graphs.

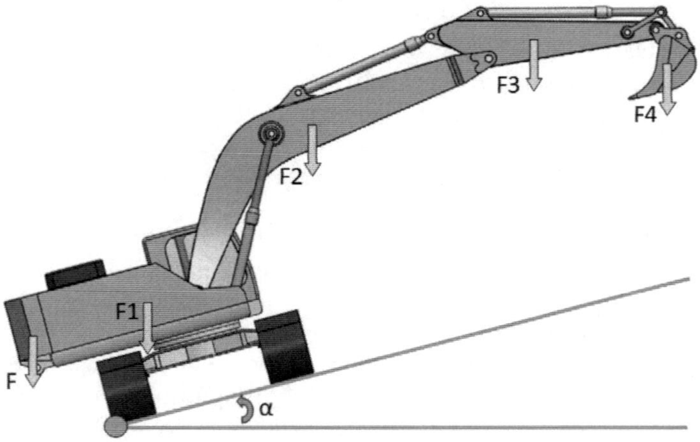

Figure 20. Schematization for calculating the maximum slope that can be overcome with the cab rotated 90° respect to the track

- Uphill load condition 2

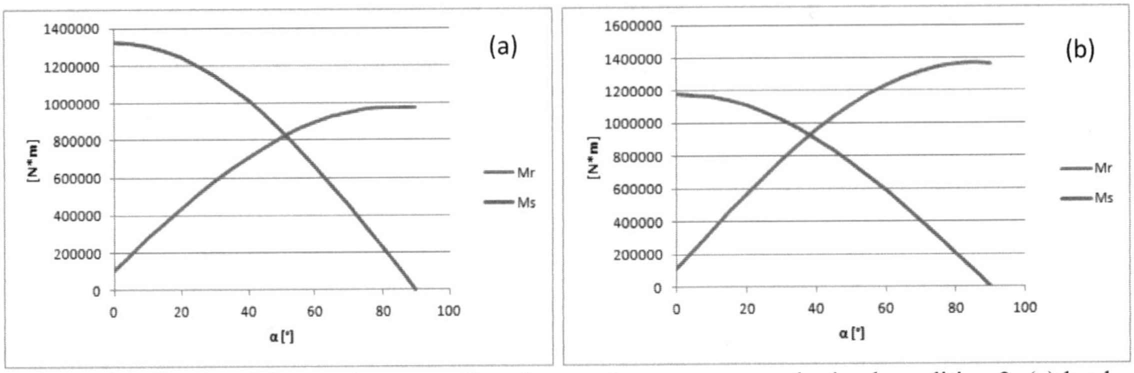

Figure 21. Graphs of the maximum gradient that can be overcome under load condition 2: (a) load condition start, (b) load condition end

- Uphill load condition 3

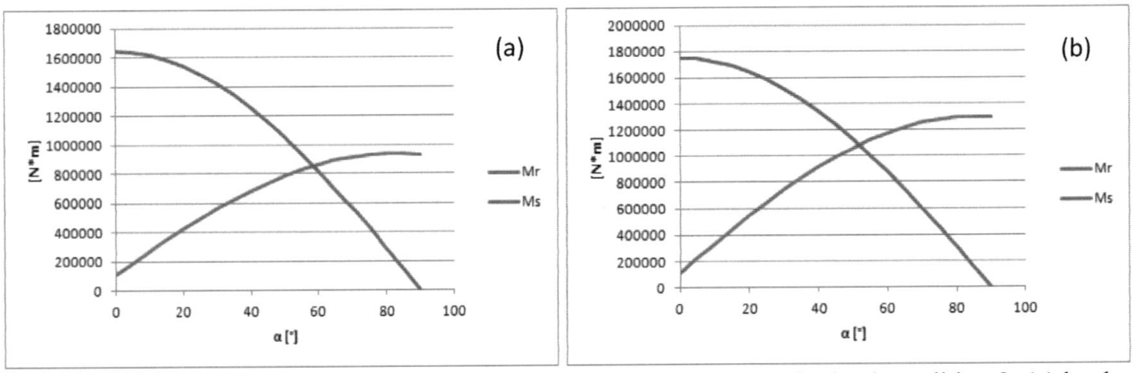

Figure 22. Graphs of the maximum gradient that can be overcome under load condition 3: (a) load condition start, (b) load condition end

9. Results

Thanks to the use of innovative materials such as aluminum alloys and in particular carbon fiber composite materials, it has been possible, as can be seen in other works [23, 24], to obtain a considerable reduction in weight [25, 26] in the main components that make up a typical excavator as can be seen from Table 6. In fact, thanks to the use of the aluminum alloy Al 6060 T5 (UNI EN 573-3) and the appropriate construction methods, it was possible to obtain a frame with structural characteristics similar to the steel frame, but with a weight reduction of approximately 56% (Table 4). This technical solution, if combined with the use of arms and hydraulic cylinders made of composite material, determines a reduction in the operating weight of the entire machine greater than 26% (Table 6).

From the graphs shown in Figure 17, 18, 20 and 21 and from those elaborated for the other load conditions, it is found that the intersection of the curves determines the limit angle value of the excavator, before its overturning. Table 7 refers to the overturning values obtained considering the classic construction configuration (all in S355 construction steel), as regards the other configurations or with different combinations of materials, the results are very similar to those in the table, so there is no difference since the counterweight value was chosen precisely to obtain the same performance as the starting machine.

IOP Conf. Series: Materials Science and Engineering 1190 (2021) 012010 doi:10.1088/1757-899X/1190/1/012010

Table 7. Maximum angle of ground inclination before overturning for two load conditions, in the start and end of movement phase

	α [°]
Load condition 2 start	50
Load condition 2 end	37
Load condition 3 start	56
Load condition 3 end	51

Thanks to the use of innovative solutions and the adoption of unconventional materials essentially composite materials, it has been possible, as seen in table 5 and table 6, to obtain undoubted advantages. First of all it was possible to reduce the force necessary to move the arms because the latter have seen a marked decrease in their weight thanks to the use of composite materials in as can be seen from Table 8, in which, moreover, the increase in the transportable mass is also reported, always obtained through the use of composite materials. Increase in load capacity was calculated for all load conditions and was evaluated both at the beginning of the movement and at the end of it. This value derives from considered the same stresses and deformations as the original configuration.

Table 8. Arm weight variation and increase of transportable weight

Material	Weight boom [N]	Weight reduction [%]	Weight stick [N]	Weight reduction [%]	Total weight [N]	Weight transportable [N]		Incremental transportable weight [%]	
						extended	closed	extended	closed
S355 (UNI EN 10025-3)	35316	-	14715	-	50031	78750		-	-
6063 T6 (UNI EN 573-3)	18640	47.2	9040	38.6	27680	90410	102070	14.8	29.6
Composite Material	12450	64.8	5850	60.2	18300	95100	111450	20.8	41.5

Secondly, from Table 6 a really important aspect emerges, in fact, by adopting the configuration: frame in Al 6060-T5, arms in composite material and hydraulic cylinders in composite material, it is possible to obtain a reduction of 26% in the operating weight of the machine. This figure can translate into a lower pressure exerted by the machine on the ground, in fact the pressure generated by the excavator in standard configuration (S355) is equal to 1.03 kg/cm^2 (taken from the catalog), while thanks to the use of a light alloy frame and above all thanks to the use of composite materials for the construction of the arms and hydraulic cylinders, the pressure value is equal to 0.76 kg/cm^2 with a reduction of about 26%. The reduction of the pressure exerted by the machine on the ground has undoubted benefits, in fact thanks to a lower pressure the excavator is able to move even in difficult to reach areas in which it is easy to sink (not very compact soils), therefore the adoption of materials innovative as well as guaranteeing benefits related to the reduction of consumption, the possibility of increasing the transportable mass, also guarantees the excavator to be able to move on land or areas not favorable to heavy vehicles.

10. Conclusions

This paper reports a systematic study of lightweight of a large excavator. The research starts from the study of the frame by developing the component also using unconventional materials for this element. The research continues from previous research focused on developing the arms and the cylinders in composite material. The global evaluation of the excavator is also conducted considering the effect of the counterweight whose value is chosen in relation to the configuration of the machine. The results show that adopting unconventional materials for these machines, such as aluminum alloy or composite materials, there is a significant reduction in the overall weight of the machine, without compromising the functionality and stability of the machine itself. In summary, therefore, with the adoption of the innovative solution, the stability of the machine is not compromised, but a much lighter excavator is guaranteed. This result can be translated in two different ways, on the one hand one could reduce the power of the propulsion and hydraulic system for the same transportable mass, thus reducing consumption (and consequently pollution) and costs, on the other hand a reduction of the operating weight of the machine with the same hydraulic system can guarantee a greater transportable mass thus leading to an increase in hourly production. A lighter excavator as obvious implies less problems for its transposition in the place where it will operate, however it must be emphasized that a reduction of the weights at stake both by the frame and by the moving parts, such as the arms and hydraulic cylinders, brings also to the undoubted advantage of being able to proceed with faster movements thus increasing the dynamics of the vehicle and consequently ensuring a saving in terms of time for each type of work performed by the excavator. Furthermore, as it has been demonstrated, having a lighter machine guarantees lower pressure on the ground, thus ensuring use even on grounds not suitable for heavy machines.

References

[1] L. Solazzi, A. Assi, and F. Ceresoli. New Design Concept for an Excavator Arms by Using Composite Material. Appl Compos Mater 2017.

[2] L. Solazzi, and A. Buffoli. Telescopic Hydraulic Cylinder Made of Composite Material. Applied Composite Materials 26(4), pp. 1189-1206; 2019.

[3] L. Solazzi. Feasibility study of hydraulic cylinder subject to high pressure made of aluminum alloy and composite material. Composite Structures, 209, pp. 739-746; 2019.

[4] G. Miscia, V. Rotondella, A. Baldini, E. Bertocchi, and L. D'Agostino. Aluminum Structures in Automotive: Experimental and Numerical Investigation for Advanced Crashworthiness. Int. Mech. Eng. Congr. Expo., 2015.

[5] L. Solazzi. Applied Research for Weight Reduction of an Industrial Trailer. FME Trans; 40:57–62, 2012.

[6] R.M. Gouveia, F.J.G. Silva. Designing a new Sustainable Approach to the Change for Lightweight Materials in Structural Components Used in Truck Industry. J Clean Prod, 115–23, 2017.

[7] F. Omar, T. Jimi, and S. Mohini. Lightweight and Sustainable Materials for Automotive Applications. 1st Editio. CRC Press; 2017.

[8] A. Michael, and J. Kara. Materials and Design. 3rd Editio. Butterworth-Heinemann; 2014.

[9] A.F. Michael, and J.R.H. David. Engineering Materials 1 An Introduction to Properties, Applications, and Design. 4th Editio. Butterworth-Heinemann; 2012.

[10] L. Solazzi, A. Assi, and F. Ceresoli. Excavator arms: Numerical, experimental and new concept design. Composite Structures, Vol. 217, pp.60-74; 2019.

[11] A.F. Michael, and J.R.H. David. Engineering Materials 2 An Introduction to Microstructures and Processing. 4th Editio. Butterworth-Heinemann; 2013.

[12] P.K. Mallick. Materials, design and manufacturing for lightweight vehicles. 1st Editio. Woodhead Publishing; 2010.

[13] P. Xiaoping, G. Xingtong, C. Jin, and W. Yabin. Structural analysis and optimized design of working device for backhoe hydraulic excavator. IFToMM World Congress Proceedings; 2015.

[14] Z. Zou, J. Chen, and X. Pang. Lightweight and high-strength optimization design for a fully parametric working attachment of a hydraulic excavator based on limiting theoretical digging capability model. Proceedings of the institution of Mechanical Engineers, Part C: Journal of Mechanical Engineering Science, Vol. 233, pp. 4819-4835; 2019.

[15] L. Solazzi, F. Ceresoli, and M. Cima. Structural analysis on lightweight excavator arms. Proceedings - European Council for Modelling and Simulation, ECMS 33(1), pp. 351-357; 2019.
[16] L. Solazzi, G. Incerti, and C. Petrogalli. Estimation of the dynamic effect in the lifting operations of a boom crane. 28th Eur. Conf. Model. Simul., p. 309–15, 2014.
[17] F. Ceresoli. Excavator with two or three arms: Dynamical behavior and structural implications. FME Transactions, Vol.48, pp.272-280, 2020.
[18] W. Jonathan. Finite Element Methods a Practical Guide. 1st Editio. Springer International Publishing; 2017.
[19] Z. Bofang. The finite Element Method: fundamentals and Applications in Civil, Hydraulic, Mechanical and Aeronautical Engineering. 1st Editio. Wiley; 2018.
[20] Y. Regassa, and H.G. Lemu. Multibody system modelling and simulation: Case study on excavator manipulator. Lecture Notes in Electrical Engineering, Vol. 484, pp.628-635; 2019.
[21] X. Yu, X. Pang, Z. Zou, G. Zhang, Y. Hu, J. Dong, and H. Song. Lightweight and high-strength design of an excavator bucket under uncertain loading. Mathematical Problems in Engineering. Vol. 2019, article number 3190819; 2019.
[22] J.S. Raju, and S. Basavaraju. Structural parameters and working range estimation of excavator backhoe mechanism. International Journal of Recent Technology and Engineering, Vol. 8, pp. 867-872; 2019
[23] M. Collotta, and L. Solazzi. New design concept of tank made of plastic material for firefighting vehicle. International Journal of Automotive and Mechanical Engineering, Vol. 14, pp.4603-4615; 2017.
[24] L. Solazzi, and R. Scalmana. New Design Concept for a Lifting Platform Made of Composite Material. Appl Compos Mater; 20:615–26, 2013.
[25] U. Hamme, and J. Henkel. New concepts in lightweight structures - Innovative telescopic boom of a mobile cran. Stahlbau, Vol. 82, pp. 246-249; 2013.
[26] K. Skonieczny, M. Delaney, D.S. Wettergreen. and W.L. Red Whittaker. Productive lightweight robotic excavation for the moon and mars. Journal of Aerospace Engineering, Vol.27, article number 04014002; 2014.

IOP Conf. Series: Materials Science and Engineering 1190 (2021) 012011 doi:10.1088/1757-899X/1190/1/012011

Stress-Strain Distribution Within Rotary Swaged Tungsten Heavy Alloy

Ludmila Krátká, Rudolf Petrmichl

Faculty of Materials Science and Technology, VŠB-Technical University of Ostrava, 17. listopadu 15, 70833, Ostrava 8, Czech republic

ludmila.kratka@vsb.cz

Abstract. This study deals with thorough investigation of stress-strain parameters within an innovatively processed tungsten heavy alloy (THA). THAs are challenging materials featuring exceptional mechanical and physical properties and they are applicable in various demanding industrial fields, such as in the aeronautics, medicine, or military. The herein studied material is a WNiCo alloy prepared from original powders via powder metallurgy. The pre-sintered material was processed via plastic deformation, in particular hot rotary swaging with the usage of induction heating directly before a swaging pass. The experimental study was supplemented with numerical simulation of the deformation process, which was performed via the finite element method (FEM). The simulation was primarily focused on prediction of the development of effective imposed strain during the swaging process performed in multiple steps, and its correlation with stress distribution during swaging. The numerically predicted effective imposed strain development was subsequently verified via Vickers microhardness measurements, whereas the predicted results of stress distribution were validated via experimental observation of residual stress distribution performed via analysing scans acquired by scanning electron microscopy (SEM, electron backscatter diffraction analysis). The results showed that the deformation technology consisting of induction heating and hot rotary swaging was well optimized as the swaged billets featured satisfactory surface roughness and no surface cracks. Also, the tungsten agglomerates within the structure exhibited deformation in the direction of the main acting swaging force, especially in the locations the effective imposed strain in which was the highest. The values of predicted effective imposed strain and microhardness correlated reliably with the experimentally acquired and predicted results of residual stress distribution. In other words, the locations the effective strain in which was the highest also exhibited a more significant presence of residual stress.

1. Introduction

Tungsten heavy alloys (THA), in our case W-Ni-Co, consist of the majority of tungsten (approx. 90% by weight), which is supplemented with a softer NiCo matrix [1,2]. Tungsten guarantees high strength (maximum strength around 1900 MPa) of the material, whereas the matrix in the material ensures plastic properties. THAs are used primarily as protection against harmful radiation or to block kinetic energy [3,4]. Given by the high density of THAs ($\approx$16 g / cm^3), the advantageous use of tungsten heavy alloys is, for example, as shielding in the manufacture of therapeutic devices in oncology, counterweights in aeronautics, or for the manufacture of kinetic penetrators used by the military.

Powder metallurgy is in most cases used for the production of THA alloys [5,6]. Mixed metal powders are usually isostatically pressed into rods. The bars are then sintered in a protective atmosphere (H, Ar, vacuum), in the temperature range between 1000 and 1500 °C. Other alloying elements with a lower melting point (Ni, Co, Mo, Re) are also added to the powder mixture with a tungsten content above 90% by weight to form the softer matrix around the tungsten agglomerates during sintering. In this way, a homogeneous material is formed [7,8]. The elements contained in the THA matrix ensure the plasticity of the material, however, their negative effect is certain decrease in strength. For optimal strength and plastic properties, for example, modification of the chemical composition, or improvement of the properties via plastic deformation, can be used [9,10].

THAs can be processed via conventional processing technologies, such as rolling, forging, extrusion, etc. [11,12]. Nevertheless, techniques of intensive plastic deformation, such as the severe plastic deformation (SPD) methods [13–17] could also be applied for their favourable effects on structure homogenization and improvement of mechanical properties. The most suitable technologies are, however, those that are continuous, such as ECAP-conform [18] or rotary swaging (RS) [19,20]. These processes take place continuously and are therefore suitable for large-scale production. Due to its great variability, the RS technology is suitable for various branches of industry. The RS technology is characterized by high-frequency hits of dies that rotate around the blank being deformed. This process can be realized under both, cold and hot conditions [21], but induction heating can advantageously be used to heat the material before processing. It is suitable for the production of solid semi-finished products, and hollow shaped parts, too. RS is an advantageous technology for processing of difficult-to-process materials (Ti, THA) and composites (Al/Cu) due to the advantageous distribution of forces during processing [22–25].

The already published studies devoted to stress-strain characteristics of THAs report differing results. For example, work by [2,20,21] characterizes the differences in the material plastic flows in cases of cold and hot rotary swaging. It is obvious that the temperature thus plays an important role not only in the case of plastic flow of material, but also in the related stress distribution. Residual stress is thus a very important indicator of the performed forming, which subsequently affects not only the mechanical characteristics of THA products [2,21]. It should be noted, however, that publications with this focus are relatively scarce. A more detailed study devoted to the currently widely studied THA (WNiCo) is thus in a great demand, as so far only partial attention has been paid to this problem.

The aim of this study is to predict stress states during rotary swaging of a THA material from original WNiCo pseudo-alloy. In the first part, the study is focused on FEM simulation of hot rotary swaging of the material. Numerical modelling was for this purpose performed using Forge NxT software. The second part of the work is focused on experimental verification of the predicted internal stress distribution in the rotary swaged pseudo-alloy.

2. Material and methods

This study is divided into two main parts. In the first part of the paper, the proposed forming process of hot rotary swaging is simulated by FEM analysis. The simulation is focused on predicting the stress state after rotary swaging. The second part of the work is devoted to practical verification of the numerically predicted stress state results using a real application of rotary swaging (Figure 1) with induction pre-heating. For these purposes, the self-developed KOMAFU S 600 forging force detection system was used. The forging itself took place on a rotary swaging machine by HMP company.

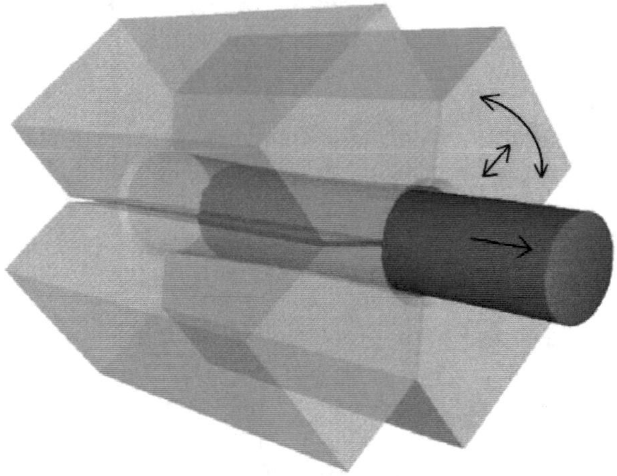

Figure 1. Schematic depiction of rotary swaging

The rotary swaging simulation was designed to predict the behaviour of the WNiCo tungsten heavy alloy. For this purpose, an elastic-viscoplastic model created by the software based on the THA tensile stress-strain curve was used. The model was determined by the Haensel - Spittel relation (equation 1), where X is the equivalent strain rate, ε is the equivalent strain, T is the temperature and A, m1, m2, m3, m4, m5, m7, m8, m9 are regression coefficients the values of which were as follows: A = 1447,002738, m1 = -0,009, m2 = 0,0895, m3 = 0,0044, m4 = -0,0069, m5, m7, m8, m9 are 0.

$$\sigma = A\exp(m_1 T)T^{m_9}\varepsilon^{m_2}\exp\left(\frac{m_4}{\varepsilon}\right)(1+\varepsilon)^{m_5 T}\exp(m_7\varepsilon)\dot{\varepsilon}^{m_3}\dot{\varepsilon}^{m_8 T} \quad (1)$$

For real experimental verification of the results, 93W7Ni3Co pseudo-alloy with the chemical composition of 90 wt.% W, 7 wt.% Ni, 3 wt.% Co. was selected. This alloy was prepared by powder metallurgy from powders with the mean grain size of 2.78 μm, by ÚJP Praha a.s. The preparation process involved mixing the powders, cold isostatic pressing at 400 MPa, and then sintering in H2 atmosphere-protected oven at 1525 ° C for 20 minutes, followed by rapid cooling in water. The minimum temperature for forming (900 ° C) was chosen based on previous experimental studies [4,21]. The mentioned works reported massive surface oxidation after exceeding the temperature of 1100 °C. Temperature control during induction heating was provided by an optical pyrometer. The WNiCo blank was gradually reduced from the original diameter of 30 mm the final diameter of 20 mm.

Structural analysis of the experimentally processed work-piece is performed on its cross-sectional sections, focusing primarily on the analysis of residual stress, and the development of structure. Observations were made using scanning electron microscopy. The samples were mechanically ground on SiC papers and subsequently polished with Eposil F substance. EBSD analyses were performed in the subsurface sample area, 1 mm from the outer edge of the rod, with the scanning step of 0.5 μm. Scanning was evaluated using ATEX software [26].

3. Results and discussion

3.1 FEM of rotary swaging

Figure 2a) shows the material flow with the projected temperature field during hot rotary swaging. The simulation shows that the material flow during rotary swaging is very complex. It can be seen that the material flows in both the directions from the neutral plane located in the reduction zone of the dies. In addition, a component of tangential plastic flow, the amplitude of which is variable during rotary swaging, is evident in the surface layers of the material. At the beginning of the work-piece, we observe that the material flows significantly in the direction of material movement during swaging. Since the material volume has the possibility to flow plastically in both the directions along the axis of the work-piece during swaging, it is necessary to determine the position of the neutral plane. Thus, the neutral plane is the location along the length of the work-piece in which the material changes the orientations of the flow vectors. From the beginning of the work-piece towards the neutral plane, the material flow gradually slows down. At both the ends of the rotary swaged work-piece, the influence of the rigid ends, which significantly affect the plastic flow of the material, can be detected. At the rear end, the rigid end is manifested by the presence of a manipulator. From the direction of the vectors can be seen that the rigid end, i.e. the presence of the manipulator, results in a deflection of the direction of the plastic flow vectors. This suggests that, during rotary swaging, the end of the work-piece tends to bend in the direction of the dominant direction resulting from these vectors.

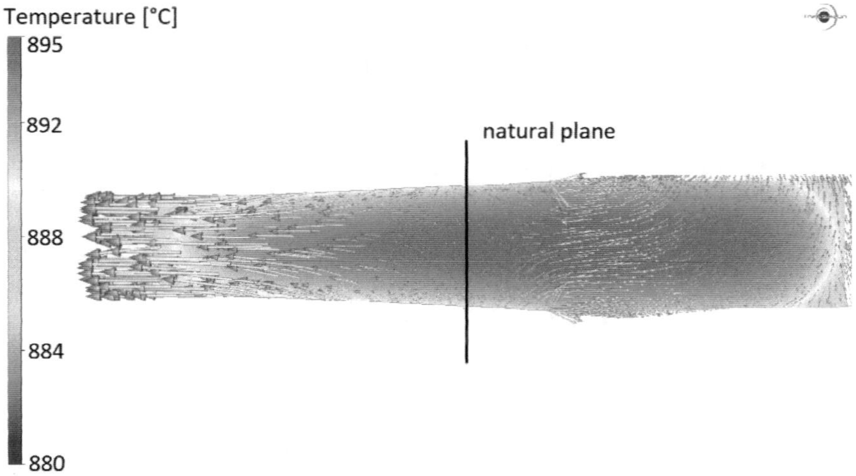

a)

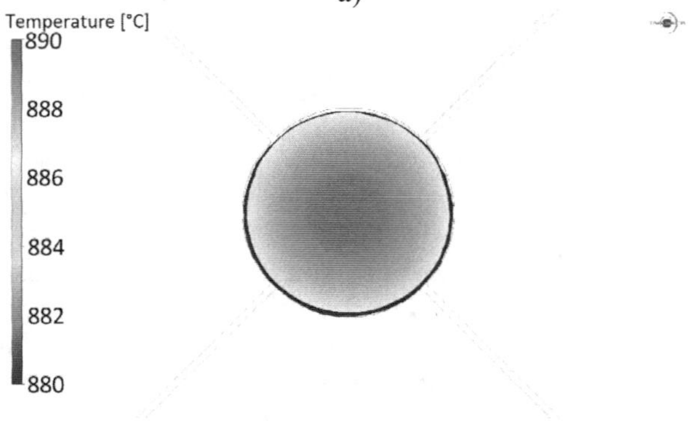

b)

Figure 2. Temperature field in the longitudinal direction a); transverse direction b)

Figure b) shows the temperature field during rotary swaging. From the figure ensues that the material cools evenly from the surface towards the axis and the temperature field visibly creates three areas; cold surface area, warmer subsurface area, and hot axial area.

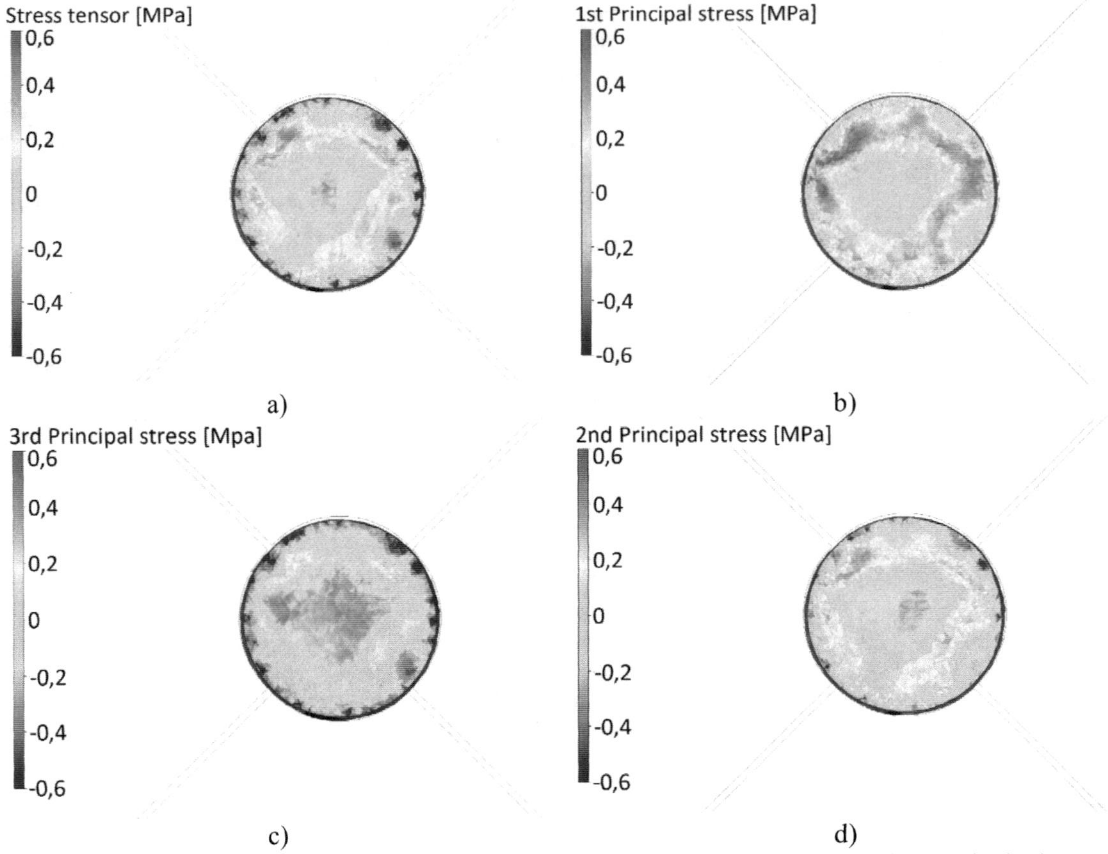

Figure 3. Stress state in the cross section of hot swaged material: stress tensor a);1st principal stress b); 3rd principal stress c); 2nd principal stress

Figure 3a) shows the stress tensor considering both the normal and shear stresses in the material during rotary swaging, where compressive stress predominate in the surface layers. This phenomenon is mainly caused by friction between the dies and the material, when during rotary swaging, the material flow in the surface layer is aggravated. Thus, this region acts more as a transmitter of plastic deformation towards the axial region of the material, by the effect of which the formation of compressive stress occurs. In subsurface areas, on the other hand, the material plastic flow is less aggravated, which is manifested by the occurrence of tensile stress. This stress point to the occurrence of significant movement of material in these locations. In the axial area, the compressive stress occurs again, this stress arises from the nature of the material flow during rotary swaging; the strain in the axial region of the work-piece is lower than in its (sub)surface regions and thus the material in this area does not have sufficient energy to flow. At the same time, significant part of the plastic flow is realized in the subsurface layers, therefore, the axial regions of the material are characterized by the compressive stress.

Figure 3b) shows the main normal tensile stress. The figure shows that the tensile stress is concentrated in the subsurface, and slightly in the surface areas. Figure 3c) shows the main normal compressive stress. This figure demonstrates the effect of compression in the middle and surface areas of the material. The resulting figure 3d) is the intersection of figure 3b) and 3c), it shows that the compressive stress is strongly suppressed in the axial region by the prevailing tensile stress.

IOP Conf. Series: Materials Science and Engineering 1190 (2021) 012011 doi:10.1088/1757-899X/1190/1/012011

From this distribution of tensile stress is evident that the stress state during rotary swaging is significantly influenced by the flow of the material, as well as the distribution of the temperature field within the material.

3.2 Structural analysis

Figure 4 a) shows the structure state after sintering. During sintering, the tungsten powder agglomerates together and agglomerated particles of tungsten are formed. This leads to the formation of agglomerates the shapes of which are very close to the shape of a sphere. Figure 4a shows large spherical agglomerates of hard tungsten surrounded by the softer NiCo matrix. The matrix ensures the cohesion of the material, as well as plastic properties. Figures 4b) and 4c) show the structural state of the work-piece after hot rotary swaging. Figure 4b) shows the subsurface area which exhibits high penetration of plastic strain. This phenomenon is evident from the pronounced deformation of the matrix and the visible deformation of the tungsten agglomerates. In many areas, the deformation of the matrix is so evident that the tungsten agglomerates developed larger contact surfaces. In Figure 4c), the structure in the axial area of the sample is shown. It is obvious that the deformation in this region is not as significant as in the subsurface area. At this point, the deformation is concentrated in the NiCo matrix. Therefore, almost no deformed tungsten grains can be seen in this location and the structural state in the axial region of the swaged material is similar to the state after sintering.

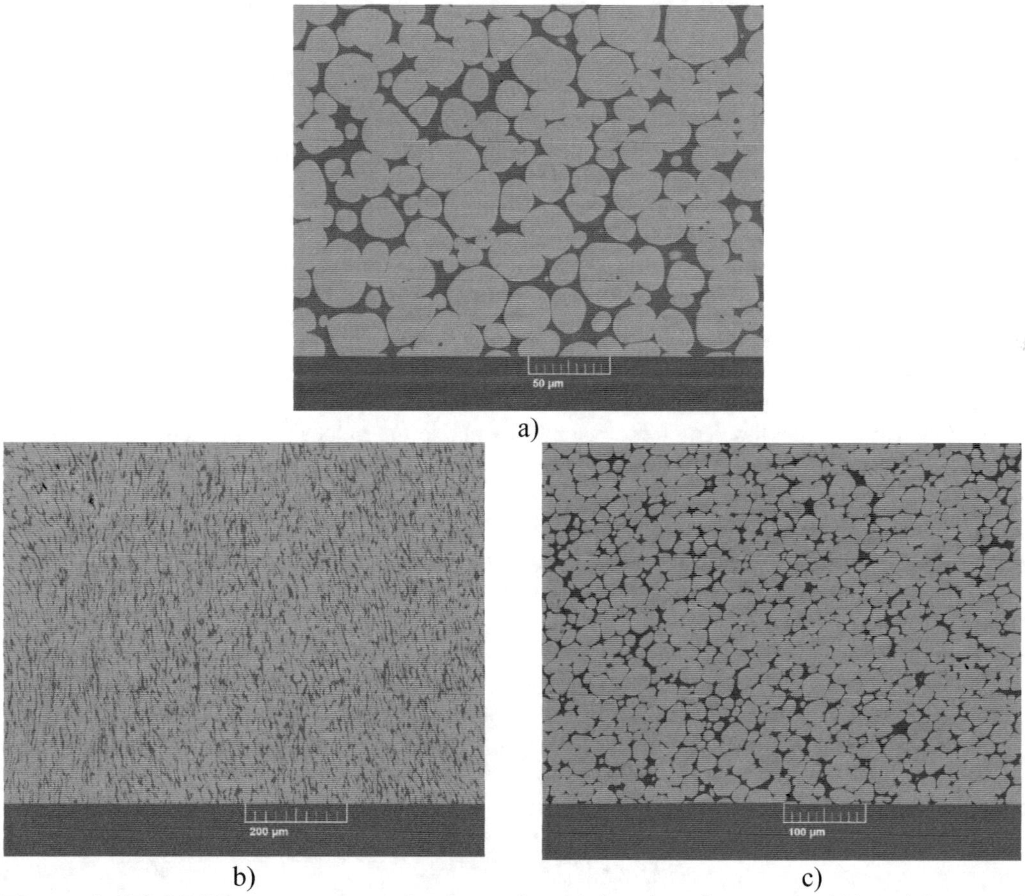

Figure 4. SEM-BSE structure scan for sintered state a); subsurface-swaged b); axial-swaged c)

3.3 Residual stress

Figure 5 shows the internal misorientation of the grains in the scale from 0° to 15 ° in the axial region of the material. These misorientation point to the distribution of residual stress. The places with the least misorientations are shown in blue in the picture. These areas largely correspond to the matrix, or to the areas with identical grains orientations. The least residual stress is manifested in these areas. This is ensured by the fact that the softer matrix allows the tension stress to relax. In the areas marked in red, there is a high concentration of grains misorientations; these are the areas with the minimum presence of matrix surrounding the tungsten agglomerates. In the red areas, the tungsten agglomerates mostly touch each other and therefore have no possibility to relax possible accumulated stress. This results into the greatest presence of residual stress. This stress development results from the deformation behaviour of the material during rotary swaging. During deformation, the orientation of the randomly arranged grains of material is changed so that the orientation of the grains is as favourable as possible. In areas in which a low volume of matrix is present, this phenomenon becomes problematic. Therefore, residual stress is most common in these places.

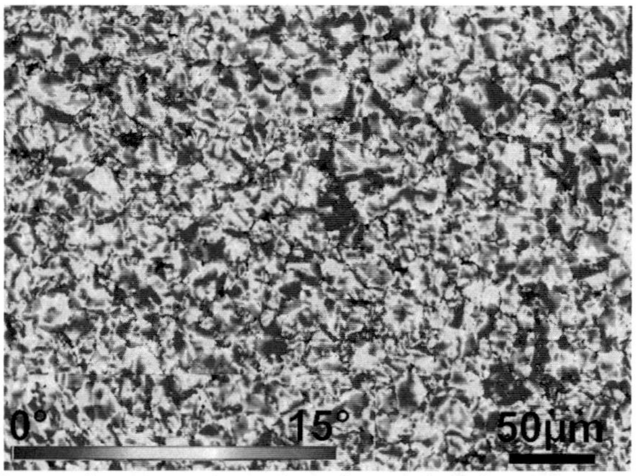

Figure 5. Internal grains orientations in axial region of swaged material

3.4 Force ratios during rotary swaging

Figure 6 shows a real record of the development of forces acting during rotary swaging in comparison with the predicted values from FEM (smooth curve). In the Figure we see the individual strokes of the dies, which reach different peaks. According to the record, it is clear that the force required to strike increases as the material continuously fills the dies. The force reaches the maximum values as soon as the working space of the dies is filled. Then follows a section in which the force remains more or less constant. In the last section of the graph, the force decreases as the material starts to empty the working space of the dies. At the end of swaging, an increase in the forces is evident. This increase is caused by cooling of the material and consequent increase in its flow stress. This trend is not significant in the predicted curve.

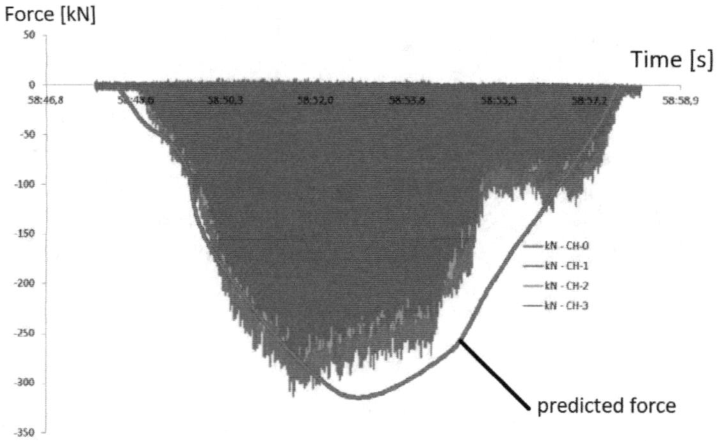

Figure 6. Force ratios acting on die

Figure 6 (smooth curve) shows that the real course differs from the predicted behaviour. The predicted curve has an upward trend up to the maximum, after the achievement of which a downward trend is evident. The predicted curve differs mainly in the last part, where the cold end of the material is swaged (in real rotary swaging). This is due to the inhomogeneity of the swaging process. The speed of feeding the material into the dies and its temperature have the major influence. The flow stress is also dependent on the temperature, i.e. it increases with decreasing temperature.

4. Conclusions

In rotary swaging, the material flow is characterized by its complexity. To characterize the material flow, it is essential to determine the neutral plane, which determines the direction of the material flow. Last but not least, the existence of rigid ends, which causes the material to bend during real forging, is also a significant phenomenon.

The stress in the material during hot forging is influenced by the temperature distribution along the cross section, and the material flow. The temperature distribution and flow of the material, according to the results from FEM, have effect on the location of tensile and compressive stress in the material. Shear stress, which occur mainly in the surface and subsurface material layers, play a considerable role in this type of forming. In general, hot forming is not characterized by high values of stress within the material due to the decreased flow stress of the material (i.e. higher plasticity).

Structural analyses confirm the effect of shear stress in the surface and subsurface layers of the material; the structure is significantly deformed in these locations. There is not such a significant deformation in the axial area; the deformation concentrates in the matrix in this location. The most significant presence of residual stress can be found in areas where there is low volume of matrix surrounding the tungsten agglomerates, and, in many cases, the agglomerates touch each other. Agglomerates with a contact surface are forced to adjust their orientations; in this way residual stress arises in the material.

The acting swaging forces acquired from the numerical simulation and experiment differed slightly. The real measurements show that the maximum experimental force acting on the dies is lower than the predicted maximum force. This fact is the effect of the inhomogeneity of the rotary forging process and is influenced by the feed rate into the dies and the temperature of the material (flow stress).

Acknowledgment

This research was supported by the 19-15479S project of the Grant Agency of the Czech Republic.

References

[1] A. Russell, K.L. Lee, Structure-Property Relations in Nonferrous Metals, 1st ed., John Wiley & Sons, Inc., New Jersey, 2005.

[2] L. Kunčická, A. Macháčková, N.P. Lavery, R. Kocich, J.C.T. Cullen, L.M. Hlaváč, Effect of thermomechanical processing via rotary swaging on properties and residual stress within tungsten heavy alloy, Int. J. Refract. Met. Hard Mater. 1–15, 2020. doi:10.1016/j.ijrmhm.2019.105120.

[3] J. Das, G. Appa Rao, S.K. Pabi, Microstructure and mechanical properties of tungsten heavy alloys, Mater. Sci. Eng. A. 7841–7847, 2010. doi:10.1016/j.msea.2010.08.071.

[4] A. Macháčková, L. Krátká, R. Petrmichl, L. Kunčická, R. Kocich, Affecting Structure Characteristics of Rotary Swaged Tungsten Heavy Alloy Via Variable deformation temperature, Materials (Basel). 12, 2019. doi:10.3390/ma12244200.

[5] P. Strunz, L. Kunčická, P. Beran, R. Kocich, C. Hervoches, Correlating Microstrain and Activated Slip Systems with Mechanical Properties within Rotary Swaged WNiCo Pseudoalloy, Materials (Basel). 13, 2020. doi:10.3390/ma13010208.

[6] S. Antusch, D.E.J. Armstrong, T. Ben Britton, L. Commin, J.S.K.L. Gibson, H. Greuner, J. Hoffmann, W. Knabl, G. Pintsuk, M. Rieth, S.G. Roberts, T. Weingaertner, Mechanical and microstructural investigations of tungsten and doped tungsten materials produced via powder injection molding, Nucl. Mater. Energy. 3–4 (2015) 22–31. doi:10.1016/j.nme.2015.04.002.

[7] L. Kunčická, R. Kocich, Z. Klečková, Effects of Sintering Conditions on Structures and Properties of Sintered Tungsten Heavy Alloy, Materials (Basel). 13, 2020.

[8] L. Ekbom, T. Antonsson, Tungsten heavy alloy: Deformation texture and recrystallization of tungsten particles, Int. J. Refract. Met. Hard Mater. 375–379, 2002. doi:10.1016/S0263-4368(02)00035-5.

[9] L. Kunčická, A. Macháčková, R. Petrmichl, Z. Klečková, M. Marek, Optimizing Induction Heating of WNiCo Billets Processed via Intensive Plastic Deformation, Appl. Sci. 10 (20AD) 8125, 2020.

[10] L. Kunčická, R. Kocich, C. Hervoches, A. Macháčková, Study of structure and residual stresses in cold rotary swaged tungsten heavy alloy, Mater. Sci. Eng. A. 25–31, 2017. doi:10.1016/j.msea.2017.07.096.

[11] O. Kovářík, J. Čech, J. Čapek, M. Hajíček, J. Klečka, J. Siegl, Mechanical properties of forged tungsten heavy alloys, in: Acta Polytech. CTU Proc., Czech Technical University, pp. 149–154, 2020. doi:10.14311/APP.2020.27.0149.

[12] J. V. Haag, D.J. Edwards, C.H. Henager, W. Setyawan, J. Wang, M. Murayama, Characterization of ductile phase toughening mechanisms in a hot-rolled tungsten heavy alloy, Acta Mater. 2021. doi:10.1016/j.actamat.2020.116523.

[13] L. Kunčická, R. Kocich, J. Drápala, V.A. Andreyachshenko, FEM simulations and comparison of the ecap and ECAP-PBP influence on Ti6Al4V alloy's deformation behaviour, Met. 2013 22nd Int. Met. Mater. Conf. 391–396, 2013.

[14] A. Nocivin, D. Raducanu, B. Vasile, C. Trisca-Rusu, E.M. Cojocaru, A. Dan, R. Irimescu, V.D. Cojocaru, Tailoring a Low Young Modulus for a Beta Titanium Alloy by Combining Severe Plastic Deformation with Solution Treatment, Materials (Basel). 14, 2021. doi:10.3390/ma14133467.

[15] L. Kunčická, R. Kocich, V. Ryukhtin, J.C.T. Cullen, N.P. Lavery, Study of structure of naturally aged aluminium after twist channel angular pressing, Mater. Charact. 94–100, 2019. doi:10.1016/j.matchar.2019.03.045.

[16] A.M. Jamili, A. Zarei-Hanzaki, H.R. Abedi, M. Mosayebi, R. Kocich, L. Kunčická, Development of fresh and fully recrystallized microstructures through friction stir processing of a rare earth bearing magnesium alloy, Mater. Sci. Eng. A. 775, 2019. doi:10.1016/j.msea.2019.138837.

[17] R. Kocich, L. Kunčická, A. Macháčková, Twist Channel Multi-Angular Pressing (TCMAP) as a method for increasing the efficiency of SPD, IOP Conf. Ser. Mater. Sci. Eng. 63, 2014. doi:10.1088/1757-899X/63/1/012006.

[18] M. Murashkin, A. Medvedev, V. Kazykhanov, A. Krokhin, G. Raab, N. Enikeev, R. Valiev, Enhanced Mechanical Properties and Electrical Conductivity in Ultrafine-Grained Al 6101 Alloy Processed via ECAP-Conform, Metals (Basel). 2148–2164, 2015. doi:10.3390/met5042148.

[19] Z. Wang, J. Chen, C. Besnard, L. Kunčická, R. Kocich, A.M. Korsunsky, In situ neutron diffraction investigation of texture-dependent Shape Memory Effect in a near equiatomic NiTi alloy, Acta Mater. 135–148, 2021. doi:10.1016/j.actamat.2020.10.049.

[20] L. Kunčická, R. Kocich, K. Dvořák, A. Macháčková, Rotary swaged laminated Cu-Al composites: Effect of structure on residual stress and mechanical and electric properties, 2019. doi:10.1016/j.msea.2018.11.026.

[21] D. Canelo-Yubero, R. Kocich, C. Hervoches, P. Strunz, L. Kunčická, L. Krátká, Neutron Diffraction Study of Residual Stresses in a W–Ni–Co Heavy Alloy Processed by Rotary Swaging at Room and High Temperatures, Met. Mater. Int. 1–12, 2021. doi:10.1007/s12540-020-00963-8.

[22] R. Kocich, L. Kunčická, P. Král, P. Strunz, Characterization of innovative rotary swaged Cu-Al clad composite wire conductors, Mater. Des. 828–835, 2018. doi:10.1016/j.matdes.2018.10.027.
[23] L. Kunčická, R. Kocich, P. Strunz, A. Macháčková, Texture and residual stress within rotary swaged Cu/Al clad composites, Mater. Lett. 88–91, 2018. doi:10.1016/j.matlet.2018.07.085.
[24] W. Pachla, M. Kulczyk, S. Przybysz, J. Skiba, K. Wojciechowski, M. Przybysz, K. Topolski, A. Sobolewski, M. Charkiewicz, Effect of severe plastic deformation realized by hydrostatic extrusion and rotary swaging on the properties of CP Ti grade 2, J. Mater. Process. Technol. 255–268, 2015. doi:10.1016/j.jmatprotec.2015.02.027.
[25] S. Hanada, N. Masahashi, T.K. Jung, M. Miyake, Y.S. Sato, H. Kokawa, Effect of swaging on Young[U+05F3]s modulus of β Ti-33.6Nb-4Sn alloy, J. Mech. Behav. Biomed. Mater. 310–320, 2014. doi:10.1016/j.jmbbm.2013.10.027.
[26] B. Beausir, J.J. Fundenberger, Analysis Tools for Electron and X-ray diffraction, ATEX - software, Available at: www.atex-software.eu, 2017.

IOP Conf. Series: Materials Science and Engineering 1190 (2021) 012012 doi:10.1088/1757-899X/1190/1/012012

Interface characterization of Al/SiC_p compacts

Mariana Ciurdas, Daniela Alina Necsulescu, Roxana Marina Solea

Faculty of Materials Science and Engineering, University Politehnica of Bucharest, Spl. Independentei 313, Bucharest, Romania

mariana.ciurdas@upb.ro

Abstract. In the last decade, several authors and scientists are interested more and more to understand the phenomena which are take place at aluminum matrix and silicon-carbide interface. This type of composite material is especially used in aerospace and automotive industries which involve high temperature applications. Aluminum metal matrix composites containing 5,15,25 weight % SiC_p were processed by powder mettalurgy route. The interaction between metallic matrix (aluminum) and particulate ceramic reinforcement (silicon-carbide) was studied. At the beginning the solid-state reaction behavior of the Al/SiC powder mixture compacts was investigated. The density, porosity, compressibility of Al-SiCp composites were calculated. Maximum densification of the composite material reinforced with 5% SiC is obtained at a compactation pressure of 420 MPa (94.5%). For 15% SiC content densification increase with higher compactation pressure given the density values of 95.66% and 96.46% for the 25% SiC composite. Conventional scanning electron microscopy revealed a faceted interface between SiC particulate and Al matrix. This chemical reaction at the interface is the result of dissolution of silicon carbide particulates in aluminum matrix during the liquid sintering process at high temperature:1100,1300,1450 and 1750°C in different sintering atmospheres. For intermetallic compounds identification was used an energy dispersive X-ray spectroscopy (EDX) investigation. The presence of compounds and phases such as: Al_4C_3, Al_4O_4C, Al_4SiC_4, SiO_2, Al_2O_3, confirmed by X-ray diffraction analysis is given by different sintering temperature. It is important to know the structure and chemical composition of the interface and near to the interface zone for study the mechanical properties of the reinforced metal matrix composites.

1. Introduction

The continuous development of aircraft and automotive industries impose new solutions for new advanced materials. Metal matrix composites represent a promising class of such materials, especially aluminum composites [1].

The composites properties depend on the metal matrix properties and on the reinforced particles form, distribution and properties. Also, the composition and physicochemical properties of the interface between metal matrix and reinforced particles must be known. From many techniques suitable to obtain the metal matrix composites the powder metallurgy is used broaden [2], [3].

The processing of composite materials with Al matrix represents an important and distinct field in the powder metallurgy. It allows us to obtain materials with complex properties, almost impossible to obtain trough classical technological processes. At the industrial level, there are several conventional techniques for obtaining aluminum sintering products and aluminum based composites using powder metallurgy, but the most commonly used are: [4], [5]

a. elementary state powder mixing (Al, Cu, Mg, etc) or binding matter - powder mixing with the ceramic particles, followed by (one direction, two direction or isostatical) warm pressing and thermal treatment; by powder mixing, cold isostatic press and warm extrusion/lamination with degasification and specific thermal treatment.
b. these techniques require specialized equipments and specific endowment. Under laboratory conditions there can be obtained composite materials by mixing the elementary state powders or by mixing the binding matter used in the pressing process with State the objectives of the work and provide an adequate background, avoiding a detailed literature survey or a summary of the results. the powder, followed by cold pressing process, controlled atmosphere sintering process and thermal treatment. The obtained composite materials can be acquired at a lower price and can have specific characteristics according to the applications used for.

The studies accomplished so far regarding the obtaining process of the composite materials trough the powder metallurgy based on the nature of the component materials and technologically parameters reveal the fact that there are still many unsolved problems with direct implications upon the properties of these materials. One of the problems encountered is the difficult pressing and sintering of aluminum powders and especially of Al/SiC powders comparing to pressing and sintering of other ferrous or copper powders. The process of cold one-direction pressing of the mixtures of metal-ceramics powders gives rise to densification problems because it requires pressing forces much higher than those used in the case of metal-metal mixtures. Aluminum sintering is difficult because each particle of Al is covered with a thin layer of a very stable aluminum oxide which cannot be reduced during the sintering process and because of that it locks the diffusion process which is necessary for the particles binding process to take place [3]. In order to solve all these problems, we have to be able to fully characterize the composite materials at the microstructural level. Structural information from X-ray powder diffraction can be obtained using the Rietveld method. Rietveld refinement of X-ray powder diffraction data can yield considerable amount of crystallographic information, in addition to quantitative phase estimation. The method itself is a whole pattern fitting least squares technique that uses the entire pattern rather than a limited number of reflections to extract the information required. In this method, the observed pattern for each phase is compared with the calculated one, and any differences between them are minimized by refining structural as well as profile related parameters. Since the method uses all lines, severely overlapping reflections are not a problem. Finally, we will give informations about weight fraction of the present phases, atoms positions in the unit cell, mean crystallite size and mean square microstrain etc.

2. Experimental details

Because of limited information concerning the interaction study taking place at the interface of the Al/SiC system, we have selected Al as a metallic matrix component and SiC powder as reinforced material. The route for processing Al/SiC$_p$ composites were powder metallurgy based on results in using the diffusion process in order to study the chemical reactions between metallic matrix and ceramic reinforcement.

The powder metallurgy processing advantages are: the obtaining of isotropic materials with complex configuration at sintering temperatures that inhibits the interaction of the matrix with ceramic particles. This process implies the restriction of forming specific breakable chemical compounds. The absence of metallic melt or the existence of metallic melt in low quantity excludes or diminish the umectability problems [6].

For their good characteristics of wear resistance, roughness, high temperature resistance was made the selection of aluminum matrix-based composites with discontinuous reinforced with rough particles of silicon carbide. We have also considered the possibility to process these materials trough conventional and economic technologies, using the existing equipment and for the low price of the raw material.

The starting materials were: green air atomized, 99.5 % purity AA300 grade aluminum metallic powder of 100-300 μm mean size, delivered by SC Alba Aluminium Zlana, Romania was used as metal matrix and fine ball milled α SiC green powder with 80-100 μm mean size, delivered by Casirom Turda, Romania. Before mixing, the SiC powder was cleaned and then washed in cold water and dried.

In order to distinguish the physical and chemical processes taking place in the processing of selected systems and to optimize the technological parameters, the ratio of SiC in the composite mixture was 5,15,25 weight % of SiC were homogenized in a TURBULA T2F mixer.

The combination of two materials is mixed in a removable container preventing cross-contamination between batches and dusting when mixing powders. The mixing time for homogenisation was 30 minute for 5 % SiC mixtures and 2 hours for 25 % SiC mixtures with rotational speed of 20 rpm.In order to avoid the agglomeration of the powder and deposition were used manganese stearate (1%) for lubrification and process controling agent (APV alcohol polyvinyl,1%). After mixing and homogenisation, the powder mixtures was presed in a cylindrical die , using and uniaxial pressing with 200, 240 and 420 MPa. The dimensions of the cylindrical die used for pressing with 200 MPa was h=84,5 mm and d=32mm, and for pressing with 240 and 420 MPa the dimensions was: h=84,5 mm and d=10 mm. Four sets of green compacts were sintered at different temperature and atmospheres as it shown in table 1.

Table 1. Sintering parameters

Sample	Sintering temperature (°C)	Sintering atmosphere
Al-SiC_{green}	1100°C	Reducing CO
Al-SiC_{green}	1300°C	protective gas (Ar)
Al-SiC_{green}	1450°C	protective gas (Ar)
Al-SiC_{green}	1750°C	air

The heat treatment of sintering was performed in CO reducing atmosphere at 1100°C, protective gas (Ar) atmosphere at 1300- 1450° C and air atmosphere at 1750° C in Siemens-Plania and Balzers IOV 16 type furnaces.

Almost all the researches upon composite materials are focused on the effects associate to the interface because these effects are crucial to the optimization of composites performances and to their processing technology.

Microstructural analysis of the composite powders was studied by scanning electron microscopy and by X-ray diffraction, a FEI XL-30-ESEM Philips electron microscope with EDS analyzer and a DRON UM 1 diffractometer in Bragg-Brentano parafocusing geometry with horizontal goniometry. A flat graphite monochromator in diffracted beam improved the pattern resolution and eliminate the Kβ line. The experimental conditions for data acquisition were: radiation type CuK_{α} (λ=1,547 Å), 35 kW, 30 mA, $2\theta initial = 20^{0}$, $2\theta final = 100^{0}$, $\Delta\theta$ step = $0{,}05^{0}$, time of acquisition per step = 8 s. For data acquisition/analysis was used specialized data acquisition software and preliminary pattern analysis software. The presence of reaction's products and component's heterogeneity leads to local variations of the charge transfer. Indifferently to the phases that bind together, the interface microstructure presents interesting particularities, having major influences upon mechanical properties of the composite [7].

3. Results and discussion

3.1. Morphological changes

The morphological characteristics of the powders it can be observing a uniform distribution of SiC particles in the Al matrix which is an important factor to be monitored in order to obtain better physical and mechanical properties of the final composites.

In order to determine some physical and technological features of the samples, the Reichart metallographic optical microscope with qualitative and quantitative imagine analysis was used. By optical microscopy the following parameters can be determined: particle's shape, distribution of particles, specific surface and apparent density. Chemical microanalysis and samples morphology were determined using ESEM FEI XL-30 (Environmental Scanning Electron Microscopy) with incorporated EDS analyzer. Figure 1 shows a relatively homogeneous distribution of polyhedral particles and porosity and also some of SiC particles are cracked.

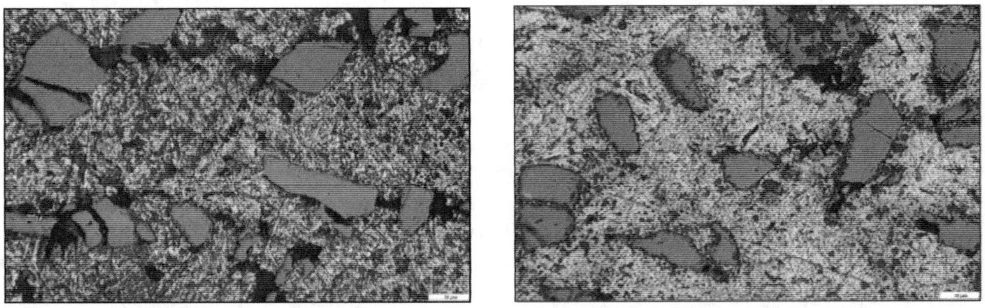

Figure 1. Optical micrographs of Al-25%SiC composite sintered at 1300°C in protective atmosphere (Ar), x200

In the case of aluminum matrix composite material reinforced with silicon carbide particles sintered in protective atmosphere (figure 1) we can observe a non-homogeneous distribution of the SiC particles with formation of an intermediate layer at the interface. According to figure 1 are parts where SiC particles are cracked and some particles are damaged.

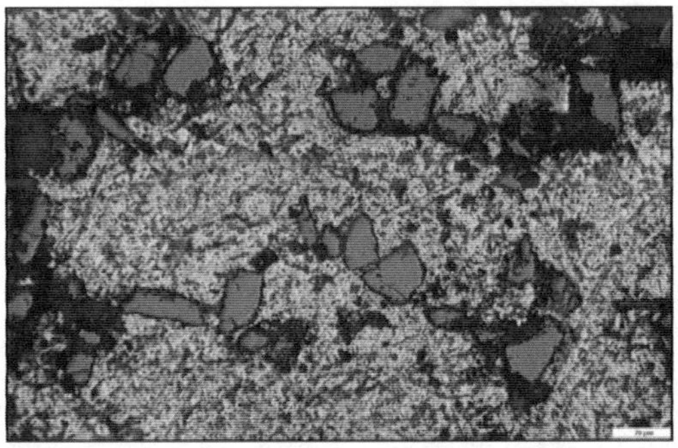

Figure 2. Optical micrographs of Al-25% SiC composite sintered at 1450°C in protective atmosphere (Ar), x200

From the optical microstructures of the composite materials sintered at temperature of 1450°C reinforced with silicon carbide particles can be observed changes in the appearance of the area of interaction (reaction) for the purpose of better adhesion at the interface with formation of a layer with relatively constant thickness (Figure 2) which confirms the existence of chemical reactions, the reaction precipitates or compounds.

3.2. Structural evolution

The characterization of Al/SiC green compacts composites is been reveals by density and compressibility changes and dimensional modification in the powder morphology. Material density is an important physical parameter being interrelated with the main mechanical parameters of materials (solids) [8-10]. To determine the effect of reinforcement content the compacts density was measured and compared with the theoretical density of the powder mixtures which represent the maximum density of material attained in the final stage, and is calculated by the rule of mixtures, after the following relationship:

$$\rho = \frac{100}{\sum_{i=1}^{n} \frac{x_i}{\rho_i}} \quad (1)$$

where: x_i is the fraction component i in mixture (i being Al, SiC, APV and Mg stearate) and ρ_i represent the density of component i (g/cm^3).

The theoretical density, calculated with relation (1) obtained for the studied samples were: $\rho_{Al+5\%SiC}$=2,725g/cm^3 : $\rho_{Al+15\%SiC}$=2,775g/cm^3 , : $\rho_{Al+25\%SiC}$=2,825g/cm^3 . The physical characteristics of the compacts were determined by measuring and weighting. Densities of the samples obtained after each processing step are given in table 2. The maximum densification it can be observed at 420 MPa pressure in the 25% SiC reinforced samples. Best values of density after compaction are recorded on sample Al-25 % SiC composites. During the compaction process the particles overlap much better; the spaces between the larger particles are filled with smaller particles, thus the volume of pressed sample is reduced. With increasing the compacting pressure was resulted a decrease on the porosity of compacts and also, the shape of particles became much more uniform, hindering the compaction process; the empty spaces among the particles of matrix and reinforcements remain unfilled properly. The processes observed during the cold unidirectional pressing operation confirm the existing data in the literature following the three stages namely: in the first stage of the pressing the particles is rearrange, allowing a better packaging of powders, then the decreasing of densification at intermediate pressures at values up to 80-85% of compaction, actually corresponds to stage II of the pressing, when occurs the contact surfaces increasing between particles by elastic-plastic deformation of the metallic particles, respectively fragmentation of ceramic particles, this increasing in density reveals from compressibility curves of our mixtures (figure 3).

Table 2. Densities, porosity and compressibility measured after compaction

Sample	Pressing pressure MPa	ρ_t g/cm^3	Apparent density g/cm^3	Compressibility %	Porosity %
Al - 5%SiC$_g$	240	2,72	2,44	89,70	10,3
	420	2,72	2,57	94,48	5,5
Al-15%SiC$_g$	240	2,77	2,47	89,16	10,3
	420	2,77	2,65	95,66	4,3
Al-25%SiC$_g$	240	2,83	2,49	86,70	12
	420	2,83	2,73	96,46	3,5

Processes corresponding to the third stage of pressing were presented in table 2, it is observed that with porosity decreasing and increasing of pressure occurs a corresponding maximum densification of massive plastic deformation and an increased number of contacts between the particles. The porosity for each composition increases with raising of SiC content reaching values of 3.5% (for samples with 25% SiC) at a pressure of 420 MPa and values of 10, 3% (for samples with 5% SiC).

IOP Conf. Series: Materials Science and Engineering 1190 (2021) 012012 doi:10.1088/1757-899X/1190/1/012012

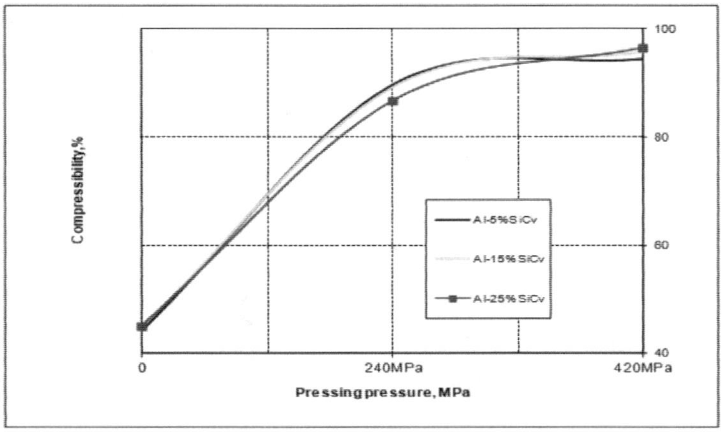

Figure 3. Compressibility curves for Al matrix reinforced with 5, 15, 25 % SiC compacts

It is known the form filling shaped unidirectional cold pressing of metal-ceramics powder mixtures it create densification problems compared with metal-metal mixtures for these reasons was studied the influence of silicon carbide proportion on the pressing behaviour of the materials, the main results of research can be summarized as follows:

- Analyzing the plotted compressibility curves (figure 3) we can conclude by the addition in the matrix of the hard and brittle SiC particles, the compressibility decreases. It's difficult to press the mixtures (higher pressing forces) with increasing of SiC amount and is explained by the fact that hard SiC particles, delay the densification by taking the load for pressing, until their broke, maximum densification taking place by repackaging SiC fragments in aluminum matrix.
- Maximum densification of the composite material reinforced with 5%SiC is obtained at a pressure of 420 MPa (94 5%). For 15% SiC content densification increase with higher pressing pressure given the density values of 95.66% and 96.46% for the 25% SiC composite.

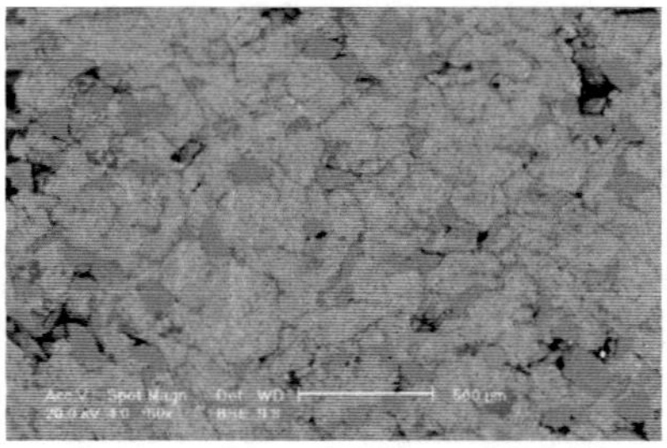

a)

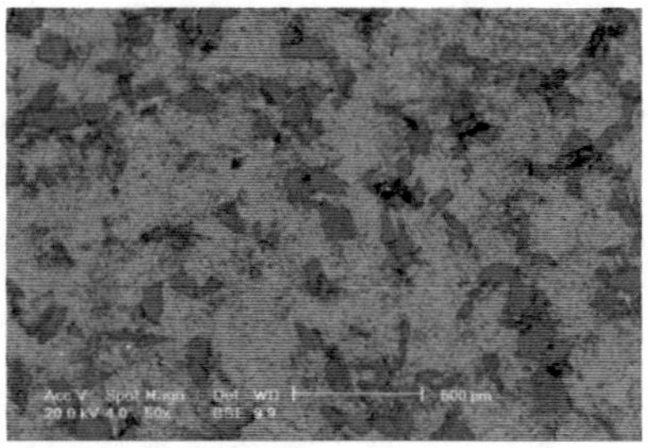

b)

c)

Figure 4. SEM images of the Al-25%SiC composites after sintering at a) 1100°C b) 1300°C c) 1450°C

From the three experimental mixtures, and based on the results one of the representative contents (Al-25% SiC) was choose to sintered and to study microstructural and compositional point of view. The structure of sintered sample was investigated by SEM, XRD and illustrated in figure 4 and 5.

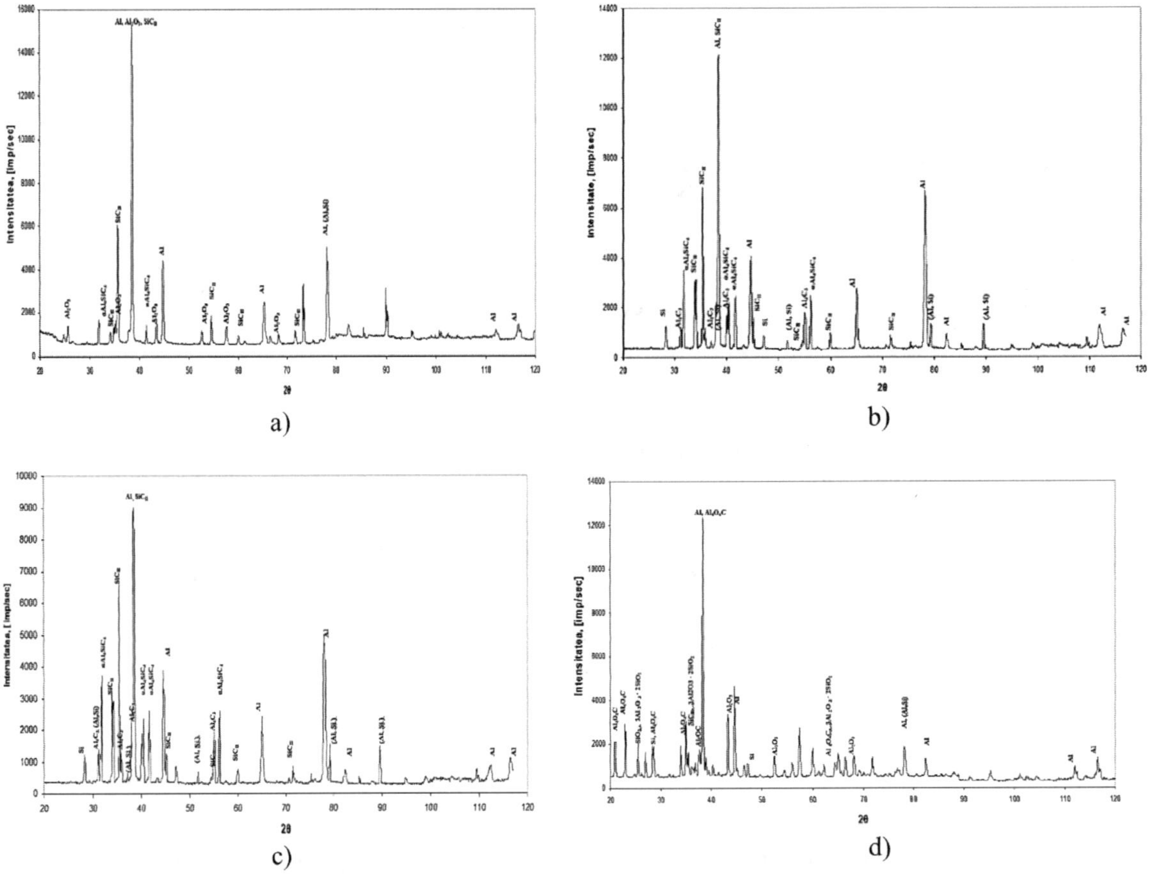

Figure 5. X-ray diffraction patterns of Al-25 % SiC composites sintered at a)1100 °C b) 1300 °C c) 1450 °C d)1750°C

In table 3 are presented the Rietveld refinement results for the analyzed XRD patterns which demonstrate the presents of reaction compounds and phases identified at the Al/SiC composites interface.

Table 3. Rietveld refinement results according to XRD patterns

No	Sintering temperature of Al/SiC composites	Phases in Al/SiC composites	Volume fraction , %	Weight fraction, %	D (Å)	(ε^2), %
1	1100 °C	**Al** (Al,fcc) PS cF4/1 cub	70.3 ±0.1	64.1 ± 0.2	539.2	0.105
		Al2 O3 PS hR10/1 trig	13.5 ± 0.1	18.3 ± 0.1	498.1	0.110
		Si C (polytype II) PS hP12/14 hexag	16.2 ± 0.1	17.6 ± 0.1	502.2	0.109
2	1300°C	**Al** (A1, fcc) PS cF4/1 cub	71.6 ± 0.2	68.7 ± 0.2	531.9	0.169
		SiC (polytype II) PS hP12/14 hexag	24.7 ± 0.1	28.2 ± 0.1	490	0.121
		Si (type A4) PS cF8/1 cub	3.7 ± 0.2	3.1 ± 0.1	502.1	0.115
3	1450°C	**Al** (A1, fcc) PS cF4/1 cub	69.6 ± 0.2	67.4 ± 0.2	538.7	0.121
		SiC(polytype II) PS hP12/14 hexag	22.6 ± 0.1	26.0 ± 0.1	496.2	0.113
		Si (type A4) PS cF8/1 cub	7.9 ± 0.2	6.6 ± 0.1	494.8	0.112
		Al (A1, fcc) PS cF4/1 cub	13.9 ± 0.0	11.3 ± 0.0	357.2	0.145
		Al2 O3 PS hR10/1 trig	42.8 ± 0.2	51.4 ± 0.2	516.8	0.100
		Al4 O4 C JCPDS 14-630 orth	17.8 ± 0.0	14.2 ± 0.0	263.2	0.175
4	1750°C	**Al4 C4 Si** (type E9.4) PS hP18/2 hex	9.6 ± 0.0	8.7 ± 0.0	495.7	0.114
		SiC (polytype II) PS hP12/14 hex	4.7 ± 0.1	4.6 ± 0.1	515.6	0.109
		Si C (type B7, polytype I) PS hR10/9 trig	7.0 ± 0.0	6.8 ± 0.0	470.6	0.125
		Si (type A4) PS cF8/1 cub	4.3 ± 0.1	3.0 ± 0.0	462.7	0.128

3.3. Interface characterization

The temperature and the sintering atmosphere were chosen as listed in the above table (table 1) and the influence the reaction between metallic matrix and ceramic reinforcement composites was studied. The chemical reactions that take place at metal-ceramic interface can be [11]: direct reduction reactions, forming compounds reactions, forming of gaseous compounds reactions.

As a result of chemical reactions between the matrix and the reinforcement material may form compounds which affect the quality of the composite, due to the emergence of an area of stress concentration; the risk of cracks is very high. In generally some areas is as compact and without fragile compounds. Since aluminum is very reactive (electropositive metal) this liquid, chemically reacts with most ceramic components. Reactions that occur between Al and SiC at 650 ° C are:

$$4Al+3SiC \rightarrow Al_4C_3 + 3Si \quad (2)$$

$$4Al+4SiC \rightarrow Al_4SiC_4 + 3Si \quad (3)$$

Some authors affirm [12], [13] the processing of pure Al – SiC that obtaining is not a simple process, which depends on temperature and time. This time-temperature dependence is given by the existing chemical reactions between Al matrix and silicon carbide. According to (2) relationship in Al-SiC melt, chemical reaction occurs:

$$Al + SiC \rightarrow [Al\text{-}Si] + Al_4C_3 + SiC \quad (4)$$

where liquid aluminum matrix will become the type Al-Si matrix, which depends on the temperature of the composite material processing. Thus, in the manufacture of composite materials Al/SiC interfacial reaction between Al and SiC depends on production parameters: temperature, atmosphere and chemical composition of the matrix and reinforcement material.

The reactions at the interface can be: the dissolution of the SiC in Al; precipitation of a phase III; segregation of the third phase at the interface. The reaction products formed at the interface are fragile and under certain conditions may affect in negative way the properties of the composite. For our experimental conditions we can observe the following reactions produced at 1300°C and 1450°C in the same sintering atmosphere (protective):

$$4Al + 3\,SiC \rightarrow Al_4C_3 + 3\,[Si] \text{ reaction at the Al – SiC interface} \quad (5)$$
$$4Al + 4SiC \rightarrow Al_4SiC_4 + 3\,[Si] \text{ reaction at the Al – SiC interface} \quad (6)$$
$$SiC + Al_4C_3 \rightarrow Al_4SiC_4 \quad (7)$$
$$Si + Al \rightarrow (Al,Si) \text{ with Si diffusion in matrix and adsorption in aluminum matrix} \quad (8)$$
$$4Al + 3O_2 \rightarrow 2Al_2O_3 \text{ reaction at Al-O interface (pores)} \quad (9)$$
$$Si + Al \rightarrow Al,Si \text{ with Si diffusion in matrix without adsorption in aluminum matrix} \quad (10)$$

Therefore, it is recommended to conduct partial reactions at the interface by controlling the thickness of the reaction zone. Increasing the reaction layer is limited by diffusion processes of the elements. the silicon carbide particles and the molten aluminum above its melting temperature (up to 1100°C).

Areas subject to EDS chemical microanalysis investigation to spot focused areas: matrix, interface, and silicon carbide particles. The aluminum content (table 4) from aluminum matrix is less than 100%.

Table 4. EDS analysis of Al-SiC composite sintered at temperatures of 1100 °C, 1300 °C, 1450 °C and 1750°C

Temperature / Sample	Zone	Elementary composition of sintered samples			
		Al, % weight	Si, % weight	C, % weight	O, % weight
1100° C Al-25 %SiC	Matrix	85,23	1,14	6,33	7,29
	Interface	67,61	2,92	0	29,48
	Particle	1,01	75,89	22,56	0,54
1300° C Al-25 %SiC	Matrix	83,05	8,93	-	8,02
	Interface	71,11	5,69	-	23,20
	Particle	0,87	75,02	23	1,12
1450° C Al-25 %SiC	Matrix	88,32	7,36	2,96	1,37
	Interface	40,74	33,86	6,33	19,07
	Particle	0,60	73,97	25,43	-
1750° C Al-25 %SiC	Matrix	90,21	5,94	-	3,85
	Interface	74,26	0,67	-	25,07
	Particle	55,52	2,50	7,46	34,52

This content can be explained by the diffusion phenomena, dissolution and / or reaction of silicon and carbon in the matrix. The amount of oxygen in the aluminum matrix can explained due to the reaction of aluminum with air provide from the pores of the material. The presence of silicon and carbon in the metal matrix cannot be explained only by the decomposition of the silicon carbide particle.

Estimated reaction products and intermetallic compounds are: Al, Al_2O_3, (Al, Si), Si, Al, Al_4O_4C, $3Al_2O_3x2SiO_2$, SiO_2, α Al_4SiC_4, SiC_{II},SiC_{III}, Al_4C_3 which are indentified by XRD analysis (figure 5). The detailed EDS / SEM analysis (figure 6-8) has revealed the interface morphology between SiC particles and aluminum matrix. As shown above, with the increasing of temperature, the composites phase structure becomes more complex, with the apparition of new interphases and a good interface zone.

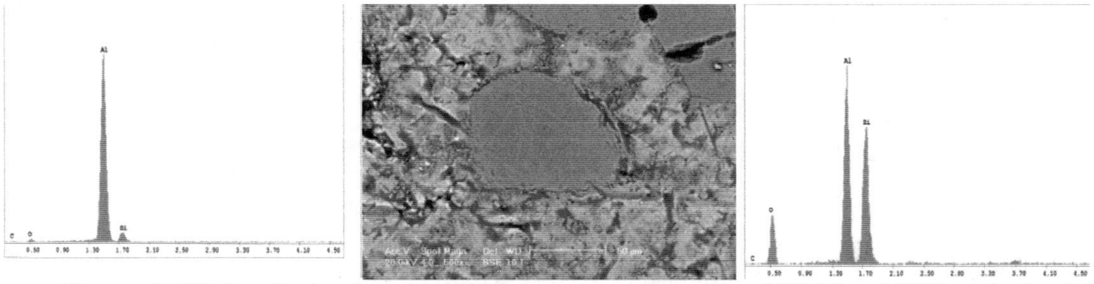

Figure 6. SEM analysis of Al-25%SiC$_p$ composites sintered at 1100°C and EDS analysis of the interested interface zone

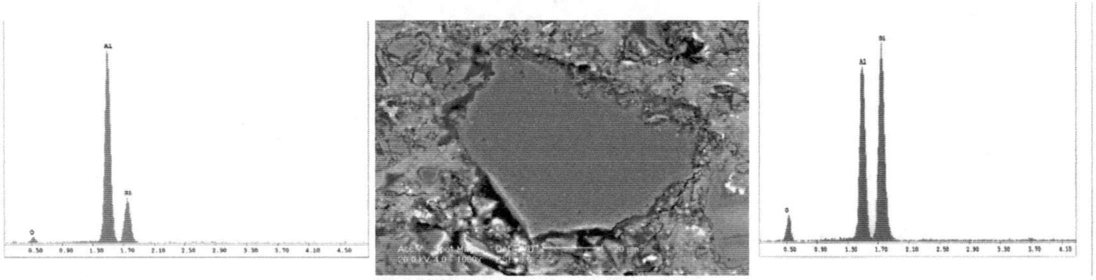

Figure 7. SEM analysis of Al-25%SiC$_p$ composites sintered at 1300°C and EDS analysis of the interested interface zone

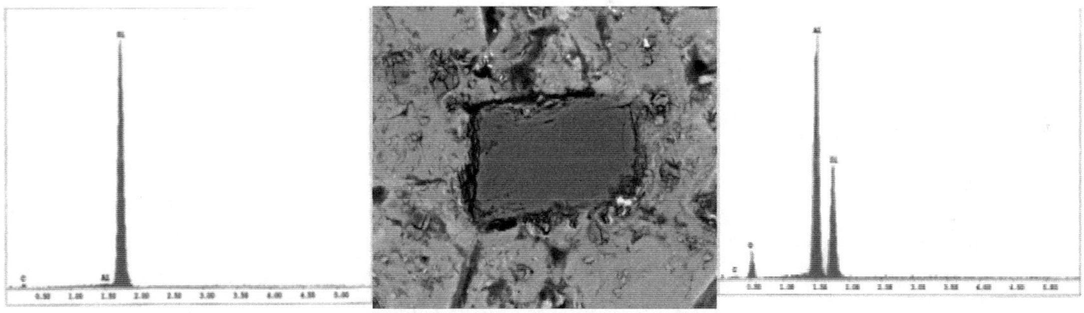

Figure 8. SEM analysis of Al-25%SiC$_p$ composites sintered at 1450°C and EDS analysis of the interested interface zone

4. Conclusions

In the paper were investigated the influence of the reinforcement content, the sintering parameters and conditions on interface between Al/SiC composites obtained by powder metallurgy route can be concluded:

- the technical applications and benefits have been studied for first time the behavior of composite materials of Al-SiC at high temperatures (1100, 1300, 1450 and 1750°C) by powder metallurgy (sintering in the liquid phase). Composites were made in different environments, namely: reducing (CO), inert (Ar) and oxidant (air). Following the characterization of the three types of composites with different contents of one polytypic state (α SiC polytype I, II) silicon carbide (5, 15, 25%) and obtained was found that the strongest interaction takes place in the composites material with composition of 75% Al and 25% SiC. By optical and electron microscopy analysis have showed the composites structure where the SiC particles are dispersed in the Al matrix.

- the EDS analysis effectuated on a concentration line has showed the soft shape the interface layer at the interface of the Al/SiC composite. The XRD analyses have established the composition of the interface layers. From the phase analysis we were able to identify the following phases: at 1100° C: Al, Al_2O_3, SiC (polytype II); at 1300° C: Al, SiC (polytype II), Si; at 1450° C: Al, SiC (polytype II), Si; at 1750° C: Al, Al_2O_3, Al_4O_4C, Al_4C_4Si, SiC (polytype II), SiC (polytype I). Sinterred powders in reducing atmosphere have led to the apparition at the interface of an Al_2O_3 layer which acted as a diffusion barrier [14], [15].

Acknowledgment(s)
The article is financed by the University Politehnica of Bucharest, through the project "Engineer in Europe", online system, registered at the Ministry of Education under no. 140 / GP / 19.04.2021, by using the fund to finance special situations that cannot be integrated in the form of financing institutions state higher education

References

[1] I. N. Popescu, "Wear behaviour of automobile disk brakes from iron based composite material using statistical analysis", *International Conference on Advanced Materials and Technologies -ROMAT 2004 - Proceedings*, pp. 64-73, 2004.
[2] M. Eriksson "Friction and Contact Phenomena of Disc Brakes Related to Squeal", *Acta Universitatis Upsaliensis, Uppsala*, 2000.
[3] M. Cojocaru, F. Gostin, "The modeling of the pressing behaviour of the high alloyed chromium-nickel- molybdenum steel powder AlSi3161L", *International Conference on Advanced Materials and Technologies-ROMAT 2004*, Proceedings, pp. 64-73, 2004.
[4] M. Y. Bal'shin, *Vestnik. Metalloprom*, vol. 418, pp. 124-137, 1938.
[5] K. Kawakita, Tsunami, *Bull. Chem. Soc. Jap., 39*, pp.1364-1368, 1969.
[6] E. Geza, A. Popa, N. Popescu, "Considerations concerning the density of metallic powders through psedo-isostatic procedures", *Second International Conference on Materials and Manufacturing Technologies, Cluj-Napoca,* 1998.
[7] R.W. Heckel, "Density-pressure relationships in powder compaction", *Trans. Metall. Soc. AIME*, 221, pp. 671-675, 1961.
[8] O.G. Neikow, S.S. Naboychenko, G. Dawson "Handbook of Non-Ferrous Metal Powders – Technologies and Applications" Oxford: Elsevier Ltd, 2009.
[9] B. Ramesh, T. Senthilvelan," Formability characteristics of aluminium based composites - a review", *International Journal of Engineering and Technology* Vol. 2, No.1, pp. 1-6, 2010.
[10] H.Z. Razavi, H.R. Hafizpour, A. Simchi. "An investigation on the compressibility of aluminum/nano-alumina composite powder prepared by blending and mechanical milling" *Material Science Engineering A*; pp 454–455: pp. 89-98, 2007.
[11] P. Scardi, M. Leoni, Whole powder Pattern Modelling: Theory and Applications, pp. 51-86, 2004.
[12] M. Zsigmond (Ciurdas), P. Moldovan, "Coating technology and processes, *International Scientific Conference, Scientific Bulletin of North University of Baia Mare, Seria D*, Vol. XVII, pp. 413-419, 2003.
[13] T. Fan, G. Yang, D. Zhang, T. Shibayanagi, M. Naka, "Temperature dependence of melt structure in SiCp/Al composites above the liquidus", Vol. 93, pp. 208-216, 2005.
[14] D. Balzar, N.C. Popa, "Crystallite Size and Residual Strain/Stress Modeling in Rietveld Refinement" In: E.J. Mittemeijer, P. Scardi (eds) *Diffraction Analysis of the Microstructure of Materials. Springer Series in Materials Science,* Vol 68. Springer, Berlin, pp. 125-144, 2004.
[15] R.K. Everett, J. Arsenault, "Metal Matrix Composites-Processing and Interfaces", Book, 1991.

IOP Conf. Series: Materials Science and Engineering 1190 (2021) 012013 doi:10.1088/1757-899X/1190/1/012013

Effect of corundum sand proportion on strength properties geopolymer mortar based on fly ash

Kinga Pławecka, Kinga Korniejenko, Patrycja Bazan

Faculty of Material Engineering and Physics, Cracow University of Technology, Jana Pawła II 37, 31-864 Cracow, Poland

kinga.plawecka@pk.edu.pl

Abstract. The main objective of this study is to investigate the possibility of replacing the traditional construction aggregate, river sand, with corundum material and to evaluate the effect of the proportion of corundum sand on the strength properties of fly ash based geopolymer mortar. Electrocorundum (white) and fly ash from a thermal power plant in Skawina were used to make the samples. Different fractions of electrocorundum were used to prepare the geopolymer mass. A geopolymer based on fly ash and river sand was used as reference material. The results showed that the average compressive strength of fly ash and alumina-based materials increased (results ranging from 25.9 to 47.0 MPa) compared to the reference samples (20.7 MPa). The flexural strength results showed a weakening of the strength properties of the electrocorundum and fly ash-based samples (results ranging from 6.1 to 4.8 MPa) compared to the reference sample (7.0 MPa). Only the geopolymer, which in its composition contained a mixture of electrocorundum with several grain gradations, showed an increase in flexural strength (7.3 MPa), compared to the reference sample. Additionally, surface topography analysis of the geopolymer samples was performed by scanning electron microscopy. The whole tests carried out on the above types of material indicate that it can be used as a substitute for sand in geopolymer mortars.

1. Introduction

The fundamental natural resource used by the construction industry is river sand, used as fine aggregate during concrete production [1]. More than 50 billion of this material is exploited annually [2], and the continuous development of the construction industry and other sectors of the economy related to natural resources makes it become a scarce commodity. Over the years, the ratio of the extraction of this raw material to the rate of replenishment of its deposits worldwide has been gradually decreasing [3]. Unfortunately, not all types of this material are suitable for use in construction mortars - desert sand, due to its streamlined grain shape is more difficult to mix and is more easily washed out by water [4]. Therefore, it is not suitable for construction applications. The increasing demand for river sand and environmental concerns about its complete depletion have prompted the investigation of other relative sources that could provide a substitute for the above raw material in construction materials [1,5].

Aggregate is a rocky loose material (coarse or fine grained) of organic or mineral origin, mainly used in the production of concrete, building mortar or as a base for road surfaces [6,7]. The geometrical features, among others, the shape or grain size of the aggregate determine the subsequent properties of the entire concrete mix. The mixture of aggregates for concrete should contain both coarse and fine fractions in appropriate proportions. The fine aggregate stabilizes the whole concrete mix (affects the viscosity and workability of the mass), and the addition of coarse aggregate is to fill the volume of the concrete mix. The optimal composition of fractions is to ensure the lowest possible demand of the concrete on the cement slurry. Aggregate grain size is also related to another property, namely water holding capacity. It depends on its surface, roughness and shape. The smaller the grain of the crumb pile, the higher the water demand of the concrete mix. In the development of concrete formulations, a balance is sought between low water-carrying capacity (to achieve the lowest possible shrinkage) and good rheological properties and surface quality of the produced elements [8].

Cement is another most widely used material in the construction industry worldwide [9]. Along with sand, it is the primary component of concrete. The construction sector is the largest explorer of natural resources, which is associated with driving the global production of cement, used as the base material in this industry [10]. The 2020 data on global cement fabrication, shows that the volume of cement production reaches values of about 4 billion tons per year [11], and the production process itself is associated with high carbon dioxide emissions. According to the International Energy Agency, the cement industry is responsible for about 7% of global emissions of this gas into the environment [12].

Research carried out on geopolymers is aimed at verifying their ability to replace Portland cement in the construction industry [13].The design process of a geopolymer mix is mainly based on its intended use and related performance criteria. The workability of the mix during its preparation and the compressive strength value of the bonded mortar are used as the main determinant [10]. Geopolymers are a class of amorphous, inorganic and aluminosilicate polymers whose properties depend on the type of base material used and the type and amount of activator used in their production [14].

Secondary raw materials or by-products from industrial processes can be used to make geopolymer materials. One of these is fly ash, which is a waste product of the coal combustion process [15,16] Currently, some of these ashes are disposed of as an admixture material for concrete production to reduce their environmental impact. The durability and mechanical properties of concrete with part of the cement replaced with fly ash have already been studied. This has been shown to produce a material with high durability and strength [17-20]. It should be noted, however, that the development of the technology of production of geopolymers on the basis of ash itself would affect the increase in the use of this raw material, which is a by-product of combustion. At the same time, it would be a positive environmental aspect.

The optimum aggregate content in geopolymers is not yet precisely specified and is mainly based on the results of laboratory scale research work [21,22]. Based on their analysis, Joseph and Mathew proved that geopolymer concrete with proper content and fixed proportion of fine aggregate can have better mechanical properties compared to those of ordinary cement concrete [23].

The most commonly proposed application for geopolymers is to obtain various types of products for use in construction, including pavers, flagstones, curbs, facade panels, or bricks. These materials are also suitable for immobilizing hazardous waste (radioactive waste, asbestos) or toxic waste (mercury, lead, arsenic) [20,24-27]. The above group of binders is characterized by, among others: high compressive strength [28-30], resistance to acids, chlorides and sulfates [31-34], thermal resistance (up to about 800°C) [35-37] or good frost resistance [38,39]. Moreover, there is no corrosion of steel reinforcement in geopolymers and the material shows good adhesion properties with steel. The synthesis process of geopolymer materials itself consumes 2-3 times less energy than the production of Portland cement [20,40].

In this article, the possibility of replacing river sand with another aggregate in the form of electrocorundum sand in fly ash based geopolymer mortar is analyzed. Strength tests were conducted on the fabricated geopolymer materials and the surface topography of the specimens was observed.

2. Materials and sample preparation

In the present work, activities related to the fabrication of geopolymers based on fly ash and an additional aggregate in the form of noble electrocorundum were undertaken. The fly ash was obtained from the Skawina CHP Plant and the abrasive material from one of the Polish suppliers - the "Herubin" company (Dobra, Poland). The table below (Table 1) shows the fractions of white electrocorundum ordered for the production of geopolymer materials. Figures 1a, 1b, and 1c show photographs of each type of abrasive, taken with a JEOL JSM-6390LV scanning microscope. This material was chosen for this study because of the shape and size of its grains (similar to river sand), its chemical composition (more than 98% pure Al2O3), and its excellent mechanical properties of high hardness and low abrasiveness.

Table 1. Grit sizes for electrocorundum according to FEPA 42-D-1984 and PN/M-59107

Grain number	Particle size [μm]	
(F)	from	to
24	710	850
80	180	212
220	53	75

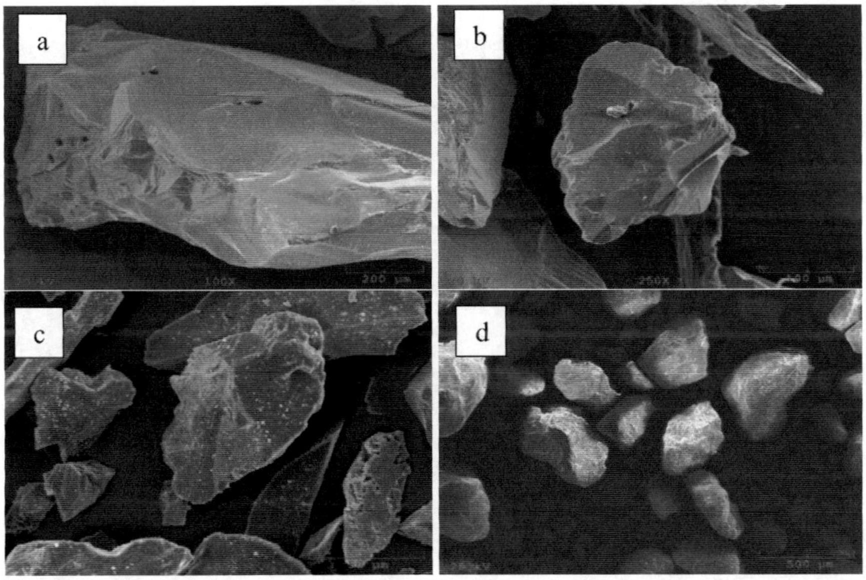

Figure 1. Aggregates used to produce geopolymers; a) F24 electrocorundum, b) F80 electrocorundum, c) F220 electrocorundum, d) river sand

In addition, one of the aggregates needed to make the reference sample mixture was river sand (Swietochlowice area). Figure 1d shows grains of the above material taken with the same scanning microscope model. River sand is a basic raw material, necessary for the production of concrete mortars and geopolymer pastes, due to the morphological nature of the grains and its general availability. Table 2 shows the proportion of the different grain fractions of the above river sand.

Table 2. Percentage of individual grain fractions in river sand

Grain diameter [mm]	>1.6	1.6-1.0	1,0-0.3	0.2-0.3	0.2-0.1	<0.1
Fraction content [%]	0.65	2.94	1.50	85.25	9.18	0.48

To prepare a suitable alkali activator for the geopolymerization process, a source of strong base, an aqueous silicate solution, and water are required. In the above work, technical sodium hydroxide in flake form with a density of about 2.13 g/cm^3, sodium water glass R-145 with a density of about 1.48 g/cm^3, and tap water were selected. The preparation of the 10M alkaline activator involved weighing an appropriate amount of technical sodium hydroxide and dissolving it in water, followed by the addition of sodium silicate. Both processes were accompanied by an exothermic reaction, so the resulting solution had to be cooled to ambient temperature (about 25°C).

The design of the geopolymer mortar was mainly based on selecting the appropriate raw materials and determining their proportions in the overall mixture. The dry ingredients were mixed with a laboratory mixer for about 5 minutes and then the alkaline solution was added, mixing everything thoroughly for 15 minutes until a homogeneous mass was obtained. On average, about 400 ml of alkaline solution was added per 1kg of dry ingredients. The obtained plastic mass was filled into sets of prismatic molds of 50 mm x 50 mm x 200 mm and cubic molds of 50 mm x 50 mm x 50 mm, and then placed on a vibrating table. The geopolymer concretes prepared in this way were cured for 24 hours in a laboratory dryer (LW 750 STD, Pol-Eko-Aparatura, Wodzisław Śląski, Poland) at 75°C. Then, the samples removed from the molds, cooled to ambient temperature, were seasoned for 28 days, after which the mechanical properties were tested and the SEM microstructure was observed. Table 3 shows the composition of the prepared samples and their mass ratios.

Table 3. Information about prepared samples

Sample	Solution	Composition
R	10 M	Fly ash +sand (1:1) - reference sample
P24	10 M	Fly ash + electrocorundum F24 (1:1)
P80	10 M	Fly ash + electrocorundum F80 (1:1)
P220	10M	Fly ash + electrocorundum F220 (1:1)
K	10M	Fly ash + electrocorundum F24 + electrocorundum F80 + electrocorundum F220 (5:1:1:1)

3. Research methods

Compressive strength tests were carried out on a testing machine MATEST 3000 kN with a speed 0.5 MPa/s, according to EN 12390-3 standard „ Testing of hardened concrete. Compressive strength of samples ". The tests were carried out on cubic samples with dimensions of 50mm x 50mm x 50mm.

The flexural strength tests were carried out similarly to the compressive strength tests, according to the EN 12390-3 on the MATEST 3000 kN testing machine with a speed of 0.5 MPa/s. These tests were carried out on 50 mm x 50 mm x 200 mm. The distance between the support points was 150mm.

The observation of the microstructure of the geopolymers was made using the scanning electron microscope (SEM) of the type JEOL JSM-6390LV with EDS on the breakthroughs of the samples after the compressive strength tests. Preparation of samples for the observation was to their previous sputtering a thin layer of gold by vacuum sprayer JEOL – JEE-4X. The samples were observed at different magnifications.

4. Results and discussions

4.1. Compressive strength

The results from compressive strength tests are presented in table 4. The measurements for each series were made on 10 samples.

Table 4. The results of compressive strength

Sample	Compressive strength [MPa]
R	20.7 ± 3.6
P24	25.9 ± 6.2
P80	50.8 ± 8.2
P220	28,9 ± 5.7
K	47.0 ± 6.2

The average compressive strength value for material R (reference specimen) is about 21 MPa. This is the lowest average value obtained compared to all types of compression specimens. Compared to sample P, an increase in average compressive strength value was recorded for all geopolymers with the addition of alumina - P24 by about 25% and P220 by about 40%. The highest fracture toughness was demonstrated for material K and P80, for which the average values were 47.0 MPa and 50.8 MPa, respectively, which corresponds to an increase in compressive strength of about 130% and 140%, compared to the reference sample. The reference sample had the lowest standard deviation, which indicates very good repeatability of the results during the experiment. The P80 sample had the highest measurement uncertainty. The standard deviation of the other test materials remained constant.

4.2. Flexural strength

The results from flexural strength tests are presented in table 5. The measurements for each series were made on 5 samples.

Table 5. The results of flexural strength

Sample	Flexural strength [MPa]
R	7.0 ± 1.8
P24	6.1 ± 0.3
P80	5.2 ± 1.5
P220	4.8 ± 0.6
K	7.3 ± 0.5

The differences between the average values of the reference sample (P) and the composite material (K) are really small, up to a limit of 10%. However, the K material has the highest flexural strength of more than 7 MPa. The reference sample came in second with a flexural strength of 7 MPa. The average flexural strength of the geopolymer material with the addition of F24 electrocorundum is about 6 MPa. In specimens P80 and P220, the addition of electrocorundum affected the deterioration of flexural strength. Compared to the reference sample, the flexural strength decreased by about 35 % for the P80 material and by about 45 % for the composite based on fly ash and F220 electrocorundum. The P24 samples had the lowest standard deviation and kept the measurement results constant. The reference sample and sample P80 had the largest spread of measurement values. Material K and P220 had a similar trend of results - the standard deviation oscillated at a similar level in both cases.

4.3. Microstructural investigation

The basis for the study of the surface topography of the specimen and its chemical composition, was the material from specimens destroyed during the performance of strength tests.

Figure 2a and figure 2b shows the morphology of the reference material sample. The fly ash and sand based geopolymer is characterized by a compact and homogeneous structure with no obvious pores. The light-colored and irregular structures visible on the surface of the material are probably remnants of unburned coal. The spherical, regular particles embedded in the geopolymer matrix are fly ash grains that have not dissolved.

IOP Conf. Series: Materials Science and Engineering 1190 (2021) 012013 doi:10.1088/1757-899X/1190/1/012013

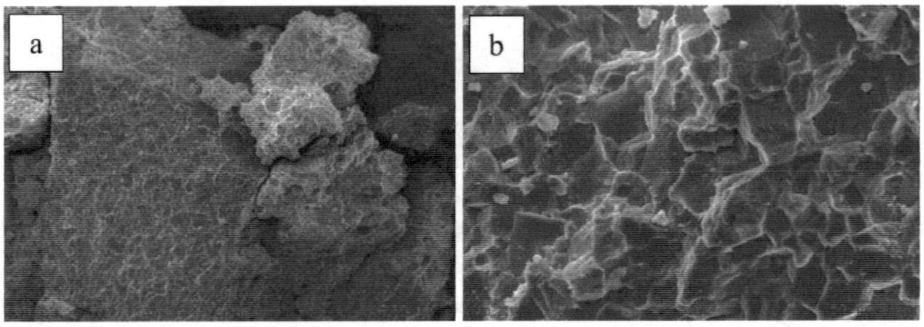

Figure 2. SEM images of geopolymer R: a) magnification 100x, b) magnification 600x

Figure 3a and figure 3b shows the microstructure of P24 geopolymer. The surface of the material is irregular, with numerous pores, whose size oscillates between 30 and 40 µm. At 600x magnification - Figure 3b, spherical, regular forms, connected with the matrix, can be seen. These are fly ash particles which have not degraded. Additionally, feathery shapes can be seen, located in numerous clusters on the surface of the material. It is likely that these are some kind of crystalline forms, but at present we do not have information on the structures formed.

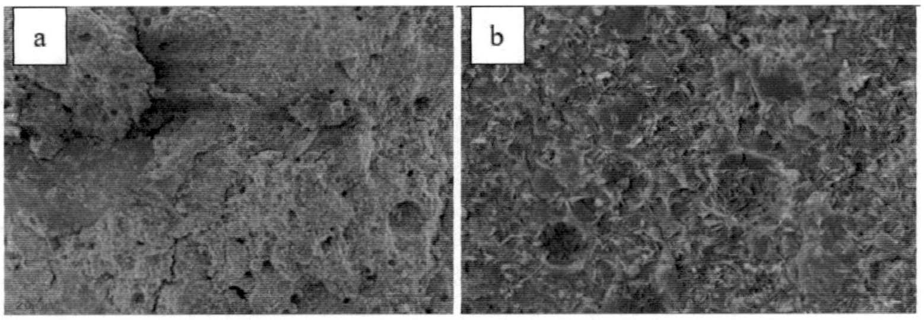

Figure 3. SEM images of geopolymer P24: a) magnification 100x, b) magnification 600x

Figure 4a and figure 4b present the surface image of P80 geopolymer. The microstructure of P80 geopolymer is characterized by numerous pores, which oscillate around 50 µm. Irregular, sharply pointed structures of several tens of micrometers in size are the grains of electrocorundum. Undissolved fly ash grains with a rounded shape tend to bond poorly with the geopolymer matrix and, consequently, to lower mechanical properties.

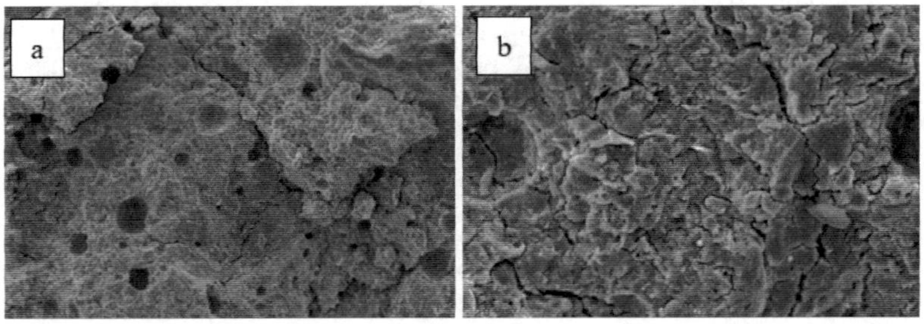

Figure 4. SEM image of geopolymer P80: a) magnification 100x, b) magnification 600x

Figure 5a and figure 5b show the surface topography of sample P220. Taking a closer look at the microstructure itself, one can see numerous small pores of 5 - 10 µm, as well as their larger counterparts of around 50 µm. In addition to the spherical ash particles, which are bonded to the material, we can also see sharp-edged corundum grains measuring approx. 30 µm. At higher magnification, the feathery structures, which appeared earlier in the case of P24 material, are revealed. However, there are far fewer of them.

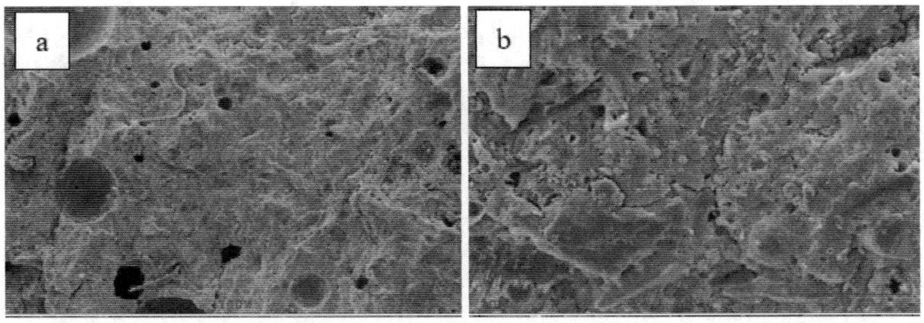

Figure 5. SEM image of geopolymer P220: a) magnification 100x, b) magnification 600x

The last type of material analyzed is geopolymer K, shown in figure 6a and figure 6b. The structure of the material is more homogeneous than in the case of P24 or P80. However, a few pores of about 40 μm in size are present. In the case of this sample, the ash particles appear to be well solidified with the geopolymer matrix, although they have not dissolved. There is no gap (lack of cohesion) around them, which compromises the mechanical properties.

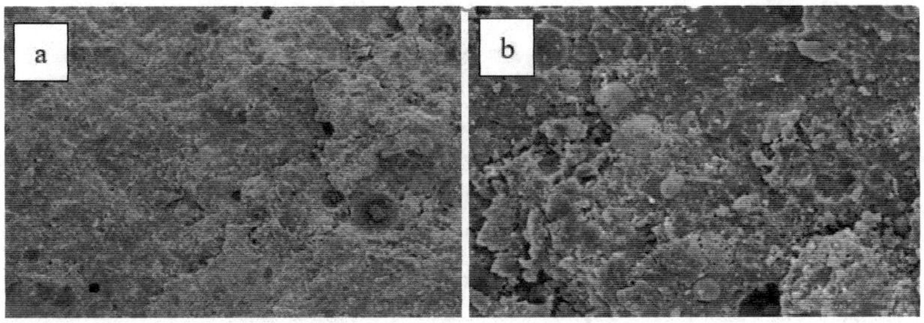

Figure 6. SEM image of geopolymer K: a) magnification 100x, b) magnification 600x

Microstructure and properties of geopolymers strongly depend on the nature of the initial raw materials even though the macrostructure characteristics of alumino-silicate-based geopolymers may appear similar, since it is related to the same silicon and aluminum bonding and the same gel phase binder [41]. Compressive strength measurements are used by many researchers as a tool to assess the success of geopolymerisation process. This is due to the low cost and simplicity of compressive strength testing, as well as due to the fact that strength properties is a primary measure of the utility of materials used in different applications of the construction industry [42].

5. Conclusions

Based on a review of the collected literature and an analysis of the results of individual experiments, the above work draws important conclusions about the properties of a fly ash-based geopolymer mortar and noble electrocorundum:

1. The good strength properties of electrocorundum and fly ash based geopolymers are related to the varying degree of aggregate grain size and grain shape.
2. The addition of aggregate in the form of homogeneous fraction of electrocorundum as a substitute for sand affects the deterioration of strength properties when tensile forces are applied to the material. This can probably be related to the failure to form a tight crumb stack in the geopolymer.
3. Addition in the form of homogeneous fraction of electrocorundum as a substitute for sand affects the improvement of strength properties when compressive forces are applied to the material. However, the best results are obtained by adding a mixture of noble aluminium oxide with different grain size (composed of F24, F80 and F220) - MK material or F80.
4. Addition in the form of differentiated fractions of electrocorundum material (composed of F24, F80 and F220 electrocorundum) contributes to a more uniform material structure. This is probably related to the formation of a tight crumb stack in the geopolymer.

Acknowledgments

The presented works were carried out as part of the project: "Geopolymer foams with low thermal conductivity produced on the basis of industrial waste as an innovative material for the circular economy", which is financed by the National Center for Research and Development under the LIDER X program, contract number LIDER / 31/0168 / L-10/18 / NCBR / 2019, implementation period: 01/01/2020 - 30/11/2022.

References

[1] P. Nanthagopalan and M. Santhanam, "Fresh and hardened properties of self-compacting concrete produced with manufactured sand," *Cem. Concr. Compos.*, vol. 33, no. 3, pp. 353–358, Mar. 2011, doi: 10.1016/j.cemconcomp.2010.11.005.

[2] M. Hernandez, S. Scarr, and K. Daigle, "The messy business of sand mining explained," [Online] 2021 [Accessed Jun. 25, 2021] Available at: https://graphics.reuters.com/GLOBAL-ENVIRONMENT/SAND/ygdpzekyavw/.

[3] X. Lai, D. Shankman, C. Huber, H. Yesou, Q. Huang, and J. Jiang, "Sand mining and increasing Poyang Lake's discharge ability: A reassessment of causes for lake decline in China," *J. Hydrol.*, vol. 519, no. PB, pp. 1698–1706, 2014, doi: 10.1016/j.jhydrol.2014.09.058.

[4] W. Kozioł and I. Baic, "Kruszywa naturalne w Polsce - aktualny stan i przyszłość.," *Przegląd Górniczy*, vol. 11 (1152), pp. 1–8, 2018.

[5] M. Guan, W. Liu, M. Lai, H. Du, J. Cui, and Y. Gan, "Seismic behaviour of innovative composite walls with high-strength manufactured sand concrete," *Eng. Struct.*, vol. 195, pp. 182–199, Sep. 2019, doi: 10.1016/j.engstruct.2019.05.096.

[6] T. I. Nelson and W. P. Bolen, "Construction Aggregates," *Min. Eng.*, vol. 60, pp. 25–26, 2008.

[7] M. Edwards, "Different Types of Aggregate Types of Concrete Each Should be Used for." [Online] 2021 [Accessed Jun. 24, 2021] Available at: https://www.diydoctor.org.uk/projects/types-of-aggregate.htm.

[8] A. Książek, "Performance properties of aggregates and their influence on concrete quality," *Surowce i Masz. Bud.*, vol. 4, pp. 70–75, 2017.

[9] P. Mehta, "Greening of the Concrete Industry for Sustainable Development," *Concr. Int.*, vol. 24 (7), pp. 23–28, 2002.

[10] N. A. Lloyd and B. V. Rangan, "Geopolymer concrete with fly ash," *2nd Int. Conf. Sustain. Constr. Mater. Technol.*, vol. 7, pp. 1493–1504, 2010.

[11] "Cement production global 2020." [Online] 2020 [Accessed May 31, 2021] Available at: https://www.statista.com/statistics/1087115/global-cement-production-volume/.

[12] "Concrete and the carbon challenge." [Online] 2020 [Accessed May 31, 2021] Available at: https://www.reuters.com/article/us-global-infrastructure-breakingviews-idUSKBN2A13KL.

[13] M. Król and T. Z. Błaszczyński, "Geopolimery w budownictwie," *Izolacje*, vol. 5, pp. 38–47, 2013.

[14] S. Alehyen, M. Achouri, and M. Taibi, "Characterization, microstructure and properties of fly ash-based geopolymer," *J. Mater. Environ. Sci.*, vol. 8, no. 5, pp. 1783–1796, 2017.

[15] B. Katarzyna *et al.*, "The fabrication of geopolymer foam composites incorporating coke dust waste," *Processes*, vol. 8, no. 9, 2020, doi: 10.3390/pr8091052.

[16] B. Kozub, P. Bazan, D. Mierzwiński, and K. Korniejenko, "Fly-ash-based geopolymers reinforced by melamine fibers," *Materials (Basel).*, vol. 14, no. 2, pp. 1–13, 2021, doi: 10.3390/ma14020400.

[17] Z. Zhang, J. L. Provis, A. Reid, and H. Wang, "Geopolymer foam concrete: An emerging material for sustainable construction," *Constr. Build. Mater.*, vol. 56, pp. 113–127, 2014, doi: 10.1016/j.conbuildmat.2014.01.081.

[18] V. M. Malhotra, "High-Performance High-Volume Fly Concrete," *Concr. Int.*, vol. 24 (7), pp. 30–34.

[19] V. G. Papadikis, "Effect of Fly Ash on Portland Cement Systems, Part I: Low-Calcium Fly Ash,"

Cem. Concr. Res., vol. 29 (11), pp. 1727–1736, 1999.
[20] J. Mikuła and M. Łach, "Geopolimery – nowa przyjazna środowisku alternatywa dla betonna bazie cementu portlandzkiego," in *Rozwiązania proekologiczne w zakresie produkcji. Nowoczesne materiały kompozytowe przyjazne środowisku*, J. Mikuła, Ed. Kraków: Wydawnictwo Politechniki Krakowskiej, 2014, pp. 13–32.
[21] A. Fernández-Jiménez, A Palomo and C. Lopez-Hombrados, "Engineering properties of alkali-activated fly ash concrete," *ACI Mater. J.*, vol. 103(2), pp. 106–112, 2006.
[22] B. V. Hardjito, D., Wallah, S.E., Sumajouw, D.M.J. and Rangan, "On the development of fly ash-based geopolymer concrete," *ACI Mater. J.*, vol. 101(6), pp. 467–472, 2004.
[23] B. Joseph and G. Mathew, "Influence of aggregate content on the behavior of fly ash based geopolymer concrete," *Sci. Iran.*, vol. 19, no. 5, pp. 1188–1194, 2012, doi: 10.1016/j.scient.2012.07.006.
[24] J. Davidovits, *Geopolymer Chemistry and Applications. 5-th edition*, no. April. France: Institut Géopolymère, 2020.
[25] K. Culkova, S. Khouri, M. Straka, A. Rosova, "Ecological and Economic Saving of Fly Ash Using as Geopolymer," *Annual Set The Environment Protection,* vol. 20, pp.73-88, 2018.
[26] A. O. Purdon, "The action of alkalis on blast furnace slag," *J. la Société des Ind. Chim.*, vol. 59, pp. 191–202, 1940.
[27] J. Davidovits, "30 Years of Successes and Failures in Geopolymer Applications. Market Trends and Potential Breakthroughs," in *Geopolymer 2002 Conference*, 2002, pp. 28–29.
[28] M. Soutsosa, A. P. Boyle, R. Vinai, A. Hadjierakleous, and S. J. Barnett, "Factors influencing the compressive strength of fly ash based geopolymers," *Constr. Build. Mater.*, vol. 110, pp. 355–368, 2016.
[29] J. Li *et al.*, "Synthesis of electrolytic manganese residue-fly ash based geopolymers with high compressive strength," *Constr. Build. Mater.*, vol. 248, p. 118489, 2020.
[30] A. Nazari, H. Khanmohammadi, M. Amini, H. Hajiallahyari, and A. Rahimi, "Production geopolymers by Portland cement: Designing the main parameters' effects on compressive strength by Taguchi method," *Mater. Des.*, vol. 41, pp. 43–49, 2012.
[31] M. A. M. Ariffin, M. A. R. Bhutta, M. W. Hussin, M. M. Tahir, and N. Aziah, "Sulfuric acid resistance of blended ash geopolymer concrete," *Constr. Build. Mater.*, vol. 43, pp. 80–86, 2013.
[32] A. Mehta and R. Siddique, "Sulfuric acid resistance of fly ash based geopolymer concrete," *Constr. Build. Mater.*, vol. 146, pp. 136–143, 2017.
[33] W. Zhang, X. Yao, T. Yang, C. Liu, and Z. Zhang, "Increasing mechanical strength and acid resistance of geopolymers by incorporating different siliceous materials," *Constr. Build. Mater.*, vol. 175, pp. 411–421, 2018.
[34] A. Özcan and M. B. Karakoç, "Evaluation of sulfate and salt resistance of ferrochrome slag and blast furnace slag-based geopolymer concretes," *Struct. Concr.*, vol. 20 (5), pp. 1607–1621, 2019.
[35] Z. Zhang, J. L. Provis, A. Reid, and H. Wang, "Mechanical, thermal insulation, thermal resistance and acoustic absorption properties of geopolymer foam concrete," *Cem. Concr. Compos.*, vol. 62, pp. 97–105, 2015.
[36] S. Luhar, S. Chaudhary, and I. Luhar, "Thermal resistance of fly ash based rubberized geopolymer concrete," *J. Build. Eng.*, vol. 19, pp. 420–428, 2018.
[37] H. Cheng-Yong, L. Yun-Ming, M. Abdullah, and K. Hussin, "Thermal Resistance Variations of Fly Ash Geopolymers: Foaming Responses," *Sci. Rep.*, vol. 7, p. 45355, 2017.
[38] R. Zhao *et al.*, "Freeze-thaw resistance of Class F fly ash-based geopolymer concrete," *Constr. Build. Mater.*, vol. 222, pp. 474–483, 2019.
[39] F. N. Degirmenci, "Freeze-Thaw and fire resistance of geopolymer mortar based on natural and waste pozzolans," *Ceram. - Silikaty*, vol. 62, no. 1, pp. 41–49, 2018, doi: 10.13168/cs.2017.0043.

[40] Z. Xu, Z. Huang, C. Liu, X. Deng, D. Hui, and S. Deng, "Research progress on mechanical properties of geopolymer recycled aggregate concrete," *Reviews on Advanced Materials Science*, vol. 60, no. 1. Walter de Gruyter GmbH, pp. 158–172, Jan. 01, 2021, doi: 10.1515/rams-2021-0021.

[41] P. Duxson, A. Fernández-Jiménez, J. L. Provis, G. C. Lukey, A. Palomo, and J. S. J. Van Deventer, "Geopolymer technology: The current state of the art," *J. Mater. Sci.*, vol. 42, no. 9, pp. 2917–2933, 2007, doi: 10.1007/s10853-006-0637-z.

[42] J. L. Provis, G. C. Lukey, and J. S. J. Van Deventer, "Do geopolymers actually contain nanocrystalline zeolites? a reexamination of existing results," *Chem. Mater.*, vol. 17, no. 12, pp. 3075–3085, 2005, doi: 10.1021/cm050230i.

IOP Conf. Series: Materials Science and Engineering 1190 (2021) 012014 doi:10.1088/1757-899X/1190/1/012014

Kinematic analysis of a hyper-redundant robot with application in vertical farming

Ciprian Lapusan, Ciprian Rad, Olimpiu Hancu

Technical University of Cluj-Napoca, Str. Memorandumului, No. 28, Cluj-Napoca, Romania

ciprian.rad@mdm.utcluj.ro

Abstract. The paper presents the forward kinematic analysis for a new hyper redundant robot with applications in vertical farming inspections and harvesting tasks. The high dexterity and the ability to adapt to various unstructured environments makes this type of robot a good solution for this kind of applications. The robot consists of seven 3-RPS (Revolute-Prismatic-Spherical) parallel modules resulting in a 21 DOF structure. The forward kinematic analysis of one module is presented first and then for the entire structure in detail. The obtained equations are then implemented in MATLAB and validated using a virtual robot model.

1. Introduction

It is estimated that about 80% of the global population will live in urban areas by 2050 [1] (more than 7 billion people). This trend will create a real challenge in producing and delivering the necessary general goods (food in particular) in these areas [2]. One proposed solution to tackle this challenge is vertical farming [1]. The concepts focus in creating dedicated buildings, in urban areas, that produce food in a controlled environment with increased efficiency [3]. One key aspect of this process is the increase automation that is implemented in order to control the plant growth, health and nutrients needs. The automation process assists the management of plants growth during the entire process starting from seeding, growing and harvesting the plants.

The assist of robotics in implementing the necessary automation systems in vertical farming is more and more present, as can be seen in the literature [4]. One area that robotics aid is the development of dedicated robotized structures for harvesting and inspection for example [5, 6]. Several structures that implement hyper redundant robots were developed with dedicated application in this area [7, 8]. The advantages of these robots (ability to operate in crowded and difficult workspaces, high robustness and the capability to perform complex movements) allow development of complex application tasks specific to the vertical farming operations.

Beside the advantages of these robots, one challenge in implementing these structures is the more complex kinematics comparing with the traditional robots [9, 10]. In implementing the direct and invers kinematics a mapping between actuator space, each module operational space and end effector (robot) operational space must be performed (figure 2). This increases the complexity of the equations and at the same time the necessary hardware performance that could compute in real time these equations are higher [11]. The path planning of these robots is also more complex to implement. For a given pose to the end effect the robot can have an infinite number of values for the actuators that allows the robot to reach that pose.

In this paper a new hyper redundant structure is proposed for use in the vertical farming inspection and harvesting tasks. The proposed structure is made out from 7 modules that give the robot a total of 21 Degree-of-Freedom (DOF). The forward kinematics of the structure is analysed, and the performance evaluated for the use in the vertical farming process for the proposed tasks.

2. Robot kinematic analysis

The developed robot has a modular structure and is made of 7 modules (figure 1a, b). Each module consists of a 3-RPS parallel structure (figure 1c) and has seven mobile elements, six kinematic joints of class 5 (three revolute joints and three prismatic joints) and three kinematic joints of class 3 (three spherical joints). The number of DOFs for one module is 3, as can be seen from mobility equation:

$$M = 6n - 5C_5 - 4C_4 - 3C_3 - 2C_2 - C_1 = 6\cdot 7 - 5\cdot 6 - 3\cdot 3 = 3. \quad (1)$$

where:

M = Degrees-of-Freedom;
n = number of mobile elements;
C_k = number of kinematic joints of class k $(k = 1 \div 5)$

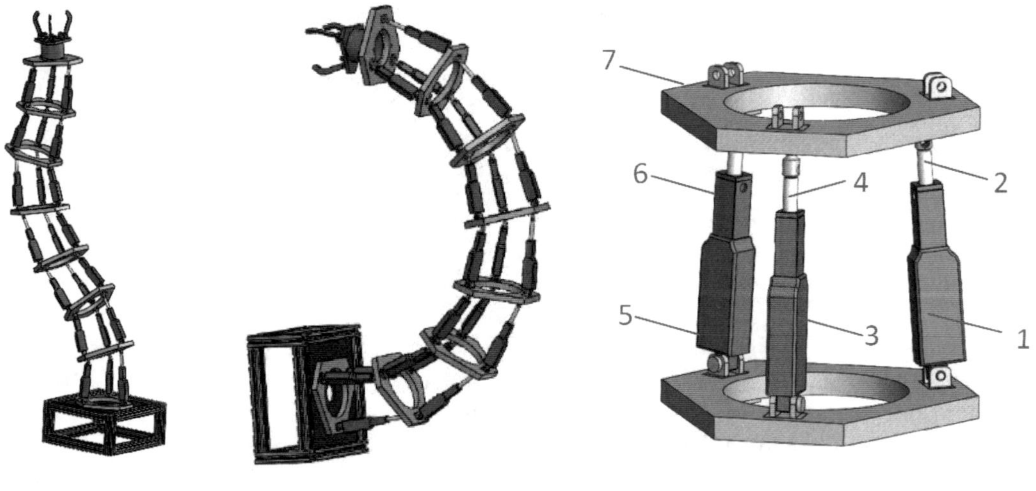

a) b) c)

Figure 1. CAD model of the new hyper redundant robot.

The actuation of the robot is done by the linear DC actuator *RA-MINI Plast* that produce a linear force of 70[N] at a speed of 3.6 [mm/s]. By multiplying the individual DOF of each module with the number of modules n_m=7 results a number of 21 DOF for the entire structure. In design process of the structure one imposed parameter was the ability for the end-effector to have amplitude in orientation of ± 180[deg] resulting the minimum number of the modules that form the structure (in this case n_m=7).

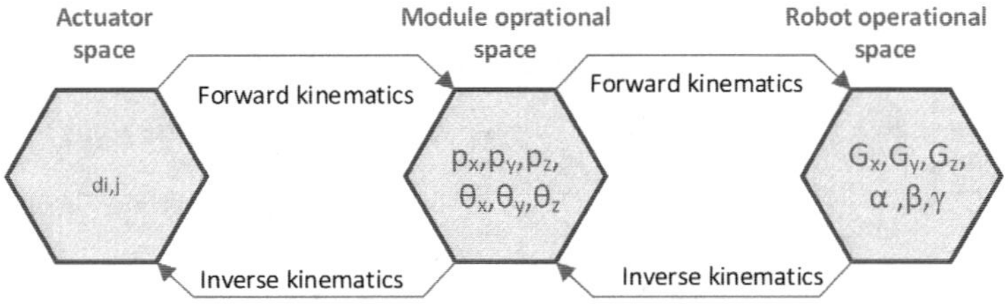

Figure 2. Kinematic mapping between robot spaces.

The kinematic of the robot can be divided in two steps that focus on the mapping process between actuator space and module operational space and the mapping between module operational space to the robot operational space (figure 2). In the first phase, based on the length of the actuators, for each module the orientation and position of each upper platform is calculated. These data are then used to calculate the position and orientation of the end effector in task space in the second phase. The propose kinematic analysis could be also extended to robots with similar architecture but with different number of modules.

2.1 Forward kinematic analysis of one module

Figure 3 presents the parallel mechanism used as a module for the proposed robot. It has three DOFs, two in orientation and one in cartesian position [12, 13]. The module consists of 3-RPS (revolute joint-prismatic joint-spherical joint) kinematic chains connected in parallel between a lower base {A} and an upper base {B}. Revolute joints R are connected at lower base at 120 degrees resulting the equilateral triangle $\Delta A_1A_2A_3$ while the spherical joints S are connected at upper base at 120 degrees resulting the equilateral triangle $\Delta B_1B_2B_3$. The three prismatic joints P are used as the inputs to the module to control the three DOFs of the upper base (two rotations about X and Y-axes and one translation about Z axis).

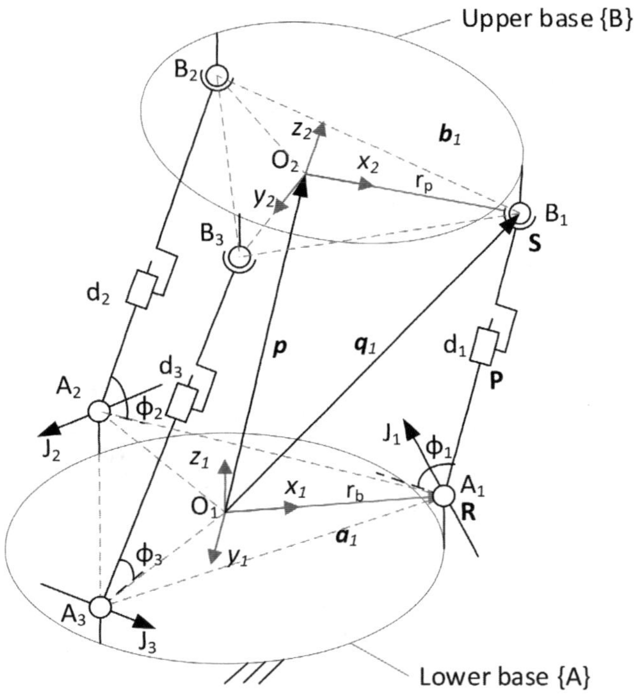

Figure 3. Kinematic diagram for a robot module

IOP Conf. Series: Materials Science and Engineering 1190 (2021) 012014 doi:10.1088/1757-899X/1190/1/012014

Each revolute joint $A_i (i = 1 \div 3)$ can rotate about its axis of rotation $J_i (i = 1 \div 3)$ in the direction described by the unit vector $\mathbf{j}_i = [j_x, j_y, j_z]^T$. Revolute joint $A_i (i = 1 \div 3)$ is described with respect to coordinate system *A(X,Y,Z)* by vector $\mathbf{a}_i = [a_x, a_y, a_z]^T$ while spherical joint $B_i (i = 1 \div 3)$ with respect to coordinate system *B(x,y,z)* by vector $\mathbf{b}_i = [b_x, b_y, b_z]^T$. Based on the previous observations, the coordinates of A_i, B_i and J_i are given as follows:

$$\begin{cases} \mathbf{a}_1 = \left[r_b, 0, 0\right]^T \\ \mathbf{a}_2 = \left[-\frac{1}{2} r_b, \frac{\sqrt{3}}{2} r_b, 0\right]^T \\ \mathbf{a}_3 = \left[-\frac{1}{2} r_b, -\frac{\sqrt{3}}{2} r_b, 0\right]^T \end{cases}, \quad \begin{cases} {}^B\mathbf{b}_1 = \left[r_p, 0, 0\right]^T \\ {}^B\mathbf{b}_2 = \left[-\frac{1}{2} r_p, \frac{\sqrt{3}}{2} r_p, 0\right]^T \\ {}^B\mathbf{b}_3 = \left[-\frac{1}{2} r_p, -\frac{\sqrt{3}}{2} r_p, 0\right]^T \end{cases}, \quad \begin{cases} \mathbf{j}_1 = \left[0, 1, 0\right]^T \\ \mathbf{j}_2 = \left[\frac{\sqrt{3}}{2}, \frac{1}{2}, 0\right]^T \\ \mathbf{j}_3 = \left[-\frac{\sqrt{3}}{2}, \frac{1}{2}, 0\right]^T \end{cases}. \tag{2}$$

Forward kinematics describes the mathematical relationship between prismatic joints P and position and orientation of the upper base {B} with respect to the lower base {A} fixed coordinate frame. That means, if prismatic joints lengths d_1, d_2, d_3 are provided, one can determine position vector $\mathbf{p} = [p_x, p_y, p_z]^T$ and rotation angles $\theta_x, \theta_y, \theta_z$ defined with respect to three successive rotation about fixed *X*, *Y*, *Z*-axes.

Let ϕ_1, ϕ_2, ϕ_3 be the angles between the extensible limbs and lower base {A} (see Figure 3). If these angles are known, the position vector $\mathbf{q}_i \left(i = 1 \div 3\right)$ of spherical joints $B_i (i = 1 \div 3)$ with respect to the lower base coordinate system *A(X,Y,Z)* can be written as follows:

$$\mathbf{q}_1 = \begin{bmatrix} r_b - d_1 \cos(\phi_1) \\ 0 \\ d_1 \sin(\phi_1) \end{bmatrix}, \mathbf{q}_2 = \begin{bmatrix} -\frac{1}{2}(r_b - d_2 \cos(\phi_2)) \\ \frac{\sqrt{3}}{2}(r_b - d_2 \cos(\phi_2)) \\ d_2 \sin(\phi_2) \end{bmatrix}, \mathbf{q}_3 = \begin{bmatrix} -\frac{1}{2}(r_b - d_3 \cos(\phi_3)) \\ -\frac{\sqrt{3}}{2}(r_b - d_3 \cos(\phi_3)) \\ d_3 \sin(\phi_3) \end{bmatrix}. \tag{3}$$

Because spherical joints S are placed in the vertexes of triangle $\Delta B_1B_2B_3$, the distance between two consecutive sides of the equilateral triangle is $|B_1B_2| = |B_2B_3| = |B_3B_1| = \sqrt{3} r_p$. Substituting (3) in the previous mathematical relationship we get a system of three loop-closure nonlinear equations (4) with three unknown variables (ϕ_1, ϕ_2, ϕ_3) and three known variables (d_1, d_2, d_3) where $\phi_i (i = 1 \div 3) \in [0, \pi]$.

$$\begin{cases} f_1 = d_1^2 + d_2^2 + 3r_b^2 - 3r_p^2 + d_1 d_2 \cos(\phi_1)\cos(\phi_2) - 2d_1 d_2 \sin(\phi_1)\sin(\phi_2) \\ \qquad - 3r_b d_1 \cos(\phi_1) - 3r_b d_2 \cos(\phi_2) \\ f_2 = d_2^2 + d_3^2 + 3r_b^2 - 3r_p^2 + d_2 d_3 \cos(\phi_2)\cos(\phi_3) - 2d_2 d_3 \sin(\phi_2)\sin(\phi_3) \\ \qquad - 3r_b d_2 \cos(\phi_2) - 3r_b d_3 \cos(\phi_3) \\ f_3 = d_3^2 + d_1^2 + 3r_b^2 - 3r_p^2 + d_3 d_1 \cos(\phi_3)\cos(\phi_1) - 2d_3 d_1 \sin(\phi_3)\sin(\phi_1) \\ \qquad - 3r_b d_3 \cos(\phi_3) - 3r_b d_1 \cos(\phi_1) \end{cases} . \tag{4}$$

There are various options to solve this system of equations such as numerical methods for nonlinear system of equations, computer algebra with software like Maple or MATLAB or algorithmic elimination methods such as Sylvester's Dialytic elimination method [12-14].

In this paper we will use the Newton-Kantorovich numerical method for nonlinear system of equations to find the numerical values of $\phi_i\,(i = 1 \div 3)$ when $d_i\,(i = 1 \div 3)$ are known. Once the angles are known, the components of position vector $\mathbf{q}_i\,(i = 1 \div 3)$ can be computed from (3). Next, the cartesian position of the upper base {B} can be determined by calculating the position vector $\mathbf{p}$ with the following equation:

$$\mathbf{p} = \frac{1}{3}(\mathbf{q}_1 + \mathbf{q}_2 + \mathbf{q}_3). \tag{5}$$

Regarding the orientation of the upper platform one can note that the position vector $\mathbf{q}_i\,(i = 1 \div 3)$ of spherical joints $B_i\,(i = 1 \div 3)$ can be expressed also as:

$$\mathbf{q}_i = \mathbf{p} + {}^{A}R_{B}\,{}^{B}\mathbf{b}_i . \tag{6}$$

where the rotation matrix ${}^{A}R_{B}$ of upper base {B} with respect to lower base {A} can be written as follows:

$${}^{A}R_{B} = \begin{pmatrix} u_x & v_x & w_x \\ u_y & v_y & w_y \\ u_z & v_z & w_z \end{pmatrix}. \tag{7}$$

Direction cosines $u_x, u_y, u_z, v_x, v_y, v_z$ are solved from (3) and (6) while w_x, w_y, w_z from the orthogonality property of the rotation matrix [14]. Rotation matrix can be defined also using the following representation [15, 16]:

$${}^{A}_{B}R_{XYZ}(\theta_x,\theta_y,\theta_z) = \begin{bmatrix} c\theta_z c\theta_y & c\theta_z s\theta_y s\theta_x - s\theta_z c\theta_x & c\theta_z s\theta_y c\theta_x + s\theta_z s\theta_x \\ s\theta_z c\theta_y & s\theta_z s\theta_y s\theta_x + c\theta_z c\theta_x & s\theta_z s\theta_y c\theta_x - c\theta_z s\theta_x \\ -s\theta_y & c\theta_y s\theta_x & c\theta_y c\theta_x \end{bmatrix}. \tag{8}$$

where: s – sine and c - cosine

Finally, by comparing (7) and (8), the three rotation angles $\theta_x, \theta_y, \theta_z$ defined with respect to fixed *X*, *Y*, *Z*-axes can be computed using the following [15, 16]:

$$\begin{cases} \theta_x = A\tan 2(v_z / \cos\theta_y, w_z / \cos\theta_y) \\ \theta_y = \sin^{-1}(-u_z) \\ \theta_z = A\tan 2(u_y / \cos\theta_y, u_x / \cos\theta_y) \end{cases} . \quad (9)$$

where: θ_x, θ_z will always be in the range [−π to +π] and θ_y will be between [−π/2 and +π/2].

2.2 Kinematic analysis of the robot

Equations obtained in Section 2.1 allows to obtain the relative orientation (angles $\theta_x, \theta_y, \theta_z$) and position (vector $\mathbf{p} = [p_x, p_y, p_z]^T$) between two consecutive platforms. Based on these parameters the pose of the end effector in respect with the global cartesian system ($O_0x_0y_0z_0$) can be calculated. For this, the transformation matrix ${}^{7}_{0}T$ is calculated using an iterative process, where the transformation matrix is obtained using (10) based on the transformation matrix associated with the previous platform ${}^{i-1}_{0}T$ that is multiplied with the current module transformation matrix ${}^{i}_{i-1}T$ (where i=1..7).

$${}^{i}_{0}T = {}^{i-1}_{0}T \, {}^{i}_{i-1}T \qquad \{i=1..7\}. \quad (10)$$

The general form of the transformation matrix of one module ${}^{i}_{i-1}T$ is presented in (11). To obtain this matrix, the orientation ${}^{i}_{i-1}R$ and position vector ${}^{i}_{i-1}P$ obtained from the parameters $\theta_x, \theta_y, \theta_z$ $, p_x, p_y, p_z$ are used. The rotation matrix is obtained using three successive rotations about the axis *Z*, *Y*, X.

$${}^{i}_{i-1}T = \begin{bmatrix} {}^{i}_{i-1}R & {}^{i}_{i-1}P \\ 0 \quad 0 \quad 0 & 1 \end{bmatrix}. \quad (11)$$

Using the ${}^{7}_{0}T$ transformation matrix the pose parameters for the end effector (gripper), could be extracted.

3. Numerical results

The validation of the obtained equations was performed using a virtual model of the robot that was imported from SolidWorks in MATLAB/Simulink. The robot model is developed using Simscape library and the CAD model from SolidWorks. The Simulink model for a module and the virtual representation of the robot are presented in figure 4.

The equations that defined the forward kinematic of the robot were implemented using an M-Function block. The inputs in the block are the values provided by the actuator sensors that define the strokes of the prismatic joints. The outputs are the calculated value for the end effector pose. The parameters for the structure are r_b = 54 [mm], r_p = 47 [mm] and the actuator length vary from 98.66 [mm] to 128,66 [mm].

Using this setup, several numerical simulations were performed. The aim was to create different work scenarios for the robot and to evaluate the responses given by the Simscape model and the values calculated using the kinematic equations.

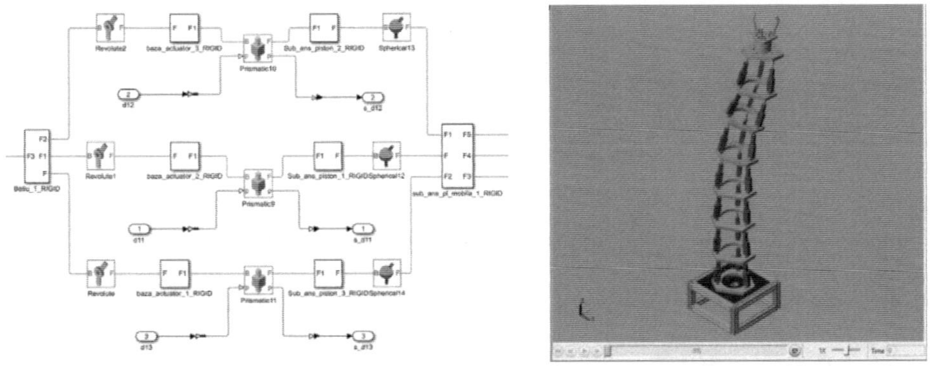

a) b)

Figure 4. Robot virtual model a) Simulink model of one module b) 3D representation of the robot in Mechanics Explorer

The obtained results are presented in figure 5. The signal measured from the end effector pose from virtual robot sensors (blue) are compared with the values calculated using the proposed algorithm (red -positions, green - angles). It can be observed that variation for the end effecter position and orientation matches the calculated results. A set of sine waves signals were used as inputs for the actuators, the amplitude of the signals was 10, 15 and 30 [mm] and a frequency of 0.1, 0.2 and 0.4 [Hz].

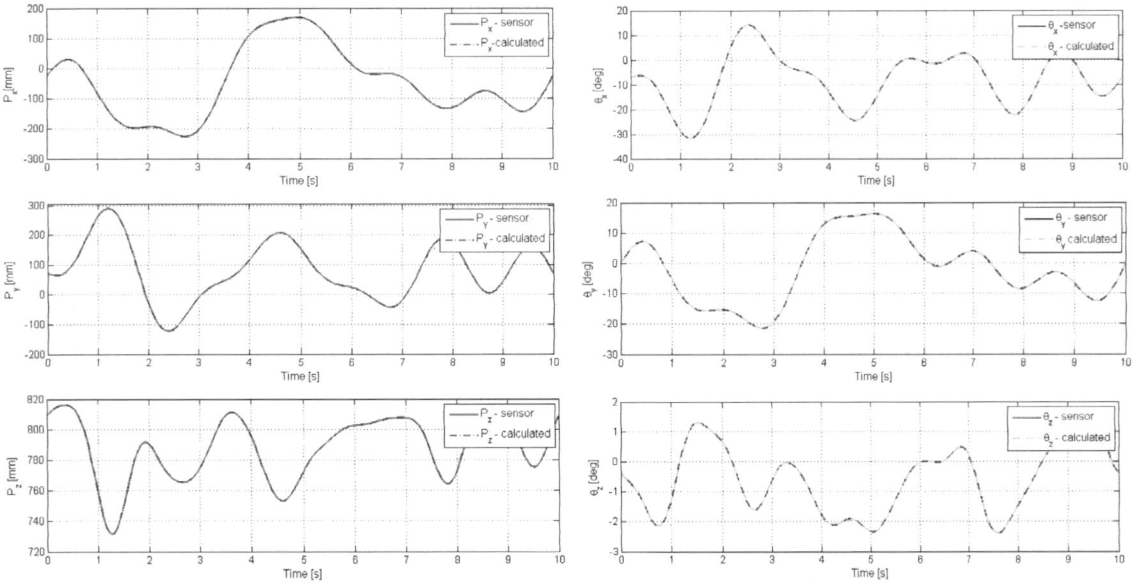

a) b)

Figure 5. Numerical results: a) end effector position b) end effector orientation

4. Conclusions

The paper presented the forward kinematic analysis for a new hyper-redundant robot that is used in applications specific to vertical farming. Due to the complex structure of the robot a hybrid method was used where both analytical and numerical methods were implemented in order to compute the end-effector pose. The obtained equations were implemented in MATLAB/Simulink and the numerical results were validated using a virtual model of the robot. The proposed method can also be used to for robot with similar architecture but with a different number of modules.

Acknowledgment(s)
„This work was supported by a grant of the Romanian Ministry of Education and Research, CCCDI - UEFISCDI, project number PN-III-P2-2.1-PED-2019-4939, within PNCDI III".
All authors contributed equally to this manuscript.

References

[1] F. Kalantari, O.M. Tahir, R.A. Joni,& E. Fatemi, "Opportunities and challenges in sustainability of vertical farming: A review," *Journal of Landscape Ecology*, vol. 11(1), pp. 35-60, 2018.
[2] D. Despommier, "Farming up the city: The rise of urban vertical farms," *Trends in Biotechnology*, vol. 31(7), pp. 388–389, 2013.
[3] K. Benke, B. Tomkins, "Future food-production systems: vertical farming and controlled-environment agriculture," *Sustainability: Science, Practice and Policy*, vol. 13(1), pp. 13-26, 2017.
[4] C. Gnauer, H. Pichler, M. Tauber, C. Schmittner, K. Christl, J. Knapitsch & M. Parapatits, "Towards a secure and self-adapting smart indoor farming framework," *e & i Elektrotechnik und Informationstechnik*, vol. 136(7), pp. 341-344, 2019.
[5] A. Krishnan & S. Swarna, "Robotics, IoT, and AI in the Automation of Agricultural Industry: A Review," *In 2020 IEEE Bangalore Humanitarian Technology Conference*, pp. 1-6, 2020.
[6] C.-R. Rad, O. Hancu, C. Lapusan, "Aspects regarding "soft" grasping in smart agricultural harvesting tasks," *Acta Technica Napocensis-series: Applied Mathematics, Mechanics, and Engineering*, vol. 63(4), pp. 389-394, 2020.
[7] S.C. Lauguico, R.S. Concepcion, D.D. Macasaet, J.D. Alejandrino, A.A. Bandala & E.P. Dadios, "Implementation of Inverse Kinematics for Crop-Harvesting Robotic Arm in Vertical Farming," *In 2019 IEEE International Conference on Cybernetics and Intelligent Systems (CIS) and IEEE Conference on Robotics, Automation and Mechatronics*, pp. 298-303, 2019.
[8] J. Baur, J. Pfaff, H. Ulbrich, T. Villgrattner, "Design and development of a redundant modular multipurpose agricultural manipulator," *Proceedings of International Conference on Advanced Intelligent Mechatronics (AIM)*, pp. 823-830, 2012.
[9] C. Lapusan, O. Hancu, C.-R., Rad, "Quaternion-based approach for solving the direct kinematics of a modular hyper redundant robot," *Acta Technica Napocensis-Series: Applied Mathematics, Mechanics, And Engineering*, Vol. 63(4), pp. 363-366, 2020.
[10] C. Lapusan, F. Fodor, A. Csiszar, A. Verl, C. Brisan, "Aspects concerning the modeling of robots with increased dexterity", *41st International Symposium on Robotics*, pp. 1-6, 2014.
[11] Z. C. Du, G. Y. Ouyang, J. Xue & Y. B. Yao, "A Review on Kinematic, Workspace, Trajectory Planning and Path Planning of Hyper-Redundant manipulators," *10th Institute of Electrical and Electronics Engineers International Conference on Cyber Technology in Automation, Control, and Intelligent Systems (CYBER)*, pp. 444-449, 2020.
[12] L.W. Tsai, "Robot Analysis: The Mechanics of Serial and Parallel Manipulator," *New York:John Wiley & Sons*, ISBN: 978-0-471-32593-2, 1999.
[13] C.-R. Rad, S.D. Stan, R. Balan, C. Lapusan, "Forward kinematics and workspace analysis of a 3-RPS medical parallel robot," *In 2010 IEEE International Conference on Automation, Quality and Testing, Robotics (AQTR)*, vol. 1, pp. 1-6, 2010.
[14] N.M. Rao, K.M Rao, "Dimensional synthesis of a spatial 3-RPS parallel manipulator for a prescribed range of motion of spherical joints," *Mechanism and Machine Theory*, vol. 44(2), pp. 477-486, 2009.
[15] J.J. Craig, "Introduction to Robotics: Mechanics and Control (3rd Edition)," *Pearson*, 2004.
[16] R. Desai, S. Muthuswamy, "A Forward, Inverse Kinematics and Workspace Analysis of 3RPS and 3RPS-R Parallel Manipulators," *Iran J Sci Technol Trans Mech Eng*, vol. 45, pp. 115–131, 2020.

Discrete element simulation of dry snow using the developed analytic bond model

Hassan Bahaloo, Johan Casselgren, Fredrik Forsberg, Mikael Sjödahl

Division of Experimental mechanics, Luleå University of Technology, Luleå, Sweden

hassan.bahaloohoreh@ltu.se

Abstract. Snow is a heterogenous, hot material which is constituted from ice particles. The bonding behavior of ice particles is an important parameter determining the macroscopic behavior of snow. Discrete Element Method (DEM) is usually used as a tool to model dry snow. The most important input data required into the DEM is bonding behavior of ice particles since ice particles can adhere to form bonds when they brought into contact. This study had two aims: first, an analytical formulation was derived to predict the bond diameter of ice-ice contacts as a function of time, compressive load, and strain rate. Using the previously published data for strain rate of ice, a solution method was developed. The results of bond diameter development with time were compared to experimental data and a good agreement was found. Second, a DEM for dry snow was developed and programmed in MATLAB and the developed bond model was employed in the simulation to study the deposition behavior of snow in a container under gravity acceleration. A specific beam element with implemented damage model was developed in implemented in the simulation using the bond data obtained from the analytical approach. The simulated parameters were macroscopic angle of repose, packing density, and surface conditions as a function of temperature and filling rate. The results showed that discrete element simulations were able to verify the existing published experimental data. Specifically, the simulation results showed that angle of repose of snow decreased rapidly with decreasing the temperature, the surface became very irregular due to the particles rotation and re-arrangement for lower falling speeds of particles, and density increased with depth of deposition. These findings were all matched with experimental observations.

1. Introduction

The DEM is one of the methods used in simulating granular materials with large number of particles [1–4]. DEM is also widely used in snow mechanic simulations [5,6], which is an open research field despite its long history [7–12].

Most particles, including snow, are naturally non-spherical; therefore, it is desired to simulate irregular particles in DEM. Generally, two avenues are paved to treat irregular particles in DEM; one is based on ellipsoids or some polygons and the other route uses agglomerates of bonded spheres/disks. For instance, snow particles can take a plethora of shapes [13], which can be modelled by including bonds between particles [5, 6]. If the second approach is chosen, bond modeling is required. To model bonded or cohesive granular materials, traditionally a compact spring model has been used [14–20]. Despite its wide application, the bending-shear coupling is neglected in most cases. A more sophisticated approach is to employ beam Finite Elements (FEs) to represent the bonds [21–23]. When employing a beam element, it is important to acknowledge that bonds can have a low length to diameter ratio, which makes the Timoshenko beam model a better solution than the Euler beam [22].

Bonds can vary with time, the so called "sintering", which has been studied extensively for metallurgical applications [24]. There has been experimental investigations into the sintering force of ice particles [25,26], and analytical modeling of diffusion sintering [27,28].

The aim of this study is to incorporate a model for short term sintering in DEM and to employ the code to simulate the effects of different parameters including friction coefficient, bonding strength, and deposition rate on the packing configuration of granular materials when short-term sintering is considered.

2. Contact law for non-bonded particles

The general nomenclature of the contact models for non-bonded and bonded particles are presented in Figure 1. In all the subsequent formulations, the left superscript indicates time or time step, left subscript indicates involved particle(s), right superscript indicates mathematical exponent of the variable, and right subscript indicates abbreviations like "r" for rotation or "s" for slip. Moreover, italic bold letters represent vectors and non-italic bold letters represent matrices.

In this study, the following damped contact laws are considered for the normal force, F_n, tangential force, F_s, and rolling moment, M_r, respectively [29–31],

$$ {}^t_I F_n = {}^t_{IJ}K_n |\min(0,\delta)|^{3/2} + \eta_n(\delta)\dot{\delta}|\min(0,\delta)| / |\delta| \tag{1} $$

$$ {}^t_I F_s = -\min\left({}^t_{IJ}K_s \int_0^{\Delta t_{cd}} {}^t_{IJ}\Delta V_s(\tau)d\tau - \eta_s {}^t_{IJ}\Delta V_s, \mu_{st} |{}^t_I F_n| \,\mathrm{sign}({}^t_{IJ}\Delta V_s) \right) \tag{2} $$

$$ {}^t_I M_r = -{}_I r\, {}^t_I F_s - {}^t_{IJ}K_r r\,\mathrm{sign}({}^t_{IJ}\Delta V_r) \int_0^{\Delta t_{cd}} {}^t_{IJ}\Delta V_r(\tau)d\tau \tag{3} $$

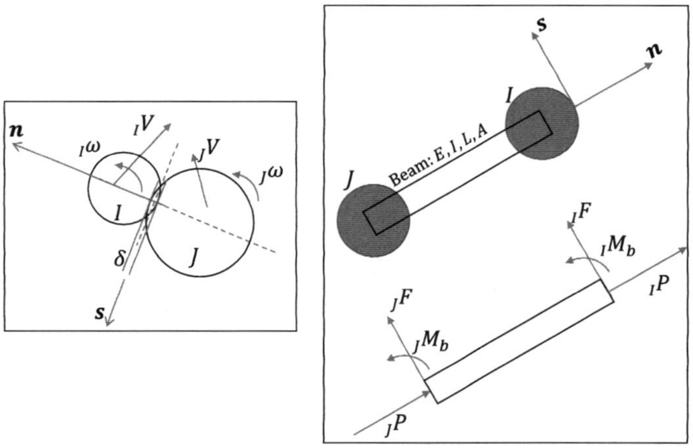

Figure 1. schematic of two particles, I and J indicating normal penetration, δ and linear and angular velocity vectors, V and ω, respectively. $\boldsymbol{n}-\boldsymbol{s}$ is the local normal-tangential coordinate system of the two non-bonded particle system (left), Beam element nomenclature used in this study for bonded particles (right)

where, "sign" is the signum function; r is the radius of the particle; η_n and η_s are the normal and tangential damping coefficient, respectively; K_n, K_s, and K_r are normal, tangential, and rolling stiffness values, respectively. The expressions for K_s, and K_r can be found in Marshall 2009 [32]; ΔV_s is the relative tangential velocity in direction s of the local $n - s$ coordinate system in Figure 1; μ_{st} is the static friction coefficient; δ and $\dot{\delta}$ are normal penetration and normal penetration speed, respectively. Finally, τ is a free integration variable.

Relative slip and angular velocities between particles I and J, at time t, are defined as, ${}_{IJ}^{t}\Delta V_s$, ${}_{IJ}^{t}\Delta V_r$, respectively, and implemented as,

$$ {}_{IJ}^{t}\Delta V_s = {}_{I}^{t}V_s - {}_{I}^{t}\omega_I r - \left({}_{J}^{t}V_s + {}_{J}^{t}\omega_J r\right), \tag{4}$$

$$ {}_{IJ}^{t}\Delta V_r = {}_{I}^{t}\omega - {}_{J}^{t}\omega \tag{5}$$

where, ${}_{I}^{t}\omega$ is the angular velocity of particle I at time t, ${}_{I}r$ is the radius of particle I, and unless otherwise stated, all units are in SI. ${}_{I}^{t}V_s$ is the velocity of particle J in direction s at time t. The damping coefficients η_s and η_n have significant effect on the behavior of granular material since they have an important role in dissipating collision energy. Consequently, it is of paramount importance to model a reasonable damping coefficient. A model followed from the work of Tsuji et al. [33] are employed in the current study in which the damping coefficient is considered as a fraction of the critical damping coefficient, $0.9\sqrt{2mK}$ or $0.1\sqrt{2mK}$, for velocities below and above the sticking velocity, respectively. The density of ice particles is considered a constant value of 917 $\mathrm{kg/m^3}$ since it does not change significantly for the temperature range considered [34].

The time step, dt, is chosen as, [21,35,36],

$$ dt = 2.9\alpha_t \left(\frac{m_{min}}{K_{max}}\right)^{1/2}, \tag{6}$$

where m_{min} is the mass of the smallest particle in the system; K_{max} is the maximum stiffness in the system, and $\alpha_t = 0.025$ or 0.2, for bonded and non-bonded cases, respectively.

The friction coefficient of ice, μ_{st} , is modelled as [37],

$$ \mu_{st} = \min\left[0.6, 0.017(T_f - T)(\Delta V_s)^{-1/2}\right] \tag{7}$$

where, ΔV_s is the sliding speed in $\mathrm{m/s}$, $T_f = 273.15\mathrm{K}$, and T is the absolute temperature.

3. Modelling of bonds

Ice particles form bonds easily and these bonds change by time. Estimating sticking velocity. The sticking velocity for small particles in which Van der Waals forces are dominant can be expressed as [38],

$$ V_{stick}^{*}(T = -10°\mathrm{C}) = \frac{73.44}{m^*}\left(\frac{R^{*4}\,\gamma^5}{E^{*2}}\right)^{1/6} \tag{8}$$

In Eq.(8), the surface tension $\gamma = 0.1\ \mathrm{N/m}$ [39], the equivalent mass $m^* = m_1 m_2/(m_1 + m_2)$, the equivalent radius, $R^* = r_1 r_1/(r_1 + r_2)$, and the equivalent stiffness $1/E^* = (1 - \nu_1^2)/E_1 + (1 - \nu_2^2)/E_2$ are utilized, where a constant Poisson's ratio, $\nu_1 = \nu_2 = 0.4$ is considered for all calculations.

Furthermore, the sticking behavior has been shown greatly affected by the temperature with increased temperature resulting in easier sticking [40,41]. In the absence of accurate experimental data, an exponential form for temperature dependence is assumed,

$$V_{stick}(T) = V_{stick}^{*} \exp\left(\frac{T+10}{25}\right) \tag{9}$$

where, T is temperature ($T < 0°\mathrm{C}$). Eq. (9) is assumed to be reasonably accurate within the temperature range $-35°\mathrm{C} < T < 0°\mathrm{C}$.

The time-dependent sintering force, $F_{sint}(t)$, can be estimated from the ultimate strength of the bond, S_{ut}, as,

$$F_{sint}(t) \approx \pi S_{ut}[a(t)]^2, \tag{10}$$

where $a(t)$ is the time-dependent bond radius.

The grain strength depends on particle and temperature, as [39],

$$S_{ut}[\mathrm{MPa}] = \left(0.05 + \frac{1.5}{\sqrt{2R^*}}\right) e^{-\frac{Q}{nR_g}\left(\frac{1}{263.15}-\frac{1}{T}\right)} \tag{11}$$

where, $Q = 31.8\frac{\mathrm{kCal}}{\mathrm{mol}}$ [42] and $R_g = 8.314\frac{\mathrm{kJ}}{\mathrm{kmolK}}$ are Boltzmann coefficient, and gas constant, respectively; T is absolute temperature; $n = 3$ [39] and R* is again the equivalent radius.

The only remaining parameter that needs to be determined in order to estimate the sintering force is the radius of the contact area, $a(t)$, as depicted in Figure 2.

The initial contact radius, $a_0 = a(t = 0, P = 0)$, when the load is zero [43] can be approximated from JKR theory as,

$$a_0 = \left(\frac{9\gamma\pi R^{*2}}{2E^*}\right)^{1/3}, \tag{12}$$

In the case of loaded particles with some external load, P, a solution for the viscoelastic case will be possible but it will depend on the loading and unloading history, which is not readily available in applications. As a result, for the initial time-step, the contact radius, $a = a_{P\neq0}(t = 0)$, of the loaded particles with load P are also assumed from JKR, as,

$$\frac{4E^*}{3R^*}a^3 = P + 3\pi\gamma R^* + [6\pi\gamma PR^* + (3\pi\gamma R^*)^2]^{1/2} \tag{13}$$

Solving for a, the initial sintering force when the particles are loaded with a load P, can simply be calculated from Eq.(10), Eq.(11), and Eq.(13), as,

$$F_{sint}(t = 0) = \pi e^{-\frac{Q}{nR_g}\left(\frac{1}{263.15}-\frac{1}{T}\right)} \left(0.05 + \frac{1.5}{\sqrt{2R^*}}\right) 10^6[a\,(t = 0)]^2 \tag{14}$$

Eq.(14) only provides the sintering force at time zero for the loaded particles. To obtain an estimate of the force development with time, the primary creep strain-stress-time relationship,

$$\varepsilon = 8.28 \times 10^{-6} \sigma^{2.43}\, t^{2/3} \quad , \tag{15}$$

is used [44], where, σ is applied contact pressure in MPa, and t is time in seconds.

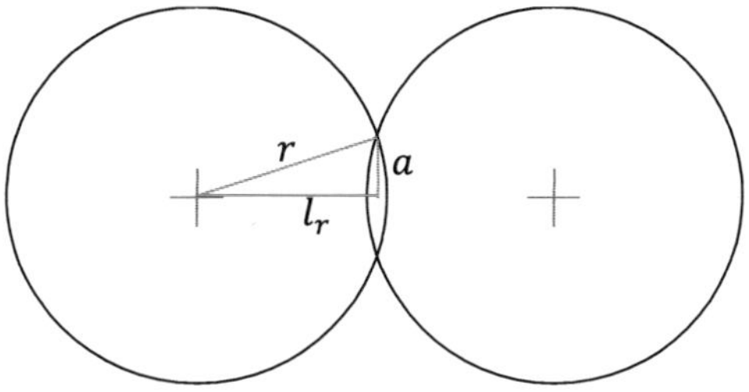

Figure 2. Nomenclature used in derivation of Eq.(16)

Assuming that the particle radius, r, as depicted in Figure 2, can be considered as a constant during a small time-step Δt, the change in the radius of the contact area can be written as,

$$\frac{\Delta a}{\Delta t} \approx \left(\frac{r^2 - a^2}{a} \right) \dot{\varepsilon} \tag{16}$$

where a is the bond radius at any time t, and $\dot{\varepsilon}$ is the time derivative of Eq.(15). In practice, this assumption means that negligible melting and diffusion mass transfer from the liquid layer to the bond neck area occurs. Therefore, with $a/r \ll 1$ for initial sintering stages, $r^2 - a^2 \approx r^2$, and,

$$a(t) = \left\{ [a_{P \neq 0}(t=0)]^2 + r^2 \left(1.656 \times 10^{-5} \sigma^{2.43}\, t^{\frac{2}{3}} \right) \right\}^{1/2} , \tag{17}$$

where $a_{P \neq 0}(t=0)$ is the initial radius obtained from Eq.(13). The maximum value of contact pressure, σ, used in Eq.(17), can be approximated as [45],

$$\sigma(t=0) = \frac{3P}{\pi [a(t=0)]^2} \tag{18}$$

Finally, the dependence of Young's modulus of the ice to temperature is in all calculations assumed as [46],

$$E[\text{GPa}] = 9.61(1 + 1.07 \times 10^{-3} T + 1.87 \times 10^{-6} T^3)^{-1} \tag{19}$$

where T is the temperature in centigrade, and E is Young's modulus.

Examples of radius of bond neck area and sintering force versus sintering time under the contact pressures 0.5MPa and 1MPa are presented in Figure 3. It is evident from this figure that with increasing contact pressure, which is equivalent to an increment of the applied load, P, the sintering force and the bond radius grow faster.

IOP Conf. Series: Materials Science and Engineering 1190 (2021) 012015 doi:10.1088/1757-899X/1190/1/012015

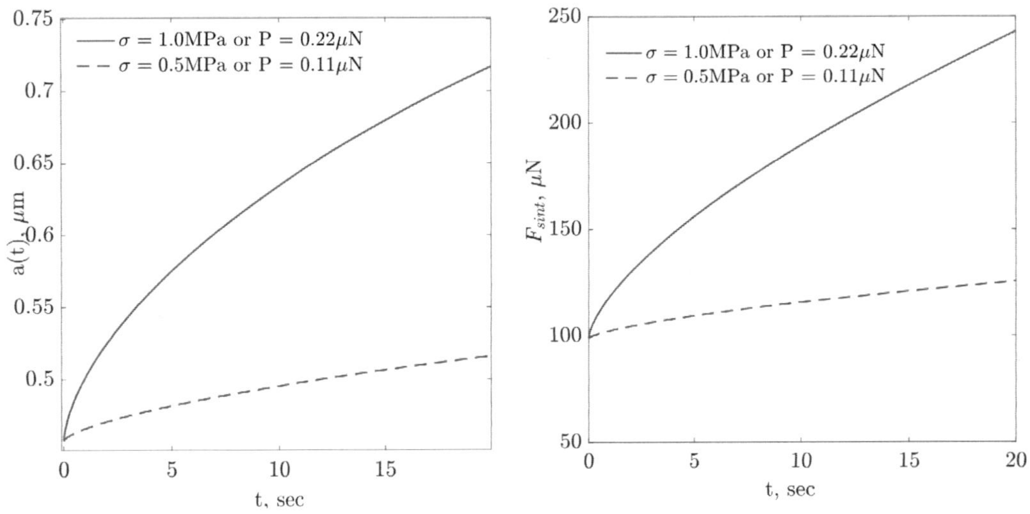

Figure 3. bond radius, $a(t)$, growth from Eq.(17) (left); equivalent sintering force, F_{sint}, (right); plots obtained for parameters $\gamma = 0.1\text{N/m}; E^* = 5.5\text{GPa}\ Eq.(19); r = 50\mu\text{m};\ T = -5°\text{C}$

For example, consider a particle of 100μm diameter. For the case with no external loading, the adhesive force using Eq.(14) will be 98μN, which is in agreement with published experimental results [25]. However, for the case where the particle diameter is 6mm, and there is no external loading, the sintering force predicted by Eq.(14) is 3mN, which is an order of magnitude less than that predicted by Szabo and Schneebeli [26]. As depicted in Figure 4(left), the sintering force increases both with time and with particle diameter. The rate of changes of the sintering force is also higher for larger particles, as seen in the figure. The increasing nature of sintering force, also fits well with published experimental results for $\sigma = 0.5$MPa [25]. The sintering force changes with the particle diameter in a nearly linear way if the maximum contact pressure is kept at a constant value, as depicted in Figure 4(right).

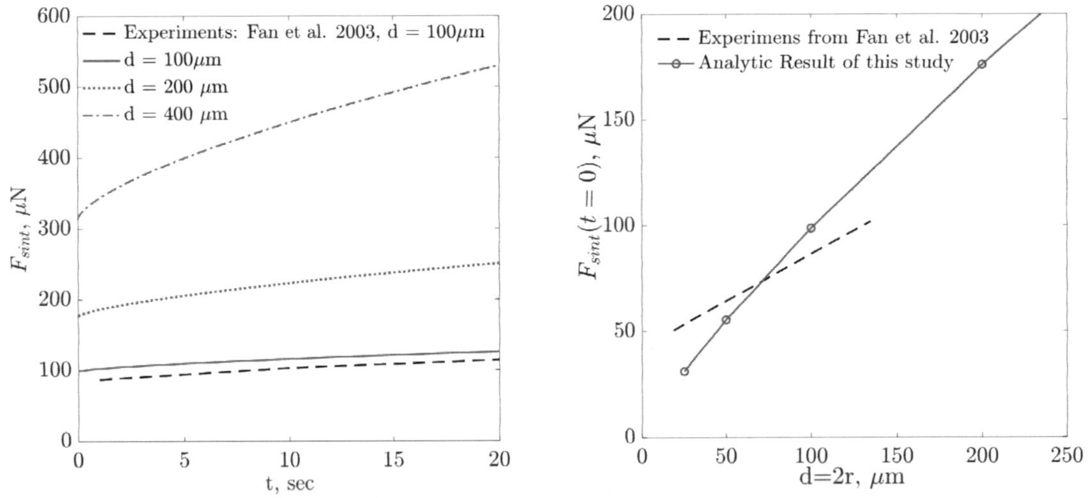

Figure 4. variation of sintering force, F_{sint}, with time and particle size (left); variation of initial sintering force with particle size (right); plots obtained from Eq.(14) for parameters $\gamma = 0.1\text{N/m}; E^* = 5.5\text{GPa}\ Eq.(19); r = 50\mu\text{m};\ T = -5°\text{C}; \sigma = 0.5\text{MPa}$

Timoshenko beam elements connecting center-center of bonded particles are used to model the bond element in DEM. With reference to the geometry in Figure 1(right), the incremental bond force is defined as,

$$^{t+dt}\boldsymbol{F}_{bond} = {}^{t}\boldsymbol{F}_{bond} + \mathbf{K}_{beam}\boldsymbol{V}_{bond}dt \tag{20}$$

where, $\mathbf{K}_{beam}$ is the ordinary Timoshenko beam stiffness element, which can be found in any finite element book [47], and $\boldsymbol{V}_{bond}$ is the beam nodal velocity vector of the beam. The Coulomb-Mohr criterion is used to identify the bond failure loads.

4. Implementation of the method

The algorithm is implemented in MATLAB 2020a® (Mathworks Inc., Natick, MA). The program have four main functions: 1) generating and filling the particles with location and velocities at top of the container to be thrown into the container, 2) finding possible collisions for each particle at each time step using a sub-cell collision detection method, 3) calculating forces and updating positions using the velocity Verlet method, as presented in Eq.(21) [48], and 4) postprocessing. Positions and velocities are updated as,

$$\begin{aligned} ^{t_i+0.5dt}\boldsymbol{v} &= {}^{t_i}\boldsymbol{v} + 0.5dt\,{}^{t_i}\boldsymbol{a} \\ ^{t_{i+1}}\boldsymbol{x} &= {}^{t_i}\boldsymbol{x} + {}^{t_i+0.5dt}\boldsymbol{v}dt \\ ^{t_{i+1}}\boldsymbol{v} &= {}^{t_i+0.5dt}\boldsymbol{v} + 0.5\,{}^{t_{i+1}}\boldsymbol{a}dt \end{aligned} \tag{21}$$

where, $\boldsymbol{x}$, $\boldsymbol{v}$, and $\boldsymbol{a}$ represent displacement, velocity, and acceleration vectors; left superscripts represent the time, and dt is the marching time from Eq.(6). Similar formulas are used for angular velocities and rotations. Ice particles are generated using a beta-distribution, with the possibility to bond the particles together in the beginning, before inserting them into the container area. At each time step, forces are calculated, and bonds and sticking of particles are checked. Accelerations are obtained simply from forces, and integrated twice to obtain the positions using Eq. (21). The required time for 1000 particles to be simulated for one-million time-steps on a Mac book Pro (CPU 2.2 GHz 6-Core Intel Core i7, 16 GB RAM 2400 MHz DDR4) was 75 minutes.

5. Numerical experiments

Numerical experiments are conducted by throwing both non-bonded and bonded particles into a container. The effects of friction coefficient, and deposition rate on the packing configuration in either case are investigated. Particles are thrown from the top of the container with an initial random velocity in horizontal direction, and without any initial velocity in vertical direction. Since falling velocity of snow particles is sharply decreasing with their mass or diameter [49] and since the velocity can be in the order of fractions of mm/s for micro-meter sized particles [50], heavy drag forces are introduced for the case of bonded-particles in the current numerical simulation to reduce the effective fall speed. For the non-bonded particles, the simulation time step is in the order of a few micro-seconds and for the bonded particles, it is in the order of a fraction of micro-seconds ($\approx 0.3 \times 10^{-6}$s). Five different numerical experiments are detailed and discussed in the following subsections.

A schematic of the numerical experiments of this study are presented in Figure 5, with the critical parameters and variables specified in the figure. In Sec.5.1. , numerical experiments on non-bonded, temperature-independent particles are detailed, while in Sec.5.2. numerical experiment on bonded, temperature-dependent particles are presented. A typical filling process for bonded ice particles, is also shown in Figure 5. Ice particles form in general single particles or 2-, and 3-particle clusters after collisions, like in reported experimental results [50]. After a few particles have filled the bottom, particles start to pile up in a sticky way. In all simulations, the friction coefficient between the ice particles and the container walls (including bottom wall) of the container is set to a constant value of

0.4. A typical plot of particle size distribution, when poly-dispersed particles are considered is presented in Figure 6 ($d = 250 + 500\beta(5,10)\mu$m), where β is the beta-distribution [51]. This size distribution is used in all the upcoming simulations considering poly-dispersed particles.

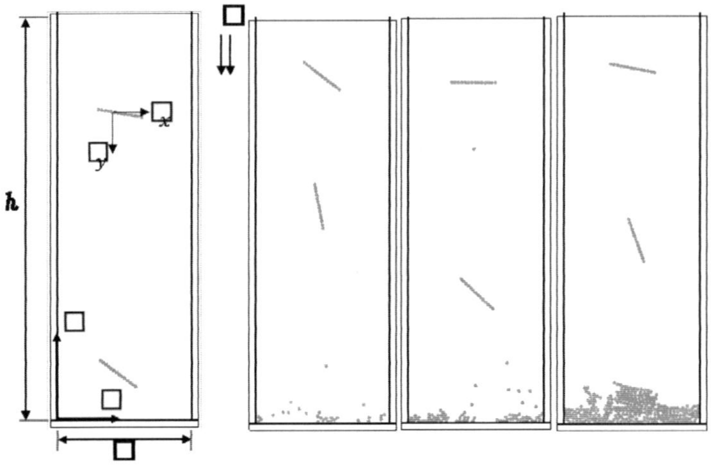

Figure 5. Schematic filling of the particles with essential parameters defined on the left image with timeframes increasing from left image to the right image

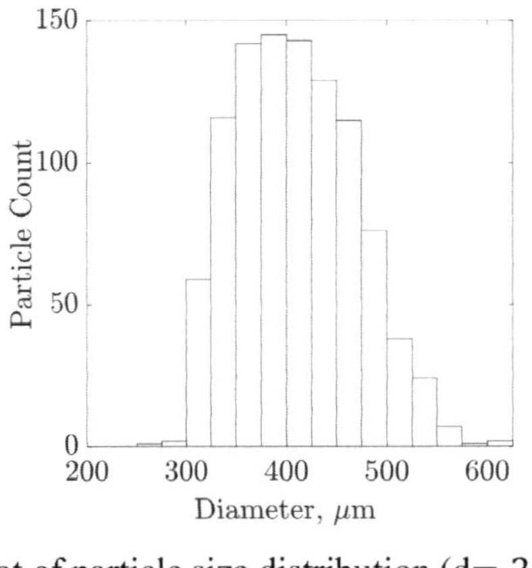

Figure 6. A typical plot of particle size distribution (d= $250 + 500\beta(5,10)\mu$m).

5.1. Non-bonded temperature-independent particles

The settings for non-bonded particles simulation are: time average time step, $dt = 1.8$ micro-seconds; container size, $w \times h = 20$mm × 60mm; particles are released with initial zero velocity; Particles are released in the form of nonbonded linear clusters as depicted in Figure 5 such that the height and horizontal location of the left particle are 54mm and 6.5mm, respectively; cluster length is 0.35 of the width of the container; the nonbonded cluster has a random inclination between 0 (horizontal configuration for the cluster) to 90° (rotated in clockwise direction); Number of particles, $N_{tot} = 500$; Young's modulus, $E = 4$GPa; Poisson's ratio, $\nu = 0.4$; Density, $\rho = 917$Kg/m^3. Gravitational acceleration and friction coefficient are varied for each experiment and expressed separately.

5.1.1. Mono-dispersed vs. poly-dispersed non-bonded, frictional particles. In these experiments, gravitational acceleration, $g = 9.81\ m/s^2$ and friction coefficient, $\mu = 0.9$ are considered and the filling process is according to Figure 5. For experiments with mono-dispersed particles, the corresponding diameter is set to 500 μm and for experiments with poly-dispersed particles the particle diameter varies according to Figure 7. As depicted in Figure 7, deposition of mono-dispersed particles shapes a "preferred"-hexagonal-order, which is the ordered-packing with a preferred highest density arrangement in two dimensional lattices. The compatibility mechanism that allowed the particles to achieve this hexagonal arrangement, is the movement of slip band (shear bands) within the material. Indeed, via geometrically necessary dislocations [52], the material adopted itself for strain gradients during the filling process to take the shape of the container. Voids are also formed as seen in Figure 7(left), which can alter the macroscopic behavior of the material significantly. Adding a perturbation in the size of particles to have nonuniform diameters results in an amorphous configuration as presented in Figure 7(right).

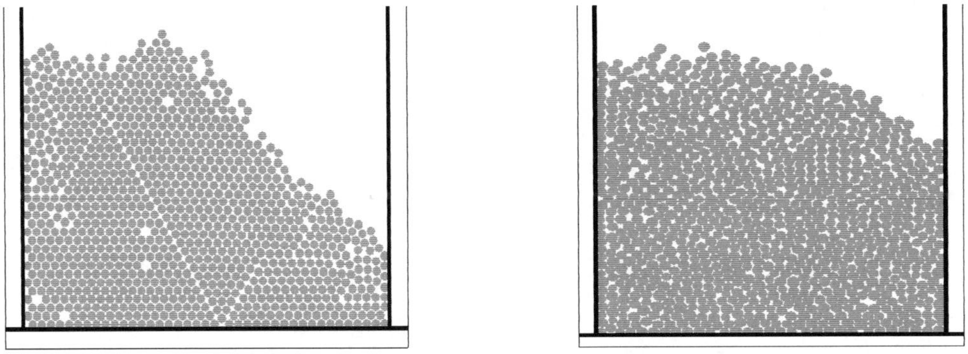

Figure 7. Mono-dispersed disks: no bonding is considered, and friction coefficient $\mu = 0.9$, slip bands and voids can be observed; particle diameter: 500 μm (left), poly-dispersed particles and friction coefficient $\mu = 0.9$(right)

5.1.2. Effects of variation in friction coefficient for non-bonded particles. Since the static friction coefficient of ice can reach to near unity [37, 53], and the related static friction force is the force responsible to allow particles to move and rearrange, an increasing friction coefficient must decrease the packing density, as observed in previous published works [54]. Moreover, it is known that the density of packing increases with the depth of deposition [55].

Here a set of simulation results for non-bonded poly-dispersed particles and various friction coefficients revealed that the density decreases with increasing friction coefficient and increasing with depth, as presented in Figure 8. Gravitational acceleration, $g = 9.81\ m/s^2$ is considered. The left image shows the simulation outcomes while the right image shows the average trends.

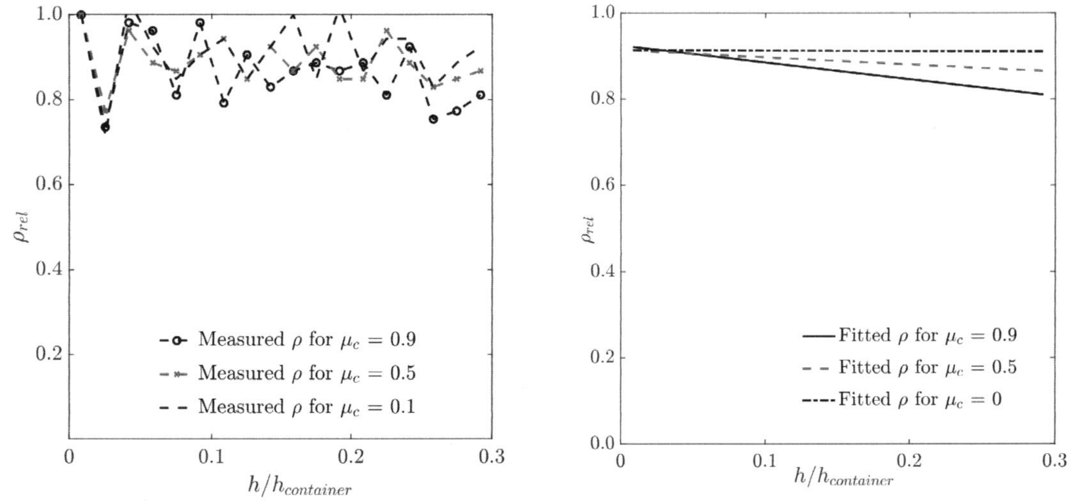

Figure 8. Simulated relative density: no bonding is considered (left) and linear fits on it (right) for different friction coefficients. No temperature dependence is considered on these plots.

5.1.3. Effect of deposition rate on the deposition density for non-bonded, frictional particles. To attain lower rate in the deposition process, it is possible to suspend the particles in a fluidic medium [56, 57]. In this case, due to the movement of particles in the fluid, several forces and moments can be exerted on the particle [32]. For smaller particles, the most important forces are drag forces, and reduced gravity [32].

Here, the effect of deposition rate is studied by reducing the effective gravitational acceleration on the particles. When $g_r \rightarrow 0$, particles hit the surface with lower kinetic energy, as presented in Figure 9. It is obvious from this figure that decreasing effective gravity decreases the effective density especially near the surface and increases the irregular pattern depth close to the surface. Gravitational acceleration, $g = \{0.490, 0.981, 9.81\}$ $\mathrm{m/s^2}$ and friction coefficient $\mu = 0.9$ are considered while the particles are kept at the constant diameter of 500μm.

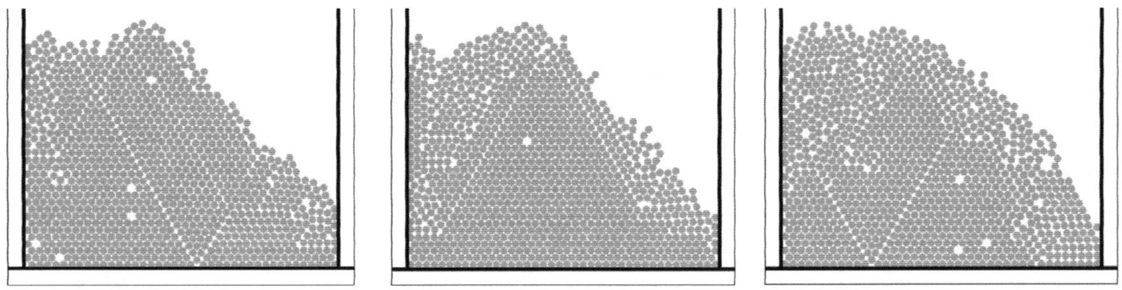

Figure 9. mono-dispersed particles: no bonding is considered, friction coefficient $\mu = 0.9$: $g = 9.81$ (left), $g = 9.81/10$ (middle), and $g = 9.81/20$ (right); particle diameter: $500\ \mu$m,

5.2. Bonded, temperature-dependent particles

5.2.1. Effects of temperature and bonding on the RLP density of snow. Random loose packing of snow is of interest since snow naturally packs under the influence of gravitational forces. Poly-dispersed particles are considered. It is seen in Figure 10 that as particles fall, they form new bonds, roll over each other and reconfigure, as previously reported [50].

The settings for this simulation are: gravitational acceleration, $g = 9{,}81\ m/s^2$; poly-dispersed particle size distribution according to Figure 7; average time step, $dt = 0.1\mu s$; container size, $w \times h = 20\text{mm} \times 60\text{mm}$; particles are released with initial zero velocity; Particles are released in the form of bonded linear clusters as depicted in Figure 5 such that the height and horizontal location of the middle particle are 54mm and 30mm, respectively; cluster length is 0.35 of the width of the container ; the bonded cluster has a random inclination between 0 (horizontal configuration for the cluster) to 90° (rotated in clockwise direction); Number of particles, $N_{tot} = 1000$; Poisson's ratio, $\nu = 0.4$; density, $\rho = 917\text{Kg/m}^3$; Friction coefficients, damping values, and Young's modulus, can be extracted based on the information provided in previous sections; drag forces are introduced as $F_{drag} = -10\pi V d$, where V and d, are velocity and diameter of the particle respectively.

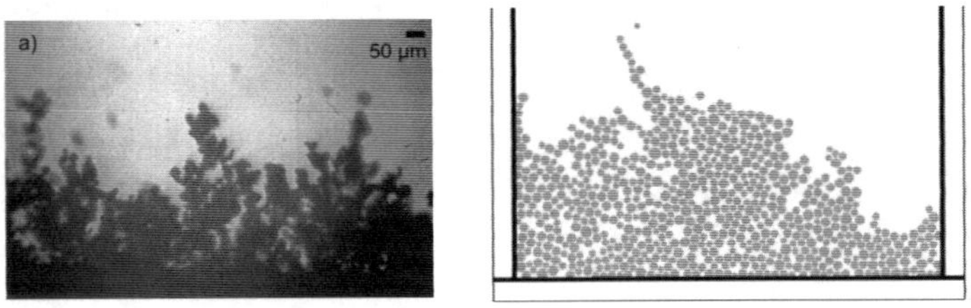

Figure 10. experimental results of sticky ice particles accumulation from experimental results [50] (left); simulated results using DEM (right) $F_{drag} = -10\pi V d$

Note: For Figure 10, which is copied here from [50], permission will be acquired from Elsevier

As a granular material, angles of repose (of snow) is a function of interparticle friction, shape, and bond forces [58]. Therefore, checking the model against an experimentally validated macroscopic model is a reliable way to discern that internal parameters of the model work righteously. A Typical repose configuration of snow at different temperatures are presented in Figure 11. The approximate measured angle of repose from the simulated packing in Figure 11 are compared with the results from Willibald et al [58] and shows good agreement with the experimental results for faceted snow. In Willibald et al. [58] the angle of repose for ice beads are significantly lower which indicates that mono-dispersed particles probably cannot capture the behavior in the best way.

The settings for this simulation are: gravitational acceleration, $g = 9.81\ m/s^2$; poly-dispersed particles are considered with an average time step, $dt = 0.1$micro-seconds; container size, $w \times h = 60\text{mm} \times 60\text{mm}$; particles are released with initial zero velocity; particles are released in the form of bonded linear clusters as depicted in Figure 5 such that the height and horizontal location of the middle particle are 54mm and 30mm, respectively; cluster length is 0.35 of the width of the container; the bonded cluster has a random inclination between 0 (horizontal configuration for the cluster) to 90° (rotating in clockwise direction); Number of particles, $N_{tot} = 1000$; Poisson's ratio, $\nu = 0.4$; Density, $\rho = 917\text{Kg/m}^3$; friction coefficients, damping values, and Young's modulus, can be extracted based on the information provided in previous section; drag forces are introduced as $F_{drag} = -3\pi V d$, where V and d, are velocity and diameter of the particle respectively. In contrast to the results in Figure 11, when the container is large enough so there are no vertical wall effects, the relative density decreases with decreasing temperature: $\rho(T = -2°\text{C}) = 0.68$, while $\rho(T = -15°\text{C}) = 0.58$. Trend-wise, these decrements are is in agreement with the results from experimental works [59]. However, the amount of density decrement is less pronounced than that reported in the literature for natural snow [59], which can be related to the variable natural conditions and wind speeds in nature, which are expected to affect the density significantly.

Interestingly, when temperature decreases, angle of repose also decreases, despite the fact that both friction coefficient and bond strength increase. This indicates the importance of bonding in the snow since with decreasing temperature, particles are less likely to stick, according to Eq. (9), and bond growth rate is also slower. In a word, it seems that, the effects of sticking and bond growth dominate the effects of friction and bond strength. To realize the sintering effect in more details, after obtaining the temperature dependent angle of repose for bonded particle, the sintering ability was turned off in the model keeping all other parameters the same, and the simulations was repeated. The results, as depicted in Figure 11(right) show that if the bonding is not considered, the angle of repose is insensitive to the temperature variations.

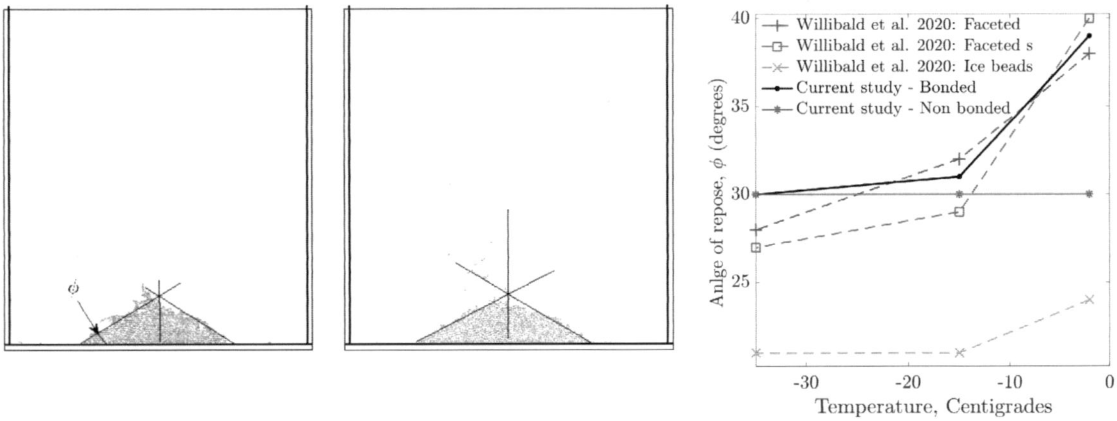

Figure 11. Effect of temperature on the packing configuration, poly-dispersed particles; container size $w \times h = 60 \times 60\text{mm}$;

$T = -2°\text{C}\ t_{fill} = 7.582\text{s}$, $\phi \approx 39°$ bonded particles (left); $T = -35°\text{C}\ t_{fill} = 7.446\text{s}$, $\phi \approx 30°$ bonded particles (middle); Angle of repose of snow as a function of temperature (faceted and faceted s represent two sizes for faceted particles in [58] with faceted s representing smaller particles) (right)

6. Conclusions

The discrete element method was used in this study to model the random loose packing behavior of non-bonded and bonded granular materials representing dry snow with different parameters. The effect of variation in friction coefficient and bond strength on random loose packing density was investigated. Bonds were modelled using Timoshenko beam element theory. With parameters described in previous sections, the developed model was verified for a few simple cases where analytic solutions are available. A time-dependent bond model was briefly presented in this study for short-term sintering of granular materials like snow, and the model was validated with previously published experimental results. This study has the following constraints: First of all, only dry snow was considered. Also, fall velocity was considered as a random variable, and, since the focus was on the material model and simpler presentation on the results, simulations were performed in 2D. Moreover, the effect of particles changing size due to the metamorphism was neglected, and the temperature gradient within the accumulated ice was not considered. Furthermore, particles crushing was not considered, and it was assumed that the clusters can break down only to the constitutive particles level. Finally, since we could not find data on the dependency of rolling friction coefficient on the temperature or velocity, it is considered as a constant in this study. If such data are provided, it is of interest to study the effects of variations in rolling friction coefficient on the macroscopic behavior of granular materials like snow.

The results showed that in the case of mono-dispersed non-bonded particles, an ordered packing was obtained upon sequential deposition and shear bands or artefacts were formed in the materials to geometrically allow for maximum density and ordered packing. However, adding a small perturbation to the particle size distribution destroyed the ordered packing and an amorphous packing was obtained. An increased friction coefficient decreased the loose packing density of the non-bonded particles. Moreover, it was found that the packing density of the non-bonded particles generally increases with the depth while reducing the deposition velocity decreases the density and generally alters the surface morphology significantly. In packing of bonded particles, deposition velocity also plays an important role. Rolling friction plays a significant role in the packing process and specially in forming the surface morphology. Angle of repose of snow, as a sample bonded material with temperature dependent properties, decreases with decreasing temperature, however becomes asymptotic below -15°C. The angle of repose is insensitive to the temperature, when the bonding ability is not considered. In the case of non-contained packing, where there are no vertical wall effects, the density of snow decreases with the temperature significantly. The results can be useful in the study of random loose packing of snow and any other granular materials and in the study of surface engineering using the deposition method.

References

[1] W. Zhong, A. Yu, X. Liu, Z. Tong, and H. Zhang, *DEM/CFD-DEM Modelling of Non-Spherical Particulate Systems: Theoretical Developments and Applications*, Powder Technol. 302, 108 (2016).

[2] P. Wang, Z. Karatza, and C. Arson, *DEM Modelling of Sequential Fragmentation of Zeolite Granules under Oedometric Compression Based on XCT Observations*, Powder Technol. 347, 66 (2019).

[3] B. Soltanbeigi, A. Podlozhnyuk, S. A. Papanicolopulos, C. Kloss, S. Pirker, and J. Y. Ooi, *DEM Study of Mechanical Characteristics of Multi-Spherical and Superquadric Particles at Micro and Macro Scales*, Powder Technol. 329, 288 (2018).

[4] G. W. Delaney, R. D. Morrison, M. D. Sinnott, S. Cummins, and P. W. Cleary, *DEM Modelling of Non-Spherical Particle Breakage and Flow in an Industrial Scale Cone Crusher*, Miner. Eng. 74, 112 (2015).

[5] J. B. Johnson and M. A. Hopkins, *Identifying Microstructural Deformation Mechanisms in Snow Using Discrete-Element Modeling*, J. Glaciol. 51, 432 (2005).

[6] C. Willibald, S. Scheuber, H. Löwe, J. Dual, and M. Schneebeli, *Ice Spheres as Model Snow: Tumbling, Sintering, and Mechanical Tests*, Front. Earth Sci. 7, 1 (2019).

[7] B. Salm, *Mechanical Properties of Snow*, Rev. Geophys. Sp. Phys. 20, 1 (1982).

[8] L. H. Shapiro, J. B. Jerome, M. Sturm, and G. L. Blaisdell, Snow Mechanics: Review of the State of Knowledge and Applications, 1997.

[9] J. J. Petrovic, *Mechanical Properties of Ice and Snow*, J. Mater. Sci. 38, 1 (2003).

[10] I. Hawkes and M. Mellor, *Deformation and Fracture of Ice Under Uniaxial Stress*, J. Glaciol. 11, 103 (1972).

[11] M. J. Percy, K. Ball, and P. B. Mellor, *An Experimental Study of the Burst Strength of the Rotating Disks, International.*, J. Mech. Sci. Pergamon Press 16, 809 (1974).

[12] S. C. Colbeck, *Classification of Seasonal Snow Cover Crystals*, Water Resour. Res. 22, 59S (1986).

[13] J. D. Locatelli and P. V Hobbs, *Fall Speeds and Masses of Solid Precipitation Particles*, J. Geophys. Res. 79, 2185 (1974).

[14] Y. Ouyang, Q. Yang, and X. Chen, *Bonded-Particle Model with Nonlinear Elastic Tensile Stiffness for Rock-like Materials*, Appl. Sci. 7, (2017).

[15] Y. He, T. J. Evans, A. B. Yu, and R. Y. Yang, *A GPU-Based DEM for Modelling Large Scale Powder Compaction with Wide Size Distributions*, Powder Technol. 333, 219 (2018).

[16] G. Lian, C. Thornton, and M. J. Adams, *Discrete Particle Simulation of Agglomerate Impact*

Coalescence, Chem. Eng. Sci. 53, 3381 (1998).
[17] J. Y. Delenne, M. S. El Youssoufi, F. Cherblanc, and J. C. Bénet, *Mechanical Behaviour and Failure of Cohesive Granular Materials*, Int. J. Numer. Anal. Methods Geomech. 28, 1577 (2004).
[18] Z. Shen, M. Jiang, and C. Thornton, *DEM Simulation of Bonded Granular Material. Part I: Contact Model and Application to Cemented Sand*, Comput. Geotech. 75, 192 (2016).
[19] D. O. Potyondy and P. A. Cundall, *A Bonded-Particle Model for Rock*, Int. J. Rock Mech. Min. Sci. 41, 1329 (2004).
[20] M. J. Metzger and B. J. Glasser, *Numerical Investigation of the Breakage of Bonded Agglomerates during Impact*, Powder Technol. 217, 304 (2012).
[21] C. O'Sullivan and J. D. Bray, *Selecting a Suitable Time Step for Discrete Element Simulations That Use the Central Difference Time Integration Scheme*, Eng. Comput. (Swansea, Wales) 21, 278 (2004).
[22] N. J. Brown, J. F. Chen, and J. Y. Ooi, *A Bond Model for DEM Simulation of Cementitious Materials and Deformable Structures*, Granul. Matter 16, 299 (2014).
[23] A. Jagota and G. W. Scherer, *Viscosities and Sintering Rates of Composite Packings of Spheres*, Journal of the American Ceramic Society.
[24] S. Luding, K. Manetsberger, and J. Müllers, *A Discrete Model for Long Time Sintering*, J. Mech. Phys. Solids 53, 455 (2005).
[25] X. Fan, P. Ten, C. Clarke, A. Bramley, and Z. Zhang, *Direct Measurement of the Adhesive Force between Ice Particles by Micromanipulation*, Powder Technol. 131, 105 (2003).
[26] D. Szabo and M. Schneebeli, *Subsecond Sintering of Ice*, Appl. Phys. Lett. 90, 1 (2007).
[27] S. C. Colbeck, *Sintering in a Dry Snow Cover*, J. Appl. Phys. 84, 4585 (1998).
[28] E. E. Adams, D. A. Miller, and R. L. Brown, *Grain Boundary Ridge on Sintered Bonds between Ice Crystals*, J. Appl. Phys. 90, 5782 (2003).
[29] H. M. Lankarani and P. E. Nikravesh, *A Contact Force Model with Hysteresis Damping for Impact Analysis of Multibody Systems*, J. Mech. Des. Trans. ASME 112, 369 (1990).
[30] W. R. Ketterhagen, J. S. Curtis, C. R. Wassgren, and B. C. Hancock, *Predicting the Flow Mode from Hoppers Using the Discrete Element Method*, Powder Technol. 195, 1 (2009).
[31] O. R. Walton and R. L. Braun, *Viscosity, Granulartemperature, and Stress Calculations for Shearing Assemblies of Inelastic,* J. Rheol. (N. Y. N. Y). 30, 949 (1986).
[32] J. S. Marshall, *Discrete-Element Modeling of Particulate Aerosol Flows*, J. Comput. Phys. 228, 1541 (2009).
[33] Y. Tsuji, T. Tanaka, and T. Ishida, *Lagrangian Numerical Simulation of Plug Flow of Cohesionless Particles in a Horizontal Pipe*, Powder Technol. 71, 239 (1992).
[34] H. Bader, Special Report 64 Density of Ice By, 1964.
[35] P. Thorsten, *Computational Granular Dynamics* (2005).
[36] F. A. Tavarez and M. E. Plesha, *Discrete Element Method for Modelling Solid and and Particulate Materials*, Int. J. Numer. Methods Eng. 70, 379 (2007).
[37] S. Sukhorukov and S. Løset, *Friction of Sea Ice on Sea Ice*, Cold Reg. Sci. Technol. 94, 1 (2013).
[38] T. Eidevåg, P. Abrahamsson, M. Eng, and A. Rasmuson, *Modelling of Dry Snow Adhesion during Normal Impact with Surfaces*, Powder Technol. (2019).
[39] V. F. Petrenko and R. W. Whitworth, *Physics of Ice* (Oxford University Press, 2006).
[40] K. D. Supulver, F. G. Bridges, S. Tiscareno, and J. Lievore, *The Sticking Properties of Water Frost Produced under Various Ambient Conditions*, Icarus 129, 539 (1997).
[41] F. G. Bridges, *Energy Loss and Sticking Mechanisms in Particle Aggregation in Planetesimal Formation*, Icarus 435, 422 (1996).
[42] J. W. Glen, *The Creep of Polycrystalline Ice*, Proc. R. Soc. A Math. Phys. Eng. Sci. 228, 519 (1955).
[43] Y. Y. Lin and C. Y. Hui, *Mechanics of Contact and Adhesion between Viscoelastic Spheres: An*

Analysis of Hysteresis during Loading and Unloading, J. Polym. Sci. Part B Polym. Phys. 40, 772 (2002).
[44] F. Azizi, *Primary Creep of Polycrystalline Ice under Constant Stress*, Cold Reg. Sci. Technol. 16, 159 (1989).
[45] R. G. Budynas and J. K. Nisbett, *Shigley's Mechanical Engineering Design*, 10th ed. (Mc Graw Hill Education, 2015).
[46] G. Dantl, *Die Elastischen Moduln von Eis-Einkristallen*, Phys. Der Kondens. Mater. 7, 390 (1968).
[47] Klaus-Jurgen-Bathe, *Finite Element Procedures*, Second (Prentice Hall, Peareson Education Inc, 2014).
[48] H.-G. Matuttis and J. Chen, *Understanding the Discrete Element Method_ Simulation of Non-Spherical Particles for Granular and Multi-Body Systems* (John Wiley & Sons, 2014).
[49] E. Barthazy and R. Schefold, *Fall Velocity of Snowflakes of Different Riming Degree and Crystal Types*, Atmos. Res. 82 82, 391 (2006).
[50] B. Gundlach, S. Kilias, E. Beitz, and J. Blum, *Micrometer-Sized Ice Particles for Planetary-Science Experiments - I. Preparation, Critical Rolling Friction Force, and Specific Surface Energy*, Icarus 214, 717 (2011).
[51] J. K. Kruschke, *Doing Bayesian Data Analysis*, 2nd ed., Vol. 81 (Academic Press, 2015).
[52] M. E. Gurtin, *A Gradient Theory of Single-Crystal Viscoplasticity That Accounts for Geometrically Necessary Dislocations*, J. Mech. Phys. Solids 50, 5 (2002).
[53] E. M. Schulson and A. L. Fortt, *Friction of Ice on Ice*, J. Geophys. Res. 117, 1 (2012).
[54] K. M. Salerno, D. S. Bolintineanu, G. S. Grest, J. B. Lechman, S. J. Plimpton, I. Srivastava, and L. E. Silbert, *Effect of Shape and Friction on the Packing and Flow of Granular Materials*, 98, 050901 (2018).
[55] L. Arnaud, J. M. Barnola, and P. Duval, *Physical Modeling of the Densification of Snow / Firn and Ice Upper Part of Polar Ice Sheets*, Phys. Ice Core Rec. 285 (2000).
[56] G. R. Farrell, K. M. Martini, and N. Menon, *Loose Packings of Frictional Spheres*, Soft Matter 6, 2925 (2010).
[57] K. J. Dong, R. Y. Yang, R. P. Zou, and A. B. Yu, *Role of Interparticle Forces in the Formation of Random Loose Packing*, Phys. Rev. Lett. 145505, (2006).
[58] C. Willibald, H. Löwe, T. Theile, J. Dual, and M. Schneebeli, *Angle of Repose Experiments with Snow : Role of Grain Shape and Cohesion*, J. Glaciol. 66, 658 (2020).
[59] M. Lafaysse, B. Cluzet, M. Dumont, Y. Lejeune, V. Vionnet, and S. Morin, *A Multiphysical Ensemble System of Numerical Snow Modelling*, Cryosph. 11.3, 1173 (2017).

IOP Conf. Series: Materials Science and Engineering 1190 (2021) 012016 doi:10.1088/1757-899X/1190/1/012016

Loose Material Filling in the Loading Trough Profile of the Belt Conveyor

Leopold Hrabovský

VSB Technical University of Ostrava, Faculty of Mechanical Engineering, Institute of Transport, 17. listopadu 2172/15, 708 00 Ostrava - Poruba, Czech Republic

leopold.hrabovsky@vsb.cz

Abstract. The paper presents a mathematical expression of the sizes of the transverse and longitudinal cross-section surfaces and of the volume of the loose material spread on the surface of the conveyor belt before the transverse partition of the belt conveyor, inclination angle of which exceeds the limit transport angle permitted for the belt conveyor of a classic design. Volume and geometric shape of the lose material; transported in the trough (two-roller bench) loading profile; are impacted by the mechanical characteristics of the transported material, width of the conveyor belt, inclination angle of the side rollers, height of the transverse ribs and especially inclination angle of the conveyor. The maximal cross-section surface of the lose material filling at the conveyor with a smooth belt can be (for conveyor belts modified based on the shape of the trough) expressed as the sum of the cap cross-section surface and the surface of the bottom part of the belt filling. The paper describes the jacket curve of the of the cap cross-section using a parabola equation, and the surface curve of the bottom part of the belt filling using a line equation. The size of the cross-section surface of the belt filling along the longitudinal plane of the conveyor belt is defined as the difference between the inclination angle of the conveyor and the discharge angle of the transported material.

1. Introduction

Due to their technical, operation as well as economic benefits, belt conveyors have been widely used for industrial purposes. Conveyors with a pulling as well as loadbearing mechanism in the form of an endless loop of the conveyor belt [1] represent transport means from the category of fluently operating devices, designated for moving lose, semiliquid as well as piece materials to a distance of up to thousand meters.

The cross-section size of the conveyor belt filling S *[m²]* depends on the shape of the loading profile (flat or trough roller bench), loading width of the belt b *[m]* and dynamic discharge angle ψ_d *[deg]* of the transported material [2]. For a flat loading profile of a conveyor belt of a horizontally installed belt conveyor (i.e., when the transport inclination angle $\beta = 0$ deg), total surface S *[m²]* of the belt filling cross-section of the transported material can be determined using formula [2, 3]. When the transported material is loaded on the inclined part of the belt ($\beta > 0$), the reduction of a part of cross-section S_1 is determined by inclination coefficient k *[-]* [2].

When the conveyor inclination angle β equals to the dynamic discharge angle of the transported material ψ_d, upper cross-section S_1 of the belt filling is cancelled and only lower cross-section S_2 can be used for the transport. This definition, which is included in technical standard [2], specifies that the limit angle, when there is no transported material on the flat loading profile of the conveyor belt anymore, acquires value $\beta_{max} = \psi_d$.

Figure 1. Conveyor belt a) grooved, b) with separated protrusions, c) profile

In real life, it is required; for example, due to a reduced built-up area by the conveyor, reduced length of the enclosed loop of the conveyor belt, etc.; that the lose or small-piece materials on the surface of the given conveyor belt are transported under the highest possible inclination angle of the conveyor. From among several methods [4, 5] of how to ensure, under a transport angle that exceeds the limit angle of a smooth conveyor belt β_m *[deg]* of a belt conveyor of a standard design, an effective transport of the grains of the transported material, a conveyor belt with transverse ribs is used, Figure 2.

2. Source of the explored topic

Various industrial fields require an increased inclination angle of the belt conveyor transport [4], or the use of belt conveyors of special designs [6-9]. The permitted inclination angle of the transport by belt conveyors of a standard design [2, 10] is particularly determined by the friction coefficient between the contact surface of the material grains (incapable of rotation) and the surface of the given conveyor belt. Material grains that are capable of rotation (grains of a spherical or cylindrical shape) must be prevented from rolling along the surface under any angle of the given inclined conveyor belt. For horizontally carried conveyor belts, we also need to prevent rotation of the spherical grains, which can occur due to the effects of the forces along the contact surface of the grains with the conveyor belt and the inertia forces. In order to increase the permitted angle of the belt conveyor transport [5], we need to:

a) increase the friction coefficient along the work surface of the conveyor belt [11, 12],

b) structurally modify the surface of the conveyor belt [13, 14], making sure that movement of the transported material grains along the belt is prevented (against the transport direction for the uphill transport and in the transport direction for downhill transport) [4],

c) increase the pressure of the transported material to the conveyor belt [6, 15, 16].

An increase of the friction (cohesive, adhesive) force of the transported material in relation to the surface of the conveyor belt is usually implemented by increasing the roughness (friction) of the surface or the operation layer of the conveyor belt [17]. Higher values of the friction coefficient can be achieved by:

a_1) applying an additional compact layer of particles on the conveyor belt operation surface after the assembly of the conveyor belt [18], or

a_2) by adding abrasive particles to the rubber cover layer when manufacturing the belt by vulcanization.

b) grooving and indenting the belt operation surface, see Figure 1a, when vulcanizing the cover rubber surface of the conveyor belt.

c) vulcanization [19] of protrusions (separated Figure 1b or profile Figure 1c) onto the belt operation surface.

Prevention of undesirable movement of the transported piece materials as well as lose material grains (capable as well as incapable of rotation); along the surface of the conveyor belts of conveyors that operate with a transport incline angle [20] that exceeds the limit angle [21, 22]; is usually ensured by implementing transverse partitions [23] (Figure 2a) with a regular span along the entire length of the endless loop of the conveyor belt.

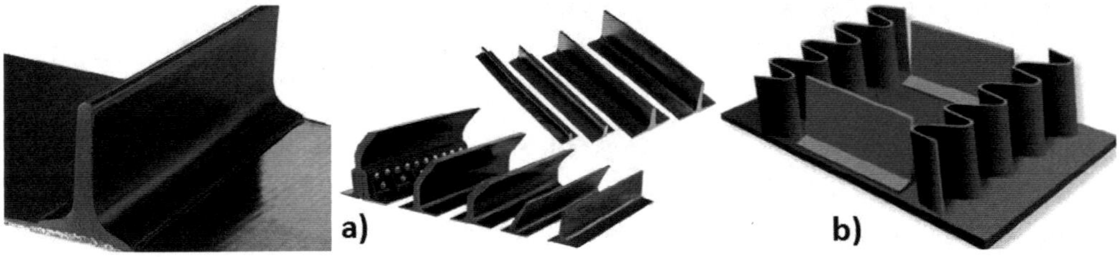

Figure 2. Conveyor belt a) with transverse partitions, b) with partitions ad side bellows

When transporting lose materials on belt conveyors under the transport angle $\beta = 90\ deg$, conveyor belts furnished (by the means of hot as well as cold vulcanization or mechanically) by partitions and side bellows [24] (Figure 2b) are used. Side bellows prevent the transported grains from falling off over the side edges or the conveyor belt and significantly increase transport performance $Q\ [t \cdot h^{-1}]$ by putting a larger volume of the lose material $V\ [m^3]$ to the area between the two mutually adjacent partitions.

3. Methodology of research

3.1. Transverse cross-section of the conveyor belt filling

The maximal cross-section surface of the lose material filling $S\ [m^2]$ at the conveyor with a smooth belt [2] can be (for conveyor belts modified based on the shape of a trough) [25] expressed as the sum of the cap cross-section surface $S_1\ [m^2]$ (3) and the surface of the bottom part $S_2\ [m^2]$ (4) of the belt filling [3]. Canonical shape of the parabola [5] with its axis in axis y, see Figure 3a, and vertex V in the distance of $h_{max}\ [m]$ from the beginning 0 of the coordinate system can be recorded using formula (1).

$$x^2 = -2 \cdot p \cdot y_{(x)} + h_{max}\ [m^2] \tag{1}$$

The value of parameter $p\ [m]$ and height of the parabola h_{max} can be determined; if we know point A_1 of the intersection of the parabola with axis x and tangent t to the parabola, which passes through point A_1; by graphic procedure (Figure 3a) [5], and it can be expressed using formula (2).

$$p = \frac{b_1}{2 \cdot tan(\psi_d)}\ [m],\ h_{max} = \frac{b_1 \cdot tan(\psi_d)}{4}\ [m] \tag{2}$$

Surface of the transverse cross-section cap $S_1\ [m^2]$ (see Table 1) of the transported material, see Figure 3b, along plane xy can be expressed using formula (3).

$$S_1 = \frac{b_1^2 \cdot tan(\psi_d)}{6} = \int_{-b_1/2}^{b_1/2} y_{(x)} \cdot dx = 2 \cdot \int_{b_1/2}^{0} y_{(x)} \cdot dx = 2 \cdot \int_{b_1/2}^{0} \frac{x^2}{-2 \cdot p} \cdot dx \tag{3}$$

Surface of the bottom part of the transverse cross-section cap $S_2\ [m^2]$ (see Table 1) of the transported material filling, see Figure 3b, can be expressed using formula (4). Distance $z_{(x)}\ [m]$ of the distribution of the material transported on the conveyor belt along plane yz, i.e., in the direction of axis z, can be expressed pursuant to Figure 3b formula (5).

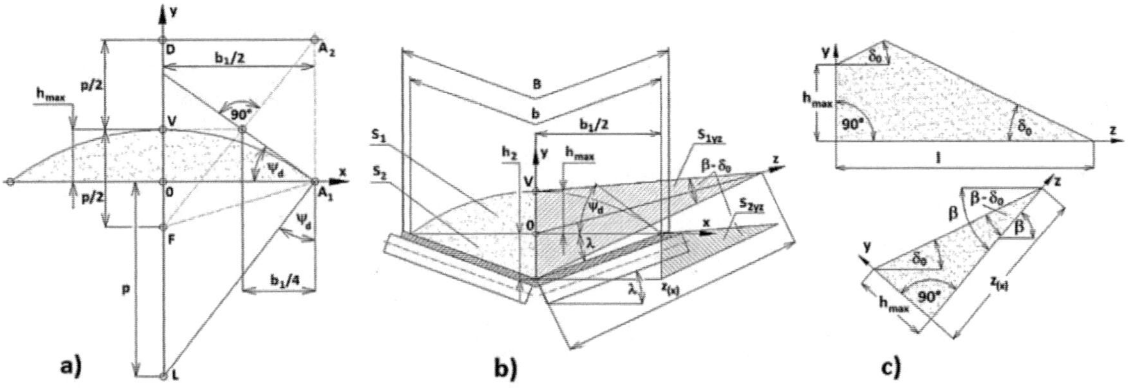

Figure 3. Surface of the a) transverse cross-section, b), c) longitudinal cross-section of the transported material

$$S_2 = \frac{b_1^2 \cdot \tan(\lambda)}{4} = 2 \cdot \int_0^{b_1/2} x \cdot tan(\lambda) \cdot dx \; [m^2] \tag{4}$$

$$z_{(x)} = \frac{y_{1(x)}}{tan(\beta - \delta_0)} = \frac{y_{(x)} + (h_2 - x \cdot tan(\lambda))}{tan(\beta - \delta_0)} = \frac{h_{max} - \frac{x^2}{2 \cdot p} + \left(\frac{b_1}{2} - x\right) \cdot tan(\lambda)}{tan(\beta - \delta_0)} \; [m] \tag{5}$$

where $y_{1(x)}$ *[m]* is the height of the lose material layer on the conveyor belt along plane *xy* [4] within the interval of distances *0* to $b_1/2$ on axis *x*.

3.2. Longitudinal cross-section of the lose material filling before the partition

The size of the longitudinal cross-section $S_{1yz(x)}$ $[m^2]$ of the transported material, see Figure 3b, can be, for the selected width of the conveyor belt *B* and inclination angle β of the conveyor, expressed by formula (6), see Table 2 through Table 4. The size of the longitudinal cross-section $S_{2yz(x)}$ $[m^2]$ of the transported material, see Figure 3b, can be, for the selected width of the conveyor belt *B* and inclination angle β of the conveyor, expressed by formula (7), see Table 2 through Table 4 (Figure 5).

Table 1. Transverse, longitudinal cross-section and lose material volume for B = 0.4÷0.65 m

B	β	S1	S2	Sxy	ψd	λ	b1	hmax	h2	δ0	S1yz	S2yz	V
m	deg		cm²		deg			mm		deg	mm²		dm³
0.4	30	45.95	77.21	123.17	18	20	291.31	23.66	53.01	20	16671.23	7969.26	1884.01
0.5		76.51	128.56	205.07			375.88	30.53	68.40		27756.47	13268.28	4047.36
0.65		136.87	229.98	366.85			502.74	40.84	91.49		49653.73	23735.71	9684.09

$$S_{1yz(x)} = \frac{1}{2} \cdot y_{3(x)} \cdot z_{3(x)} = \frac{1}{2} \cdot \left(y_{(x)} + \frac{b_1}{2} \cdot tan(\lambda)\right) \cdot \frac{y_{(x)} + \frac{b_1}{2} \cdot tan(\lambda)}{tan(\beta - \delta_0)} = \frac{\left[h_{max} - \frac{x^2}{2 \cdot p} + \frac{b_1}{2} \cdot tan(\lambda)\right]^2}{2 \cdot tan(\beta - \delta_0)} \tag{6}$$

$$S_{2yz(x)} = \frac{1}{2} \cdot y_{4(x)} \cdot z_{4(x)} = \frac{1}{2} \cdot (h_2 - x \cdot tan(\lambda)) \cdot \frac{h_2 - x \cdot tan(\lambda)}{tan(\beta - \delta_0)} = \frac{\left[\left(\frac{b_1}{2} - x\right) \cdot tan(\lambda)\right]^2}{2 \cdot tan(\beta - \delta_0)} \tag{7}$$

4. Results

4.1. Volume of the lose material batch on the conveyor belt

The volume of the lose material $V_{(i)}$ $[m^3]$ spread on the conveyor belt with a width of B before the transverse partition; provided the height of the transverse partition is at least $h_{max} + h_2\,[m]$ (where $h_2 = tan(\lambda)\cdot b_1/2$) see Figure 4a.; can be, for inclination angle β of the belt conveyor, expressed by formula (8).

$$V_{(i)} = \int_{-b_1/2}^{b_1/2} \frac{1}{2}\cdot y_{1(x)}\cdot z_{(x)}\cdot dx = 2\cdot\int_{0}^{b_1/2} \frac{1}{2}\cdot y_{1(x)}\cdot z_{(x)}\cdot dx = 2\cdot\int_{0}^{b_1/2} S_{yz(x)}\cdot dx =$$

$$= 2\cdot\int_{0}^{b_1/2}\left[h_{max} - \frac{x^2}{2\cdot p} + \left(\frac{b_1}{2} - x\right)\cdot tan(\lambda)\right]^2 \cdot \frac{1}{2\cdot tan(\beta - \delta_0)}\cdot dx \tag{8}$$

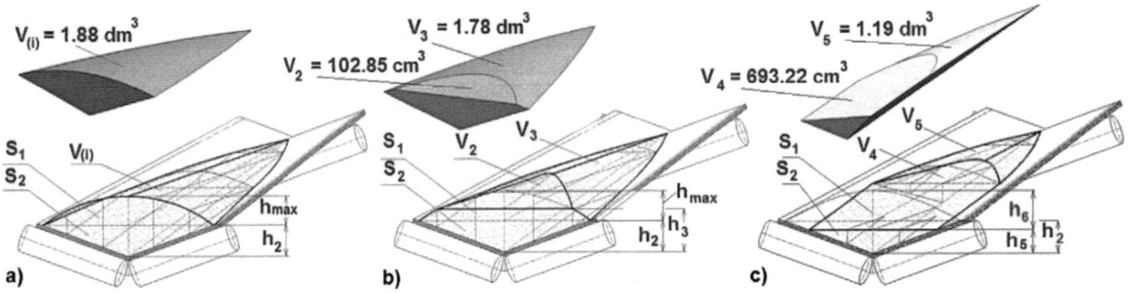

Figure 4. The volume of the transported material for a partition height a) $\geq h_{max} + h_2$, b) $> h_3$ a $\leq h_{max} + h_2$, c) $< h_2$

Table 2. Longitudinal cross-section and lose material volume for B = 0.4 m for β = 30÷60 deg

i		1	2	3	4
β	deg	30	40	50	60
B	mm	400			
$z_{(x)}$	mm	434.85	210.66	132.81	91.38
$S_{1yz(i)}$	cm²	16671.23	8076.45	5091.52	3503.26
$S_{2yz(i)}$	cm²	7969.26	3860.74	2433.87	1674.65
$V_{1(i)}$	cm³	246.67	119.50	75.34	51.84
$V_{(i)}$	cm³	**1884.01**[*1]	912.72	575.39	395.9

Table 3. Longitudinal cross-section and lose material volume for B = 0.5 m for β = 30÷60 deg

i		1	2	3	4
β	deg	30	40	50	60
B	mm	500			
$z_{(x)}$	mm	561.10	271.83	171.36	117.91
$S_{1yz(i)}$	cm²	27756.47	13446.74	8477.03	5832.7
$S_{2yz(i)}$	cm²	13268.28	6427.88	4052.23	2788.17
$V_{1(i)}$	cm³	529.90	256.72	161.84	111.35
$V_{(i)}$	cm³	4047.36	1960.79	1236.11	850.52

Table 4. Longitudinal cross-section and transported material volume for B = 0.65 m for β = 30÷60 deg

i		1	2	3	4
β	deg	30	40	50	60
B	mm	650			
$z_{(x)}$	mm	750.47	363.57	229.20	157.70
$S_{1yz(i)}$	cm^2	49653.73	24054.97	15164.61	10434.15
$S_{2yz(i)}$	cm^2	23735.71	11498.87	7249.06	4987.78
$V_{1(i)}$	cm^3	1267.90	614.25	387.24	266.44
$V_{(i)}$	cm^3	9684.09	4691.50	2957.59	2034.98

When the height of the transverse partition installed on the conveyor belt reaches h_3 *[m]* (which is lower than $h_{max} + h_2$, but higher than h_2, see Figure 4b), the volume of the given lose material V_3 *[m^3]* spread on the conveyor belt with a width of *B* before the transverse partition with a height of h_3 *[m]*, can be expressed by formula (10).

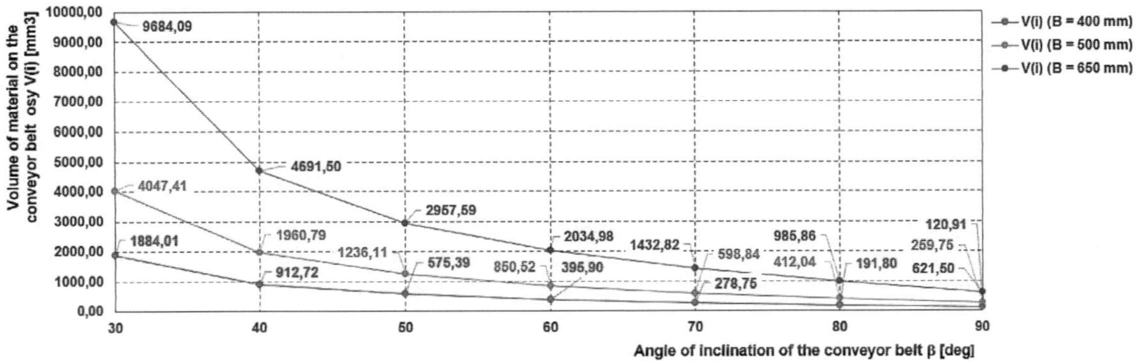

Figure 5. Volume of the lose material for $B = 0.4 \div 0.65$ *m* for $\beta = 30 \div 60$ *deg*

Pursuant to formula (9) and Figure 4b, we can express height h_4 *[m]* and length b_2 *[m]* of the parabolic surface along plane *xy*, which is higher than the transverse partition with a height of h_3. The loss of the transported material (i.e., volume V_2 *[m^3]*, Figure 4b); which falls over the upper edge of the partition due to an insufficient height of the partition h_3 from the original volume of $V_{(i)}$ (see Figure 4a) of the transported material (8); can be expressed by expression V_2 *[m^3]*, which is specified in formula (10).

Table 5. Volume of the transported material V_2 *[m^3]* for h_3 *[m]* and $\beta = 30 \div 60°$ at $B = 0.4$ m

B	**β**							
mm	deg	**h_3**	mm	55	60	65	70	75
400	30	V_2	cm^3	198.10	**102.85**[*2]	42.19	10.43	0.33
400	40	V_2	cm^3	95.97	49.82	20.44	5.05	0.16
400	50	V_2	cm^3	60.50	31.41	12.88	3.19	0.10
400	60	V_2	cm^3	41.63	21.61	8.87	2.19	0.07

[*2] see Figure 4b, $V_{(i)} - V_2 = 1884.01 - 102.85 =$ **1781.16** cm^3

Volume of the lose material V_2, falling over the upper edge of the partition with a height of h_3, for the given width of the conveyor belt *B* and inclination angle of the conveyor belt is showed in Table 5 through Table 7.

When the height of the transverse partition installed on the conveyor belt reaches h_5 *[m]* (which is lower than h_2, see Figure 4c), the volume of the given lose material V_5 *[m^3]* spread on the conveyor belt with a width of *B* before the transverse partition with a height of h_5 *[m]*, can be expressed by formula (12).

Table 6. Volume of the transported material $V_2\ [m^3]$ for $h_3\ [m]$ and $\beta = 30 \div 60°$ at $B = 0.5$ m

B	β								
mm	deg	h3	mm	70	75	80	85	90	95
500	30	V2	cm³	463.33	288.37	160.52	74.59	24.56	3.16
	40			224.47	139.70	77.77	36.13	11.90	1.53
	50			141.51	88.07	49.02	22.78	7.50	1.00
	60			97.37	60.60	33.73	15.67	5.16	0.66

Table 7. Volume of the transported material $V_2\ [m^3]$ for $h_3\ [m]$ and $\beta = 30 \div 60°$ at $B = 0.65$ m

B	β										
mm	deg	h3	mm	95	100	105	110	115	120	125	130
650	30	V2	cm³	1012.74	706.92	464.45	280.24	148.69	63.48	17.29	0.98
	40			490.65	342.47	225.00	135.77	72.03	30.75	8.38	0.48
	50			309.31	215.90	141.85	85.59	45.41	19.39	5.28	0.30
	60			212.83	148.55	97.60	58.90	31.24	13.34	3.63	0.21

$$h_4 = h_{max} + h_2 - h_3,\ b_2 = \sqrt{8 \cdot p \cdot h_4}\ [m] \tag{9}$$

$$V_3 = V_{(i)} - V_2 = V_{(i)} - 2 \cdot \int_0^{b_2/2} S_{3yz(x)} \cdot dx = V_{(i)} - 2 \cdot \int_0^{b_2/2} \left(h_4 - \frac{x^2}{2 \cdot p} \right)^2 \cdot \frac{1}{2 \cdot tan(\beta - \delta_0)} \cdot dx\ [m^3] \tag{10}$$

Pursuant to Figure 4c, we can express height $h_6\ [m]$ and length $b_3\ [m]$ of the parabola along plane *xy*, which is higher than the transverse partition with height of h_5, using formula (11). The loss of the transported material (i.e., volume $V_4\ [m^3]$, Figure 4c); which falls over the upper edge of the partition due to an insufficient height of the partition h_5 from the original volume of $V_{(i)}$ (see Figure 4a) of the transported material (8); can be expressed by formula (12).

Table 8. Volume of the transported material $V_{4i}\ [m^3]$ for $h_5\ [m]$ and $\beta = 30 \div 60°$ at $B = 0.4$ m

B	β							
mm	deg	h5	mm	25	30	35	40	45
400	30	V41/V42	cm³	971.77/168.19	913.55/96.69	802.54/48.05	655.69/18.76 **693.22**[*3]	491.81/4.53
	40			470.78/81.48	442.57/46.84	388.79/23.28	317.65/9.09	238.27/2.20
	50			296.78/51.37	279.00/29.53	245.10/14.67	200.25/5.73	150.21/1.38
	60			204.21/35.34	191.97/20.32	168.64/10.10	137.79/3.94	103.35/0.95

[*3] see Figure 4c, $V_{(i)} - (V_{41} + 2 \cdot V_{42}) = 1884.01 - 693.22 = \mathbf{1190.79}$ cm³

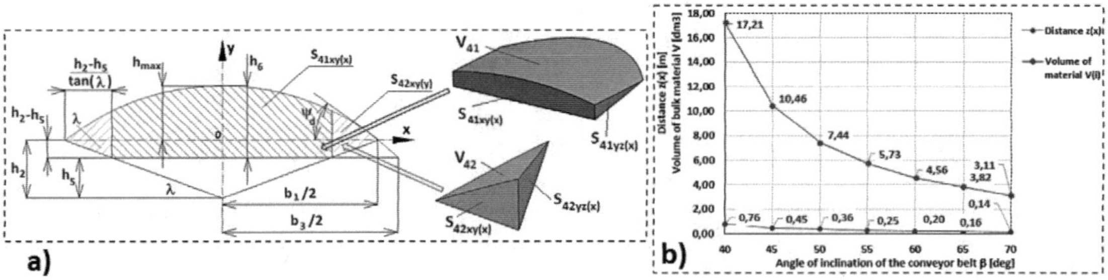

Figure 6. a) overfalling volume $V_{4i}\ [m^3]$, b) distribution $z_{(x)}\ [mm]$ and volume $V_{(i)}\ [m^3]$, of the lose material

$$h_6 = h_{max} + h_2 - h_5,\ b_3 = \sqrt{8 \cdot p \cdot h_6}\ [m] \tag{11}$$

Figure 6a shows the surfaces of the transverse cross-section ($S_{41xy(x)}$ a $S_{42xy(x)}$), longitudinal cross-section ($S_{41yz(x)}$ a $S_{42yz(x)}$), and volumes (V_{41} a V_{42}) of the lose material that falls over the transverse partition with a height of h_5 in relation to increasing inclination angle β of the conveyor belt. (Figure 7)

$$V_5 = V_{(i)} - V_4 = V_{(i)} - V_{41} - V_{42} = V_{(i)} - 2 \cdot \int_0^{\frac{b_1}{2} - \frac{h_2 - h_5}{tan(\lambda)}} S_{41yz(x)} \cdot dx - 2 \cdot \int_{\frac{b_1}{2} - \frac{h_2 - h_5}{tan(\lambda)}}^{b_1/2} S_{42yz(x)} \cdot dx =$$

$$= V_{(i)} - 2 \cdot \int_0^{\frac{b_1}{2} - \frac{h_2 - h_5}{tan(\lambda)}} \left(h_6 - \frac{x^2}{2 \cdot p} \right) \cdot \frac{1}{2 \cdot \tan(\beta - \delta_0)} \cdot dx - 2 \cdot \int_{\frac{b_1}{2} - \frac{h_2 - h_5}{tan(\lambda)}}^{b_1/2} \left(h_6 - \frac{x^2}{2 \cdot p} - \left(x \cdot tan(\lambda) - h_5 \right) \right) \cdot$$

$$\cdot \frac{1}{2 \cdot \tan(\beta - \delta_0)} \cdot dx \quad (12)$$

Figure 7. Loose material distribution on the surface of the conveyor belt before the transverse partition

Volume of the lose material V_4, falling over the upper edge of the partition with a height of h_5, for the given width of the conveyor belt B = 0.4 m and inclination angle of the conveyor belt β is demonstrated by Table 8.

5. Conclusion

One of the several practical ways of transporting lose material grains on the surface of a conveyor belt, the inclination angle of which exceeds the limit angle of a classic design of belt conveyors, is to prevent the transported grain from moving. Grains are prevented from moving (against the direction of the transport for uphill transport and in the transport direction for downhill transport) by the implementation of transverse ribs, fastened to the surface of the conveyor belt.

The transverse cross-section of the belt filling increases with increasing transport capacity (volumetric or weight), i.e., with the material volume that needs to be transported by the given belt conveyor over a certain time. Increasing transverse cross-section surface of the belt filling results in an increasing length of the base area of the transported material, spread on the surface of the conveyor belt. The maximal length of the surface base of the transverse cross-section of the belt filling, which must not exceed the value of the so-called used loading width, results in the need to use a wider conveyor belt when the use of a flat profile of the conveyor belt and identical transport speed are required for transporting an increased volume of the given material. When it is possible to use a trough loading profile of the conveyor belt, the surface of the transverse cross-section of the belt filling can be increased (while the width of the conveyor belt remains the same) by selecting a bigger inclination angle of the side rollers (two- as well as three-roller bench) or by adjusting the length of the middle roller (three-roller bench).

The paper describes a mathematical method for determining the surface of the transverse as well as longitudinal cross-section and volume of the lose material spread on the surface of the conveyor belt before the transverse partition, provided we know mechanical characteristics of the transported lose material. Because of the limited space, the paper only states the values that apply to a trough loading profile (two-roller bench) of the conveyor belt, along which transverse partitions of the appropriate height are installed with a regular span. The height of the transverse partition can exceed the height of the given lose material layer in the middle of the conveyor belt in the transverse cross-section of the belt conveyor. Alternatively, the height of the transverse partition may not exceed the height of the given lose material layer in the middle of the conveyor belt in the transverse cross-section of the belt conveyor. The paper specifies formulas for both these variants. Using the formulas, you can calculate the surface of the cross-sections and volume of the lose material batches, spread before the transverse partition for the given inclination angle of the conveyor belt.

The experimental tests conducted in the Laboratory of Research and Testing were supposed to verify the measured maximal length of the spread of the lose material on the conveyor belt for the known inclination angle, with theoretical values, calculated pursuant to the above-stated mathematical formulas. The laboratory tests also included measurements of the batch volumes of two types of lose materials (dry sand and ground marble chips), which were caught by a transverse rib. The values were then compared with the corresponding, theoretically determined values.

Mathematical formulas that express the volume and distance of the lose material layer distribution along the length of the conveyor belt with a direct support and a conveyor belt with undulated side rails or without them can be found in [5]. The surface of the transverse as well as longitudinal cross-sections and the volume of the lose material spread on the surface of the conveyor belt (with a trough loading profile, three-roller bench) before the transverse partition can be determined pursuant to [4].

Acknowledgements
This work has been supported by The Ministry of Education, Youth and Sports of the Czech Republic from the Specific Research Project SP2021/53.

References

[1] I. Tarpay, “Steep incline and vertical conveyors-advantages, challenges and applications”, *3th International Conference on Bulk Materials Storage, Handling and Transportation (ICBMH 2019)*, p. 330, Engineers Australia, 2019.
[2] ČSN ISO 5048, “Continuous mechanical handling equipment. Belt conveyors with carrying idlers. Calculation of operating power and tensile forces”, ÚNI, Praha, pp. 15, 1994.
[3] ČSN 26 3102, “Belt conveyors. Calculation bases”, ÚNM, Praha, pp. 48, 1988.
[4] J. Perten, “Krutonaklonnyje konvejeri”, (In English: High Angle Conveyors), Leningrad, *Mashinostrojenie*, ISBN 621-867-212-7, 1977.
[5] L. Hrabovský L, “Pásové dopravníky převyšující mezní úhel slonu dopravy”, (In English: Belt conveyors exceeding the limit angle of the transport elephant), *VSB - Technická univerzita Ostrava*, ISBN 80-248-0779-3, pp. 127, 2005.
[6] S. Guo, W. Huang and X. Li, “Normal Force and Sag Resistance of Pipe Conveyor”, *Chinese Journal of Mechanical Engineering*, vol. 33(1), 2020.
[7] V. Molnar, G. Fedorko, S. Honus, M. Andrejiova, A. Grincova, P. Michalik and J. Palencar. “Research in placement of measuring sensors on hexagonal idler housing with regard to requirements of pipe conveyor failure analysis”, *Engineering Failure Analysis*, vol. 116, 2020.
[8] E. Bombasaro and R. Oddera, “Field measurement of contact forces on rollers for a large diameter pipe conveyor”, *Open Engineering*, vol. 11(1), 2020.
[9] J. A. Dos Santos, “Theory and design of Sandwich Belt High Angle Conveyors according to the Expanded Conveyor Technology”, *Bulk Solids Handling*, vol. 20(1), pp. 27-38, 2000.

[10] D. Marasova, J. Saderova and L. Ambrisko, "Simulation of the use of the material handling equipment in the operation process", *Open Engineering*, 10(1), pp. 216-223, 2020.

[11] E. Ross, "Using Cleats to Extend Belt Conveyor Capabilities", *Plant Engineering*, vol. 37(7), pp. 58-63, 1983.

[12] B. Fullard, "Improve Operations Through the Use of Conveyor Belt Cleats", *Industrial Engineering*, vol. 24(3), pp: 32-40, 1992.

[13] B. Stehlikova, V. Molnar, G. Fedorko, P. Michalik and A. Paulikova, "Research about influence of the tension forces, asymmetrical tensioning and filling rate of pipe conveyor belt filled with the material on the contact forces of idler rolls in hexagonal idler housing", *Measurement*, 156, 107598, 2020.

[14] M. Wang, Q. Liu, Y. Ou and X. Zou, "Experimental Study of the Seed-Filling Uniformity of Sugarcane Single-Bud Billet Planter", *Sugar Tech*, 1-11, 2021.

[15] Dos Santos, Joseph A. Sandwich Belt High Angle Conveyors Coal Mine to Prep Plant and Beyond–2016. In: *XVIII International Coal Preparation Congress*. Springer, Cham, 2016. p. 111-117.

[16] A. V. Glebov, "Innovative engineering design of the high-angle conveyor for mining of deep-seated mineral deposits", *Aspects in Mining and Mineral Science*, vol. 2(2), pp. 535-535, 2018.

[17] L. Hrabovský, "Stream Lined Emission Particles Handling For Civil Engineering Purposes", *IOP Conference Series: Materials Science and Engineering*, WMCAUS 12-16 June 2017, Prague, Czech Republic.

[18] K. Mazurek, M. Szyguła, K. Perutka, K. Turczyński and K. Stankiewicz, "Acquiring electric energy from the transport conveyor roller movement for distributed sensors network", *Journal of Electrical Engineering*, 71(2), pp. 87-95, 2020.

[19] P. Michalik, J. Dobranský, L. Hrabovský and M. Petruš, "Assessment of the manufacturing possibility of thin-walled robotic portals for conveyance workplaces", *Advances in Science and Technology. Research Journal*, 12(1), 2018.

[20] V. Molnár, G. Fedorko, B. Stehlíková, P. Michalik and M. Weiszer, "A regression model for prediction of pipe conveyor belt contact forces on idler rolls", *Measurement*, 46(10), pp. 3910-3917, 2013.

[21] G. Fedorko, V. Ivančo, V. Molnár and N. Husáková, "Simulation of interaction of a pipe conveyor belt with moulding rolls", *Procedia engineering*, 48, pp. 129-134, 2021.

[22] V. Taraba, D. Marasová, D. Kubala, and V. Semjon, "Mathematical modelling of the process of conveyor belt damage caused by the impact stress", *Applied Mechanics and Materials*, vol. 683, pp. 153-158, 2014.

[23] E. Ross, "How to Attach and Apply Belt Conveyor Cleats", *Plant Engineering*, vol. 37(8), pp. 132-139, 1983.

[24] J. W. Paelke, R. C. Emerton and J. A. Williams, "Wyoming coal mine truck load-out facility carries 65,000 tons per month on FLEXOWELL steep angle conveyors", *Bulk Solids Handling*, vol. 6(2), 1986.

[25] ČSN ISO 2406, "Continuous mechanical handling equipment. Mobile and portable conveyors, Constructional specifications", ÚNI, Praha, pp. 4, 1993.

IOP Conf. Series: Materials Science and Engineering 1190 (2021) 012017 doi:10.1088/1757-899X/1190/1/012017

Mechanical Friction Retainers Used for Braking and Locking Pallets in Place

Leopold Hrabovský[1], Martin Mantič[2], Gustav Kotajný[3], Václav Žáček[3]

[1]VSB - Technical University of Ostrava, Faculty of Mechanical Engineering, Institute of Transport, 17. listopadu 2172/15, 708 00 Ostrava - Poruba, Czech Republic
[2]Technical University of Košice, Faculty of Mechanical Engineering, Institute of design machine and process engineering, Letná 9 Street, 042 00 Košice, Slovak Republic
[3]KOMA Industry s.r.o., Ruská 514/41, 706 02 Ostrava - Vítkovice, Czech Republic

leopold.hrabovsky@vsb.cz

Abstract. During 2020, a total of 140,336 new passenger cars were registered in the Czech Republic. Thus, at the beginning of 2021, the number of registered passenger cars in the Czech Republic reached 6,129,874. Even though the number of parking areas and parking spaces associated with individual types of buildings in the Czech Republic is determined pursuant to ČSN – Czech Technical Standards (ČSN 73 6056 - Parking areas for road vehicles and ČSN 73 6110 - Design of urban roads), most cities suffer acutely from an insufficient number of parking spaces. In the Czech Republic, dimensions of parking spaces are set by ČSN Standard 73 6056, or ČSN 73 6058 - Small, multi-story and mass garages. In densely populated urban and panel-block housing areas, it is almost a daily occurrence that drivers cannot find a place to park properly. One of the possible solutions that could cut down parking problems is "mass garages", single or multi-story, which efficiently make use of the built-up area and space. Parking in mass garages is based on the definition given in ČSN 73 6056. The basic length/width of perpendicular parking is 2.5 m/5.0 m (according to ČSN 73 6056), to which it is necessary to add area for spacing (depending on whether the parking is on an edge, next to a pillar, etc.). The next most efficient solution is "automated parking structures", which don´t require additional area for spacing around vehicles. In automated parking structures, vehicles are handled on steel pallets. The pallets are vertically transported by the computer control software to the designated floor by a lifting device, where they are horizontally moved to a transport device, which transports them to a parking rack cell. The pallet is inserted into / removed from the rack cell by a handling platform installed on the frame of the steel construction of the transport device. The pallet located in the rack cell must be fixed in the required position to prevent spontaneous movement of the pallet. The pallet is fixed by a locking device, the so-called mechanical brake. The paper describes three types of locking spring retainers, which can in some cases replace the existing structurally more complex brakes used in existing parking structures.

1. Introduction

The basic conceptual variants of automatic parking systems (hereinafter only APS) implemented by KOMA-Industry s.r.o. and KOMA Parking [1] (hereinafter referred to as KOMA) use steel pallets to move vehicles in the interiors of parking garages.

Since the APS can be designed as above ground, underground or combined, pallets with vehicles are moved vertically between the individual floors of the APS using freight elevators.

For horizontal shifting of pallets on individual floors of the APS, handling platforms are used, which (in the most complex conceptual variant of APS "MULTI TOWER" [1]) are moved forwards or backwards in one direction (on the given APS floor) by cylindrical wheels of transport equipment on fixed tracks.

Two variants of mechanical locking brakes, which lock the pallet in the cell of the parking rack of the APS, are described in papers [2], [3], [4]. A third version of a mechanical locking brake is presented in [5]. Mechanical locking brakes used in KOMA parking buildings do not always work entirely without malfunctioning [2], [6]. The method of locking the pallet in the desired position using electromagnets is presented in [7].

Steel pallets are inserted into/removed from the cells of parking racks by chain manipulators, which are installed on the handling platforms of the APS equipment [8]. The limitation of the high amount of shear friction [9] in the contact surfaces of the inserted/extended pallet and the rack beam is ensured by rolling the pallet on sheaves, see Figure 1a. The sheaves, in a predetermined number according to the maximum load of the pallet, are attached to the lower part of the pallet or are mounted on pallet-driven beams of the steel structure of the rack. The function of the mechanical locking brakes is to fix and lock the pallet with or without a parked vehicle (i.e., the pallet is empty, without being loaded with the vehicle's weight) in the exact required position in the depth of the rack cell. The resting state of the parked pallet in the required position for the duration of the pallet's parking in the rack cell is necessary for trouble-free attachment of the pallet at the moment when it is necessary to relocate this pallet within the parking structure.

Figure 1. Pallet locking brake in a parking building in the city of Slaný

The relatively complex construction of the technical equipment [5], [4], the so-called locking brakes (Figure 1b) [2], [3], used in KOMA parking garages, whose task is to arrest the movement of the pallet at rest and to prevent unwanted movement of the pallet, can be replaced by a "mechanical retainer" with a friction segment (Figure 2, Figure 3 and Figure 6).

2. Mechanical Retainer with Friction Segment

Construction designs for three of numerous other possible design variants of mechanical retainers are shown in Figures 2 to 4. The prevention of horizontal movement of a pallet in the cell of the parking rack is accomplished by means of the friction force F_T (1) acting in the horizontal contact surfaces of the friction segments of the mechanical retainer. The magnitude of the friction force F_T is directly proportional to the magnitude of the coefficient of friction f *[-]* of the materials used in the design of the friction segments. It is required that the coefficient of friction f at the contact surfaces of the friction segments acquires the highest possible value. The friction segments can therefore be made of technical plastics (e. g. PolyTAN [10] - polyurethane in green or blue), which have high abrasion resistance and a high friction coefficient. ($f = 0.3 \div 0.4$ for PolyTAN).

$$F_T = F_{S(i)} \cdot f = F_S \cdot i_S \cdot f = k_S \cdot s_S \cdot i_S \cdot f \ [N] \quad (1)$$

IOP Conf. Series: Materials Science and Engineering 1190 (2021) 012017 doi:10.1088/1757-899X/1190/1/012017

An empty pallet with a mass of m_p *[kg]* is locked in the parking position in the rack cell by two pieces of mechanical retainers (for example, Figure 2a, c or Figure 3a, c). Each retainer is fitted with two compression coil springs 4 [11], of the R203-514 type, $D_H/D_d = 25/12.5$ mm, (Figure 2b) of open length $L_0 = 89$ mm and a stiffness $k_s = 24$ N·mm^{-1}, with a rectangular cross-section wire $b \cdot h = 5.4 \cdot 2.7$ mm (Figure 2b).

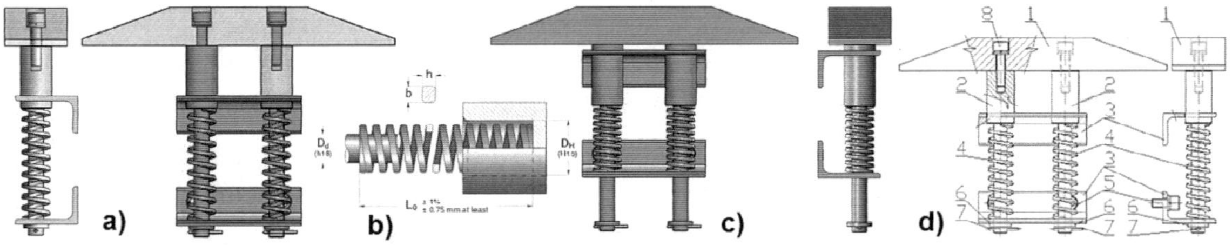

Figure 2. Locking mechanical retainer a), c) 3D model, b) compression spring R203-514 [11], d) 2D construction design

The friction segment 1 of the stopping mechanical retainer (Figure 2d) is fastened to the pins 2, by two screws 8. Cylindrical compression springs 4 are threaded on the cylindrical part, 12 mm in diameter, of both pins 2. Two ϕ 26 mm holes are drilled in the upper L-profile 3. Two ϕ 13 mm holes are drilled in the lower L-profile 3. Washers 6 are threaded on the end cylindrical part of both pins 2 (protruding from the L-profile 3). The pins 2 are secured against sliding out of the L-profiles 3 by cotter pins 7.

If detachment of the pallet rollers from the travel track of the rack is to be prevented, then the magnitude of the resulting sum of compressive forces $F_{s(i)}$ of all springs used must be less than the magnitude of the weight of the empty pallet G_p *[N]*. The maximum spring compression s_s *[mm]* [11] can be determined according to relationship (2).

$$s_S \leq \frac{G_p}{k_S \cdot i_S} = \frac{F_{S(i)}}{k_S \cdot i_S} \; [mm] \tag{2}$$

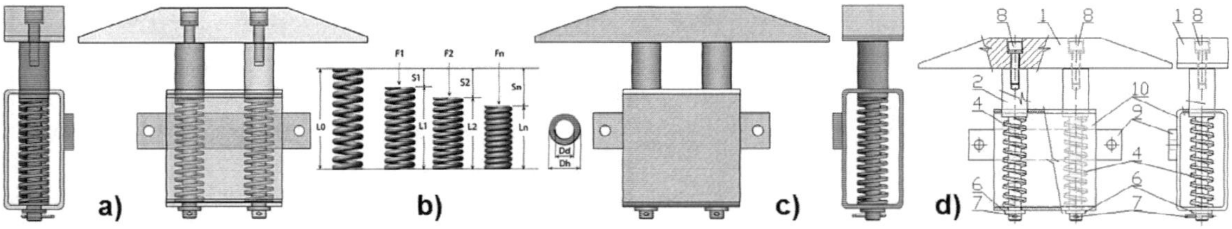

Figure 3. Mechanical locking retainer a), c) 3D model, b) compression spring, d) 2D construction design

The friction segment 1 of the locking mechanical retainer (Figure 3d) is fastened to the pins 2, by two screws 8. Cylindrical compression springs 4 are threaded on the cylindrical part, 12 mm in diameter, of both pins 2. Two ϕ 26 mm holes are drilled in the upper surface of the closed profile with a rectangular cross-section 10. Two ϕ 13 mm holes are drilled in the lower surface of the rectangular, closed profile. Washers 6 are threaded on the end cylindrical part of both pins 2 (protruding from the profile 10). The pins 2 are secured against sliding out of the profiles 10 by cotter pins 7.

When inserting the pallet into/removing the pallet from the rack cell, the compressing force *F [N]* of the chain feeder overcomes the moving resistances of the pallet as well as the frictional resistance forces F_{FT} *[N]* (3) (Figure 4b) and F_{ST} *[N]* (4) (Figure 4c) in the chamfered (at an angle ε *[deg]*) leading surfaces of the friction segments 1.

$$F_{FT} = F_1 \cdot f = F \cdot sin\varepsilon \cdot f\ [N] \tag{3}$$

$$F_{sT(i)} = F_{s1(i)} \cdot f = F_{s1} \cdot i_s \cdot f = F_{s(i)} \cdot cos\varepsilon \cdot f = F_s \cdot i_s \cdot cos\varepsilon \cdot f\ [N] \tag{4}$$

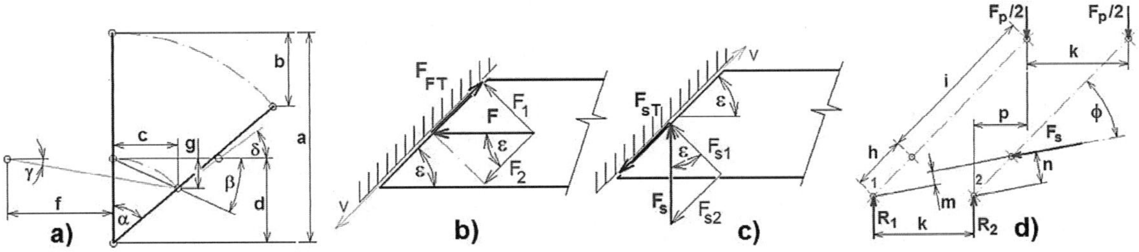

Figure 4. Acting forces, kinematics of activity and geometric dimensions of the arms of the mechanical retainer

The instantaneous values of the compressive force $F_{s(i)}$, the components of the normal forces F_{s1} and the tangent F_{s2}, and the values of the friction force F_{sT} *[N]* in the contact surfaces of the friction segments of one (two pieces are shown in Figure 2 and Figure 3) and two $F_{sT(i)}$ (4) of the compression springs 4 are shown in Figure 5.

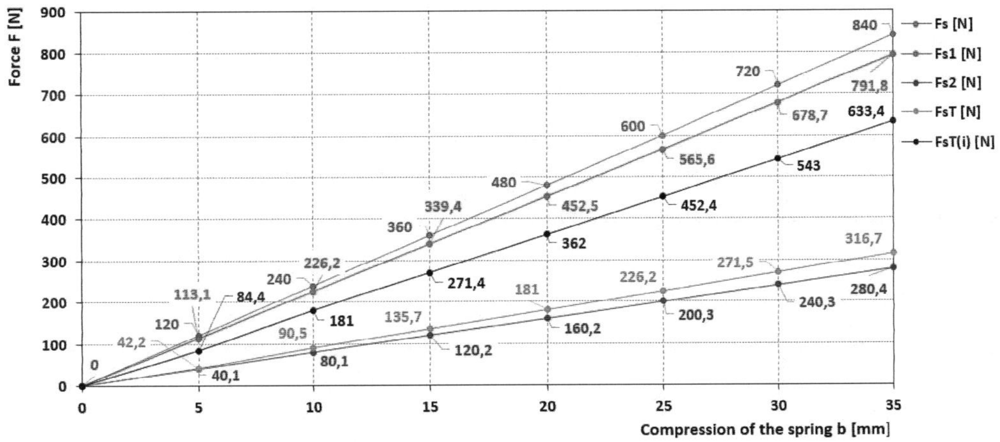

Figure 5. Forces acting in the stopping and locking mechanical retainer induced by compression of cylindrical springs

Table 1. Friction force $F_{sT(i)}$ *[N]* during vertical movement (compression of spring 4) of the friction segment 1 of the mechanical retainer

s_s	mm	0	5	10	15	20	25	30	35
$F_{sT(i)}$	N	0	84.4	181.0	271.4	362.0	452.4	543.0	633.4

If it is considered that the coefficient of friction in the contact surfaces of the friction segments acquires the value $f = 0.4$, then the value of the instantaneous friction force $F_{sT(i)}$ (4) can be subtracted, depending on the compression s_s of the two compression springs 4 (see Figure 2 and Figure 3) from Table 1.

Locking in place of the pallet in the cell of the APS parking rack can be ensured by a mechanical retainer see Figure 6.

IOP Conf. Series: Materials Science and Engineering 1190 (2021) 012017 doi:10.1088/1757-899X/1190/1/012017

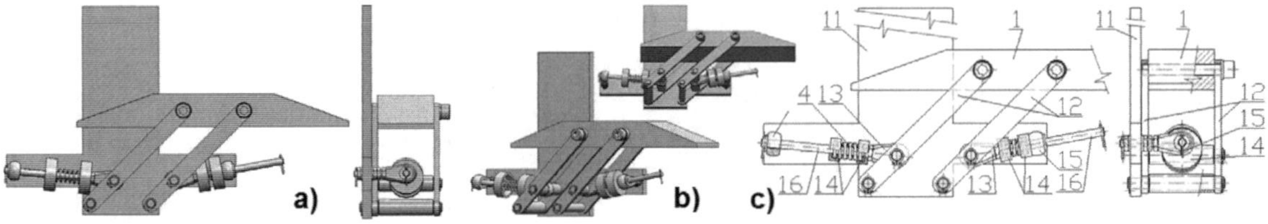

Figure 6. Mechanical retainer a), b) loaded with a pallet 3D model, c) 2D construction design

In the case of a mechanical retainer (Figure 6c), the arms <u>12</u> are fastened to the friction segment <u>1</u> with two screws. Compression springs <u>4</u> are threaded on the cylindrical part of both pins <u>16</u>, and their ends are inserted into the cups <u>14</u>. The pins <u>16</u>, are provided at one of their ends with an external M6 thread, and these are screwed into the holes of the eyes <u>15</u> (type EF-E006Z [12]), which are provided with an internal M6 thread. Deflection of the arms <u>12</u> from the vertical position results in the compression of the spring <u>4</u>. The strength of the induced pressing force F_p, corresponding to the angle of inclination α of the arms <u>12</u>, is given in Table 3.

The mechanical retainer, see Figure 6, uses ISO device springs <u>4</u> [12] (red, for heavy loads) type R205-204, D_h/D_d = 12.5/6.3 mm, L_0 = 25 mm, s_s = 7.5 mm ($L_N = L_0 - s_s = 25 - 7.5 = 17.5\ mm$), $k_s = 42.2\ N{\cdot}mm^{-1}$. If the retainer is in the locking position (Figure 6a, b), that is, locking the pallet in the parked position in the rack cell, then the spring <u>4</u> (see Figure 6d) receives the compression force $F_s = s_{s4}{\cdot}k_s = 5.6{\cdot}42.2 = 236.32\ N$.

According to Figure 4a it is possible, for the lowest possible position, b = 35 mm, of the upper surface of the friction segment and for the selected construction dimensions a = 120 mm, d = 30 mm, f = 70 mm, expressed as: $\alpha = arccos[(a - b)/a] = 44.9\ deg$, $c = d{\cdot}sin\alpha = 21.18\ mm$, $g = d - d{\cdot}sin\alpha = 8.75\ mm$, $\beta = arctan(g/c) = 22.45\ deg$, $\gamma = arctan[g/(c + f)] = 5.48\ deg$, $\delta = arctan[g/(f - c)] = 10.16\ deg$. The values of c, g, α, β, γ, δ for the position of vertical shifting $b = 0 \div 35\ mm$ of the upper surface of the friction segment are listed in Table 2.

Table 2. Angles of incline and distance of points, see Figure 4a in the mechanical retainer (Figure 6), corresponding to the vertical shift of the upper surface of the friction segment <u>1</u>.

b	mm	0	5	10	15	20	25	30	35
α	deg	0	16.6	23.56	28.96	33.56	37.66	41.41	44.9
c	mm	0	8.57	11.99	14.52	16.58	18.33	19.84	21.18
g	mm	0	1.25	2.5	3.75	5.0	6.25	7.5	8.75
β	deg	0	8.3	11.78	14.48	16.78	18.83	20.7	22.45
γ	deg	0	0.91	1.75	2.54	3.31	4.05	4.77	5.48
δ	deg	0	1.17	2.47	3.87	5.35	6.9	8.5	10.16

To determine the strength of the vertical force $F_p\ [N]$ (8), by which the friction segment of the mechanical retainer is pressed against the friction segment attached to the lower surface of the steel pallet, the sum equation of the forces acting in the vertical direction (5) and the torque equations (6) and (7) are expressed according to Figure 4d.

$$\Sigma F_y = 0{:}\ R_{1y} + R_{2y} - \frac{F_{py}}{2} - \frac{F_{py}}{2} - F_{sy} = 0\ N \tag{5}$$

$$\Sigma F_x = 0{:}\ R_{1x} + R_{2x} + \frac{F_{px}}{2} + \frac{F_{px}}{2} - F_{sx} = 0\ N \tag{6}$$

$$\Sigma M_1 = 0: F_{py}\cdot\frac{k+p}{2} + F_{py}\cdot\frac{k+p+k}{2} + F_{px}\cdot a\cdot\cos\alpha + F_S\cdot m - R_{2y}\cdot k = 0 \Rightarrow$$

$$\Rightarrow R_{2y} = \frac{F_{py}\cdot(p+1.5\cdot k) - F_S\cdot m + F_{px}\cdot a\cdot\cos\alpha}{k} = \frac{F_{py}\cdot A - F_S\cdot m + F_{px}\cdot a\cdot\cos\alpha}{k}\ [N] \quad (7)$$

$$\Sigma M_2 = 0: F_{py}\cdot\frac{p}{2} + F_{py}\cdot\frac{p+k}{2} + R_{1y}\cdot k + F_{px}\cdot a\cdot\cos\alpha - F_S\cdot n = 0 \Rightarrow$$

$$\Rightarrow R_{1y} = \frac{F_S\cdot n - F_{py}\cdot(p+0.5\cdot k) - F_{px}\cdot a\cdot\cos\alpha}{k} = \frac{F_S\cdot n - F_{py}\cdot B - F_{px}\cdot a\cdot\cos\alpha}{k}\ [N] \quad (8)$$

where, as per Figure 4d, for the lowest possible position $b = 35$ mm of the upper friction surface of the friction segment and for the chosen construction dimensions $h = d\ [mm]$, $i = 90\ mm$, $k = 55\ mm$ it applies that: $\phi = \pi/2 - \alpha - \delta = 34.94\ deg$, $p = a\cdot sin\alpha = 29.71\ mm$, $n = h\cdot sin\phi = 17.18\ mm$, $\delta = \pi/2 - \alpha - \phi = 10.16\ deg$, $m = n - k\cdot sin\delta = 7.48\ mm$, $A = 1.5\cdot k + p = 112.21\ mm$, $B = 0.5\cdot k + p = 51.21\ mm$, $F_{sy} = F_s\cdot sin\delta\ [N]$. The values p, n, m, A, B for the position of vertical shifting $b = 0\div35\ mm$ of the upper surface of the friction segment are listed in Table 3.

Inserting the relationships (7) and (8) into the expression (5) and using mathematical modification, we obtain the relationship for calculating the vertical force $F_{py}\ [N]$ (9).

$$F_{py} = \frac{F_S\cdot(m - n + k\cdot sin\delta)}{A - B - k}\ [N] \quad (9)$$

The instantaneous value of the compressive force F_s of the coil spring 4 can be found, for the dimensions of the mechanical retainer (Figure 4a, d) given in Table 2, depending on the vertical displacement b of the upper surface of the friction segment 1 in Table 3.

Table 3. Angle of incline, for distance of points, see Figure 4d and the forces generated in the mechanical retainer (Figure 6) derived from the vertical movement of the friction segment 1.

b	mm	0	5	10	15	20	25	30	35
ϕ	deg	90	72.24	63.98	57.18	51.1	45.44	40.09	34.94
p	mm	-55.0	-20.72	-7.04	3.09	11.33	18.31	24.37	29.71
n	mm	30.0	28.57	26.96	25.21	23.35	21.38	19.32	17.18
m	mm	30.0	16.06	14.81	13.47	12.05	10.58	9.05	7.48
A	mm	27.5	61.78	75.46	85.59	93.83	100.81	106.87	112.21
B	mm	-27.5	6.78	20.46	30.59	38.83	45.81	51.87	57.21
s_s	mm	0	0	0	0	1.55	3.15	4.49	5.6
F_s	N	0	0	0	0	65.41	132.93	189.48	236.32
F_p	N	0	0	0	0	17.01	33.23	45.48	51.99
F_T	N	0	0	0	0	6.80	13.29	18.19	20.80

The values of the compressive force F_s of the coil spring 4, the normal force F_p (10) and the friction force F_T in the contact surfaces of the friction segments (coefficient of friction $f = 0.4$), depending on the compression s_s of the coil spring 4, are shown in Figure 7.

$$F_p = \frac{F_S\cdot n}{a\cdot sin\alpha} = \frac{F_S\cdot n}{(h+i)\cdot sin\alpha}\ [N] \quad (10)$$

IOP Conf. Series: Materials Science and Engineering 1190 (2021) 012017 doi:10.1088/1757-899X/1190/1/012017

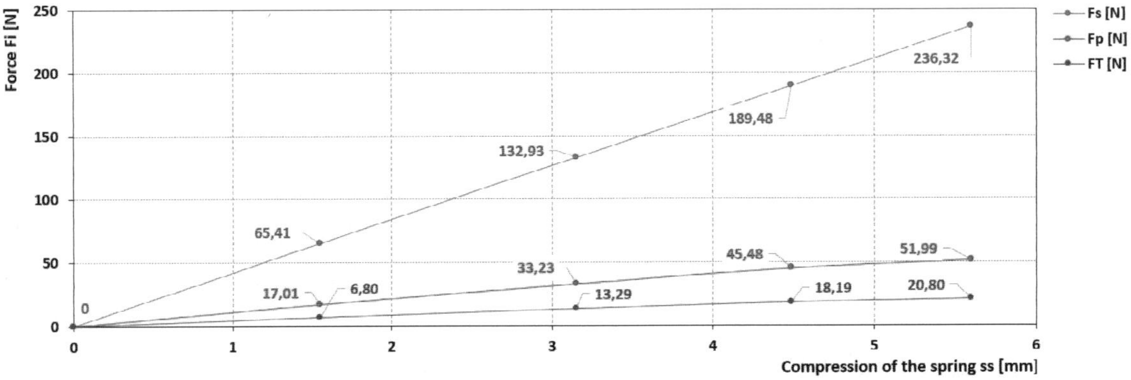

Figure 7. Compression force of the spring, normal and friction force in the contact surfaces of the friction segments of the mechanical retainer

Changing the vertical position of the upper surface of the friction segment 1 *b* (in the range of 0 to 35 mm) when inserting the pallet into the rack cell causes compression of the spring 4 of strength s_s (in the range of 0 to 5.6 mm) in the mechanical retainer, see Figure 6a. By compression s_s, the spring exerts a compressive force F_s, the instantaneous strength of which is directly proportional to the stiffness k_s. The arms of the mechanical retainer *I* transmit the compressive force of the spring F_s to the friction segment *II*, the upper surface of which, when the pallet is inserted into the rack cell, contacts the lower surface of the opposite segment *III* (Figure 8). The contact surfaces of the segments are pressed against each other by a force F_p, when the pallet is moved horizontally, a frictional force F_T is generated on the surfaces of the friction segments.

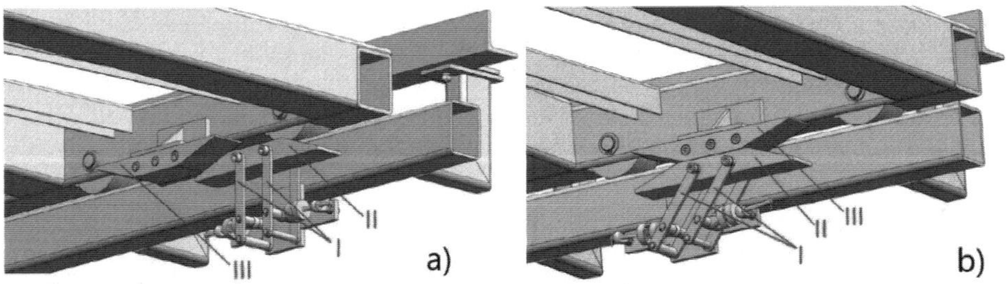

Figure 8. Steel pallet inserted in the rack cell in position a) unlocked, b) fixed

For trouble-free catching of the pallet by the automatically operating chain feeder (roller chain 16-B 1 [13], dimension $p{\cdot}b_1 = 25.4{\cdot}17.02\ mm$, sprocket with $z_k = 13$ as the number of teeth, a pitch circle diameter of $d_k = 106.14$ mm) it is required that at the time of requesting the pallet to be ejected from the rack cell, that the pallet does not change its position [6] for the duration of the parking time. Locking of the pallet in the parking rack cell can be ensured by using a mechanical brake [2], [5] or a mechanical retainer with a friction segment (Figure 2 or Figure 6).

3. Conclusion

The desperate lack of parking spaces, caused by the ever-increasing number of vehicles, leads to the need for an acute increase in the number of parking spaces. A partial solution to increase the number of parking spaces where possible is a more appropriate way of arranging parked cars on the streets, for example by replacing horizontal parking with vertical parking. A more efficient way is to build parking terminals, multi-story garages and especially multi-story parking structures with fully automated operation. In automatic parking garages, vehicles are handled on steel pallets, which must be fixed in the required position in the cells of parking racks. Pallet fixation is currently provided by mechanical brakes, which use the compressive forces of the longitudinally deformed coil spring. The introduction of this paper refers to two types of mechanical brakes that are used in APS implemented by KOMA.

Three types of mechanical retainers, which are presented in the paper as possible structural alternatives of fixing and locking elements of steel pallets, use the compressive forces of the coil spring to achieve enough pressure between the contact surfaces of the friction segments, with a high coefficient of friction. The frictional force in the contact surfaces of both friction segments makes it is possible to keep the pallet with the vehicle in the desired position in the cell of the parking rack.

The text of the paper states that for high loads, of the R205-204 type, the mechanical retainer shown in Figure 6 uses springs. However, this type of spring exerts a compressive force during compression, which is too small for practical application in fixing pallets with vehicles in a real parking structure. This spring (stiffness 42.2 N/mm) was used for the purposes of laboratory measurements, which are described in [3], [4]. According to relationship (10), it is possible to express the normal force by which the friction segment of the mechanical retainer is pressed against the friction segment attached to the beam of the steel structure of the parking rack. The pallets are fixed in the required position by a friction force in the contact surfaces of both friction segments. The magnitude of the friction force in the contact surfaces of both friction segments can be influenced by the coefficient of friction (material of friction segments with a high coefficient of friction) and the magnitude of the pressing (normal) force. The compressive force is therefore directly proportional to the compressive force (stiffness) of the spring. According to the catalog of springs [12], a spring, or R205-504, could be used in the design for practical use, with a stiffness of 375.6 $N \cdot mm^{-1}$.

Acknowledgment(s)
This work has been supported by The Ministry of Education, Youth and Sports of the Czech Republic from the Specific Research Project SP2021/53.

References

[1] Automated parking systems (APS), [Online] 2003. Available at: www.komaparking.cz/automated-parking-system/.
[2] L. Hrabovsky and D. Dluhos, "Calibration of Transducers and of a Coil Compression Spring Constant on the Testing Equipment Simulating the Process of a Pallet Positioning in a Rack Cell", *Open Engineering*, vol. 9, pp. 631-640, 2019.
[3] L. Hrabovsky and T. Mlcak, "Vertical Forces Acting on the Lock-Off Element of the Pallet Brake System", *Advanced in Science and Technology-Research Journal*, vol. 14(1), pp. 116-126, 2020.
[4] L. Hrabovsky, T. Mlcak and G. Kotajny, "Forces Generated in the Parking Brake of the Pallet Locking System", *Advanced in Science and Technology-Research Journal*, vol. 13(4), pp. 181-187, 2019.
[5] L. Hrabovsky, V. Molnar, G. Kotajny and J. Kulka, "Pallet Fixing Lock, Determination of Shear Resistance and Spring Stiffness", *Advanced in Science and Technology-Research Journal*, vol. 14(2), pp. 190-197, 2020.
[6] L. Hrabovsky and V. Zacek, "Forces in the lock-off device during the rolling of the brake pulley roller over the brake ramp in the built automatic parking house" *IOP Conference Series: Materials Science and Engineering*, vol. 960(2), 2020.
[7] L. Hrabovsky, G. Fedorko, L. Mlynek and P. Michalik, "Electromagnetic Locking Devices of Car Handling Units", *Scientific Journal of Silesian University of Technology-Series Transport*,. vol. 107, pp. 73-83, 2020.
[8] J. Kulka, M. Mantic, M. Kopas and E. Faltinová, „Locking the movement of persons on the bridge crane", *Advances in Science and Technology- Research Journal*, vol. 12(2), pp. 260-265, 2018.

[9] J. Fries, I. Onderkova and M. Prochazka, "Comparison of Shearer Sumping Methods", *16th International Multidisciplinary Scientific Geoconference (SGEM 2016)*, pp. 141-148, 2016.
[10] PolyTAN (polyuretan), [Online] 2021. Available at: www.polyplasty.cz/material-polyuretan.html.
[11] Nástrojové pružiny dle ISO 10243, [Online] 2021. Available at: www.pruziny-fevos.cz/nastrojove.pdf.
[12] Tažné pružiny - Pozinkované, [Online] 2021. Available at: www.sodemann-pruziny.cz/produkty/tazne-pruziny/rada-c-pozinkovane.
[13] Válečkové řetězy podle DIN 8187, [Online] 2021. Available at: www.tyma.cz/produkty/retezy/valeckove-hnaci/din-8187/.

17-4 PH SS Manufactured via Selective Laser Melting: High Cycle Fatigue Properties

Lorenzo Fraccaroli [1], Franco Concli [1]

[1] Free University of Bolzano/Bozen, Faculty of Science and Technology, piazza Università 5, 39100 Bolzano (I), Italy

franco.concli@unibz.it

Abstract. In the recent years, Additive Manufacturing (AM) is becoming an emerging technology and more and more material powders are available on the market. The knowledge of their specific mechanical behaviour is a fundamental aspect for their large-scale adoption, also considering the lack of reliable data and dedicated standards.In this work the High Cycle Fatigue characterization of a 17-4 PH Stainless Steel (SS) is presented. The cylindrical samples used were manufactured via Selective Laser Melting (SLM) technology using an EOS M280 machine. Two separate series of samples differing in terms of surface finishing were tested. The first series of samples was used in the as-build condition while the second series was machined in order to obtain a better surface finishing.The effect of the surface finishing on the fatigue behaviour of AM materials is fundamental considering that in most of the applications the parts – reticular or lattice structures with complex geometries – are set into operation in the as-build condition because a surface finishing is not feasible. The comparison of the results obtained for the two different series of specimens allows the quantification of the reduction of the mechanical performances due to the actual limits of the SLM technology. The fatigue limit was obtained with the short stair-case approach according to the Dixon statistical method. The maximum number of cycles (run-out) was set equal to 5000000. The left part of the Wöhler diagram was also studied by means of additional tests at higher stress levels.Additional quasi-static tests were performed in order to characterise the static behaviour.

1. Introduction

Laser sintered 17-4 PH SS, is a material with a very good corrosion resistance and high strength guaranteed by its martensitic structure [1]. The high strength is due to the precipitation of fine Cu-rich face centred cubic phase. A dendritic structure with body centred cubic martensite and about 50% of reserved austenite is favoured by the SLM process. According to [2], the mechanical properties of additive manufactured materials are comparable with those produced with classic technologies. Additive Manufacturing (AM) is characterized by high solidification speeds, leading to a metastable austenite phase in the matrix. For this reason, this material is generally hardened. Nevertheless, in literature are present several works in which this type of material shows a low fatigue resistance. Internal and surface defects like porosities, inclusions and a high roughness in the as-build condition, can act as crack nucleation sites [3]–[5].

For this reason, in addition to a first series of as-build samples which surface roughness is the one that can be obtained with the SLM process, a second series of samples was produced with the same technology and successively machined.

Tests were performed on both series of samples in order to understand the harmful effect of the low surface quality achievable with SLM. This comparison is particularly significant considering that the AM technologies are suitable for producing complex geometries such as reticular or lattice structures, that cannot be successively surface finished.

2. Chemical composition and microstructure

The chemical composition of 17-4 PH SS is shown in Table 1. The particles of the powder have a mean size of 44.454 μm. Chromium interacts with Carbon, forming chromium carbides along grains [6].

Table 1. Chemical composition of the 17-4 PH SS.

C	Si	Mn	P	S	Cr	Ni	Mo
0.036	0.78	0.33	0.009	0.004	16.2	4.02	0.002

Co	V	Ti	Cu	W	Sn	Al	Nb
0.012	0.018	0.002	0.372	0.009	0.004	0.001	0.29

In order to prevent the creation of $Cr_{23}C_6$ and to increase strength, Niobium is used. It promotes the formation of NbC. Silicon decreases the mechanical properties promoting the formation of ferrite, but it is useful for fluidizing the material during the casting. To compensate this effect, Chromium and Nickel are used. They reduce the formation of ferrite in favour of austenite. Nickel and Manganese are helpful to avoid the formation of $Cr_{23}C6$ and FeS phases respectively, forming Cr_2N and MnS instead. Cuprum helps increasing the total strength of the material forming precipitates.

Despite the chemical analysis, Scanning Electron Microscopy (SEM) measurements have been made to better characterize the size of the defects that was found to be up to 100 μm.

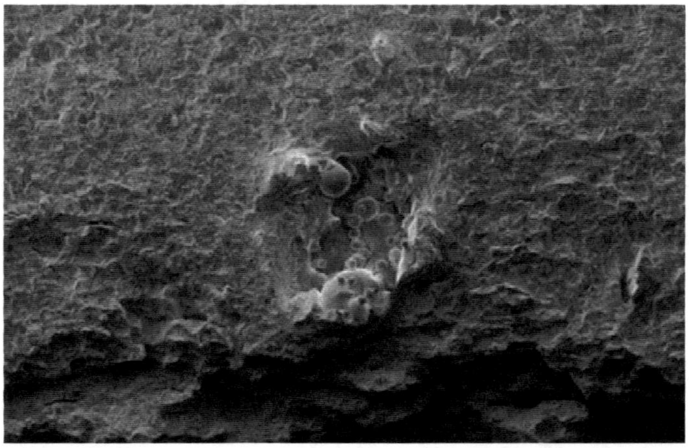

Figure 1. SEM analysis of the fracture surface

3. Experimental tests

High-cycle-fatigue (HCF) tests were performed on cylindrical samples (Figure 1) having a geometry according to the ASTM E466 standard [7]. The apparatus used for testing is a STEPLab-UD04 fatigue testing machine capable to apply up to 5 kN at maximum frequency of 35 Hz already used for the Low-Cycle-Fatigue characterization of the same material [8].

In addition to the HCF tests, a quasi-static tensile characterization was performed. The geometry of the adopted samples is shown in Figure 2. Tensile tests were performed at room temperature with a crosshead speed set to 0.1 mm/min.

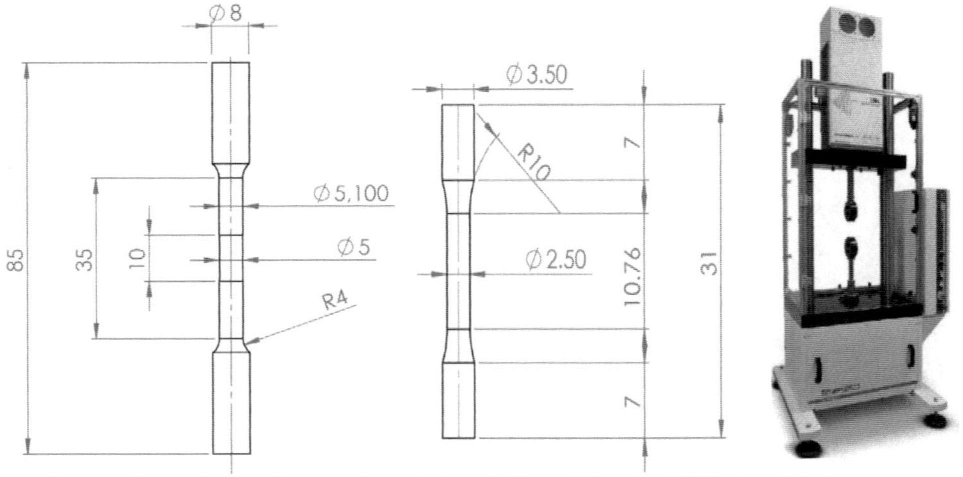

Figure 2. Standard specimen: for quasi-static (up-left) and for the HCF (up-right) tests according to ASTM and testing machine (down)

The tested samples present the typical cone-cup shaped fracture surface (ductile failure). Plastic deformation was also confirmed by the presence of a fibrous and irregular surface (necking region). The Quasi-Static (QS) stress-strain curve is shown in figure 3.

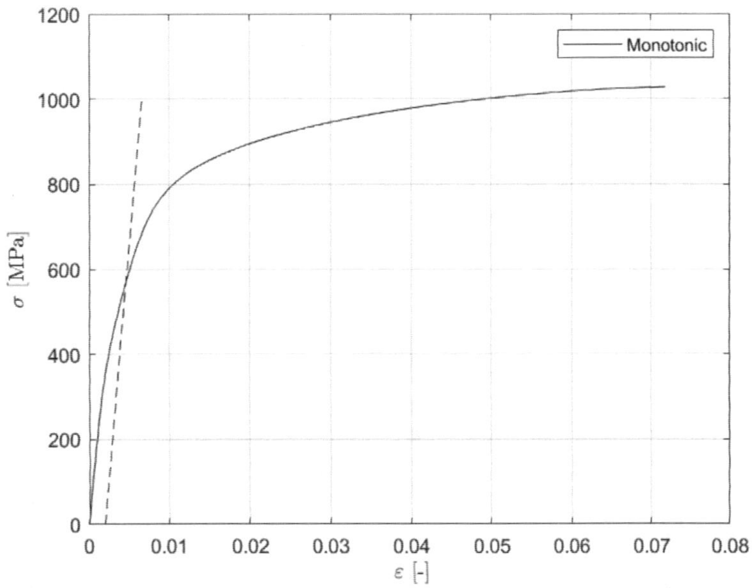

Figure 3. Quasi-static monotonic curve

According to the test results, the yielding point was found to be $\sigma_Y = 570MPa$ ($\varepsilon = 0.2\%$). This data has been used for defining the left part of the Wöhler ($S-n$) diagram.

HFC tests were performed according to the ASTM E466 standard [7].

The fatigue limit σ_F was calculated with the short stair-case approach [9]–[12] on samples having the geometry shown in figure 2 (right).

IOP Conf. Series: Materials Science and Engineering 1190 (2021) 012018 doi:10.1088/1757-899X/1190/1/012018

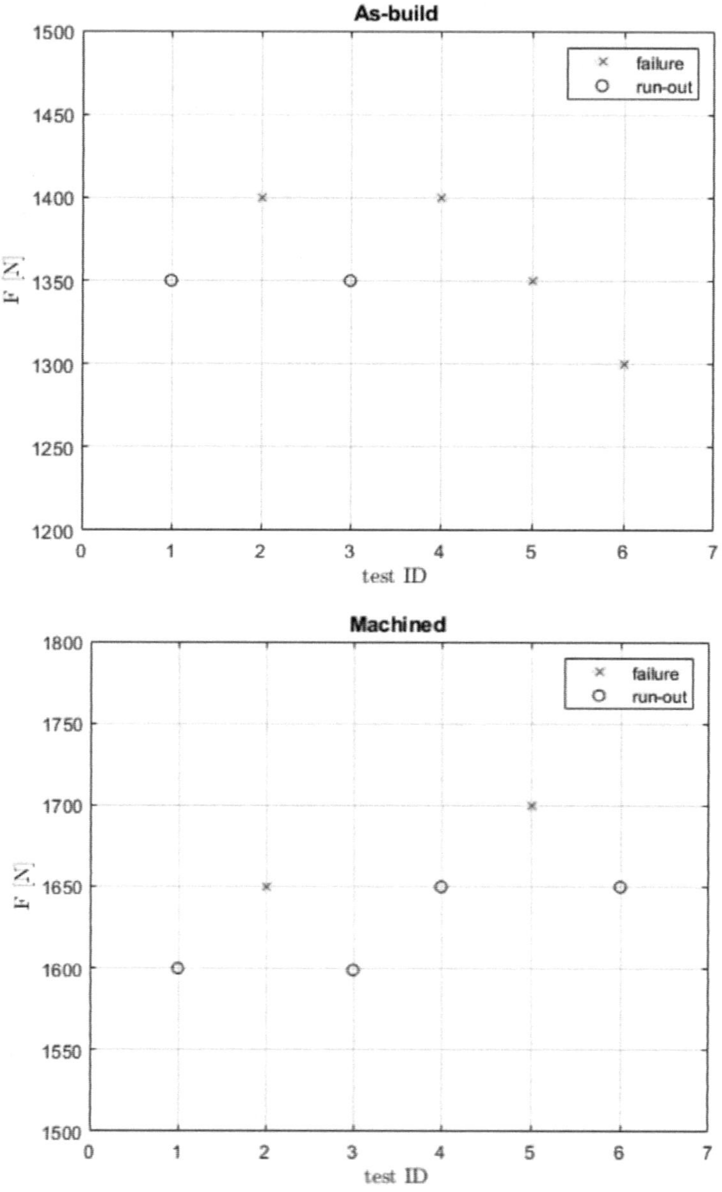

Figure 4. Short stair-case sequences for the as-build and for the machined samples

Several tests at different load levels were performed.

The stair-case method prescribes to define a force increment ΔF (that is directly related to the accuracy of the results that can be achieved). The actual ΔF was equal to 50 N corresponding to a stress interval $\Delta\sigma$ equal to 10 MPa.

If the initial test, performed at a force level F_i, reaches the run-out condition (5000000 cycles without breaking), the next test has to be performed on a force level increased by ΔF i.e. $F_{i+1} = F_i + \Delta F$

Otherwise, if the specimen broke, the next force level must be decreased by ΔF i.e. $F_{i+1} = F_i - \Delta F$.

Figures 4 and 5 show the short stair-case sequences (6 tests) for the two series of samples, as-build and machined.

The number of tested specimens was, for both series of samples, below 15 as suggested by the standards.

Therefore the Dixon statistical approach was used [13]. This approach allows the calculation of the fatigue limit taking into consideration a number of tests between 2 and 7. Statistical significance of the results is ensured by the correction coefficients.

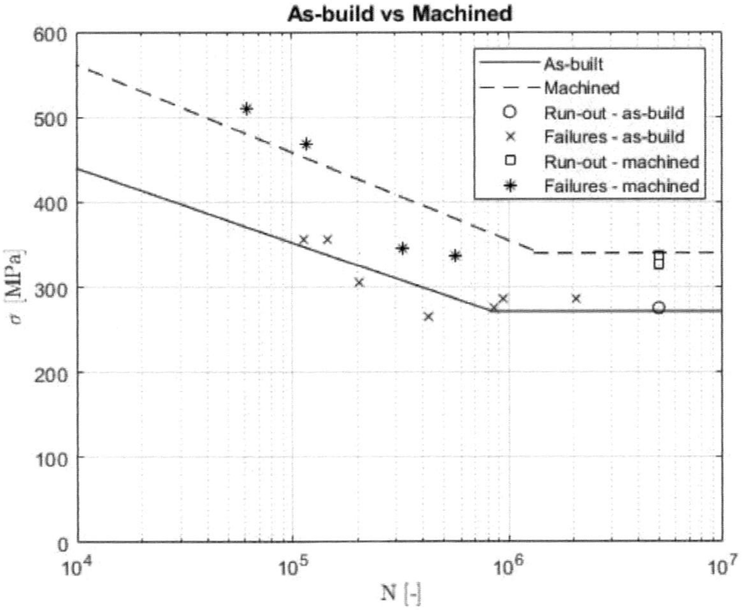

Figure 5. S-N curve and comparison of as-build and machined fatigue properties

According to the Dixon approach, the sequence order of runouts (O) – failures (X) is also considered. The fatigue limit can be calculated by means of the following relation

$$\sigma_F = x_f + k \cdot \Delta\sigma \tag{1}$$

in which σ_F is the fatigue limit at a 50% probability, x_f is the stress level of the last test (ID=6 with reference to figure 4) and k a coefficient depending on the sequence order.

For the as-build testing campaign, the tests order results in O-X-O-X-X-X ($x_f = 265$ MPa). Having a k factor equal to 0.661, the fatigue limit results $\sigma_F = 271$ MPa.

For the machined samples, the test order was O-X-O-O-X-O ($x_f = 336$ MPa). The Dixon coefficient k was 0.372 resulting in a fatigue limit equal to $\sigma_F = 340$ MPa.

Additional tests were performed for both the sample groups at higher levels of stress in order to obtain also the left part of the Wöhler diagram.

Figure 4 shows the S-N curve, Run-outs are identified by (○ or □) while failures with (× or ∗). Continuous line refers to the as-build condition, the dashed line to the machined one.

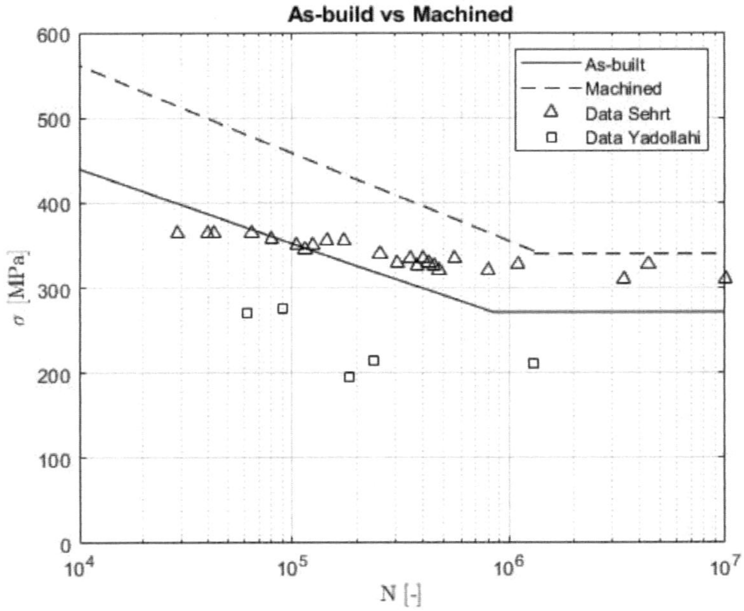

Figure 6. Comparison between the actual S-N curves and the data from literature [14][3][15].

Figure 6 compares the actual results with those obtained by other authors on as-build samples.

4. Conclusions

A 17-4 PH SS was characterized with quasi-static tensile and fatigue tests. The yielding point was found to be 571 MPa and the ultimate strength limit (UTS) at 1027 MPa that is coherent with literature data.

Fatigue teste were performed on as-build as well as machined samples differing in the surface finishing only. The fatigue limit was obtained exploiting the short star-case method and the Dixon's statistical approach. Fatigue limit results equal to 271 MPa for the as-build samples and 340 MPa for the machined ones. Surface defects act as crack initiation sites reducing the fatigue strength of the material.

Acknowledgment(s)

The authors would like to thank the Free University of Bozen-Bolzano for the financial support given to this study through the projects M.AM.De (TN2092, call CRC2017 Unibz PI Franco Concli franco.concli@unibz.it).

References

[1] S. Cheruvathur, E. A. Lass, and C. E. Campbell, "Additive Manufacturing of 17-4 PH Stainless Steel: Post-processing Heat Treatment to Achieve Uniform Reproducible Microstructure," *JOM*, 2016.

[2] A. et al Yadollahi, "Fatigue Behavior of 17-4 PH Stainless Steel Fabricated Using Selective Laser Melting," in *26th International Solid Freeform Fabrication Symposium - An Additive Manufacturing ConferenceAt*, 2015.

[3] G. Sehrt J.T., Witt, "Dynamic Strength and Fracture Toughness Analysis of Beam Melted Parts," in *Proc. 36th Int. MATADOR Conf.*, 2010.

[4] A. Yadollahi, N. Shamsaei, S. M. Thompson, A. Elwany, and L. Bian, "Mechanical and microstructural properties of selective laser melted 17-4 ph stainless steel," in *ASME International Mechanical Engineering Congress and Exposition, Proceedings (IMECE)*, 2015,

vol. 2A-2015.
[5] C. Wu, J., Lin, "Tensile and Fatigue Properties of 17-4 PH Stainless Steel at High Temperatures," *Met. Mater. Trans.*, 2002.
[6] R. J. Hamlin, "Microstructural Evolution and Mechanical Properties of Simulated Heat Affected Zones in Cast Precipitation Hardened Stainless Steels 17-4 and 13-8 + Mo," 2015.
[7] ASTM, "Standard Practice for Conducting Force Controlled Constant Amplitude Axial Fatigue Tests of Metallic Materials," *ASTM E466-15*.
[8] L. Maccioni, E. Rampazzo, F. Nalli, Y. Borgianni, and F. Concli, "Low-Cycle-Fatigue Properties of a 17-4 PH Stainless Steel Manufactured via Selective Laser Melting," in *Key Engineering Materials*, 2020.
[9] C. Gorla, F. Rosa, F. Concli, and H. Albertini, "Bending fatigue strength of innovative gear materials for wind turbines gearboxes: Effect of surface coatings," in *ASME International Mechanical Engineering Congress and Exposition, Proceedings (IMECE)*, 2012, vol. 7, no. PARTS A, B, C, D.
[10] C. Gorla, E. Conrado, F. Rosa, and F. Concli, "Contact and bending fatigue behaviour of austempered ductile iron gears," *Proc. Inst. Mech. Eng. Part C J. Mech. Eng. Sci.*, vol. 232, no. 6, 2018.
[11] F. Concli, "Austempered Ductile Iron (ADI) for gears: Contact and bending fatigue behavior," *Procedia Struct. Integr.*, vol. 8, pp. 14–23, 2018.
[12] C. Gorla, F. Rosa, E. Conrado, and F. Concli, "Bending fatigue strength of case carburized and nitrided gear steels for aeronautical applications," *Int. J. Appl. Eng. Res.*, vol. 12, no. 21, 2017.
[13] W. J. Dixon, "The Up-and-Down Method for Small Samples," *J. Am. Stat. Assoc.*, vol. 60, no. 312, pp. 967–978, 1965.
[14] Yadollahi, A., Shamsaei, N., "Additive manufacturing of fatigue resistant materials: Challenges and opportunities," *Int. J. Fatigue*, vol. 98, pp. 14–31, 2017.
[15] L. Maccioni, L. Fraccaroli, and F. Concli, "High-Cycle-Fatigue Characterization of an Additive Manufacturing 17-4 PH Stainless Steel," in *Key Engineering Materials*, 2020.

IOP Conf. Series: Materials Science and Engineering 1190 (2021) 012019 doi:10.1088/1757-899X/1190/1/012019

Innovative Solution for the Transport of Wire Ropes of Large Size and Length

Luigi Solazzi [1], Federico Ceresoli [1]

[1] Department of Mechanical and industrial Engineering, Via Branze 38, 25123 Brescia, Italy

f.ceresoli002@unibs.it

Abstract. This paper presents the study about the development of an innovative solution for road transport of high dimensions and lengths wire ropes. Currently these operations are limited by various constraints related both to the maximum weight and to the dimensions, even when using special vehicles, the passage of the load on the road becomes impossible. The proposed solution involves the use of two reels positioned on two separate trailers. In this case, the overall dimensions are reduced and the maneuverability of the load is considerably increased. The research is based on a concrete case in which the results relating to the sizing of the reels and the wire rope winding system are also reported, for the transport of a wire rope with a diameter of 125 mm and a weight of 400000kg. The research therefore shows that through this solution, which can also be extended to several vehicles connected to each other, the transport of wire ropes of large dimensions and length is achievable partially releasing the problem of dimensions and road travel.

1. Introduction

The subject of this paper is the study of an innovative method of transporting wire ropes of large dimensions and lengths, usually used for structural uses, like the mining sector and the off-shore industry. In fact, compared to the transport of other types of goods and materials, the transport of wire ropes, it has some peculiarities attributable largely to the shape of the reels, which have a cylindrical geometry and therefore are more bulky if compared, with the same weight, to squared geometries used, for example, for the construction of electrical transformers. The road transport, by industrial vehicles, also presents problems related to legislative aspects concerning transport (e.g. the maximum weight per axle) and the presence of possible obstacles present during the journey to transport the reel from one place of interest to another, for example from the place of production to the port to then be taken to the place of use. For these reasons, the type of vehicles for the transport of reels and their number is not efficient and plausible for the transport of large diameter and long length wire ropes. With these problems, the aim of this work is to present a possible alternative solution for the transport of large reels, this solution is based on the subdivision of a single large reel into two or more reels of smaller dimensions so as to be able to be transported with more readily available and maneuverable vehicles. At the production site, the wire rope is wound on several reels and positioned on different industrial vehicles; once they reach their destination, the reels must be rewound to form a single reel. The present study, after an evaluation of the problem, also based on the state of the art, finds concrete application in the transport of a 125 mm diameter wire rope, 4000 m long and with a total weight therefore equal to about 400000 kg.

2. Conceptual development of a new reel transport system

In order to develop a new transport solution, first of all, we proceeded with the study of possible methodologies and solutions, considering: the maximum load height, the overall dimensions and finally the availability of a trailer suitable for both weight and to the volumes related to the transported load, it has also been seen that the optimization of the chassis does not significantly reduce the total weight [1] [2]. Figure 1 show the two different construction solutions adopted for the transposition of large reels.

(a)

(b)

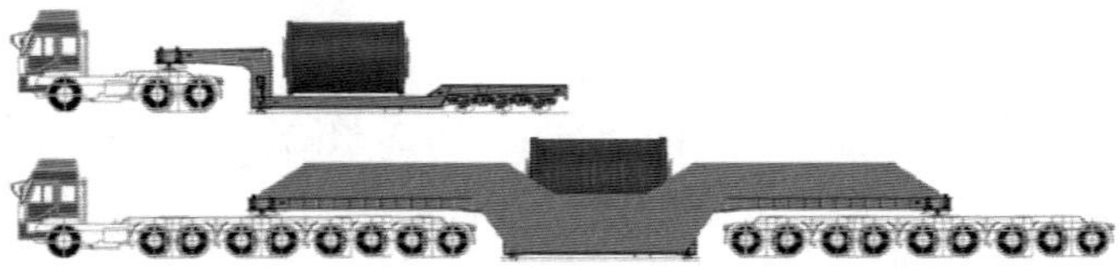

Figure 1. Possible transport configurations.

(a) Semi-trailer trolley available for loads of less than 200 tons, not usable for large reels as the imposed height limits would be exceeded.
(b) Trolley for 300/400 ton transport, the problem related to height is solved, but the longitudinal dimensions and the turning radius are increased (solution adopted for the transport of large reels).
In this work the two innovative solutions shown in figure 2 will be developed.

(a)

(b)

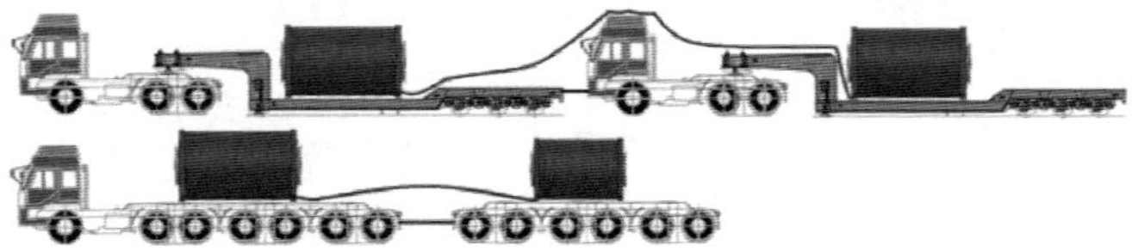

Figure 2. Possible innovative configurations for the transport of reels.

The first solution of figure 2 (a) refers to the transport of two coils of the same size via two trailers towed by two different trucks, in this way it is possible to use trucks with a maximum capacity of 200 tons. The second solution differs because the transport is for two reels of different sizes, in this way it is possible to use two trailers towed by the same truck. The solutions of figure 2 are two innovative solutions which have many advantages compared to the traditional solutions of figure 1 which are currently used for the transport of large diameter reels. The two proposed solutions, given by the wire rope wound on two or more reels, are the most convenient considering the use of equipment with smaller dimensions, in fact exceptional vehicles must be specially designed [3] and guaranteeing greater maneuvering capacity than, for example when the solution (b) is adopted, as can be seen from figure 3, which shows the comparison of the turning radius of the two different solutions.

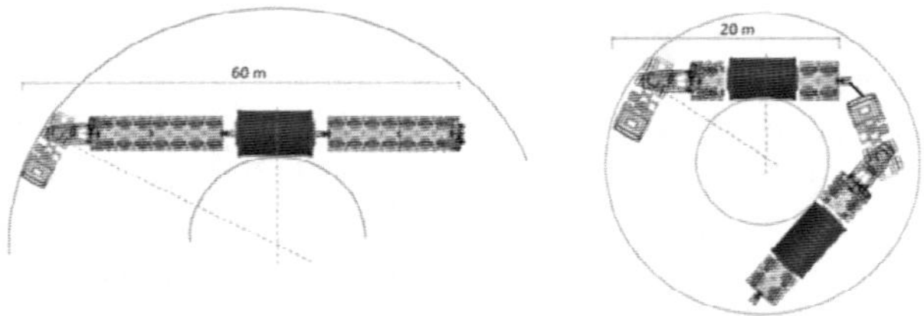

Figure 3. Turning radius comparison.

3. Development of innovative reel for transport

In order to use the chosen solutions, it is necessary to develop a new configuration of the transport reel system, a solution to reduce the weights of the reels and therefore take advantage of less performing transport systems could be use reels made with different materials [4], but this road is not feasible because the actions that the wire rope exerts would be impossible for plastic materials. The new configuration of the reels must be composed of a reel capable of containing the entire wire rope, but with the particularity of being a decomposable reel and a temporary reel on which part of the wire rope will be wound. The reel must be decomposable, or, demountable. The two external discs must be reduced, when only a portion of the wire rope is wound, in order to reduce the total height of the system reel and vehicle. The reel must be reassembled when the height (or even width) constraints are no longer present, such as transport by ship. The size of a reel also depends on the maximum diameter of the wound wire rope, its length and its weight. In general, a coil is made up of different parts, as can be seen from figures 4 (1: drum, 2: square (to support the flanges); 3: flange; 4: drum head; 5: bush for rotation; 6: spindle support for winding. For the sizing and verification of the wire rope, the Eurocodes [5] have been adopted, in particular the standards relating to Eurocode 3 [6] [7].

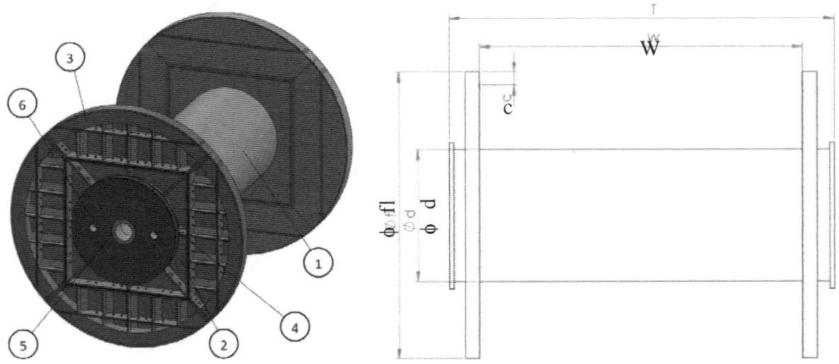

Figure 4. Main parts and characteristics of the reel.

The development of the project concerns the concrete case referred to the Usha Martin company which needs to transport a wire rope of large dimensions, both in diameter and length, from the production site to the port. Currently the solution that the company adopts is a) of figure 1, however this solution is not feasible for the wire rope being transported, so it should be passed to solution b) of figure 1, but it is not feasible for a series of problems connected to the road route which should include, in addition to the removal of some road obstacles (traffic lights, signs, etc.) and the temporary closure of some motorway sections, the construction of special roundabouts to allow the transit of the vehicle b) of figure 1, which is absolutely impractical. For this reason, the solutions of figure 2 and in particular the solution (a) of figure 2 have been developed. The main characteristics of the wire rope to be transported are: nominal diameter dwire rope = 125mm; maximum allowable diameter dadm = 130mm; nominal length Lwire rope = 4000m; maximum admissible length Ladm = 4060m; maximum wire rope weight 400000 kg.

4. Geometric sizing of the reel

Based on these characteristics, it is possible to estimate the size of the reel through three different methods: volume equivalence, average length per loop and incremental method. We therefore opted for the incremental method, as it allows to calculate for each layer the exact length of wire rope that can be wound on the drum, remembering that at each layer the winding diameter increases allowing to contain more wire rope. It is important to underline that the diameter of the drum on which the wire rope is wound derives from various considerations table 1. A smaller wire rope winding diameter, determines greater stresses present in the wire rope, a greater force required for the winding and a smaller winding diameter of the whole wire rope, effectively reducing the height dimensions of the vehicle transport with coil. Based on these aspects, and primarily on the stresses induced in the wire rope as a result of the winding, a winding diameter of approximately 2000 mm has been adopted. In this case the ratio D/d = 2000/125 = 16.

Table 1. Table for calculating the flange diameter.

Layers	ϕ layer [mm]	Length [mm]	L. tot [mm]	n° loop	W [mm]	L. wire rope max [m]	c [mm]	ϕ fl [mm]	Choice criterion
0	1910		0						
1	2170	6409	6	633.5	82355	4060	1441	5052	82510
2	2430	7226	14	298.5	38805	4070	679	3788	38989
3	2690	8042	22	187.5	24375	4064	427	5453	24631
4	2950	8859	31	133.5	17355	4077	304	3557	17716
5	3210	9676	40	101.5	13195	4082	231	3672	13696
6	3470	10493	51	80.5	10465	4082	183	3836	11146
7	3730	11310	62	65.5	8515	4062	149	4028	9420
8	3990	12127	74	55.5	7215	4115	126	4243	8370
9	4250	12943	87	47.5	6175	4137	108	4466	7621
10	4510	13760	101	40.5	5265	4084	92	4694	7054
11	4770	14577	115	35.5	4615	4097	81	4932	6754
12	5030	15394	131	31.5	4095	4121	72	5173	6598
13	5290	16211	147	28.5	3705	4190	65	5420	6565
14	5550	17027	164	25.5	3315	4183	58	5666	6565
15	5810	17844	182	22.5	2925	4093	51	5912	6596

- Layers: number of layers of wire rope subsequently wound on the drum;
- ϕ layer: outer diameter of the nth layer of wire rope;
- Length: length of a loop of wire rope belonging to the nth layer;
- L.tot: sum of the development of all the previous layers (for ease of reading it is expressed in meters for each loop);
- n° loop: is the ratio between the nominal length of the wire rope in meters and the total length of one turn in meters, we proceeded to round up to the upper one and a half turn using the following algorithm

$$n^\circ loop = INT\left(\frac{L.wire\ rope}{Length} + 0{,}499\right) + 0{,}5 \quad (1)$$

- W: is given by the product of the number of turns for the admissible diameter of the wire rope;
- L.wire rope max: is the maximum length that can be wound on the reel, it is calculated based on the number of loop for the length of each loop;
- c (clearance): safety margin to prevent the wire rope from coming into contact with the ground $c = 0{,}0175\ W$;
- ϕ fl: is the effective diameter of the coil flange to be made, it is given by:

$$\emptyset_{fl} = \phi\ layer + 2 * c. \quad (2)$$

- choice criterion: extrapolation of what is generally used for the design of the stranding machines. In practice it is the diagonal of the coil seen in plan and is given by:

$$\sqrt{\phi\ fl^2 + W^2} \quad (3)$$

Among the various dimensional combinations, the one with the choice criterion of lower value is considered preferable for its greater "compactness". By keeping the reel diameter and the wire rope length constant, it is possible to determine: the diameter of the flange, the number of layers and the value of the diameter relative to the small flange, this is possible through the use of the following system.

$$\begin{cases} n^\circ\ layers = INT\left[\frac{\phi_{fl}-2c}{2\phi_{adm}}\right] \\ D_{mid} = n^\circ\ layers * \phi_{adm} + \phi_d \\ L\ wire\ rope\ wound = \frac{n^\circ\ loop * n^\circ\ layers * D_{mid}}{1000} \end{cases} \quad (4)$$

Through iteration, by entering different values of ϕ_{fl}, the minimum value is identified which allows to contain a wire rope with a length of 2030 m (half of the total length of the wire rope). Going to consider the cases in which there are 12, 11 and 10 layers, the results obtained are reported in table 2.

Table 2. Flange diameters reduced as a function of the number of layers.

Layers	W [mm]	ϕ fl [mm]	ϕ fl reduced [mm]
12	4095	5173	4150
11	4615	4932	3900
10	5265	4694	3660

Considering that, the height of the platform of a generic trailer varies from 850mm to 1200mm, the solution has been adopted which provides 10 layers of wire rope on the reel. Thanks to this it is possible to obtain a reel of reduced dimensions during transport as can be seen in figure 5.

The complete reel therefore has the following dimensions:

- $\emptyset_d$ = 1910 mm;
- $\emptyset_{fl}$ = 4700 mm;
- W = 5270 mm;
- c = 92 mm;
- $\emptyset_{fl\ reduced}$ = 3660 mm.

The reduction of the flange size (from 4700mm to 3660mm) is possible thanks to the decomposition of the same as can be seen in figure 5.

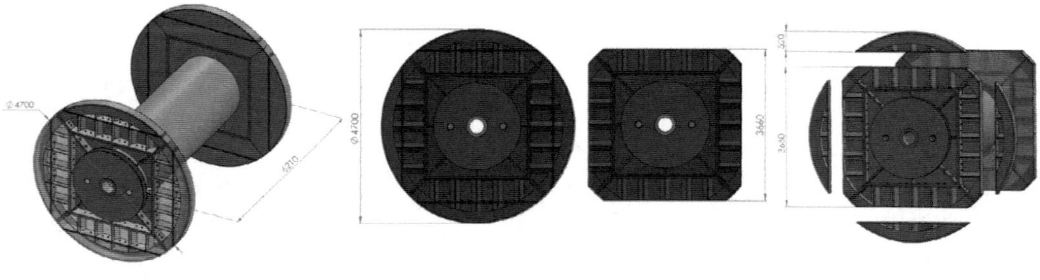

Figure 5. Decomposable reel geometry.

5. Structural sizing of the reel

In addition to problems related to instability [8] [9] [10], the reel is subject to different loading conditions that essentially derive from the way the reel is used/positioned during the transport phases by road and by ship. The different loading conditions are shown in figure 6 they refer to the conditions of transport and handling of the reel, obviously the analyzes were carried out considering all the actions acting on the structure or the own weight and also the various dynamic coefficients induced by the handling of the load [11] [12] [13] [14]. The safety coefficient with which the structure was designed is approximately 10 by adopting S275 UNI EN 10025-2 steel whose main characteristics are: breaking load 430-580MPa and yield strength 275MPa. This fairly high value is justified by the high stiffness that the structure must have both on the drum and on the flanges. The pressure of the wire rope acting on the drum and on the flanges must not induce high values in the elongations and/or bends in order to avoid incorrect winding of the wire rope.

IOP Conf. Series: Materials Science and Engineering 1190 (2021) 012019 doi:10.1088/1757-899X/1190/1/012019

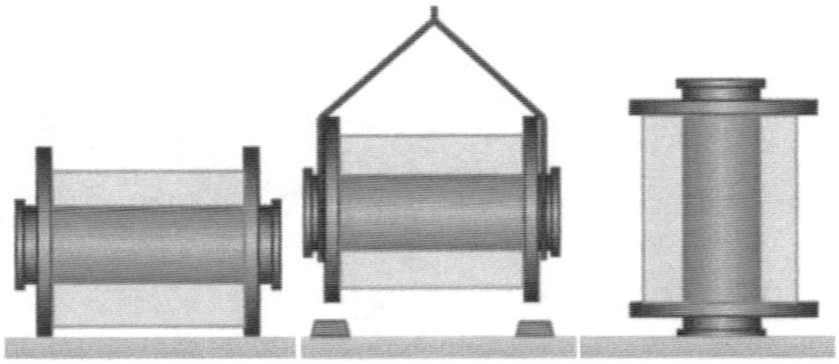

Figure 6. Load conditions diagram to which the reel is subjected (a) resting on the ground; b) lifting the reel with wire ropes; c) vertical support.

Figure 7 shows the results of the finite element analyzes carried out on the reel only for the load conditions identified in the figure 7, a more correct evaluation would be possible by carrying out experimental tests in the field such as [15].

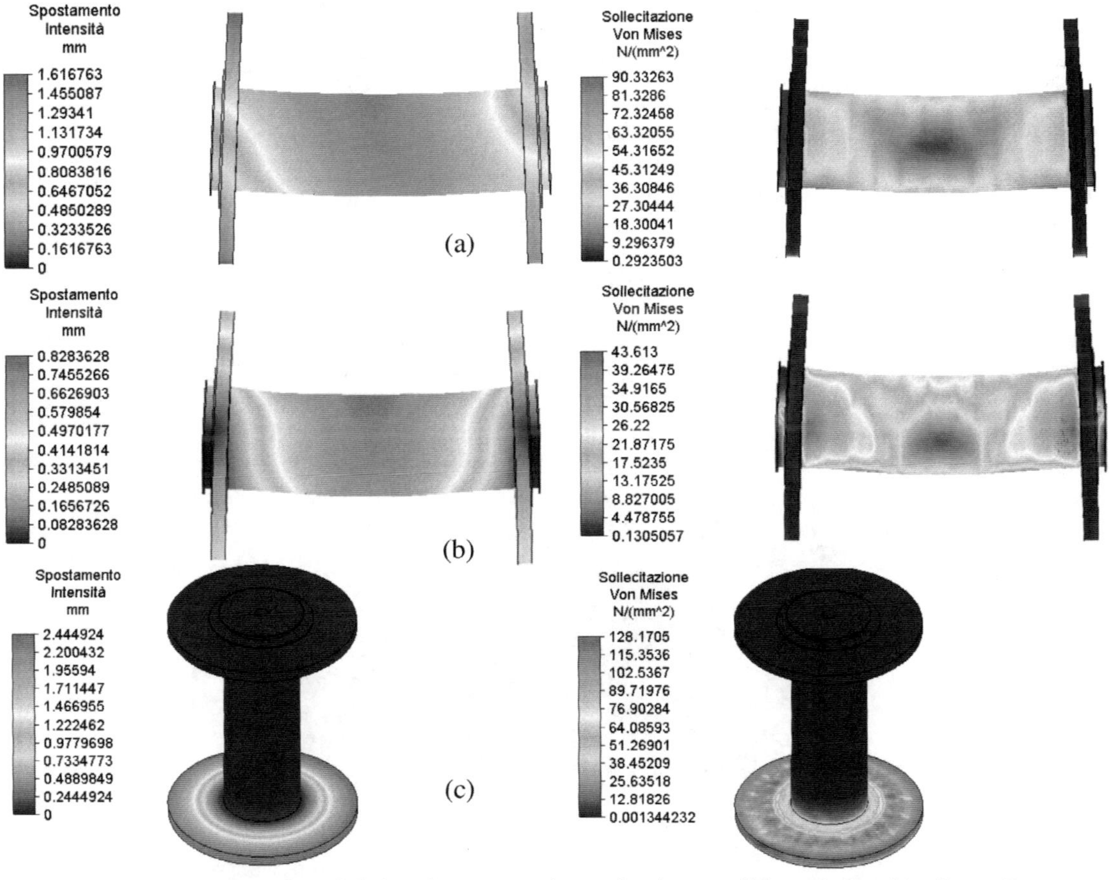

Figure 7. Results of finite element analyzes for the conditions defined in figure 7.

6. Realization of the reel, and winding it

Figure 8 shows a step during the construction of the reel as well as the winding phases of the wire rope [16]. After winding the wire rope just made on the main reel on the production site, we move on to the partial unloading of the main reel to start winding the wire rope on the auxiliary reel, through the use of specially designed equipment, as can be seen from figure 9.

Figure 8. Construction and winding of the wire rope on the main reel.

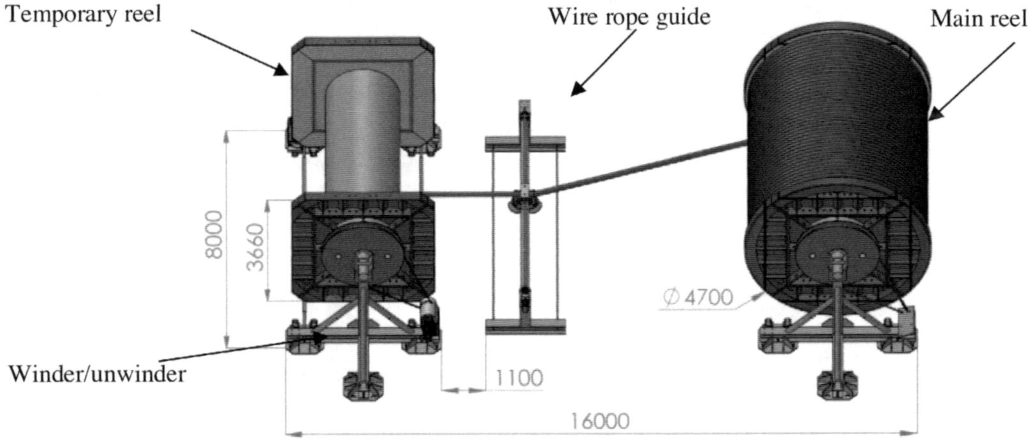

Figure 9. Wire rope winding scheme.

In particular, the motorized winder/unwinder (currently not available for the capacity indicated), is composed of a stand support that can be modified from winder to unwinder or vice versa through a minimum number of modifications, this system is equipped with a motor to 2.2 kW. It should be remembered that between the two reels there must be a free wire rope length in order to allow the necessary movements between the two vehicles used for transport. Once the unwinding and winding phases have been completed, the two reels will appear as in figure 10 where they are partially full and free of circular segments in order to reduce transport dimensions.

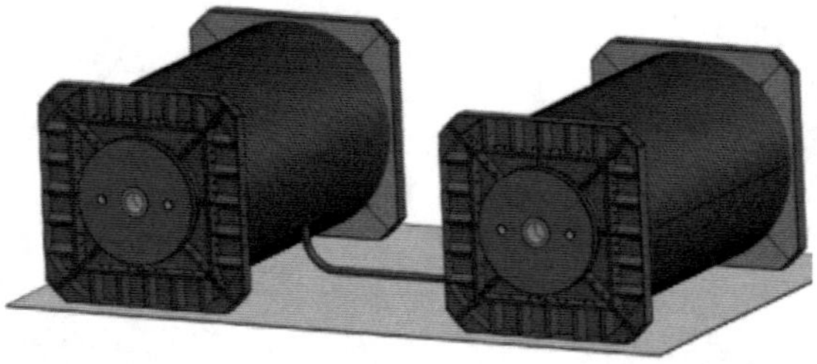

Figure 10. Reels ready for transport.

Thanks to the removal of the circular sectors from the two reels, it is possible to obtain a reduction in space, maximizing the maximum occupation of the volume in the loading area of the trailer. Furthermore, considering the minimum height of the subways and overpasses of 5m, it is sufficient to have trailers with a maximum height of the loading platform of less than 1140 mm available, so as to maintain a distance from the overpass of 200mm. If the classic solution had been adopted, the reel would have had a diameter of 4700m and a weight of about 400 tons and the transport would have taken place with a 300/400 tons truss trolley of figure 1 (b), (it is the trailer conventionally used for the transport of large reels of wire rope, it usually has a length of 60 m). As mentioned in chapter 2, this solution was not absolutely viable as this type of trailer allows to solve the problem of the height of the load, but it increases its longitudinal dimensions and the turning radius, effectively preventing circulation on the roads. For the transport it was therefore decided to adopt the solution (a) of figure 2, for this purpose two trucks equipped with 200-ton trailers are therefore necessary. In choosing the trailer, several table 3 solutions were considered, and then decided to use the trailer N3.0-3 produced by the Titan company (titantrailers.com), whose main dimensions are shown in figure 11.

Table 3. Trailers dimensions.

Type	Width Platform [mm]	Wheelbase [mm]	Height Platform [mm]	Height Minima [mm]
N3.0-1	2990	1550	1070±300	770
N3.4	3400	1600	1070±210	860
N3.0-2	2990	1550	1080±325	755
G3.0	3000	1500	1175±300	875
N3.0-3	2990	1550	1070±300	770

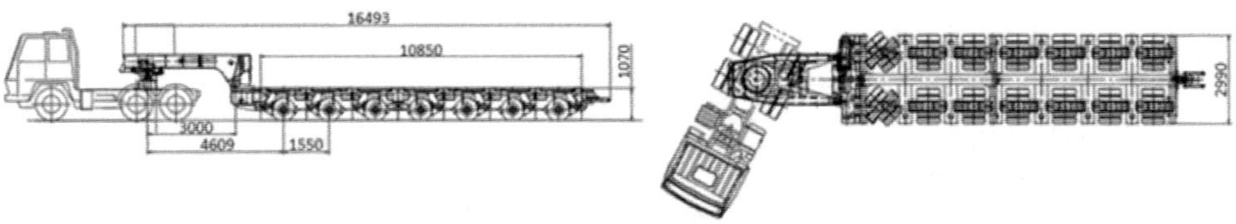

Figure 11. Main dimensions of the trailer under consideration.

If we consider the dimensions of the reels transported, the overall dimensions are given by:

- overall height: 3660mm;
- overall width: 3660mm;
- length: 6200 mm occupied by the reel on the trailer, the remaining space on the trailer can be used by additional equipment.

Table 4. Dimensions and weights of the various parts transported.

n°	Part	Weight [kN]
1	Drum	43.25
2	Drum square frame	122.18
3,4,5	Sectors of the flange	14.34
6	Flange	57.36
7	Reel	236.91
8	Winder	85.02
9	Unwinder	80.54
10	Saddle	55.43
11	Wire rope guide system	15.16
	Total structure to transport	749.20

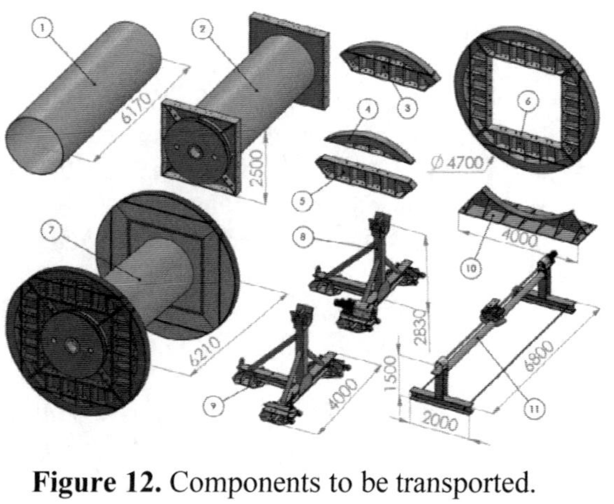

Figure 12. Components to be transported.

The two trucks must also be rigidly connected to each other, this connection is defined as a rudder, this to prevent the wire rope from being subject to unwanted pulls during transport. In addition, to keep the wire rope in the correct position and avoid excessive movements and oscillations, the cabin of the second truck can be equipped with a support system capable of supporting and guiding the wire rope, allowing only the longitudinal movements necessary to face the curves during the journey. In addition to the reels on the trailers, as previously highlighted, there is also the system for winding and unwinding the reels. Table 4 and figure 13 show the elements to be transported with the relative dimensions and weights. Once at the destination it will be possible to unload the reels and rewind the reel on a single reel after the related circular sectors have been installed.

7. Conclusions

With this work an innovative system was presented to solve the problems related to the transport of reels for large wire ropes. The innovative solution of adopting two vehicles and a decomposable reel is the only one that has allowed the transport of a wire rope with a diameter of 125 mm and a weight of 400000 kg. If we had not chosen this system, a smaller wire rope would have had to be used, which would have involved, in this case, lifting the load with several sections of wire rope, the presence of winches, their control, etc., considerably increasing handling costs. In this case, the costs for both construction and transportation would have increased. With the proposed solution, which was then implemented, advantages have been obtained linked to the fact that transport presents fewer problems, above all relating to traffic, traffic on the route and the benefits it can bring to the road surface [17] [18] [19], without considering the greater ease of finding suitable vehicle for transport. This system can be convenient if compared with the current available solutions (not feasible for this situation which presents a wire rope of considerable size and length). The transport system studied, which can be generalized by considering more "light" vehicles bound for the transport of wire ropes, could allow a reduction in the cost of transporting goods up to a third of the initial value, even in the face of an investment for the construction of the reels and the winding system (elements that can obviously be reused for subsequent transport and therefore depreciated over time thus guaranteeing greater savings) [20] [21]. For these reasons, the study conducted presents an innovative, efficient and economic solution to traditional transport systems or impossible as regards coils for large wire ropes, trying not to overstress the road surface [22]. Research is still under development by studying a system consisting of several reels designed to transport a single wire rope of large size and length, which is currently not specifically made for problems related to transportation.

Acknowledgment

The authors wish to thank Eng. Franco Clerici and Eng. Davide Clerici for support in carrying out this applied research.

References

[1] L. Solazzi, "Applied research for Weight Reduction of an industrial Trailer", *FME Transactions*, vol. 40, pp. 57-62, 2012.
[2] L. Solazzi, "Design of a truck cab torsion bar", *International Journal of Heavy Vehicle Systems IJHVS*, vol. 26, No 5, pp 692-706, 2019.
[3] N. Esmaeili, R. Kazemi, and S. H. Tabatabaei Oreh, "Design of a new integrated controller (braking and steering) to maintain the stability of a long articulated vehicle", *Proceedings of the Institution of Mechanical Engineers, Part C: Journal of Mechanical Engineering Science*, vol. 234, pp3 981-1013, 2020.
[4] J. Kim, H. Y. Kang, Y. G. Song and C. J. Kim, "Feasibility structural analysis of engineering plastic reel module for carrying wound high-voltage eletric transmission line", *Lecture Notes in Eletrical Engineering*, vol. 554, pp. 778-788, 2020.
[5] L. S. da Silva, R. Simoes, H. Gervasio, "Design of Steel Structures 2nd Edition - Eurocode 3: Design of Steel Structures Part 1-1: General rules and rules for buildings", *Ernst and Sohn* ISBN-13: 978-3433031636.
[6] Eurocode 3: "Design of steel structures - Part 1-1: General rules and rules for buildings", *EN 1993-1-1*.
[7] W. Segui, "Steel Design" *Cengage learning Inc*, 2017.
[8] E. Corona, L. H. Lee, and S. Kyriakides, "Yeld anisotropy effects on buckling of circular tubes

under bending", *Internstional Journal of Solids and Structures*, vol. 43, pp. 7099-7118, 2006.

[9] L. Solazzi, "Innovative bolted junction with high ductility for circular tubular element", *Journal of constructional Steel Research*, vol. 112, pp. 175-182, 2015.

[10] E. Corona, and S. Kyriakides, "On the collapse of inelastic tubes under combined bending and pressure", *International Journal of Solids and Structures*, vol. 24, pp. 505-535, 1988.

[11] L. Solazzi, G. Incerti, and C. Petrogalli, "Estimation of the dynamic effect in the lifting operations of a boom crane", *Proceedings of the 28th Euwire ropean Conference on modelling and simulation ECMS 2014*, 27-30/05/2014, Brescia, Italy, pp. 309-315.

[12] G. Incerti, L. Solazzi, and C. Petrogalli, "Lumped parameter models for numerical simulation of the dynamic response of hoisting appliances", *ICSEMA 2015, 17th International Conference on Systems Engineering Modelling and Analysis*, November 9-10/2015, Venice, Italy.

[13] L. Solazzi, and N. Zrnić, "Design of a high capacity derrick crane considering the effects induced by load application and release", *Journal of Applied Engineering Science*, 15 (2017) 1, 409, 15-24; doi:10.5937/jaes15-11930, Published 03 April 2017. ISSN: 1451-4117, E-ISSN: 1821-3197.

[14] L. Solazzi, "Experimental and analytical study on elevating working platform", *Procedia Engineering*, vol. 199, pp. 2597-2602, 2017.

[15] G. Donzella, S. Matteazzi, and L. Solazzi, "Experimental and numerical analysis of the body of a refuse collecting vehicle", *International Journal of Computer Applications in Technology*, vol. 22, pp. 89-97, 2005.

[16] L. Yuan, and S. Kyriakides, "Liner buckling during reeling of lined pipe" *International Journal of Solids and Structures,* vol. 185-186, pp. 1-13, 2020.

[17] A. Martinez, and M. Campos, "Load aggressivenes on different pavement structures due to the weight increase of heavy vehicles from 40 tonnes to 44 tonnes", *Baltic Journal of Road and Bridge Engineering*, vol.14, pp. 405-424, 2019.

[18] J. Ren, R. G. Thompson, and L. Zhang, "Impact of Payload Spectra of Heavy Vehicles on Pavement Based on Weight-in-Motion Data", *Journal of Transportation Engineering Part B: Pavements*, vol. 145, issue 2, 2019.

[19] T.C. Martin, "Estimating heavy vehicle road wear costs for bituminous-surfaced arterial roads", *Journal of Transportation Engineering*, vol. 128, pp. 103-110, 2002.

[20] M. Ziyardi, H. Ozer, S. Kang, and L. I. Al-Qadi, "Vehicle energy consumption and an enviromental impact calculation model for the transportation infrastructure systems",*Journal of Cleaner Production*, vol. 174, pp. 424-436, 2018.

[21] J. Wang, and H.A. Rakha, "Fuel consumption model for heavy duty diesel trucks: Model development and testing", *Transportation Research Part D: Transport and Enviroment*, vol. 55, pp.127-141, 2017.

[22] R.V. Siddharthan, P.E. Sebbly, M. El-Desouky, D. Strand, and D. Huft, "Heavy off-road vehicle tire-pavement interactions and response", *Journal of Transportation Engineering*, vol. 131, pp. 239-247, 2005.

Wheelchair Climbing Construction for Industrial Logistical Applications

Bernhard Heiden[1,3], Daniel Kattnig[1], Julia Pechmann[1], Volodymyr Alieksieiev[1,2], Bianca Tonino-Heiden[3]

[1] Carinthia University of Applied Sciences, Industrial Engineering Studiengang, Europastrasse 4, 9524 Villach, Austria
[2] Institute for Mechanical Engineering and Transport, National Technical University 'Kharkiv Polytechnic Institute', Kyrpychova 2, Kharkiv, Ukraine
[3] University of Graz, Heinrichstraße 26/V, 8010 Graz, Austria

b.heiden@cuas.at

Abstract. The human condition is - ever since becoming, in an evolutionary sense, human-associated with running on two legs. Only in the latest human technical development, stairs have become increasingly present as they crystallize to be more suitable as an artificial environment. Although stairs are suitable for humans without walking handicap, they are not for wheel-driven vehicles, like Automated Guided Vehicle (AGV) in industry or personal accompanying support vehicles as an example. In this mixed human-machine environment a sort of stair climbing is essential for many logistic applications, as this is an, from the mechanical engineering point of view, essential step in mechanization and their automation. Only recently, a stair-climbing device has been issued a patent, called the "Malteserrollstuhl", which is designed for stair climbing for handicapped people. This patented mechanism could be further used for industrial applications. The goal of this work is to investigate the possible future directions and strategies that are most likely to be implemented, on the base of this patent, as the next steps towards logistical industrial applications.

1. Introduction

The human condition is - ever since becoming, in an evolutionary sense, human-associated with running on two legs. Only in the latest human technical development, stairs have become increasingly present as they crystallize to be more suitable as an artificial environment. Although stairs are suitable for humans without walking handicap, they are not for wheel-driven vehicles, like Automated Guided Vehicle (AGV) in industry or personal accompanying support vehicles as an example. In this mixed human-machine environment a sort of stair climbing is essential for many logistic applications, as this is an, from the mechanical engineering point of view, essential step in mechanization and their automation. In times where postal and delivery service providers are constantly struggling with lowering margins in the market, it is getting more important to design processes lean and efficient. Furthermore, the availability of reliable and capable workforce in this sector more often proves a bottleneck because, though it is a physically demanding task, not well paid. Therefore, the substitution of the labor force with technology could be interesting for companies operating in this sector. Hence, the possible application of the "Malteserrollstuhl" (Malteser-wheelchair) in logistics processes will be investigated within this paper.

The objective of this work is to investigate, the systemic novelties of the patent application "Malteserrollstuhl" [1], and to give a prospect of how this could be integrated into logistical industrial applications.

First in the section "Systemic Analysis" we go deeper into what are machines, what are they for in general, and what of some in the context special important properties can be seen for transport-machines. Second, in the section *Malteserrollstuhl* (Malteser-wheelchair), we investigate the Malteserrollstuhl, its development-steps, and its basic function as well as how it can be embedded in a systemic environment of stair-climbing and its relation to general system properties, as is with forces of construction variants. Third, in the section "Industrial Logistical Applications" we look at the actual arising industrial applications and markets that can be impacted by the Malteserrollstuhl-principles and its further transport vehicle integration. Finally, we make a summary and give an outlook for future application and mechanical engineering pathways for wheel-chair vehicle automation in the context of this work.

2. Systemic Analysis of the question: "What is a machine for?"

In this section, it will be analyzed what a machine is for and how it can be accomplished for it to be a defined goal for somebody.

2.1 Precondition – The All-Around Point of View Point (AAVP)

What is an All-Around Viewpoint? A standpoint from which everything is seen, or can be seen or could be seen. Wittgenstein introduces a thought [2,§352], to the sentence of the excluded third party in logics, that "we cannot turn our gaze away from this image." He means that once we believe that this sentence exists and that the principle of the excluded third party applies, and we believe this, we cannot turn our gaze away. But there are other views and points of view possible and these are then just different theories and then e.g. with these this sentence is not applicable yet. So we have to revise our first theory, or we look at it from the All-Around View Point - AAVP.

2.2 A Machine Regarded as a System – Conficiency and the All-Around Point of View.

A machine can be regarded as a kind of tool a means like a hand is a means to build a tool. The tool itself is then a means for further tools ad infinitum. What we regard in human evolution is that together with the development of the brain about threefold in mass in about 3 Mio years from the ape brain, humans evolutionary natural tool, the hand, coevolved by differentiating his fine motoric. This is now called the *pinch grip* - two fingers are used like with a gripper - of humans, which allows him to hold a pencil and write and many other useful things (q.v. [3]).

A machine - an application - is an instrument, is a means, is a limit, a border, which can be interpreted according to Spencer Brown [4]. But is this really true? Let us make a thought experiment: "I see the world without a machine. Then I see the world with a machine. E.g. I drive a car and see simultaneously the world as I drive." This makes the difference, with and without the machine. The machine changes my subject view of the world, in this example. Therefore it is a border that I am looking through like through a window. It is between me and the world. Therefore it is a means.

"I see the world with a machine and so the world is changed", one might say. "The machine is a means by which I see the world." another could say. So the machine itself is a limit. The machine is a border to the world. Another machine is then another border to the world.

A machine that is more permeable to the world is more *conficient* than other machines (compare Figure 1). That can be also interpreted by giving more specific possibilities, concerning persons, subjects, and the environment.

IOP Conf. Series: Materials Science and Engineering 1190 (2021) 012020 doi:10.1088/1757-899X/1190/1/012020

The universal virtual classroom, or the practical application of remote teaching, can be regarded as another machine [5]. It is more permeable, more conficient than other machines. It appears transparent as in the extended picture in Figure 2. This gives the transparent orgiton image. An orgiton is here an elementary cybernetic unit of mass, energy, and information [6]. "Look at the class as parents, then the teachers, the school looks at your private". The transparency applies in both directions. The machine the "virtual classroom" is a tool that increases the "conficiency".

That everything is connected is the prerequisite. The edges – according to graph theory, a system that can be described by edges and nodes [7, 8] - are the precondition that conficiency happens. The edges, in this case, are e.g. the flow across the boundary (compare also Figure 1). They are a necessary condition. Like a lower generation orgiton is the precondition for the next, as is the case in an information feed-back cycle [9].

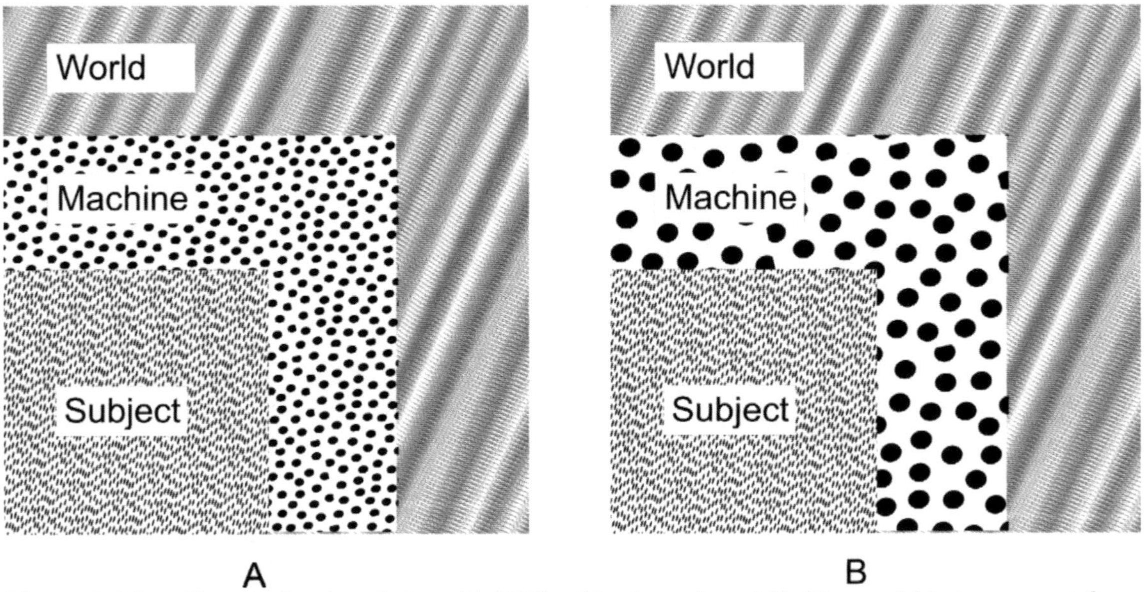

Figure 1. The *all-around point of view* (AAVP) with views A and B. The world is transparent for a subject using a machine, means, application. It is differently transparent compared to B. The fine structure of the machine produces a changed perception. The subject is always in his world with his point of view. The border to the world is the means. Different means have different configurations. To be an all-round point of view, the world must become more transparent, the machines must guarantee this in a form. Is A more transparent than B? That depends. If transparency is the transmission of linearly directed rays, then A is more mediated than B. But it is also possible to imagine a "pathway" which is curved and therefore less mediated than in B. This means that the machine is a means that is differently efficient in relation to different subjects and their will to take a pathway. A machine is therefore a means of enabling a particular subject to find its way to the world. It is a potential space. Regarding it, there are many possible machines. Fewer of them are conficient in the sense, that they allow people to interact.

The different orgitons (compare also [6]) express the interlocking. An orgiton is the combination of material, energy, and information into one elementary unit, that entails himself to the following generations. In the virtual classroom the transparency increases (see also Figure 1), which can be described with different transparent images (q.v. Figure 2).

Figure 2. Transparent orgiton model as an extension to [5]. Due to the transparency properties, the order structures are visible. Predecessor, successor, and the interlocking - the interlocking is the confluence in the material sense and the *conficiency* in the sense of social-togetherness. Each form (triangle, circle, square) can be regarded as properties related to matter, energy, and information. The borders depict the borders of the forms. Properties of conficiency can be accomplished by transparency or transmittance through borders as a picture of the process potentials, allowing for increasing interactions on different orgiton generation levels.

What decisive is, is the interaction, the efficiency. The "causa conficientis". Through the interaction, a new order emerges which has only become possible through transparency. Parents, teachers, children work together in a new village on a path for the future. This is how one could describe the principle of conficiency.

2.3 Transport Orgiton (TO) or General Elementary Transport System (GETS).
In the all-around point of view (AAVP) a transport orgiton can be defined as a set of sentences – axioms - which allow for further extension:

1.) A transport orgiton is a machine M that moves in the World (W).
2.) M is a border relative to a subject S in AAVP.
3.) S is connected to M in a fixed way.
4.) Redundant and multiple-type inter-, intra-, and trans-border-connections increase *osmotic* (nonintentional subjects (objects)) and *conficient* properties of GETS potentially.

This system depends strongly on the subject's intentions. This can be interpreted as personal machines or user-centered machine-adaptations.

It is important to get these osmotic or conficient properties of the border-connections. Applications to these are flexibility in usage, and stability due to redundancy, or stability. When applying e.g. stair climbing, then M is the border between W and S. More possible interactions mean broader distribution of how the points of center of mass are distributed, leading to the stabilization of vehicle transport properties.

3. Malteserrollstuhl (Malteser-wheelchair)
In this section, there will be provided a short history of the Malteserrollstuhl.

3.1. First Version of Malteserrollstuhl
The first version of the Malteserrollstuhl is depicted in Figure 3. There is an instability point for the last stair.

IOP Conf. Series: Materials Science and Engineering 1190 (2021) 012020 doi:10.1088/1757-899X/1190/1/012020

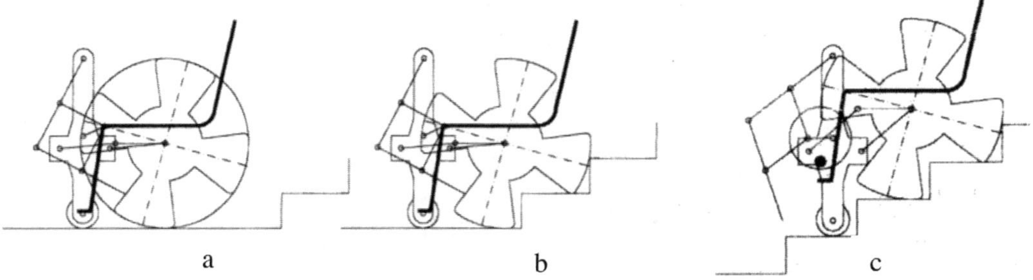

Figure 3. Malteserrollstuhl in its first version according to [10]. In (a) the unfolded Malteserrad (Malteser-wheel) for the use in the plane is shown. In (b, c) the Malteserrollstuhl is climbing on the stair.

3.2. Second Version of Malteserrollstuhl

In Figure 4, the second version of the Malteserrollstuhl is depicted. It solves compared to the first version of the stair climbing by inversing Malteser-wheel by a "Schwinge" (wing).

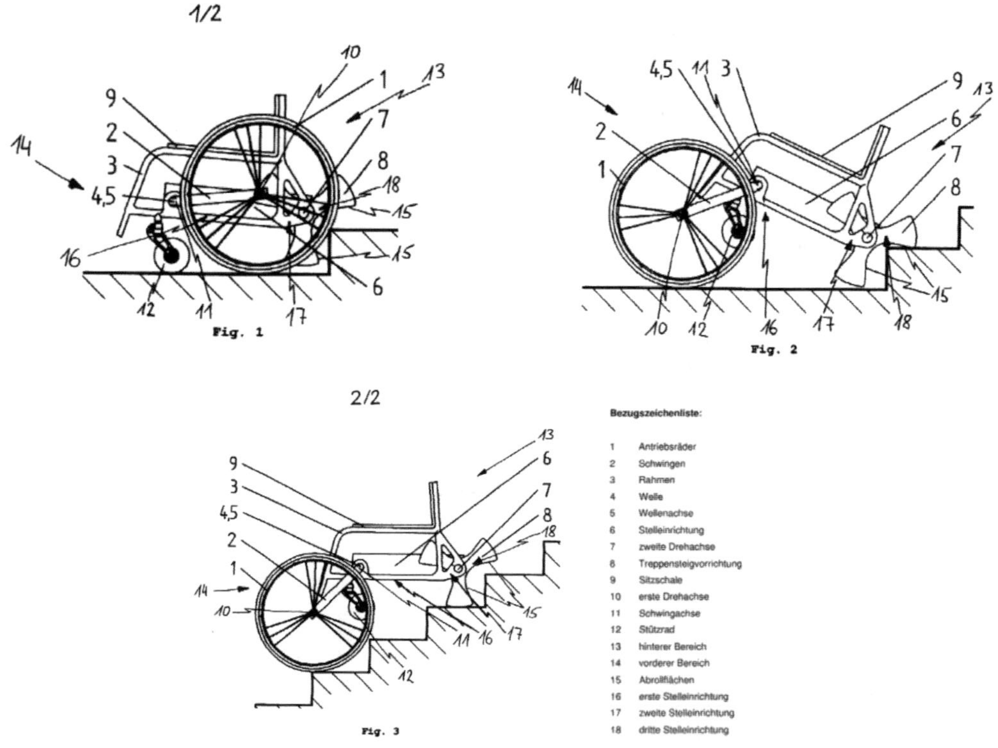

Figure 4. Malserrollstuhl in its second version according to [1]

Figure 5. Malteserrollstuhl built as a model in the scaling 1:4 [11].

This leads to the fact, that the big Malteser-wheel has not to be transformed like in version one. The only remaining "move" is the switching of the Wing. This then leads to an overall stable movement along the stairs which was demonstrated with a 1:4 model (see Figure 5).

3.3. Malteserrollstuhl – Aspects concerning Wheel Forces and Transportation Properties
One of the most important tasks of Malteserrollstuhl is a minimization of the human load. In addition to its use for people with walking handicaps, some of them can also be used for logistical purposes in industry, e.g. for the lifting of packages upstairs. The goal of this section is to investigate, how human load can be minimized using a distribution of mechanical forces.

When the Malteserrollstuhl goes upstairs, it performs a complex combined motion with variable velocity (see also [12]). In this case, the most important requirement for minimization of the human load is a minimization of peak loads. For this purpose, the dimensions and proportions of the chairs' mobile parts have to be optimized as well as forces, which are involved in the lifting process, have to be considered. For this purpose, the building of trajectories of wheels' teeth centers of mass and their subsequent analysis have to be done. From the statics law-axioms, points of force applications with the biggest tangential angle to the horizontal correspond to the maximum force value and vice versa. (sum of forces is constant). The solution is to create a bigger amount of trajectories for wheels centers of mass with variable wheels parameters (diameter, number of teeth) to determine its minimal angle to the horizontal. This investigation is shown in Figure 6 (instead of teeth a different amount of wheels was used).

In Figure 6 the trajectories of transposition of wheels centers of mass are shown, where the five "Malteser-wheels" block demonstrates the smallest angle to the horizontal in comparison to the three "Malteser-wheels" block. Hence is to conclude, that such a construction allows a relative minimization of the human loads mechanically.

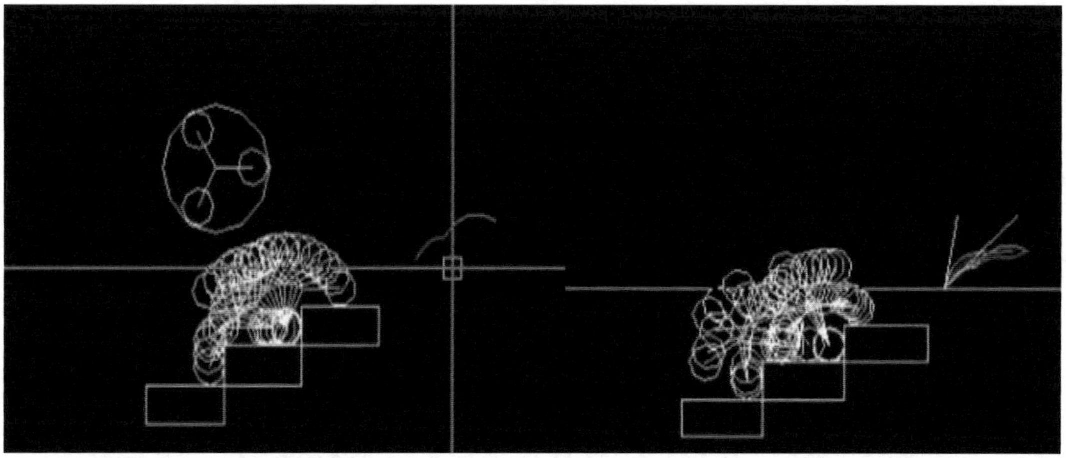

a b

Figure 6. Trajectories of transposition of Malteser-wheel centers of mass: a – for the wheel-chair with three “Malteser-wheels” block (three teeth), b – for the wheel-chair with five “Malteser-wheels” block (five teeth) [12]

4. Industrial Logistical Applications

In the following, some future industrial applications are reviewed. First, a critical analysis of the patent applications and related work [1, 10, 11] is done concerning potential future applications, indicating technical missing steps. Second a more detailed view of the situation of current Automated Guided Vehicles is given with more specific needs in this field to apply for stair climbing and similar gravity overcoming efforts for those transport applications.

4.1 Industrial Stair Climbing – Some Potential Future Applications – Critical Analysis.

Stair climbing technologies are to be found in the context of the transportation of heavy cargo through vertical borders with a person as operator [13]. However, several further projects and patents explore the application of this technology in other fields of operation, such as military and security applications [14] or rescue applications [15]. This technology could furthermore be of special interest for logistics service providers operating in the sector of small goods transport and/or parcel delivery. Especially in urban areas with a grand density of high buildings, the application of an autonomous stair-climbing robot could decrease the need for labor force within a fully automated process. The feasibility and efficiency of that use case are subject to the Master-thesis “Stair-Climbing Autonomous Driving Device for Small Goods Transportation”, currently being written by Daniel Kattnig [16], with parts of the expected outcome being displayed in the following.

Logistics processes demand numerous requirements for autonomous application devices for parcel delivery, such as high reliability, weather-resistance, compliance with safety regulations, or the ability to navigate through angled stairwells. Heterogeneous construction of stairs in both, dimension and material add further to the challenges in designing such a device. However, it was found that a robot constructed based on the “Malteser-wheel”-technology [1, 10, 11] could fulfill the task, at least in an operational sense.

The process that could be deployed currently would see the operating personnel navigating the package delivery vehicle to the end customer. The device would match the packages with the location and take the right one(s) accordingly. Another intervention on part of the labor force is needed though, as the device is not able to unlock the main doors of the tenements. Environmental scans ensure that the package is delivered to the right flat, given they are marked with the address number.

In terms of efficiency, the device lacks compared to the current manual delivery as well as alternative, new technologies, such as drone delivery (see e.g. [17]), which is currently being tested e.g. by Amazon [18]. This is due to high expected costs for construction, slow operating speed, and the infeasibility of operating the system fully automated as long as autonomous vehicles are not being used broadly and no solution is found for unlocking doors with the robot. Further development in automation and associated construction (see also [17]) in the elementary transport units (transport-orgitons) leads to further integration of this type of vehicle into transport logistics.

4.2 Industrial Transportation – Automated Guided Vehicles (AGV) – Industrial State of the Art
The automated guided vehicles (AGVs) were invented in America in the course of the 1950s and are now an important part of intralogistics. As the first application, a so-called tug, based on very simple tracking technology, was used to transport recurring group transports over a greater distance.

Intralogistics is the main field of application for AGVs. This means the control, implementation, organization, and optimization of internal material, as well as the goods, flow, information flows, and the handling of goods in industry, trade, and public institutions [19].

To this day, the AGVs are continuously developed and can be used in a variety of ways. The following criteria are decisive to be able to define which is the right design and technology for the respective application:

- Area of application and local conditions
- System complexity and number of vehicles required
- Navigation and control options
- Size and weight of the cargo to be transported

The functionality of *navigation* and the importance of *safety* play an important role in the selection of the right AGV and navigation technology. The interaction of the elements, which play a role in functionality and use, must be coordinated with the security systems so that the best suitable solution can be chosen. This solution must not cause any damage to the material, the environment, or the people surrounding the system. The navigation of the AGV depicts three very important areas. The first point is the current location of the AGV. The second point is the definition of the chosen target, which is directly related to the third point and that is to find the optimum way to the desired target. The navigation can either take place via a physical or virtual line or without any line. The respective manufacturer can decide which type of navigation to choose. Regarding security, the requirements must be strictly observed. Differences are depending on the system, but especially within Europe, relevant European standards and country-specific requirements have to be considered. The manufacturers should comply with these specifications, but the operators also have the responsibility to ensure correct use [19].

The key task of all AGVs is to transport a transport unit from a source to a target. This task is coordinated by the AGVS control system. An AGVS control system is defined by the VDI as follows: "*An AGVS control system consists of hardware and software. The core is a computer program that runs on one or more computers. It serves to coordinate several driverless transport vehicles and / or takes over the integration of the AGV into the internal processes* " [20].

For the system architecture, it is fundamental that the control system also known as a fleet manager communicates with the respective AGV and then with the control system either based on a host or PLC interface. The automation mode is the main operating mode, which enables the AGVs to be used in 24/7 mode without interaction. Depending on the version of the respective vehicle, the charge cycles must be included in the calculation of the number of shuttles. However, the decisive criteria are:

"What should be transported? Where are the transfer points? What are the distances? How often should something be transported?" A general statement about the number is therefore not serious. All aspects must always be considered and analyzed. Learning in the course of the installation is also important here since one often encounters framework conditions that were not known in advance but have a direct influence on the system [19].

4.3 *Thoughts About the Current Problems in the Industry Regarding Stairs and Floor Requirements*
Generally, all suppliers have issues with uneven floors and stairs. This has got several reasons. First is that the shuttles are only allowed to have a small floor clearance due to safety reasons. If a shoe or something similar could get under the shuttle it may lead to injury. Second, the shuttles are constructed to be efficient in driving and have rather small wheels. If one of the wheels gets into a bigger gap it may get stuck. On the other hand, the shuttle usually carries some kind of load unit, which can have over 100kg depending on the supplier. If this load is being shaken and not secured enough it may fall and get damaged or worse, leading to an injury.

Stairs are mostly an issue because they can´t be detected by the shuttles. Safety systems are usually created to detect obstacles at a low height like legs from workers. This safety scanner can not check the floor for stairs usually.

5. Results, Discussions and Conclusions

The results of this works are an investigation about the systemic properties of machines with the special application to transport vehicles and their relation to generic development. First-principles are formulated, which can be further explored. The Malteserrollstuhl as a practical application was introduced. Future industrial applications were sketched, among others also especially in the field of AGVs, based on the Malteserrollstuhl patent applications.

In conclusion, there can be stated that there is still a big gap between the present patent applications and future industrial suitable applications. This gap can only be closed by further development of detailed mechanical constructions and prototyped-models that take into account the special needs of all stakeholders: Use in the logistical business application, solving the detailed transport tasks, solving the special need of the designed material that is to be transported, special needs of the customers who use or are delivered by means of the transport vehicles, the geographic environment, the sustainable lightweight construction, can be seen as only some of the future challenges for those kinds of applications.

Acknowledgment(s)

The authors wish to acknowledge the support of the Carinthia University of Applied Sciences for this work, namely Claudia Pacher, Franz Riemelmoser, Robert Hauser, and Erich Hartlieb. Part of the text was translated with the open DeepL Software.

References

[1] B. Heiden and T. Frohnwieser, "Treppensteigender Rollstuhl," (in German), Austria Patent 520885 Patent Appl. A 50107/2018, 15.02.2020, 2020. [Online]. Available: http://seeip.patentamt.at/Publication/PdfDocument/?DocumentId=b70adc49-43e5-41ad-b23d-544c95ccb678
[2] L. Wittgenstein, *Philosophische Untersuchungen*. Suhrkamp Verlag AG, 2019.
[3] J. C. Eccles, *Evolution of the brain creation of the self*. London etc.: Routledge, 1989, pp. XV, 282 S.
[4] G. Spencer-Brown, *Laws of form Gesetze der Form*, 2. Aufl. ed. Lübeck: Bohmeier, 1999, p. 202 S.

[5] B. Heiden, B. Tonino-Heiden, and M. Decleva, *TOWARDS A WITTGENSTEINEAN LADDER FOR THE UNIVERSAL VIRTUAL CLASSROOM (UVC)* (Proceedings of SMART LIVING FORUM 2019 - 14 November 2019, Villach, Austria). Villach: BoD, Norderstedt, Germany, 2020, pp. 71-77 https://forschung.fh-kaernten.at/aal/files/2020/05/11-heiden.pdf.

[6] B. Heiden *et al.*, "Orgiton Theory (unpublished work)," 2019.

[7] P. Läuchli, *Algorithmische Graphentheorie*. 1991.

[8] S. Dasgupta, C. Papadimitriou, and U. Vazirani, *Algorithms*. Boston: The McGraw-Hill Companies, 2008.

[9] B. Heiden and B. Tonino-Heiden, "Key to Artificial Intelligence (AI) (unpublished work, accepted for publication)," 2020.

[10] B. Heiden, "Malteserrollstuhl," Austria Patent Appl. AT Patent App. A1253-2012, 2012. [Online]. Available: http://seeip.patentamt.at/Publication/PdfDocument/?DocumentId=2ee1cd02-3b0f-4843-8eeb-84a54220067e

[11] T. S. Frohnwieser, "Konstruktion und Bau eines Funktionsmodells eines treppengängigen Rollstuhls auf Basis des Malteserradkonzepts," Bachelor Thesis, Carinthia University of Applied Sciences, 2016.

[12] V. Alieksieiev and O. Turchyn, "Studie der geometrischen Parameter der Handfahrwerke fürs Ladungsheben auf Stufen," in *Thesen der Fünften Studierendenkonferenz zur Präsentation von Jahres- und Abschlussarbeiten*, 2019, vol. 48: NTUU „KPI“, Kyjiw, pp. 5-7.

[13] G. Patents. Treppensteigende Sackkarre [Online] Available: https://patents.google.com/patent/DE4409807A1/de

[14] U. o. Minnesota. Stair Climbing Robot for Military and Security Applications, Technology #z08062 [Online] Available: http://license.umn.edu/technologies/z08062_stair-climbing-robot-for-military-and-security-applications

[15] M. Eich, F. Grimminger, and F. Kirchner, "A Versatile Stair-Climbing Robot for Search and Rescue Applications," presented at the 2008 IEEE International Workshop on Safety, Security and Rescue Robotics, 2008. [Online]. Available: https://ieeexplore.ieee.org/document/4745874.

[16] D. Kattnig, "Stair-Climbing Autonomous Driving Device for Small Goods Transportation (ongoing work)," Masterthesis, Carinthia University of Applied Sciences, Villach, 2020.

[17] X. Mingxing, "Analyse möglicher Anwendungen von Gütertransportdrohnen (GTD), Entwicklung und Demonstration eines zugehörigen funktionellen Prototyps für ein automatisiertes Gütertransportgestell (GTG) und Modellierung und Simulation eines elementaren Gütertransportnetzes (GTN) von GTD im wirtschaftlichen Kontext," MSc Masterthesis, Carinthia University of Applied Sciences (CUAS), Villach, 2019.

[18] Amazon. (Info) Amazon startet Drohnen-Lieferung in Österreich! [Online] Available: https://www.preisjaeger.at/deals/info-amazon-startet-drohnen-lieferung-in-osterreich-202986

[19] T. A. G. Ullrich, *Fahrerlose Transportsysteme, Wiesbaden*. Springer Fachmedien Wiesbaden GmbH, 2019.

[20] *VDI 4451-7 "Kompatibilität von Fahrerlosen Transportsystem (FTS) - Leitsteuerung für FTS"*. Berlin: Beuth-Verlag, 2015.

IOP Conf. Series: Materials Science and Engineering 1190 (2021) 012021 doi:10.1088/1757-899X/1190/1/012021

A Method to Optimise Semi-Trailers with the Use of FEM Analysis and Durability Tests with a Road Simulator

Marcin Januszka [1,2], **Maciej Kaczor** [2,1]

[1] Silesian University of Technology, Department of Fundamentals of Machinery Design, ul. Konarskiego 18a, 44-100 Gliwice, Poland
[2] Wielton S.A., ul. Rymarkiewicz 6, 98-300 Wieluń, Poland

marcin.januszka@polsl.pl

Abstract. This paper presents a method developed by Wielton company – a manufacturer of trailers and semi-trailers – for the optimization of semi-trailer's construction, supported by FEM analysis and the modern road simulator – a test stand included as part of the method for optimising semi-trailers. The aim of the research was to optimize the process of developing a technical measure in terms of shortening the time of implementing new products while guaranteeing high durability and functionality of the products. As a result of the optimization of the construction with the use of FEM analysis and the road simulator, the semi-trailer as a new product for sale is delivered quickly and possible defects of construction are eliminated on early stage of new product development. The method assumes the following stages of the development process: determination of input data to the process, development of a 3D model, FEM numerical analysis (static and dynamic), development of an optimized 3D model based on numerical analyses, construction of the first prototype, prototype validation tests using the road simulator and other test stands. Important feature of the developed method to optimise construction is connected with specific data achieved from FEM analysis and used in tests with the road simulator. The effect of using presented method is the elimination of 98% errors in construction and technology. These errors concern not only the key parts of the semi-trailer, i.e. the body and the frame, but also less important components, i.e. fixing of tanks and boxes, ladder, spare wheel basket, extinguisher mounting. The method to optimise semi-trailers developed in Wielton have many advantages, such as: limiting the risk of launching breakthrough solutions on the market, faster increase in vehicle competitiveness, optimizing vehicle construction, reduction of production time and costs, and as a consequence - faster implementation of new products, reducing the number of complaints and service costs, increasing the value and prestige.

1. Introduction

Providing new, innovative construction solutions in the field of manufacturing commercial vehicles for transporting, among others trailers and semi-trailers is a difficult and long-lasting process. New vehicles must meet a number of requirements in terms of safety, durability and reliability [1][2]. The final stage associated with the introduction of a new vehicle on the market are validation tests, which carried out in a traditional way on roads can consume a lot of time and generate additional costs. Wielton company developed and propose alternative way for final validation of product. A special method of validation has been developed, which bases on FEM analysis and a mechatronic test stand for validation of prototypes. A road conditions simulator, which is located in the Wielton's R&D Center, enables validation testing of prototype products in a much shorter time.

Mechatronic test stands for vehicle structures, the so-called road simulators enable fatigue and durability tests of vehicles and their components. Road simulators are widely used research devices in the automotive industry [1][3][4]. However, they are not widespread when tests are carried out on vehicles such as semi-trailers and trailers. On the base of our research Authors would like to state that this type of research equipment bring tangible benefits to manufacturers of commercial vehicles, i.e. trailers and semi-trailers. With reflecting the real road conditions on the test stand, after registering road profiles on the measuring devices Authors of this paper would like to show how to accelerate the research process, which in turn will cause faster introduction of the new product to the market. The most important thing is to check, identify as fast as possible and eliminate potential threats and design problems that may occur during the long-term use of semi-trailers and trailers in various road conditions and different "culture" of vehicle use. It seems obvious that there is no better method to detect defects or weaknesses in the construction of complete vehicles than to simulate reality. Combination of FEM analysis and the MTS 320 road simulator [5] make it possible to map the vehicle movement observed while driving on different roads, regardless of the quality of their surface. The technical possibilities of the road simulator allow to simulate extremely different road conditions with high efficiency, from small vibrations to very strong impacts caused by road potholes. However such approach is not easy, because of complexity of the problem [6][7].

2. Simulators of road conditions in the automotive industry

The use of road simulators when introducing cars to the market is generally popular. With such test devices vehicle manufacturers are able to carry out the following tests (Figure 1) [3][5]:

- Vehicle durability testing through fatigue and durability tests allowing to assess the complete vehicle structure, components and subassemblies already at the prototype construction stage, improve reliability and performance while maintaining high quality, reduce costs and reduce the time of new implementations.
- Quality testing at the end of production stage, the "first road" in a controlled and repeatable environment (established road profiles) where, after successful testing, the vehicle is handed over to the end user (customer). Without a road simulator, a manufacturer who wants to sell products that meet the quality requirements must perform expensive, unique tests carried out in changing weather conditions on special test tracks, which will not allow for the elimination of potential irregularities.
- Driving comfort test that is conducted to compare with competing vehicle models. During the tests, it is possible to monitor the effects of changing test parameters, which cannot be carried out during tests on special test tracks, including in addition subjective opinions from the driver and passenger, which are insufficient for a correct assessment. The road simulator allows for high accuracy, repeatability and simplicity of driving comfort assessment. It is possible to recreate exactly the same conditions for each vehicle multiple times, stopping, visual inspection and restarting the test. Engineers are able to observe and evaluate, and then introduce changes to a design.
- Assessment of noise (buzzing, squeaking, rattling noise), which is perceived as a very important feature, has nowadays become the basic indicator of vehicle evaluation for both manufacturers and customers. Temperature, humidity, age and road conditions are the main factors that have the greatest impact on vehicle wear, which lead to the occurrence of the above defects, so some manufacturers decide to place the road simulator in special chambers that almost 100% reflect the real conditions in which vehicles are moving.
- A racing vehicle test, in which it is possible to very accurately simulate the dynamics of a vehicle that occurs during competition by adjusting the height, roll, inclination of the vehicle and allows to precisely simulate the aerodynamic load effects on the vehicle.

IOP Conf. Series: Materials Science and Engineering 1190 (2021) 012021 doi:10.1088/1757-899X/1190/1/012021

Figure 1. Different types of tests for vehicles [3][5][8]: vehicle testing with road simulators; special test tracks, road simulator placed in a special climate chamber, testing vehicles under extreme loads for the needs of sport.

However, an important research problem is the development of a test stand that allows testing of large and long (more than 12 meters of length) vehicles such as trucks, buses, semi-trailers, and truck trailers. Therefore Wielton, as a producer such type of long and big vehicles, had to elaborate own mechatronic road simulator for specific conditions. Development process have been realised in cooperation with MTS company, and a final results are presented in this paper.

3. The road simulator for trailers and semi-trailers in Wielton company - construction and application

The stand for testing commercial vehicles, i.e. semi-trailers and trailers at Wielton, known in the literature as "road simulator", is currently the most modern test stand that allows to determine various static and dynamic characteristics and, most importantly, to precisely determine the durability of the entire structure of the vehicle and its components in real conditions road loads. The stand for durability tests (understood as fatigue tests on a real object) at Wielton consists of six hydraulic cylinders placed on an active damping mass (Figure 2). The cylinders create a vertical force on all the tires and the saddle plate of the trailer. The configuration of the stand enables the testing of single-axle vehicles with a coupling, two-axle vehicles (passenger cars, agricultural trailers), and three-axle vehicles (buses, semi-trailers, military vehicles). It is possible thanks to the steel structure of the floor with interchangeable, configurable elements, which enable different settings of the distance between: axles, vehicle wheels, king pin (coupling) and the axle of a semi-trailer/trailer.

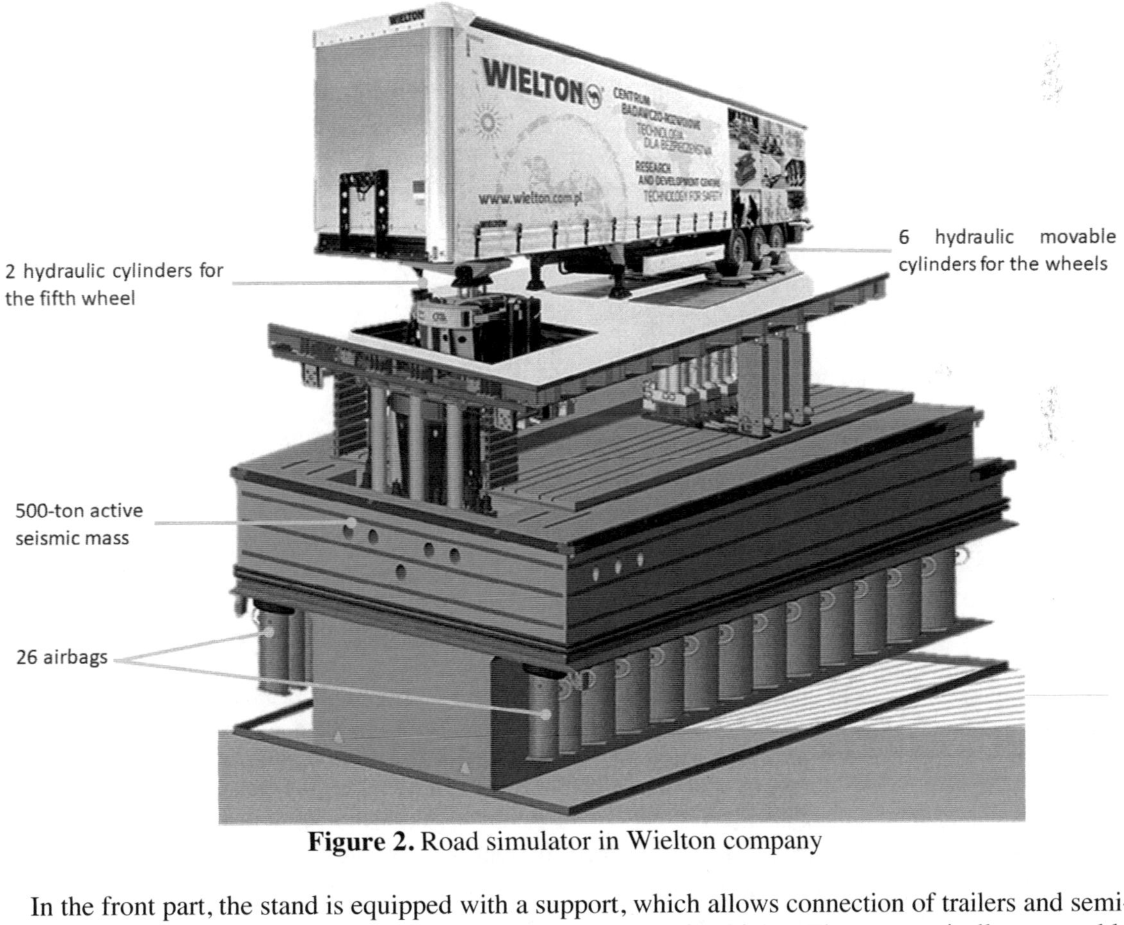

Figure 2. Road simulator in Wielton company

In the front part, the stand is equipped with a support, which allows connection of trailers and semi-trailers. This support can be hidden for testing other types of vehicles. The automatically retractable support attached to the damping mass, consisting, among other things, of two actuators, responsible for reflecting the recorded inputs in the front part of the semi-trailer or trailer, can be equipped with a standard fifth wheel or coupling, which are used in trucks depending on the tested vehicle, semi-trailer or trailer. Active damping mass with a mass exceeding 500 tons is suspended on pneumatic airbags.

IOP Conf. Series: Materials Science and Engineering 1190 (2021) 012021 doi:10.1088/1757-899X/1190/1/012021

Such a construction of the stand allows to limit the vibrations to the test field itself, damping the vibrations of the tested object, the weight of which may exceed 40 tons.

A specialized software package RPC (Remote Parameter Control) [9][10] is used for durability tests. It allows for a comprehensive analysis of the recorded loads of the vehicle in various road or terrain conditions and the selection of loads that will allow for the quickest determination of the actual durability of the vehicle.

At Wielton S.A., the road simulator is used for durability tests of curtain semi-trailers, tipper semi-trailers, container semi-trailers, center-axle trailers, trailers with a turntable and others. The use of this type of device offers the following advantages:

1. Reducing the risk associated with the introduction of breakthrough solutions to the market,
2. Faster growth of vehicle competitiveness,
3. Optimizing product design, reducing production time and costs and, consequently, faster implementation of new products,
4. Limitation of complaints and servicing costs,
5. Increasing the value and prestige of the Wielton brand.

4. New product development and optimisation method

Ensuring high-quality products offered on the road transport vehicle market (which is growing year-by-year [11]) requires comprehensive control throughout the design process. Improvements to currently offered products and the introduction of completely new products must be properly verified before they reach the customer. The final design of a new product is approved on the basis of positive results of verification tests (FEM analysis) and final validation tests (with the use of road simulator). Generally it is a long process, therefore at a continuous improvement during a design stage, the aim is to maximize the use of existing proven and correct solutions. However, after carrying out a certain number of such redesign iterations, a situation appears that further improvement of the technical means is associated with the introduction of significant changes, which in turn obliges to carry out a comprehensive re-verification of the product.

As mentioned, the implementation of a new product is a complex and time-consuming process. New product development and optimisation process at Wielton is presented on Figure 3. It begins with traditional market research. Its purpose is to define the characteristics of a new vehicle. At the stage of defining the design assumptions, customer expectations as to the durability of the new product, its durability, maintenance costs and purchase price are taken into account. At the same time, activities are carried out to determine the possibility of using a new type of materials, design solutions and manufacturing technologies. In order to be approved for road traffic, the designed vehicle must also meet certain legal standards and approval requirements. They define certain functional characteristics of the vehicle, such as maximum overall dimensions, generated permissible axle loads, required safety systems. Taking into account the above requirements and based on over 20 years of experience in Wielton in the construction of semi-trailers, the first product model is prepared in the form of a CAD file. Then it is verified in numerical strength analysis (FEM). Their main purpose is to detect overstrained areas. Various load cases that may occur during the use of the semi-trailer in accordance with its intended use are investigated. Developed method for new product development bases in this step on results of previous validation tests on the road simulator MTS 320 for similar products. Places, where some abnormalities have been detected after tests on the road simulator in the past, are listed. Such list of places with increased risk is used for FEM analysis. In such places more detailed FEM static analysis is realised. In safe places less careful FEM static analysis is performed. However, if a too high stress distribution in the structure is detected, the model is redesigned until satisfactory analysis results are obtained. Results from tests with the use of the road simulator are used as well for FEM dynamic analysis.

Dynamic analysis is performed as last type of FEM analysis, just before first prototyping. As an input to dynamic FEM analysis, a reference 60-second mileage is implemented from exactly the same road profile which is used on the road simulator at the Wielton. FEM dynamic analysis is performed only for steel frame of semi-trailers. The simulation takes into account 8 input signals: 6 signals of vertical movement of the semitrailer wheels, vertical movement and torsional movement of the fifth wheel around the axis defining the vehicle's driving direction. A reference simulation of 60 seconds is responsible for the cumulative fatigue damage of the material equivalent to a 15 km mileage. Finally these result are related to 400 000 km semi-trailer mileage, which is an equivalent of min. 3 years of a real exploitation of a semi-trailer. The results of FEM dynamic analysis are presented in the form of the damage accumulation factor. The accumulation factor of damage for welded joints is determined based on the SN fatigue curve according to PN-EN 1993-1-9-2007 Eurocode 3 for fatigue category 36 and the notch type corresponding to the weld with incomplete penetration (fillet weld). The adopted curve corresponded to the probability of failure of 95% for failure at the weld root. The SN fatigue curve of Strenx 700MC (which is typical material used for frames of semitrailers produced by Wielton) corresponds to a failure probability of 97,7%.

Positive results of FEM analysis provide the basis for manufacturing of prototype of a semi-trailer. It is being built to carry out further necessary tests, especially with the use of the road simulator, but not only. Functional tests check the performance characteristics of body elements (front wall, doors, side tarpaulins, roof) and chassis elements (bumper, support legs, electro-pneumatic system, boxes, buffers, etc.). This stage also controls the weight of the vehicle, the pressure generated on the fifth wheel and individual axles, and the tightness of the front wall and rear portal. Tensometric tests together with photogrammetric tests provide information on the strength properties of a new product and its components. Based on the recorded displacements, both methods determine the stress values in the structure. Such results are also used to define input boundary conditions for FEM analysis for next products. Tests are carried out for various load cases depending on the intended use of the product. Dropside and curtain semi-trailers are tested with a forklift simulating the loading and unloading process. In this way, the stability of the semi-trailer is checked, the strength of the floor and the functionality provided during loading (the possibility of opening/closing the tarpaulin, opening the door, lowering the sides). In turn, for example, tipper semi-trailers are tested in places such as gravel pits, quarries. On a semi-trailer equipped with appropriately placed sensors, the displacement of the side walls is measured depending on the type of the loaded material or the behavior of the entire structure at the time of the so-called tipping, i.e. dumping the load from the load box. The results obtained from such actual tests give an objective picture as to the correct functioning of the new product.

The last stage of the verification tests are durability tests (in the literature also referred to as endurance tests, fatigue tests or mileage tests [12]). Conducted on the MTS road conditions simulator, they finally verify the new product in terms of durability of both the complete vehicle and its individual components. The entire implementation process, from the design stage to validation research, is iterative and continuous. The detection of structural defects at any stage causes the work to be returned to the design stage in order to introduce the necessary structural changes. After their application, 3D CAD model and then the prototype are tested again. Only a positive result of all tests guarantees that the product is free from defects and is ready to be placed on the market. The general scheme of the new product development process in Wielton is presented on Figure 3.

The developed method for product optimisation, which base on FEM analysis as well as on a road simulator allows to obtain a high convergence of the results of numerical analysis with measurements during tests on a test stand. Table 1 presents comparison of average values of normal stress for the same points on FEM analysis model and real prototype (measured during tests on the road simulator). In Wielton the acceptable level of difference is 6%. The main reason of such high convergence is that

IOP Conf. Series: Materials Science and Engineering 1190 (2021) 012021 doi:10.1088/1757-899X/1190/1/012021

data from a road simulator is used as an input for FEM analysis and dangerous areas detected during FEM analysis are identified and inspected during durability tests on the road simulator. Such integration is very beneficial and allows to eliminate 98% errors in new products. The remaining 2% errors that occur in the products concern only minor components (not connected with safety or the functionality of the product), supplied by partners and not tested (e.g. additional equipment attached to the trailer).

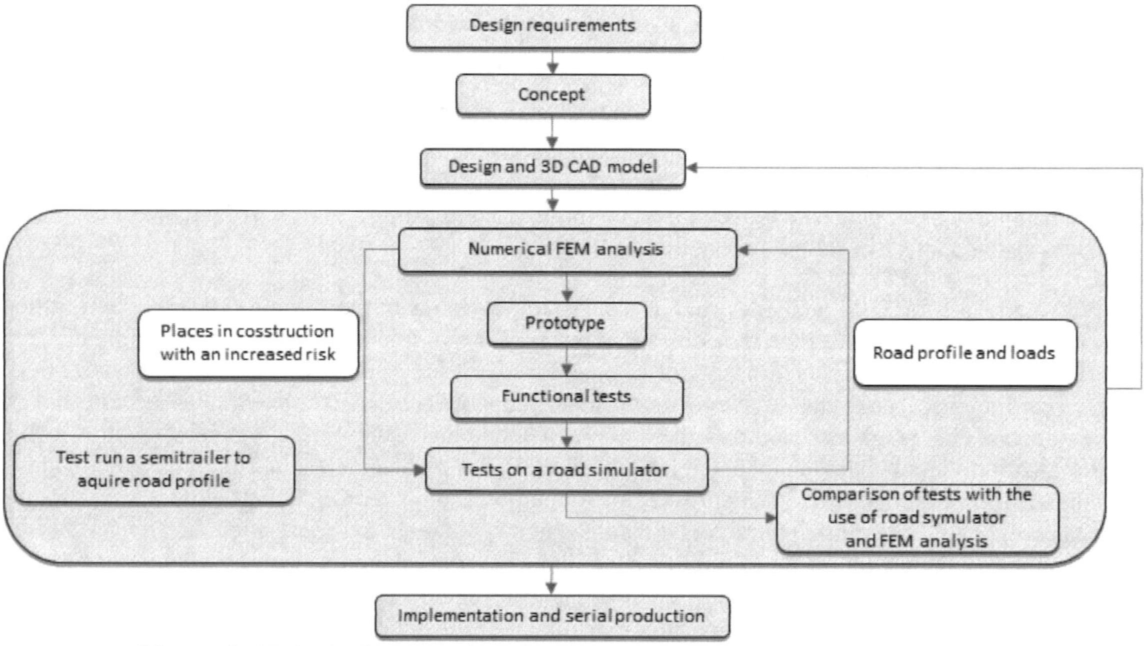

Figure 3. Method of new product development and optimisation used by Wielton

Table 1. The comparison of results of the numeric and experimental research

Strain gauge	Normal stresses measured from experiment on a road simulator [MPa]	Normal stresses obtained from numerical FEM analysis [MPa]	Difference [MPa]
1	410	387	23
2	389	413	24
3	381	398	17

5. Product validation and verification using the road simulator

The durability of the product has always been one of the key properties of each technical requirement. More than once, it played the role of the main criterion that guided the customer when buying a new machine or tool. There has also been a perception on semitrailers market that a "durable product" is a "heavy product". This reasoning is due to the fact that in the past, high durability was most often obtained by oversizing some key components of the machine. As a consequence, these actions increased the weight of the final product. Nowadays, the approach to product durability is a bit different. During the design process, the goal is to achieve the "golden mean" between a durable and a lightweight and cheap product. These criteria are used in optimisation process. The product lifetime has also been significantly shortened compared to similar products offered a dozen or so years ago.

However, one important fact should be noted. The products currently offered by manufacturers, also in the transport industry, are prepared for more effective work than the corresponding structures from previous years. The new generations of trailers are a good example. Thanks to the use of high-strength steels, such products are much lighter, but at the same time still durable. Hundreds of kilograms saved thanks to mass reduction of semi-trailers can be effectively used in the form of additional cargo that will be transported. The new technologies used for the production of semi-trailers, such as robotic welding, roll-forming, hydroforming, significantly affect the quality of the products offered. In order to maximize the potential of all these new solutions, it is necessary to use specialized tools to support the design process. It is possible to use CAD programs or environments used to numerically verify the durability of a new product. Modern tools used at various stages of the process of developing a new technical objects make it possible to determine the durability of the designed product with a certain probability. For this reason, validation tests are necessary to verify this feature of the new product.

Validation is a very important stage in the implementation process of a new product. The Polish standard defines it as a process that provides objective evidence confirming compliance with the requirements for a given use or application. The expression "objective evidence" refers to the test result, calculations or specific documents such as specifications, standards, procedures, etc. [13]. At Wielton, the validation process was carried out from the beginning on the basis of the results of a number of verification tests, including strain gauges, functional or utility. The latter were carried out at the customer's site during the trial period of operation. During this time, the recipient provided his own observations and comments regarding the product used. Additionally, after exceeding the set mileage, the semi-trailer underwent a detailed service inspection. They were inspected i.a. wear of the undercarriage components, the condition of the paint coating, the general condition of the main frame or the efficiency of power pneumatic systems. The time needed to test the trailer ranged from a few to several months. Nowadays, the key is to quickly introduce a durable and reliable technical measure. In order to meet the market requirements and to be able to compete with the largest trailer manufacturers, such as Schmitz Carbobull or Krone, it was necessary to implement modern tools that would allow for the elimination of defects at the early stages of the technical means development process. One of the basic tools used for this purpose are numerical strength analyzes. Basing on the method presented in this paper, thanks to the quick localization of excessively stressed areas, it is possible to properly modify the structure before building the prototype. Semitrailers and trailers are used in difficult conditions. Freight forwarding companies cannot afford to "waste" cargo space, which is why the trailers almost always move with the maximum permissible load. Their annual mileage reaches an average of 120 000 km, while for a passenger car they usually do not exceed 30 000 km. Loads occurring while driving constantly cause strenuous work of the entire structure. For this reason, it is crucial to correctly detect and remove even small, excessively loaded areas during the design process, as they may lead to later damage resulting from, among others, from the occurrence of material fatigue.

Correctly carrying out numerical analysis of complex systems such as semi-trailers is a difficult task. For this reason, it is necessary to conduct additional actual tests that allow to verify the correctness of the FEM model. The tool used for this purpose is the road conditions simulator MTS 320. The durability tests performed on this stand are the last tests carried out as part of the new product validation process. However, simulation with the use of road simulator must be properly prepared. In the first phase, sensors are installed on a prototype vehicle. Strain gauges providing information about displacements are placed in places that showed increased stress values in the numerical strength analysis. The accelerometers mounted on the trailer measure the parameters of the movement of the axles and the kingpin pivot. Additionally, sensors controlling the pressure in the suspension system and a GPS antenna are installed. Signals from all sensors collected during the reference run give a picture of the response of the trailer to real traffic forces. As part of the reference run, the prototype product travels a total of approx. 2000 km. Cases of driving without a load and with the maximum permissible load are investigated. Data is collected from highways, main and local roads. This allows the behaviour of the semi-trailer to be examined in the event of various types of extortion. The next step is to prepare a road profile, i.e. a set of forces that will be applied on the simulator. Such information from test run is used also in FEM dynamic analysis. The main purpose of using the MTS stand in the product validation process is to precisely determine the durability of the trailer over a specified period of time and confirm correctness of previous analysis. At this stage, the key task is to prepare an effective road profile. Only such extortion allows the actual wear process of the semi-trailer to be accelerated, which translates into faster determination of its durability. A specialized RPC software package [9][10] is used for this purpose. Remote Parameter Control is an advanced simulation technique used to simulate and analyse repetitive inputs using a dynamic mechanical system in a controlled laboratory environment. The essence of the operation of the entire system is based on the continuous imposition of such extortion that the response of the tested object to this extortion is consistent with its response recorded in the reference run. Thanks to this, the correct mapping of road conditions is ensured, and several years of vehicle operation are achieved during several weeks of simulation on the stand. The use of the MTS stand as a validation tool allows for precise determination of the durability of the entire product and the detection of damage that may occur during normal operating conditions.

6. Conclusions

Market requirements force manufacturers like Wielton to shorten the time needed to implement new products as much as possible. Unfortunately, the prevailing trend of reducing the weight of vehicles while maintaining the required durability and reliability, additionally extends the design and implementation process of the new product. For this reason, there was a need to systematize the optimization processes of new products developed at Wielton.

The requirements for transport vehicles are higher each year. For this reason, the currently designed technical measures must be constantly optimized. Optimization includes economic, design (including strength), technological, operational and other criteria. Optimization of complex technical measures such as commercial vehicles, including semi-trailers, is a complex problem that requires systematization. There are different methods used in the various stages of the new product development process (from determining the need to the design and construction process, verification testing, etc.). These methods make it possible to optimize solutions with greater or lesser success. However, prototypes are often built in the final stages of the development of a complex technical objects (including semi-trailers) and subjected to further validation tests on advanced research stands. Only validation tests most often allow to confirm the correctness of the adopted solutions, especially from the point of view of the durability criterion.

In Wielton company a method for product optimisation within better durability has been developed. The method consists of some stages of the development process: determination of input data to the process, development of a 3D model, FEM numerical analysis (static and dynamic), development of an optimized 3D model based on numerical analysis, construction of the first prototype, prototype validation tests using the road simulator and other test stands. The most important feature of this method is an integration of data used for FEM analysis and validation test with the use a road simulator and real prototype of a product. Such approach allows the elimination of 98% of errors in construction and technology in faster way.

Acknowledgment(s)

Research co-financed from the funds of the Ministry of Science and Higher Education (Poland) No. 12/DW/2017/01/1 dated November 7, 2017 in the amount of PLN 4 433 520;

References

[1] E. Azrulhisham, Y.M. Asri, A.W. Dzuraidah, N.M. Nik Abdullah, C.H. Che Hassan, A. Shahrom, "Application of Road Simulator Service Loads in Automotive Component Durability

Assessment", The Open Industrial & Manufacturing Engineering Journal, No. 4, pp. 1-7, 2011.

[2] H. Fu, J. Gao, "Vehicle durability test based on user survey", Computer Modelling & New Technologies 18(10), pp. 479-483, 2014.

[3] C. Dodds, A. Plummer, "Laboratory Road Simulation for Full Vehicle Testing: A Review", SAE Technical Paper 2001-26-0047, 2001.

[4] S. You, S. Joo, "Virtual Testing and Correlation with Spindle Coupled Full Vehicle Testing System," SAE Technical Paper 2006-01-0993, 2006.

[5] MTS. Model 320 Tire-Coupled Road Simulators, dostępne pod adresem: https://www.mts.com/cs/groups/public/documents/library/dev_002219.pdf, (available: 25.11.2021r.).

[6] M. Dilip, G. Gangwani, "Design Optimization of Truck Body", IJIRST – International Journal for Innovative Research in Science & Technology, Vol. 3/02, pp. 383-385, 2016.

[7] S. Medepalli, R. Rao, "Prediction of Road Loads for Fatigue Design- Sensitivity Study", International Journal of Vehicle Design, 23 (1-2), pp. 161-175, 2000.

[8] Website of BharatzBenz, https://www.bharatbenz.com/about-us (available: 27.04.2021

[9] MTS RPC Pro Software, https://www.mts.com/cs/groups/public/documents/library/cm3_002017.pdf (available: 27.04.2021).

[10] Manual: cRPC Pro Fundamentals. MTS Systems

[11] „Transport of goods and passengers in 2020", Statistics Poland, https://stat.gov.pl/obszary-tematyczne/transport-i-lacznosc/transport/przewozy-ladunkow-i-pasazerow-w-2020-roku,11,9.html (available: 27.04.2021).

[12] P. Simiński, Research durability in terms of special vehicles their safety (in Polish), Logistics, No 4/205, pp. 8288—8291, 2015

[13] PN-EN ISO 9000:2015-10 (ISO 9000:2015) Quality management system – Fundamentals and vocabulary.

IOP Conf. Series: Materials Science and Engineering 1190 (2021) 012022 doi:10.1088/1757-899X/1190/1/012022

Design and Manufacturing Method of GFRP Blades for Vertical Axis Wind Turbine

Paul Bere [1], **Radu Ciobanu** [2], **Oleg Ciobanu** [2], **Marin Guțu** [2]

[1] Technical University of Cluj-Napoca, Bd. Muncii 103-105, Cod 400641, Cluj-Napoca, România
[2] Technical University of Moldova, Bd. Ştefan cel Mare, 168, MD-2004, Chişinău, Republica Moldova

marin.gutu@pmai.utm.md

Abstract. Today, environmental protection and the rational use of natural resources have become growing challenges. One of these aspects is energy producing with less impact on the environment. For some applications the small wind turbines can be a solution. In this paper are presented some aspects regarding the design and manufacturing technology of 500 W vertical axis wind turbine blades. The turbine will be installed in the urban environment so that requirements regarding rotor aesthetics and noise level will be considered. The optimal geometric parameters of the turbine rotor were determined by specialized MathCAD and elaborated in 3D modelling software. For the blades the NACA 0018 airfoil was used with the chord length of 110 mm and is constant throughout the length of the blade. The blades are helical and are 1800 mm long. To estimate the aerodynamic performance of the turbine rotor, a computational fluid dynamics model was developed using ANSYS CFX finite element analysis software. The authors proposed a new method for rapid prototyping of blades using fused deposition modelling procedure. In order to increase the mechanical properties, the blades were covered with several layers of glass-fibres reinforced polymer. Estimation the blade strength to operating loads was performed using Ansys Workbench software. The modelling of the composite material architecture on the blade surface was performed using the special ANSYS Composite PrepPost module. Several simulations were performed with different number of layers and stacking sequences. In order to prevent rotor over engineering the maximum rotation speed of 400 min^{-1} was accepted (equivalent to ≈12 m/s wind speed). Simulation of the blade strength to operating loads indicate values of around 40 MPa for shear stress (50% less than failure shear stress) and around 80 MPa principal stress (60% less than failure tensile stress). The maximum blade deformations of 3 – 4 mm shows that the blade is stiff enough at imposed operation conditions. For the blade manufacturing, based on the numerical analysis results, the configuration of the strength structure composed of 5 layers of GFRP was selected, which forms a total blade weight of 1895g.

1. Introduction

Environment protection is one of the most important factors of sustainable development. Energy production without harming the environment is an ongoing concern of research centers. The wind turbine in last period has a very impressive development. Different authors traded in papers research regarding the evolution, design or behavior of wind turbine generators. Using different airfoils the positioning of attack angle of the blades were obtained different performance and efficiency. Numerical modeling and simulation of blades rotation are a challenge and different authors proposed some solutions in order to obtain the best blades geometry, airfoils, curved-bladed and helically twisted configuration [1, 2]. In [2] the authors presented analysis for H-rotor Darrieus turbine as a low speed wind energy converter. The numerical analyze presented in [3] can be adapted like solution in order to simulate the blades motion. The vertical axis of wind turbine generators are studied in [3-4] from numerical analyses point of view. In this paper are presented aspects regarding the design, Computational Fluid Dynamics (CFD) analysis, blade strength structure evaluation, and manufacturing method for vertical axis wind turbine blades. To obtain the blade prototype an additive manufacturing methodology was used [7-10]. Using a Fused Deposition Modeling (FDM) solution a plastic prototype from ABS was manufactured. The prototype of the blade was covered by Glass Fiber reinforced Polymer (GFRP) in order to improve the mechanical properties of the blades.

2. Design and manufacturing

2.1. Design of vertical axis wind turbine rotor

The parameters required to develop the wind turbine rotor with aerodynamic blades are the following: the airfoil, airfoil chord (which defines the rotor solidity), aspect ratio, rotor diameter and its height. For this rotor the symmetrical NACA 0018 airfoil was chosen which is among the most used one for VAWTs. The airfoil selected is good one for low values of Reynolds numbers (Re < 200 000) and show reasonably constant performance, in terms of lift and drag coefficients Cl/Cd ratio, for a large range variation of the angle of attack and unsteady wind conditions, according to *airfoiltools* publicly available database. The VAWT's solidity is the ratio between the blade's area and turbine's swept area. The blade's area is considered the area of one side of it. For a straight bladed VAWT, the blade's area is calculated as chord's length and blade's height, thus the solidity σ is:

$$\sigma = (N \cdot c \cdot L)/A = (N \cdot c)/D \quad (1)$$

Where: N – the number of blades; c – the chord length; L – blade's length; D – turbine's diameter.

Solidity directly influences the turbine's performance depicted by the power coefficient as well as tip speed ratio TSR. The TSR is the ratio between the blade's tangential speed and unperturbed wind speed. The optimal length of the airfoil chord was determined with the special applet QBlade. Several rotor simulations were performed for decent wind speed - 7 m/s (Re <100,000). At this speed the rotor should generate about 30% of its rated power. The results are presented in the diagram of the dependence of the power coefficient and the tip speed ratio, figure 1. From the diagram 1 we can observe that high solidity rotors have an improved self-starting ability, but their tip speed ratio (TSR) is lower compared to those with a low solidity.The rotor diameter and its height were determined approximately from the power calculation relation. The power output P of a wind turbine is directly proportional to its swept area:

$$P = \frac{1}{2} \cdot Cp \cdot \rho \cdot A \cdot U^3 \quad (2)$$

Where: ρ – air density; U – wind speed, A – the swept area of the rotor; Cp – the power coefficient which in efficiency terms can takes values ranging from 0 to 0,593 (Betz limit).

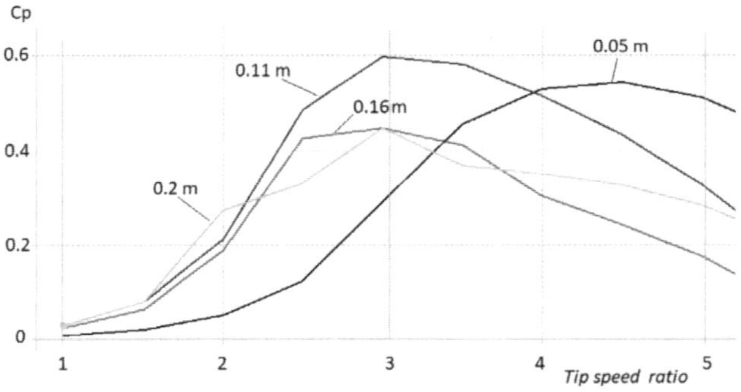

Figure 1. Power coefficient vs TSR dependence on chord length for Re < 100000.

To determinate the nominal power of a wind rotor, the wind speed equal to 11 m/s was accepted. This value fits in the range of wind speeds between 10 – 12 m/s used by most manufacturers of wind turbines. According to the parameters of the vertical axis wind turbines with aerodynamic blades existing on the market, the power coefficient does not exceed the value of 0.3. In our case, in order to select the power coefficient close to the real one, several works were analyzed, of which the most relevant was considered [1]. According to this paper the power coefficient was accepted equal to 0.22. All the calculations were performed in MathCAD software. For the rotor design the aspect ratio between height and diameter was accepted close to 1. Since the rotor is an experimental prototype, determining the optimal aspect ratio will be achieved under real test conditions. So, the rotor diameter was obtained 1600 mm and its height - 1800 mm. From the graph presented in figure 1 the airfoil chord length of 110 mm was accepted. The rotor designed in SolidWorks is presented in figure 2.

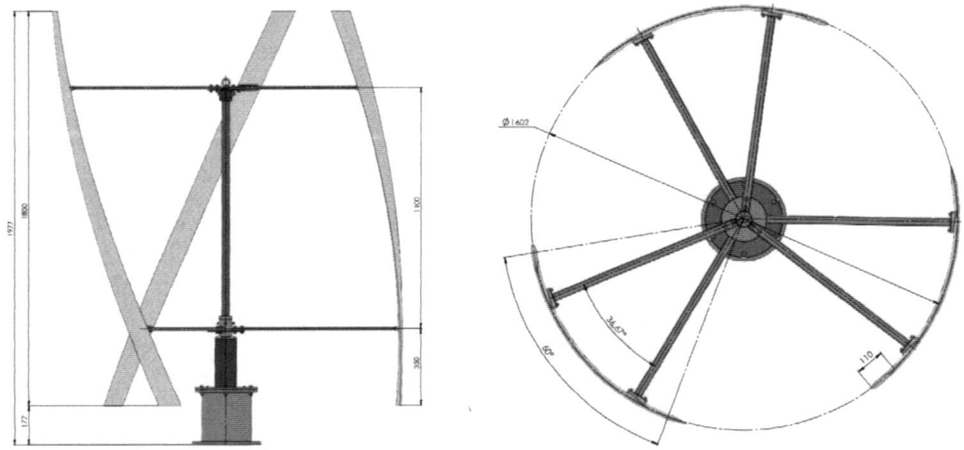

Figure 2. Wind turbine rotor geometry

The blade is designed in a helical shape according to recommendations in [2] and it is twisted to 60°.

2.1.1. CFD analysis of the rotor model. In order to verify the rotor conversion efficiency or the power coefficient a CFD model was developed using finite element analysis software ANSYS CFX. The rotor geometry, designed using SolidWorks, was then imported into the ANSYS Design Modeler software. In order to reduce the time and computing resources, only blades geometry was considered. The dimensions of the computational fluid domain were chosen taking into account the recommendations of the authors [3] so that the boundaries of the field do not influence the free flow of the air. The simulated fluid domain was divided into two subdomains: the Stator (static) subdomain and the Rotor subdomain inside of it (of cylindrical form, which rotates around its axis).

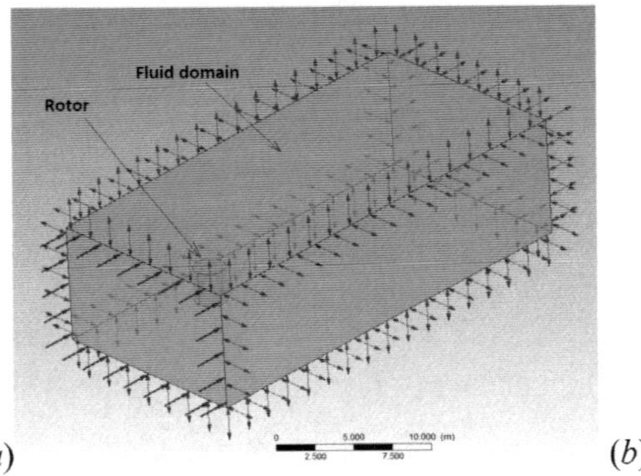

(a)

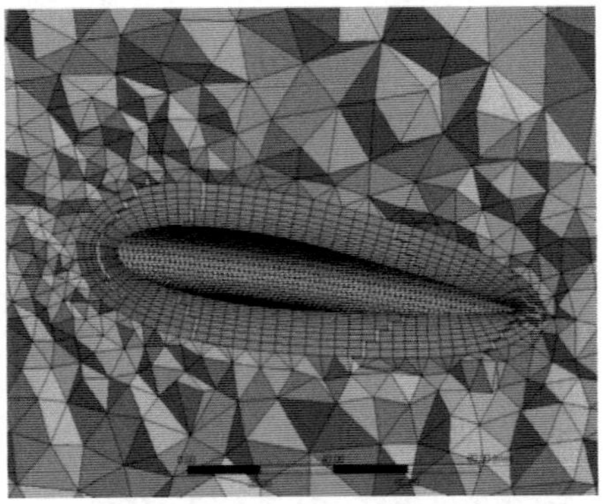
(b)

Figure 3. Computational model: (*a*) computational fluid domain; (*b*) mesh details around the blade.

After uploading the geometric model the following regions were defined: (Inlet – the face with black arrows on the perimeter), (Outlet – the opposed face of the Inlet), (Openings – the four side faces), and the common regions between *Fluid_stator* and *Rotor* (Fluid-Fluid). Figure 3 (*a*) illustrates the considered fluid domains. The mesh used for finite element analysis of the rotor was generated using the ANSYS Meshing Workbench. In order to capture the effects of the boundary layer on the surface of the blades it was fine meshed in 9 layers using the function *inflation*. The transition from the fine-meshed areas to the gross meshed ones was done by specifying the Growth Rate = 1.1 expansion factor, figure 3 (*b*). The maximum variation of the characteristic dimensions of two adjacent elements is not bigger than 5%. The entire domain was meshed into approx. 5 000 000 finite elements.

In order to perform CFD simulation of the rotor the following input parameters are necessary: wind speed and rotation speed. Optimal rotor speed depends on suitable tip speed ratio. These aspects were well analyzed in the paper [1], on the basis of which the tip speed ratio value equal to 3 was accepted. Several simulations were performed according input parameters indicated in table 1.

Table 1. Input parameters for CFD Simulation

Wind speed, m/s	4	6	7	8	10	12	14
Rotational speed, min^{-1}	142	213	263	283	354	425	496
Tip speed ratio				3			

Using ANSYS CFX Expression Language (CEL) the following output parameters were defined: the torque appearing on the rotor shaft (*Mz = torque_z()@Rotor*) and the power (*P = Mz*pi*n/30*). These parameters are necessary to draw preliminary rotor power curve.

2.1.2. Wind turbine rotor blade strength structure design and analysis. Because of blades complex geometry and relatively low dimensions their support structure will be made of plastic ABS by 3D printing technology then covered by several layers of Glass Fiber Reinforced Polymer (GFRP). Thus, these materials were analysed for modelling the strength structure of the blade. Blade support structure was modeled in SolidWorks from ABS giving suitable wall thickness in order to achieve a mass of maximum 1 kg. The thickness of 2.4 mm ABS wall was accepted for the equivalent mass of approximately 0.94 kg. Modeling the stacking sequence of the composite material of the blade structure was performed using the ANSYS Composite PrepPost (ACP) application. An orthotropic elastic material model was used to model 0.3 mm - 280 g/m^2 and 0.16 mm - 160 g/m^2 2×2 Twill, E-glass woven fabric composite material while the ABS was modeled as an isotropic elastic material. The section of the meshed blade with 400000 elements is presented in figure 4. For the preparation of the blade analysis model some recommendations for boundary conditions settings presented in the paper [5] were used. Blade model boundary conditions are depicted in figure 5. Several simulations were performed according the input parameters presented in table 2. During operation, the wind turbine rotor is charged at variable wind speeds causing aerodynamic and centrifugal forces in the blades leading to cyclic deformations and fatigue. For centrifugal load evaluation, in order to prevent rotor over engineering, the maximum rotation speed of 400 min^{-1} was accepted (equivalent to ≈12 m/s wind speed).

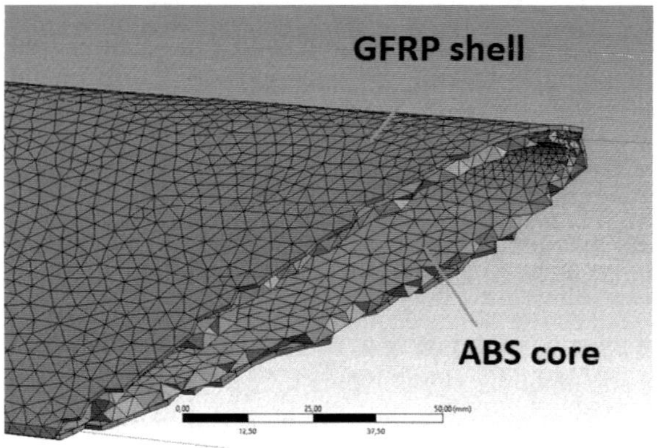

Figure 4. Mesh details of composite blade

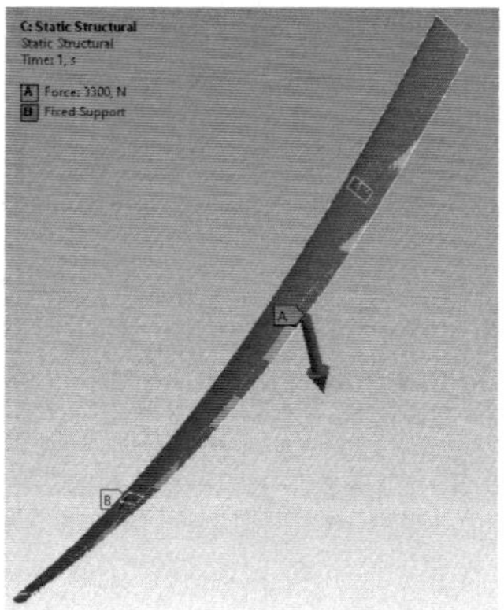

Figure 5. Blade model boundary conditions

Table 2. Input parameters for strength structure simulation.

Number of layers	[2][a]+1[b]	[3][a]+1[b]	[4][a]+1[b]	[5][a]+1[b]	[6][a]+1[b]
GFRP stacking sequence	[0/90/±45] +[0/90]	[0/90/±45/0/90] +[0/90]	$[0/90/\pm45]_2$ +[0/90]	[0/90/±45/0/90/±45/0/90] +[0/90]	$[0/90/\pm45]_3$ +[0/90]
GFRP thickness, mm	0.76	1.06	1.36	1.66	1.964
Total mass, kg	1.40	1.64	1.88	2.12	2.36
Centrifugal force, kN	1.97	2.3	2.64	2.97	3.3

[a] Composite layup of 280 g/m^2 Twill fabric (1 layer mass - approx. 240 g);
[b] +1 is the final layer of Twill fabric 163 g/m^2.

All the values of centrifugal forces were calculated using 400 min-1 rotation speed. Therefore, the following constraints were imposed for the preliminary design of the blade strength structure: the total mass of the blade must not exceed 2 kg; maximum shear stresses [σ] = 40 MPa (shear failure stresses are around 80 MPa). Shear stresses are critical for composite materials because they cause resin degradation and delamination of the layers. The value of the allowable shear stresses was accepted based on the own research presented in the paper [6].

3. Manufacturing of the blades

3.1. Rapid prototyping of the blades

After the complex blades geometry analysis was decided to use an additive manufacturing procedure. A FDM method was use to obtain a plastic prototype. The blades length was 1800 mm. Because of limited working volume of the printer, the blades were sectioned in 8 segments and assembled finally in one piece. The FDM procedure greatly reduces manufacturing time and ensures good accuracy. The parts surface obtained have certain roughness. That is proper to create a good adhesion of the glass fiber reinforced polymer layers and the plastic prototype. For this, an ABS thermoplastic material was chosen to be used in order to make the blades prototype. The FDM printer used in this case is type *LeapFrog*, the Creart HS XL model, its print size is 280x270x590. Close manufacturing conditions were used in [7].

The authors used the same printer in the previews work. They used the different material (PLA) like master model and different parameters of the printer. To create the G slicing code used to make the blade prototype Ultimaker software Cura 4.0 was used. The used material was *ABS LEAPFROG MAXX Essentials* by the printer provider. The printing temperature was 245 °C for the material. The printer heated plate temperature was 80°C. The used filament thickness is 1.75 mm, and extruder head nozzle diameter 0.4 mm. The layer thickness was 0.2 mm, and as the internal structure of the piece we have: wall thickness - 1 layer; for the exterior blade infill (filling) - 85%; structure of infill "grid". The printing speed was 60 mm/s. For the exterior shell printing, a lower speed of 40 mm/s was adapted in order to achieve better dimensional accuracy and roughness. Considering the weight evaluation of the blade segments finally the grid internal structure was chosen. The segments were assembled by bonding using a structural adhesive type Scotch-Weld™ Epoxy Adhesive from 3M Company. In order to have a good connection and centering of the ABS segments plastic joints were designed on each of face where the segments were assembled by bonding.

3.2. Reinforcement of the blades with composite materials

In order to increase the mechanical properties of the blades the ABS obtained prototype were covered by GFRP composite [7, 8]. Before prototyping the blade geometry was reduced by 1 mm in order to compensate the GFRP coating.For the reinforced materials were used 5 layers glass fiber type Twill with follow stacking sequence:

- $[0/90/\pm45]_s$ - 4 layers, 2X2 Twill, E-glass Woven Fabric, 280 g/m^2.
- [0/90] 1 layer, 2X2 Twill, E-glass Woven Fabric, 163 g/m^2.

The matrix used was epoxy type Epiphen RE/DE 4020 (France). The mixing ratio resin/hardener was 100:30 according to provider data sheet. Hand lay-up technology was used to applying the composite layers on plastic prototype. The last layer applied on the composite was Peel-ply by 163 g/m^2. by Nylon textile (figure 6).

Figure 6. GFRP layers applied on the blade plastic prototype.

The curing time were done 24 hours at 22° C and 8 hours at 75° C in the oven. At the final the peel-ply were rejected from the blade and the end border was cut at final dimension. The exterior surface was traded with sprayed filler and at the final the surfaces were covered by epoxy gel coat.

4. Results and discussions

4.1. Blade design results

The authors decided this design due to the following advantages: - a section of the rotor blade is always positioned optimally in relation to the wind direction, this allows obtaining a power coefficient relatively uniform, thus reducing fatigue loads; - the rotor is quieter and it has an aesthetic appearance. The rotor was designed with the possibility of easy disassembly of the blades in order to facilitate the change of blades with different geometry. For each wind speed was obtained a diagram as presented in figure 8 (*a*). The power diagram is variable because of variable torque appearing on the rotor shaft. For power curve construction, the average power values from each simulation were considered. In order to check flow field around the rotor, the velocity distribution was analyzed. Velocity contour section for 7 m/s wind speed is depicted in the figure 7 (*b*).

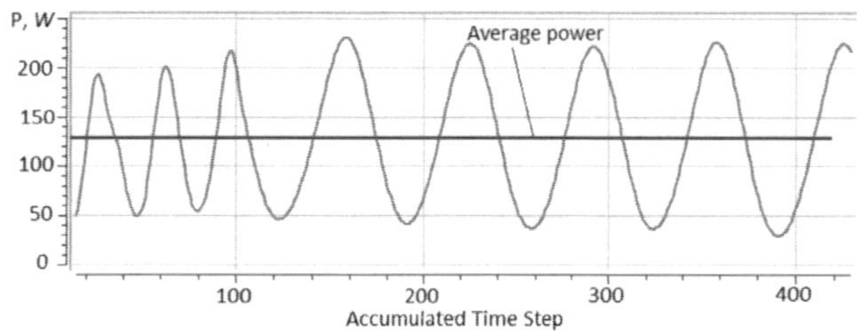

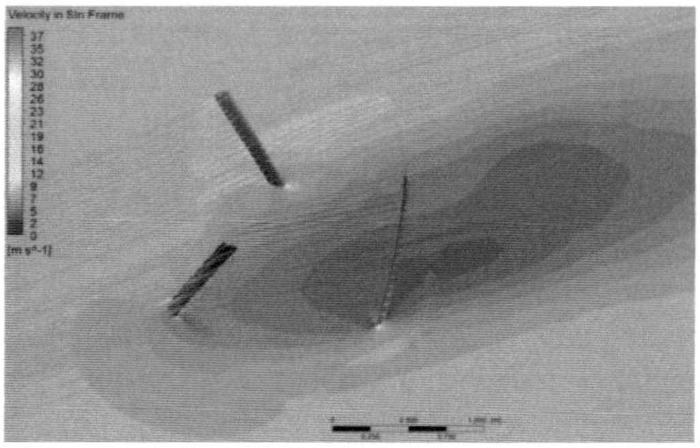

(a) *(b)*

Figure 7. CFX simulation results at 7 m/s wind speed: (a) generated power diagram, (b) velocity contour.

As was mentioned above the power curve was created using simulation obtained values and for comparison, another one was drawn based on the calculations in the Mathcad application using the power coefficient Cp = 0.22. The results are synthetized in the figure 8. It can be noticed from the figure that the power estimated by ANSYS CFX is lower because of 3D aerodynamic effects. The blades experiences turbulences that are formed at the trailing edge and at the tips. The last ones have the most negative impact on the performance of the blade. The simplest method for diminishing the effect of the vortexes at the tips on the performance is endowing the blades with endplates.

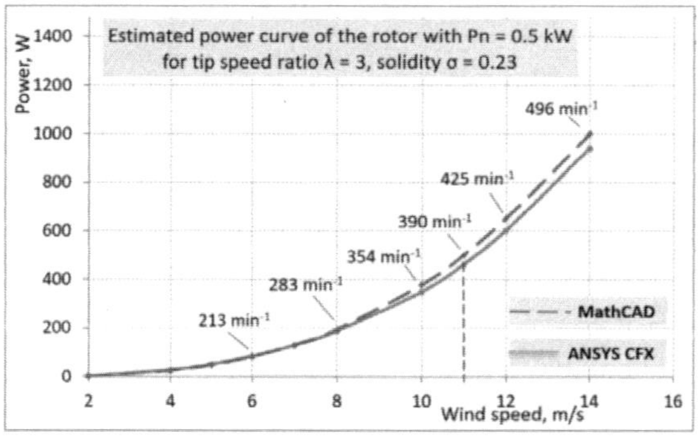

Figure 8. Wind turbine rotor power curve

Validation of the CFD model will be performed after manufacturing the experimental prototype of the rotor and testing it in real operating conditions. The results regarding blade strength structure analysis are the following. The main output parameter was considered the Shear stress. The value obtained is around the maximum allowed value and was obtained because, for reasons of the blade geometry simplification, only the elements on the blade surface were constrained, figure 9. In fact, the real blade will be fixed with metal elements and bolts. If we check the principal stresses, they have values around 85 MPa. All results are presented in the table 3.

Table 3. Simulation results according input data.

Total blade mass, kg	1,40	1,64	1,88	2,12	2,36
Centrifugal force, kN	1,97	2,30	2,64	2,97	3,30
Shear stress, MPa	36	39	41	42	43
Normal stress, MPa	43	46	50	52	53
Principal stress, MPa	73	78	82	85	86
Total deformation, mm	3,35	3,44	3,5	3,6	3,7

IOP Conf. Series: Materials Science and Engineering 1190 (2021) 012022 doi:10.1088/1757-899X/1190/1/012022

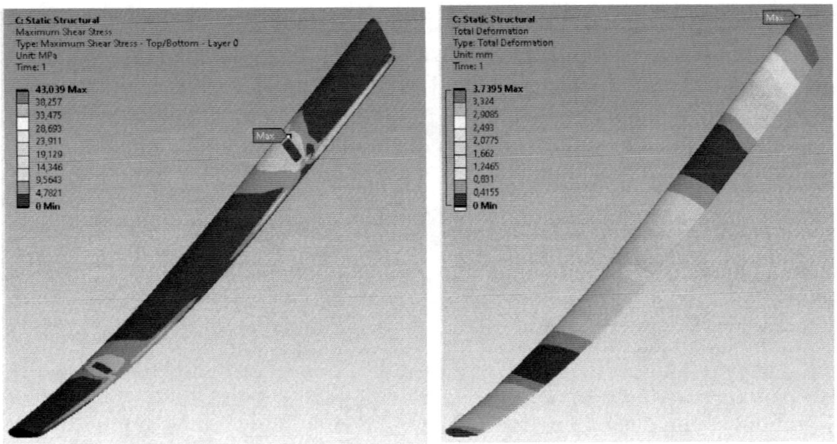

Figure 9. Blade simulation results: shear stress and deformation

Tensile stresses for the weakest bidirectional fiberglass composite materials are over 200 MPa. Total deformation of the blade indicates that it is robust enough for the imposed conditions.

4.2. Blade manufacturing results

The FDM process is very suitable to obtain the complex prototype. The obtained segments of the blades prototype using ABS material have a good surface. The surface roughness Ra of the blade segments measured by a Namicon TR-220 device was 21±5μm. Very close results are obtained and mentioned by the authors in [9] using FDM 3D Printer. The ABS material was a good option for manufacturing a prototype support of the blades. This material is sensitive to solvents from epoxy resin and makes very good coupling between these two materials. In this case the GFRP applied on the ABS prototype create a good adhesion between composite layers and Plastic porotype. Different solution was applied by the authors in [7] which used a PLA prototype. Manufacturing of the parts from PLA using FDM method is more easily and the obtained prototype is more accurate but in this case the epoxy resin does not adhere so well to the surface of the PLA material. In this case delamination layers can happen. In Fig. 10 is presented the obtained segments from ABS prototype. In figure 10 (c). the ABS master model of the blade is covered by GFRP layers.

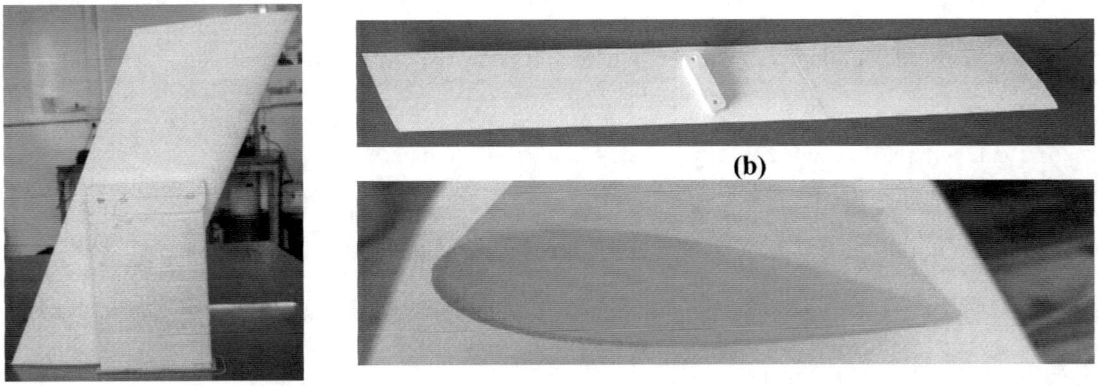

(b)

(a) (c)

Figure 10. ABS segments of the blades obtained by FDM method. **a.** Connection points area of the blade. **b.** Three segments assembled at the end of the blade. **c.** The blade prototype covered by GFRP

The weight of the plastic ABS prototype was 910 g. The cost of plastic ABS filament is 11 €/100 g.

In our case this represented around 100 euros/ blade. The obtained weight fraction ratio for GFRP material was 50%. The final weight of the blade was 1895 g. A tolerance of maximum 12 g main value was found between the blades. To balance the weight blades the supplementary plumb piece were applied to lighter blades to have the same weight. Obtained blade is presented in figure 11.

Figure 11. Obtained blade obtained from ABS and GFRP 1800 mm length.

This solution to reinforce the plastic prototype by GFRP is suitable to obtaining the master models for the blades. This can be a cheap option in order to obtain a functional prototype. In order to produce many parts for GFRP material the indicate solution is to manufacture a mold. The composite parts are obtaining into the mold and we can use couple manufacturing solution. In the prototyping phase in order to eliminate costs and manufacturing time of the molds this adopted solution can be used.

5. Conclusions

At the design stage, the aim was to develop an experimental prototype of the vertical axis wind turbine rotor to serve as a pilot station for testing blades with different geometry. As the wind turbine will be installed in the urban environment, the requirements regarding the aesthetics of the rotor and the noise level were considered. The estimated power of the rotor is obtained at a number of rotations compatible with those of the available electric generators in this range (200 - 400 min^{-1}).

Analyzing the comparative wind turbine rotor power curves it can be concluded the following. At the wind speed of 11 m/s for determination of the nominal power of the wind rotor, a difference of 10% was obtained between the values of the power calculated in the application MathCAD and the one obtained in ANSYS CFX. Therefore the value of the *Cp* for the simulated rotor model is 10% lower than the *Cp* value accepted in the initial calculations. This result can be considered as a partial validation of the computing model.

Simulation of the blade strength to operating loads indicate values of around 40 MPa for shear stress (50% less than failure shear stresses) and around 80 MPa principal stress (60% less than failure tensile stress). The maximum blade deformations of 3 – 4 mm shows that the blade is stiff enough at imposed operation conditions.

The manufacturing complexity of the blades was solved by using FDM method. The solution to use a plastic prototype from ABS covered with GFRP payers eliminate the mold of the blades. The FDM technology has a lot of benefices and can be mentioned: plastic material pieces can be produced in a short time, the internal structure of the blades can be designed in initial stage and manufactured according to the design, for the bigger piece that can be obtained by small sections and assembled, the cost production are low, reduced manufacturing time. If consider that the mold of the blades at this dimension (1800 mm) obtained by CNC milling machine from epoxy block, cam be around 5000 euros we can observe a large cost reduction for the manufacture of a blade. The cost/one blade is around 150 euros. This is for functional prototype and the mold is not included. For production blades is useful to use a mold and a GFRP sandwich material can be solution for the blades. In this case the cost production of the blades can be reduced by 3 times.

The mechanical resistance of the blades increases by GFRP composite cover. In the same time the stiffness of the blades are bigger. Using the GFRP and plastic materials the blades can be lighter like traditional material (wood or aluminum). It can be obtained some specific internal structure design which it was impossible to obtain by classic technology or materials.

Acknowledgment

The authors are grateful for the opportunity to conduct this research within the State Program Project „Study of wind and solar energy potential of the Republic of Moldova and development of conversion systems for dispersed consumers", cipher 20.80009.7007.10, funded by the Ministry of Education, Culture and Research of the Republic of Moldova.

References

[1] D. Hilewit, E. A. Matida, A. Fereidooni, H. Abo el Ella, F. Nitzsche,"Power coefficient measurements of a novel vertical axis wind turbine", *Energy Science & Engineering* Vol. 7/ 6 p. 2373-2382, 2019.

[2] F. Scheurich, T. M. Fletcher, „The influence of blade curvature and helical blade twist on the performance of a vertical-axis wind turbine", *American Institute of Aeronautics and Astronautics. Meeting AIAA,* eISBN: 978-1-60086-959-4, 2010 DOI:10.2514/6.2010-1579

[3] A. Alaimo, A. Esposito, A. Messineo, C. Orlando, D. Tumino, „3D CFD Analysis of a Vertical Axis Wind Turbine". *Energies. 8(4)*: p. 3013-3033, 2015.

[4] P. Sabaeifard, H. Razzaghi, A. Forouzandeh, Determination of vertical Axis wind turbines optimal configuration through CFD simulations, *International Conference on Future Environment and Energy, Singapore,* p. 109-113, 2012

[5] K. A. Brown & R. Brooks, „Design and analysis of vertical axis thermoplastic composite wind turbine blade", *Plastics, Rubber and Composites*, 39:3-5, p. 111-121, 2010.

[6] M. Guțu, „Correlation of composite material test results with finite element analysis" *IOP Conference Series: Materials Science and Engineering, Volume 147, 7th International Conference on Advanced Concepts in Mechanical Engineering,* 2016, Iasi, Romania 147(1):012004, DOI: 10.1088/1757-899X/147/1/012004.

[7] A. P. Chiriță, P. Bere, et al., "Aspects regarding the use of 3D printing technology and composite materials for testing and manufacturing vertical axis wind turbines", *Mat. Plast.* 56, no. 4, pp. 910-917. 2019

[8] M. Bordoni, A. Boschetto, Thickening of surfaces for direct additive manufacturing fabrication, *Rapid Prototyping Journal*, 18, 4, p. 308–317, 2012

[9] G. Krolczyk, P. Raos, S. Legutko, "Experimental analysis of surface roughness and surface texture of machinedand fused deposition modelled parts". *Teh. Vjesn.*, 21, pp. 217–221, 2014

[10] R. Roşculeţ, C. Fetecău,"Techniques for rapid prototyping of scale models of blades for wind turbines", *Journal of Engineering Sciences and Innovation,* Vol. 3, Issue 1/, pp. 15-24, 2018

IOP Conf. Series: Materials Science and Engineering 1190 (2021) 012023 doi:10.1088/1757-899X/1190/1/012023

FEM Modelling of Tool Wear in Hard Turning Operations

Cristian Cappellini [1], Franco Concli [1]

[1] Free University of Bolzano, Piazza Università 1, 39100 Bolzano, Italy

cristian.cappellini@unibz.it

Abstract. Hard turning is a machining operation of finishing performed on material with a hardness higher than 45 HRC and it has been demonstrated to be a profitable alternative to grinding process allowing, respect to this latter, better surface quality, improved process control reliability, and faster production rates. The high pressure and sliding velocity at the tool-chip interface, in addition to the deformation energy generated, lead to the formation of high temperature zones localized in the cutting region. These thermal loads enhance harmful aspects on both cutting tool and machined material. On the tool, the abrasive-diffusive wear mechanism is promoted. On the workpiece, a local surface heat treatment of quenching is observed, upgrading the formation of a thin layer of over-quenched martensite, called white layer, due to its appearance under microscopical observation, that is undesirable due to its brittleness and related cracks generation. In addition to this, hard turning is usually performed in dry condition further augmenting machining temperature and tool wear rate. Because of the extremely high hardness of the machined materials, the employment of expensive ultra-high hardness cutting tools is mandatory. Thus, to make the process valuable, it is essential to optimize the tool usage by avoiding catastrophic failure and being able to forecast the tool wear evolution during cutting time as a function of the applied process parameters. One of the possible approaches for predicting the wear behaviour is finite element (FE) simulations of the turning process. In this work, the wear of polycrystalline cubic boron nitride (PCBN) tools when turning AISI 52100 hardened steel is investigated. An abrasive-diffusive wear model has been developed and implemented in a FE engine, by means of a dedicated subroutine able to update the geometry of the worn tool as a function of the cutting parameters, with the intent of simulating the tool wear behaviour, in terms of crater and flank wear extension. For the calibration of the model, the tool wear measurements achieved from an experimental campaign of orthogonal cutting tests have been used. The influence of process parameters on the wear rate has been examined, underlining that the most affecting parameter on tool wear is the cutting speed: the higher the cutting speed the higher the tool wear. Furthermore, it has been observed that the tool breakage was due to an excessive depth of the crater wear that modified the initial negative rake angle to a positive one. This modification changed the stress state of the tool from compressive to tensile, not well tolerated by PCBN tools, bringing to catastrophic failure. The tool wear simulation results are in good agreement with the experimental ones, validating the usability of the proposed computational methodology for crater and flank wear prediction. Therefore, the application of this technique allows the optimization of cutting parameters from a preventive FE simulation analysis avoiding the need of exploiting new costly and time-consuming experimental tests.

1. Introduction

Hard turning is a machining process that deal with the cutting of materials whose hardness is higher than 45 HRC [1]. Compared with traditional grinding process, this technique allows a better surface finish, an improved dimensional accuracy, and a production time reduction [2]. Hence, in the last years, hard turning has been developed as a valuable alternative to grinding [3]. Due to its widespread in the industrial filed, a good knowledge of the effects of the technological aspects, such as process parameters, physical and chemical properties of materials, and tool wear, on the final product quality is mandatory. With this intent, several studies have been accomplished, leading to the development of empirical and analytical models for the optimization of cutting parameters [4], or for the assessment of tool life [5] and tool wear [6]. This latter affects geometrical accuracy, roughness, residual stresses, and surface modifications, that are related to the localized formation of white and dark layers [7-9], thus the possibility to forecast tool wear represents a profitable way to reduce production costs [10]. Because of the hardness of the worked material, hard turning is performed with extremely hard materials, and among these, polycrystalline cubic boron nitride (PCBN) is one the most employed. The wear mechanisms of PCBN tools are mainly related to abrasion, adhesion, and diffusion [6]. Their occurrence is a function of tool chemical composition [11], namely the ratio between binder and hard particles contents (CBN), and cutting parameters [12], such as cutting speed, feed rate, and depth of cut. Aimed to predict the evolution of PCBN tool wear, analytical models [13] and artificial neural networks (ANNs) [14] have been utilized in the past, while poor are the contributions to finite element method (FEM) modeling that demonstrated to be a good methodology to characterize the tool wear behavior during the cutting time [15]. In the present work, a FEM strategy for simulating the wear of PCBN tools in hard turning of AISI 52100 steel is presented. An abrasive-diffusive tool wear model, concerning the wear on the tool rake face, defined as crater wear (*KT*), and on the tool flank (*VB*), has been developed. This model evaluates the wear rate as a function of influencing parameters, namely sliding velocity, contact temperature, and interface pressure, at the chip-tool contact interface. Its implementation in the FEM software has been achieved by the generation of a dedicated subroutine able to move the nodes of the tool mesh for updating the geometry of the worn tool. For the collection of the tool wear measurements, an extensive experimental campaign has been performed by means of several hard turning tests. The proposed methodology has been calibrated and validated by the comparison of simulated results with the experimental ones. The outcomes show good agreement between experiments and simulations, underlining the suitability of the proposed methodology for forecasting tool wear growth.

2. Materials and methods

In order to obtain the tool wear measurements, subsequently used for the calibration of the developed wear model, an experimental campaign consisting of orthogonal turning tests has been performed by the variation of the process parameters, namely cutting speed (V_C) and feed rate (f). Their values are shown in figure 1, where it can be observed that two levels have been assigned to both cutting speed and feed rate, and the relative test name. During the tests, workpiece and tool material, dry condition, depth of cut, and geometries, have been kept constant.

2.1. Workpiece preparation

The workpieces consisted of disks made of AISI 52100 hardened steel with a thickness of 1.2 mm, also representing the depth of cut, and an initial diameter of 150 mm. The selection of AISI 52100 steel material was due to its industrial relevance and hardenability. Its chemical composition, expressed in percentage in weight, is reported in table 1.

Table 1. AISI 52100 chemical composition.

Element	Fe	C	Cr	Mn	P	Si	S
% in weight	97.0	0.98-1.10	1.45	0.35	<0.025	0.230	<0.025

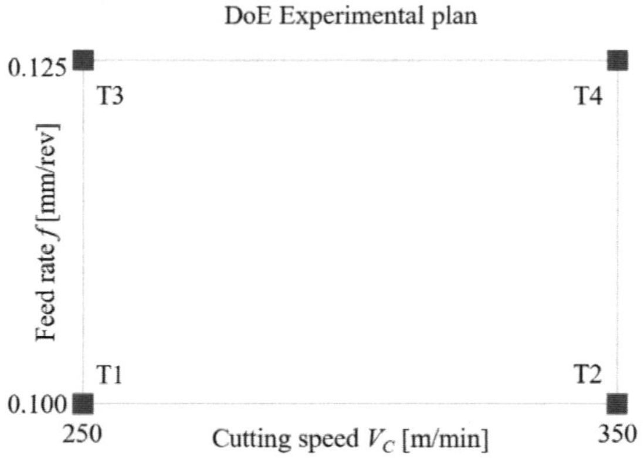

Figure 1. Values of the process parameters implemented in the experimental tests.

The specimens have been prepared by means of the heat treatment cycle of quenching and tempering described in figure 2. The application of the heat treatment was necessary in order to obtain values of mechanical properties and hardness comparable with the ones detected in common industrial applications, as reported by the standard UNI EN ISO 683-17:2014 for technical characteristics of furniture of ball bearing steels. The thermal cycle contemplates a normalization at 550 °C followed by a quenching treatment with an austenitization temperature of 860 °C. Both normalization and quenching are performed in a salt furnace. This technique is mandatory to allow a uniform heating of the disk and to prevent excessive deformation since, the low thickness increases the thermal distortion effect. For the same reason, a salt bath maintained at 175 °C is used as quenching medium. After quenching, the cooling at environment temperature, is completed in air, giving a hardness that ranges between 63 HRC and 65 HRC. To achieve the desired hardness, the specimen is then tempered in furnace at 390 °C. At the end of the tempering treatment, the disk is left to cool in air, leading to the desired hardness of 57 HRC. Finally, the disks are gently ground to obtain a flat surface of the requested thickness.

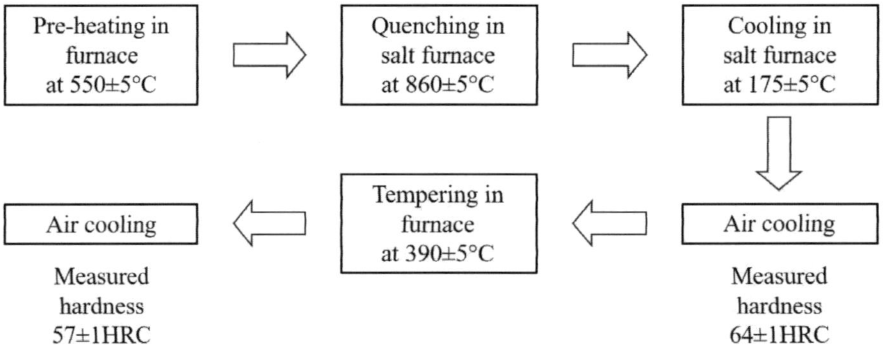

Figure 2. Heat treatment cycle of the specimens.

2.2. Cutting tests and tool wear measurements

The orthogonal cutting tests has been carried out in dry condition on a computer numerical control (CNC) lathe furnishing a nominal power of 7 kW. The tool employed was a PCBN insert ISO TNGN 110308S-01020 CBN100 by Seco Tools, where cubic-boron-nitride particles are dispersed in a TiC binder matrix. The properties of the insert are reported in table 2. To provide a negative rake angle (γ) of -8°, an inclination angle (λ) of 0°, and an entering angle (χ) of 90°, a toolholder

ISO CTFNR3225P11-PL was adopted. A specifically designed equipment, visible in figure 3, was used to fix the disks on the mandrel.

Table 2. PCBN tool properties.

CBN content	Binder	Hardness	Young's modulus	Poisson's ratio	Thermal expansion	Thermal conductivity
50%	TiC	27.5 HK	588 GPa	0.153	6.2 10^{-6} 1/K	40 W/mK

Figure 3. Experimental setup for orthogonal cutting tests.

As indicated by the tool-life testing standard ISO 3685:1993, throughout each test the flank wear (*VB*), and crater wear (*KT*) have been measured at regular time intervals. The catastrophic tool breakage has been chosen as the end of each tool-life test. In order to determine the time intervals for each test, the estimation of the final tool life is needed. For the evaluation of the tool life, an analytical model developed in [5] has been employed. This model correlates the process parameters with the tool life and is reported in equation (1):

$$V_C \cdot T_L^G \cdot f^E \cdot d_C^F \cdot \left(\frac{H}{H_0}\right)^D = C \tag{1}$$

where V_C is the cutting speed in [m/min], T_L is the tool-life in [min], f is the feed rate in [mm/rev], d_C is the depth of cut in [mm], H is the workpiece hardness in [HRC], H_0 is a reference hardness equal to 60 HRC, C is a constant equal to 172, G, E, F, and D, are exponents calibrated as a function of workpiece and tool material, respectively equal to 0.285, 0.335, 0.112, and 1.07. The tool life values calculated by equation (1) were found in accordance with the experimental ones. The cutting time intervals have been calculated dividing the total tool life approximately in 10 sub-steps except for the tests concerning a cutting speed of 350 m/min, where a lower number of sub-steps were used. For detecting the flank wear at each sub-step, the insert was removed from the toolholder and measured by means of an optical coordinate measuring machine CMM (Mitutoyo QS200) with a resolution of 0.5 μm and an accuracy lower than 2.5 μm. A typical flank wear measurement is visible in figure 4a. Crater wear depth was acquired by means of a profilometer (Mitutoyo SJ-301) with a resolution of 0.01 μm. Concerning the measurement of the crater wear depth, several profiles on the worn tool rake

IOP Conf. Series: Materials Science and Engineering 1190 (2021) 012023 doi:10.1088/1757-899X/1190/1/012023

face have been acquired at regular spatial intervals of 0.1 mm, as shown in figure 4b, and among these the maximum value of *KT* has been selected as the representative crater wear depth.

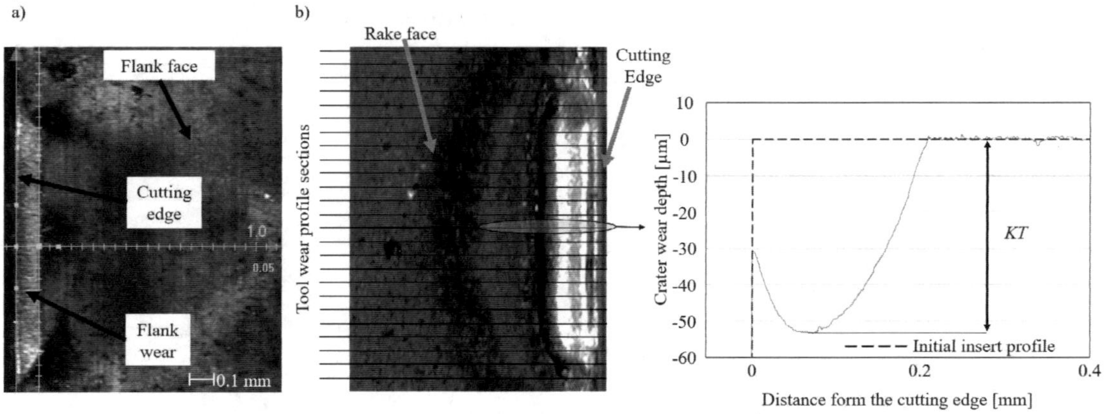

Figure 4. a) Flank wear measurement technique, and b) crater wear measurement technique (test T1, cutting time = 150s).

2.3. The developed tool wear model

The development of a coupled abrasive-diffusive tool wear model is determined to consider the different wear mechanisms acting on rake and flank face of the insert. In particular, the part of the model concerning crater wear is a combination of abrasion and diffusion contribution, while, due to the high hardness of worked material, the flank wear is modelled by pure abrasion, as observed in [12]. Because of the low CBN content of the employed tool, the adhesive contribution in the proposed model is neglected [6]. This assumption has been also confirmed by the observation of the worn tool (figure 5), where no adherent workpiece material is visible.

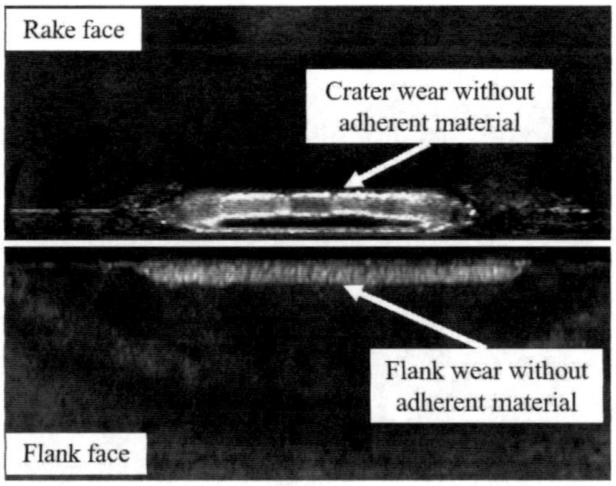

Figure 5. Rake and flank face of the tool without adherent material (test T3, cutting time = 130s).

Equation (2) reports the expression for the implemented tool wear model; Where $\partial W_{KT}/\partial t$ and $\partial W_{VB}/\partial t$ are the tool wear rate for rake and flank face respectively, H_{wp} [N/mm2] and H_t [N/mm2] are temperature hardness functions for workpiece and tool respectively, $K_Q = 20460$ is a constant related to the activation energy for diffusion, v_S [mm/s] is the sliding velocity at tool-chip interface, σ_n [MPa] is the contact pressure at tool-chip interface, T_{int} [°C] is the temperature at tool-chip interface, V_C [mm/s] is the cutting speed, $K = 0.416$ and $n = 7.0$ are hardness constant and hardness exponent whose values are dependent from the workpiece-tool hardness ratio. The coefficients $K_{KT\_abr} = 0.735$, $K_{KT\_diff} = 24602$, and $K_{VB\_abr} = 0.085$ that are respectively related to the abrasive crater wear, the diffusive crater wear, and the abrasive flank wear, have been calibrated by comparing the worn tool geometry simulation results with the experimental measurements of crater and flank wear of the first cutting sub-step for all the tests. The simulation software provides the values of contact pressure (σ_n), sliding velocity (v_S), and interface temperature (T_{int}) in their local values. For the flank wear model, the sliding velocity is assumed equal to the cutting speed (V_C). The tool wear model calculates the value of wear rate for each mesh node of the tool and subsequently, the implemented subroutine moves the tool nodes, to update the worn tool geometry.

$$\begin{cases} \frac{\partial W_{KT}}{\partial t} = K_{KT_{abr}} \cdot K \cdot \left(\frac{H_{wp}^{n-1}}{H_t^n}\right) \cdot v_S \cdot \sigma_n + K_{KT_{diff}} \cdot \sqrt{v_S} \cdot e^{-K_Q/(T_{int}+273)} & for\ crater\ wear \\ \frac{\partial W_{VB}}{\partial t} = K_{VB_{abr}} \cdot K \cdot \left(\frac{H_{wp}^{n-1}}{H_t^n}\right) \cdot V_C \cdot \sigma_n & for\ flank\ wear \end{cases} \quad (2)$$

$$H_{wp} = 11760 \cdot exp(-16.3 \cdot 10^{-4} T_{int})$$

$$H_t = 45000 - 4.324 T_{int}$$

2.4. FEM simulations

The cutting experiments have been faithfully reproduced, in terms of cutting parameters, materials, lubrication condition, in the FE environment. Figure 6 shows the FEM model with tool and workpiece at the beginning (figure 6a) and during the simulation (figure 6b) where the chip formation is evident.

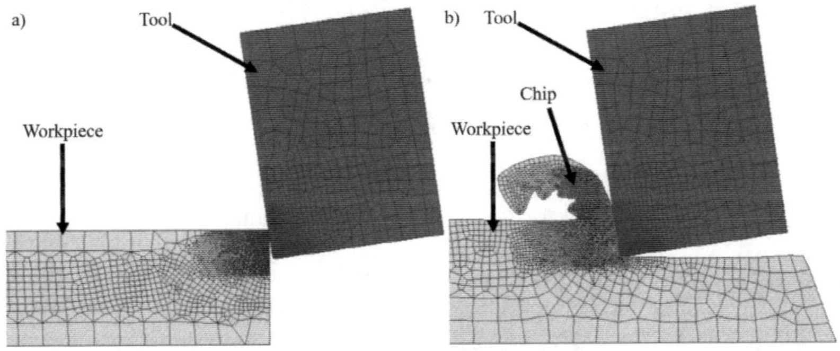

Figure 6. FEM model, a) initial simulation step, b) chip formation during simulation.

The workpiece has been considered as a rigid-plastic object and discretized with a mesh of more than 5000 quadrilateral linear elements. The rigid-plastic behavior of the workpiece material has been characterized by the modified Johnson-Cook's model of equation (3) proposed in [16], where ε is the strain, ε_r is the strain rate, T is the local temperature, $B(T)$ is a polynomial function of temperature, $J(HRC)$ and $I(HRC)$ are functions of the hardness, while $A = 0.0567$, $L = 1092$ MPa, $p = 0.083$, $m = 0.1259$, $M = 0.3$, $r = 5$, and $S = 20$ are constants.

$$\sigma = B(T)[L\varepsilon^p + J(HRC) + I(HRC)\varepsilon][1 + (ln(\varepsilon_r^m) - A)]\left[M + (1 + M)\left[tanh\left(\frac{1}{\varepsilon^r}\right)\right]^S\right] \quad (3)$$

The tool has been modelled as a rigid object and meshed with more than 1000 quadrilateral linear elements. A finer mesh, in the correspondence of the tool-chip contact area, has been created, achieving a minimum element size of 3 μm for both the objects, and allowing a correct evaluation of generated and exchanged heat between them. The initial temperature of both tool and workpiece has been set equal to 20 °C. The heat transfer coefficient (*htc*) of the two objects with the environment, applied on their free surfaces, has been set equal to 20 W/m^{2}K, while at the tool-chip interface *htc* has been imposed equal to 45000 W/m^{2}K. In order to represent the dry cutting conditions, the Zorev's hybrid frictional model [17] has been implemented. This model subdivides the tool-chip contact area in two zones: a sticking zone from the tool edge up to two times the feed, where the frictional stresses are equal to the shear flow stress of the workpiece, and a sliding zone in the remaining contact length where the frictional stresses are proportional to the contact pressure. In the proposed model the proportionality constant was set equal to 0.35.

To correctly simulate T_{int}, σ_n, and v_S at the tool-chip interface, it is mandatory to reach the thermo-mechanical steady-state before the application of the tool wear subroutine. With this intent, the simulation strategy reported in figure 7 has been adopted. In the first stage of the procedure an incremental Lagrangian (IL) solver is applied to reach the mechanical steady-state. The subsequent thermal steady-state is then achieved by the exploitation of an arbitrary Lagrangian-Eulerian (ALE) solver. After this, the subroutine for tool wear rate calculation and the geometry update of the worn tool is called. Here, the tool wear rate is calculated for each tool node in accordance with the rake or flank tool wear model of equation (2). Based on the time interval and estimated wear rate, the tool wear is determined, and the tool geometry is updated. The update of the geometry is performed by the movement of the nodes of a quantity equal to the estimated wear in the opposite direction respect the outgoing normal of the elements at which the nodes are connected. Once the tool geometry is updated, the IL solver is employed again to reach a new thermo-mechanical steady-state considering the worn tool topology. The described technique is iteratively applied until the total machining time is reached. An example of simulated tool wear evolution is visible in figure 8.

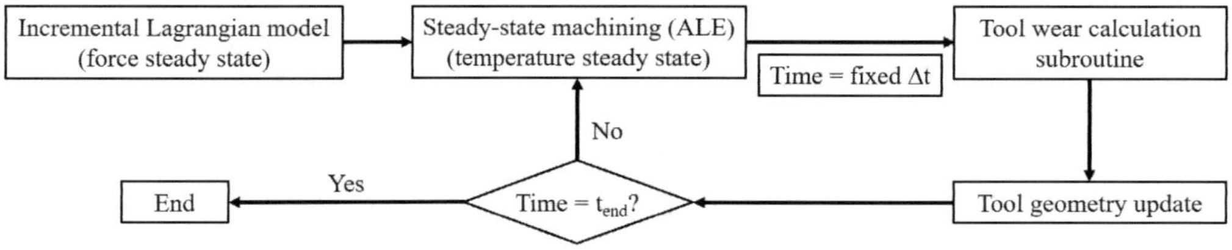

Figure 7. Applied simulation strategy.

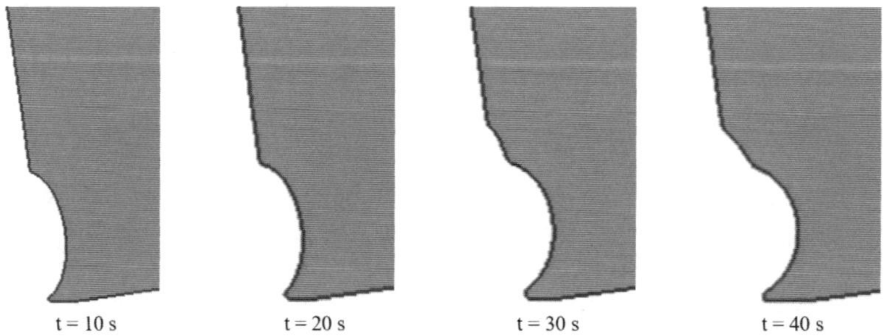

Figure 8. Example of simulated tool wear evolution (test T2).

3. Results and discussion

3.1. Experimental results

In figure 9 the evolutions of flank (*VB*) and crater (*KT*) wear during the cutting time of each test are visible. It is evident that both flank (figure 9a) and crater (figure 9b) wear are deeply influenced by the cutting speed: the higher the cutting speed the higher the tool wear rate. Moreover, the feed rate does not reveal considerable effects of the flank wear rate. In contrast, the influence of feed rate on the crater wear evolution is more prominent especially at lower cutting speed. This behavior is due to the diffusive wear mechanism that is higher with the raise of temperature. An increased feed rate leads to a higher chip section area that requires greater cutting forces. At constant cutting speeds, this means that more power is required for the machining operation, thus the deformation energy augments too. The deformation energy in the chip is converted in heat that increases the cutting temperature and consequently the diffusion mechanism is enhanced. The analysis of *KT* evolution indicates that the insert breakage occurs when the crater depth is slightly higher than 60 μm. The severe crater wear on the PCBN inserts leads to a modification of the tool rake angle from an initial negative value to a positive one (figure 10a). This geometric alteration reduces the resistant section area of the tool and increases the tensile stresses at which the resistant section is subjected. The low strength of PCBN tools to tensile stresses, brings to the premature failure of the insert. This behavior has been also verified by a simulation of the mechanical stresses on the tool during the cutting (figure 10b).

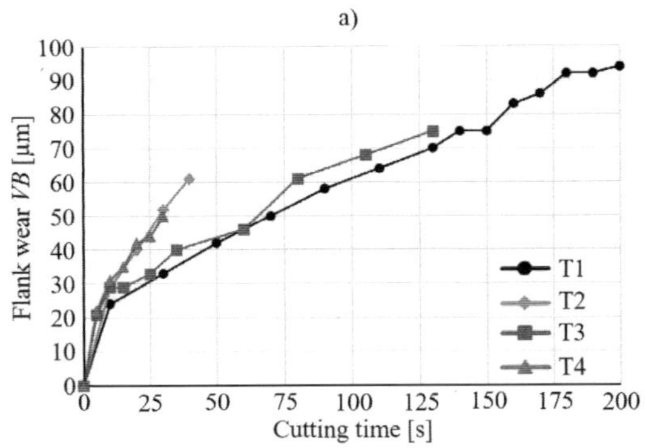

Figure 9. Example of simulated tool wear evolution (test T2).

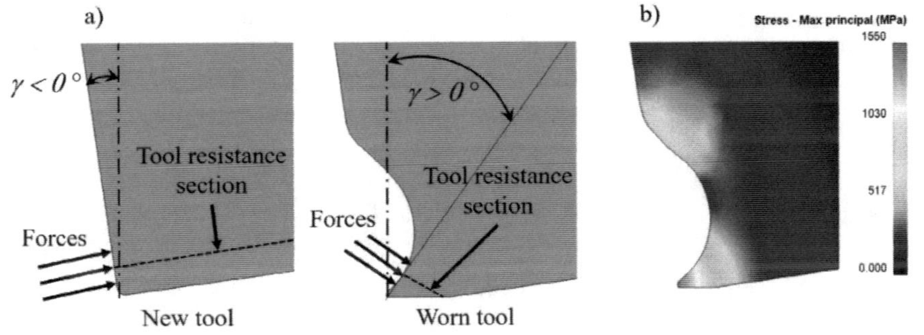

Figure 10. a) Effect of wear on effective rake angle, b) Tensile stresses on a worn tool.

3.2. Experimental results

The implemented wear model has been validated by the comparison of the simulated tool wear with the experimental one. The evolution of experimental and simulated crater wear (*KT*) and flank wear (*VB*) during the cutting time for all the tests is depicted in the plots of figure 11. The good match amongst the different results is clearly visible. Therefore, the employment of the combined rake-flank tool wear model allows to correctly simulate the wear behavior as a function of the cutting parameters, avoiding the need of performing costly and time-consuming experimental tests and permitting an optimization of process parameters. Moreover, the implementation of this new tool wear model grants the prediction of the effective rake angle, due to the growing of crater wear. Figure 12 shows the evolution of simulated effective rake angle during time, normalized respect the tool life of each test. The increase of the crater wear changes the effective rake angle from a negative to a positive value.

It is important to supervise the value of this angle since PCBN tools present extremely high resistance to compressive stresses but poor to tensile ones. Consequently, the application of a negative rake angle is recommended to enhance compressive stresses on them. Conversely, when the rake angle is positive tensile stresses increases. Figure 12 underlines the independency of rake angle limit leading to tool breakage from process parameters, in fact its value ranges between 25° to 30° for all the tests.

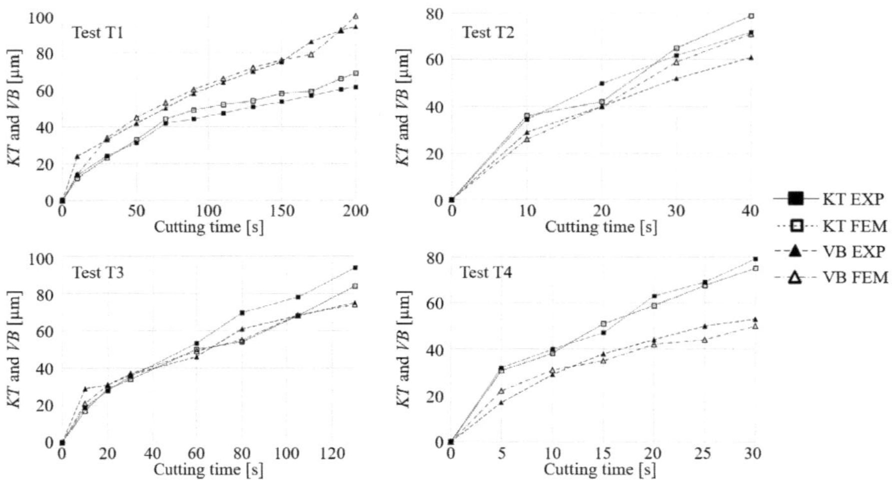

Figure 11. Comparison between experimental and simulated *KT* and *VB* values.

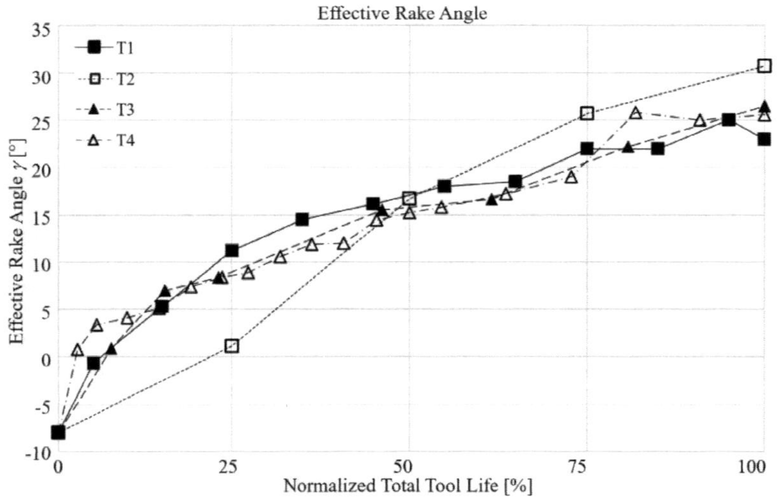

Figure 12. Evolution of effective rake angle as a function of the normalized tool life.

4. Conclusions

In the present paper the results achieved from a study on tool wear simulation of PCBN inserts when cutting AISI 52100 hardened steel, aimed to improve the quality and the reliability of FEM analysis of machining operations, has been presented. Simulations replicating experimental tests has been performed. Moreover, a coupled abrasive-diffusive wear model able to update the geometry of the worn tool during machining simulations, have been implemented into a FEM engine by means of a suitable subroutine. Experimental and simulated wear parameters have been compared in terms of crater depth and flank wear width. This comparison demonstrated the capability of the FEM model to closely reproduce the tool wear throughout the cutting time. The validated FEM model has been also employed to analyze the effect of tool wear on the effective rake angle of the worn insert. This underlined how the increase of crater depth leads to a modification of the rake angle from a negative to a positive value, reducing the tool resistance section. The combination of this with the high brittleness of PCBN material, has been individuated as the mechanism of failure of the tool, bringing to premature breakage. Hence, the proposed FEM strategy can be applied to optimize the process parameters by continuously monitoring crater and flank wear evolution, preventing catastrophic rupture, and allowing a correct replacement of the tool.

References

[1] G. Bartarya, S.K. Choudhury, "State of the art in hard turning", *International Journal of Machine Tools & Manufacture*, vol. 53, pp. -14, 2012.
[2] W.F. Sales, J. Schoop, L.R.R. da Silva, A.R. Machado, I.S. Jawahir, "A review of surface integrity in machining of hardened steels", *Journal of Manufacturing Processes*, vol. 58, pp. 136-162, 2020.
[3] C. Cappellini, A. Attanasio, G. Rotella, D. Umbrello, "Formation of white and dark layers in hard cutting: Influence of tool wear", *International Journal of Material forming*, vol. 3, pp. 455-458, 2010.
[4] A.V. Pradeep, D.L. Raju, S. Ramakrishna, "Cutting force analysis in hard turning of AISI 52100 steel using multi-layer coated carbide inserts – RSM approach", *Materials today: Proceedings*, https://doi.org/10.1016/j.matpr.2019.07.369
[5] G. Poulachon, A. Moisan, I.S. Jawahir, "Tool-wear mechanisms in hard turning with polycrystalline cubic boron nitride tools", *Wear*, vol. 250, pp. 576-586, 2001.
[6] Y. Huang, Y.K. Chou, S.Y. Liang, "CBN tool wear in hard turning: a survey on research progresses", *The International Journal of Advanced Manufacturing Technology*, vol. 35, pp. 443-453, 2007.
[7] M.A. Yallese, K. Chaoui, N. Zeghib, L. Boulanova, J-F. Rigal, "Hard machining of hardened bearing steel using cubic boron nitride tool", *Journal of Materials Processing Technology*, vol. 209, pp. 1092-1104, 2009.
[8] A. Attanasio, D. Umbrello, C. Cappellini, G. Rotella, R. M'Saoubi, "Tool wear effects on white and dark layer formation in hard turning of AISI 52100 steel", *Wear*, vol. 286-287, pp. 98-107, 2012.
[9] F. Zemzemi, H. Khochtali, W.B. Salem, B. Alzahrani, M-L. Bouazizi, "Analytical multi-physics model of microstructure changes in hard turning of AISI 52100 steel: prediction of thicknesses of white and dark layers", *The International Journal of Advanced Manufacturing Technology*, vol. 112, pp. 2755-2771, 2021.
[10] Y. Huang, T.G. Dawson, "Tool creater wear depth modeling in CBN hard turning", *Wear*, vol. 258, pp. 1455-1461, 2005.
[11] S. Gordon, P. Phelan, C. Lahiff, "The effect of high speed machining on the crater wear behaviour of PCBN tools in hard turning", *Procedia Manufacturing*, vol. 38, pp. 1833-1848, 2019.
[12] S. Chinchanikar, S.K. Choudhury, "Predictive modeling for flank wear progression of coated carbide tool in turning hardened steel under practical machining conditions", *The International Journal of Advanced Manufacturing Technology*, vol. 76, pp. 1185-1201, 2015.
[13] D. Singh, P.V. Rao, "Flank wear prediction of ceramic tools in hard turning", *The International Journal of Advanced Manufacturing Technology*, vol. 50, pp. 479-493, 2010.
[14] U.M.R. Paturi, S. Cheruku, V.P.K. Pasunuri, S. Salike, "Modeling of tool wear in machining of AISI 52100 steel using artificial neural networks", *Materials Today: Proceedings*, https://doi.org/10.1016/j.matpr.2020.06.581
[15] A. Attanasio, E. Ceretti, A. Fiorentino, C. Cappellini, C. Giardini, "Investigation and FEM-based simulation of tool wear in turning operations with uncoated carbide tools", *Wear*, vol. 269, pp. 344-350, 2010.
[16] D. Umbrello, J. Hua, R. Shivpuri, "Hardness based flow stress for numerical modeling of hard machining AISI 52100 bearing steel", *Mater. Sci. Eng.–A*, vol. 347, pp. 90-100, 2004.
[17] N.I.N. Zorev, "Metal Cutting Mechanics", Pergamon Press, Oxford, New York, xv, pp. 526, 1966.

IOP Conf. Series: Materials Science and Engineering 1190 (2021) 012024 doi:10.1088/1757-899X/1190/1/012024

Analysis of Longitudinal Beam Damage of a Container Chassis Semi-Trailer Using Finite Element Method – Case Study

Kazimierz Drozd[1], Aleksander Nieoczym[2]

[1]Department of Materials Engineering, Faculty of Mechanical Engineering, Lublin University of Technology, Lublin, Poland
[2]Department of Machine Design, Faculty of Mechanical Engineering, Lublin University of Technology, Lublin, Poland

k.drozd@pollub.pl

Abstract. This study investigates the damage of a container chassis loaded with a container having a capacity of 22 thousand litres. The chassis was damaged during full-load operation. Based on the observed effects of damage induced in a longitudinal beam (longeron) of the container chassis, a hypothetical course of events was elaborated. An attempt was made to recreate the actual course of events by eliminating the reasons due to which the effects of damage significantly differed from those observed post factums. To this end, numerical finite element method (FEM) simulations were performed using the Abaqus software. The damaged longitudinal beam of the container chassis was modelled using three-dimensional (3D) solid elements. It was assumed that the beam had the same profile cross-section over its entire length. In the numerical model, the container chassis had a pinned support on the fifth wheel. Elastic elements were used to model the axle-supported regions of the container chassis with the wheels. The following three events were identified as the potential causes of container chassis damage: the container chassis was supported by the bumper on the ground when the wheels on the right side were in contact with the bottom of the road surface cavity; the container chassis was supported by the bumper on the ground when the wheels on the right side were not in contact with the bottom of the road surface cavity; the bumper was subjected to point impact during vehicle reversal. The numerical results demonstrated that for each of the proposed damage hypothesis, the highest reduced stresses were located in the very region where the container chassis damage was actually observed. It was found that the observed damage was most probably induced when the container chassis was supported on the bumper and there was no contact between the wheels and the road surface due to the presence of a cavity in the road surface. Moreover, analytical calculation results demonstrated that the force acting on the structural members of the container chassis in the analysed load case exceeded the force resulting from the static load of the structure by more than 2.5 times. The load-induced stress in the beam material significantly exceeded the strength of the structure.

1. Introduction

The frame is the main structural member of a semi-trailer, and it transfers both loads from the transported cargo weight and road surface reactions. The correct vehicle design and its execution in compliance with requirements specified in the vehicle approval are key factors ensuring the durability and reliability of the semi-trailer. Even if these requirements are satisfied, there are cases when a vehicle is excluded from further use due to its structural damage. Potential damage – whatever its cause – is modelled and evaluated via simulations. Nowadays technical objects are modelled by the finite element method (FEM) in order to achieve basically two aims. One is related to the analysis of stresses in the current design of a semi-trailer or its design optimization. The other concerns the identification and verification of potential causes of damage.

In FEM strength calculations, the main task is to calculate load forces and torques and to simulate road conditions in such a way that the numerically modelled conditions would reflect the real road conditions. The FEM analysis begins with calculating strength under static load from the weight of the cargo and the structure itself. The cargo load can be evenly distributed over the top surface of the longitudinal beam as shown by Han et al. [1] or Zhang et al. [2]. As a result, it is possible to determine bending stresses. If the model does not consider the support simulating reactions of the left front wheel, the load causes the longitudinal beam to undergo bending and torsion at the same time. The semi-trailer's motion on an uneven road is simulated by eliminating support in the parts of the structure in which the road wheels are mounted [2]. In the simulations, the container chassis was loaded with dynamic forces for the following drive modes: driving on a rectilinear track with constant speed; driving on a curved track; acceleration; and braking [3]. The dynamic force required to apply load was calculated as the product of static load and dynamic coefficient.

The aim of the modal analysis of semi-trailer frame vibration is to investigate the effect of vibration frequency on the structural deformation mode. Zhang et al. [4] determined the ranges of vibration frequencies and their impact on the deformation of a semi-trailer frame resulting from bending and torsion of specified structural members. The problem of transport system vibration, in combination with the determination of stiffness and damping coefficients of its structural members, was solved by Abdelkareem et al. [5]. The authors of studies [4, 5] also proposed important frame design modifications by the use of additional stiffening elements or local increase in the cross-sectional area of the frame profile.

Numerical results are sometimes verified via bench tests. Napierała [6] investigated the strength of welded joints connecting longitudinal and transverse beams in a semi-trailer frame. Numerical results showed good agreement with those obtained in the experiment that was conducted with the use of resistance strain gauges and displacement sensors mounted on the frame elements. In [7] Działak and Karliński described the procedure for measuring the deflection of a lifted semi-trailer by the photogrammetric method. The method based on the use of the MTS road simulator, described by Kaczor et al. [8], involves fatigue verification of entire vehicles and their individual parts. Load distribution on several vehicle axles can be measured on a test stand equipped with a deformable beam [9].

2. Object of the study and research methods

The object of a FEM numerical analysis was a damaged three-axle semi-trailer for transporting standard 20' (feet) containers (figure 1a). Loaded with a liquid cargo tank, this semi-trailer having an actual weight of 22 tons took part in two events. One was a collision at a stop situation, which resulted in rear bumper beam damage. The other event took place when the right wheels of the semi-trailer hit a pothole, which led to the right side of the rear bumper resting on the road surface. The analysed object had visible fractures of the right-side longeron located in relief holes (figures 1b-c) and visible deformation of the semi-trailer structural members (figures 1d-f).

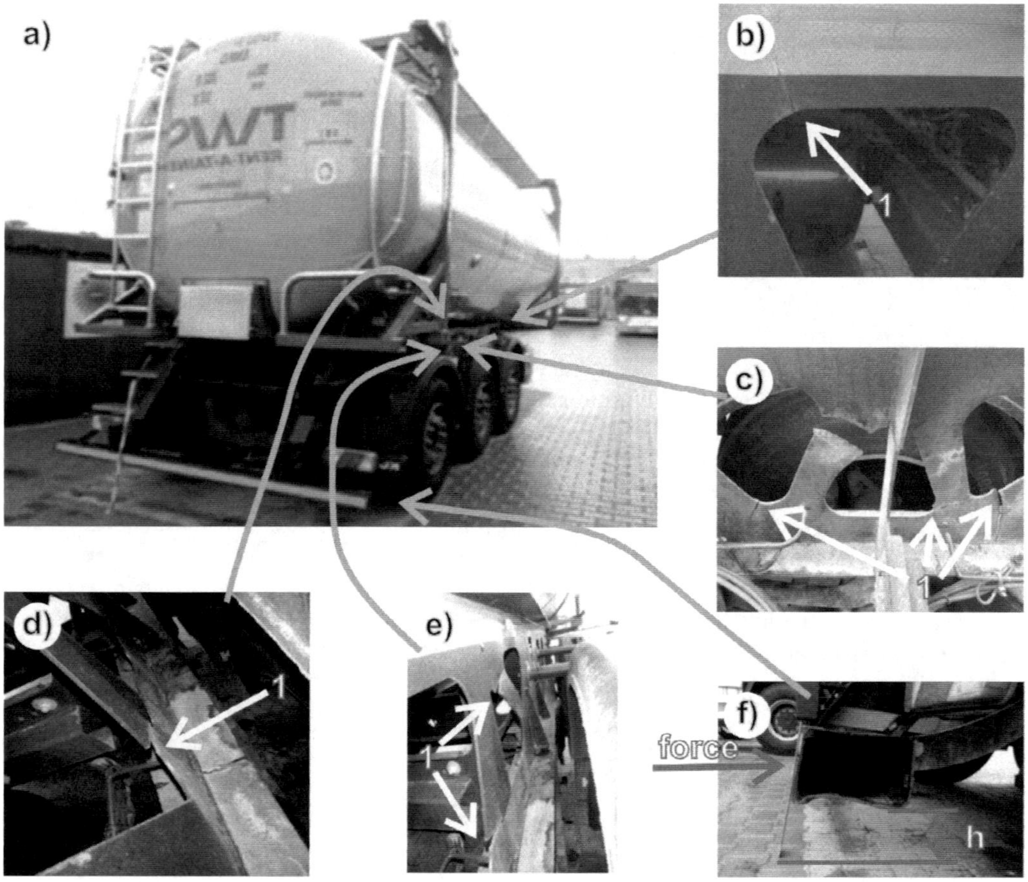

Figure 1. Semi-trailer under analysis – green arrows show the place and direction of photographing: a) general view, b) – e) longeron damage location, f) deformed rear bumper, 1 –fracture initiation location.

The analysis has shown the presence of a 20 mm deep indentation located on the outer surface of the rear bumper (force in figure 1f). The lower plane of the bumper is deformed too (figure 1f); however, the arch-shaped element connecting the bumper with the semi-trailer frame is undamaged. The observed mode of deformation of the lower bumper shell indicates that the deformation could have occurred as a result of the bumper's resting on the road surface.

The semi-trailer frame longitudinal beam (figure 1e) has a total length of approximately 8,300 mm. This I-beam has the overall cross-sectional dimensions of 360x100 mm and a wall thickness of 11 mm. In the rear part of the frame, only in the web, six triangular holes were factory-made (figure 1e). The frame damage took the form of material cracking of different lengths on the web between the triangular (relief) holes (figures 1b-e). As a result of the damage, the web fragments were separated by a distance of up to 25 mm (figure 1c). Given the deformation mode of the beam elements, it was concluded that the fracture initiated at Point 1 (figures 1b-e), where the greatest distance between the separated fragments could be observed. It was also observed that the fracture always initiated at the transition between the arch-shaped profile and the rectilinear edge of the triangular hole in the beam web.

Given that apart from the fracture of the right-side longeron and the rear bumper, no other structural deformation was observed, the model was simplified in terms of geometry and kinematics.

The geometric simplification consisted in analysing a single beam model, with its profile and transverse dimensions in the front (undamaged) end being unchanged, without the support and the rear bumper. The kinematic simplification was related to considering only the kinematic connection between the rear bumper and the beam end. This meant treating the bumper support as a non-deformable element. The above simplifications made it possible to investigate the quasi-static problem in a linear-elastic range and to shorten the simulation time considerably. This applied simplifications also resulted from the fact that – given an unknown driving speed – an attempt at simulating dynamic phenomena would have to involve optimization of the simulation parameters for several variants that are described later in the manuscript.

To determine the cause of semi-trailer frame damage, FEM strength simulations were carried out to verify three hypotheses of the potential course of events. In two of these hypotheses, the semi-trailer drove over a pothole, which resulted in the bumper's resting on the road with the right side wheels of the vehicle having (hypothesis 1) or losing (hypothesis 2) contact with the bottom of the road defect. Under hypothesis 3, it was assumed that the semi-trailer damage was caused by point impact to the rear bumper of the semi-trailer.

Numerical analyses were carried out for three hypotheses, and for each of these hypotheses two variants (A and B) of beam loading were tested. The scope of the numerical simulations was extended to include these variants because the literature of the subject does not provide any recommendations with respect to loading a semi-trailer with a container mounted thereon. In the A variant, the beam with only continuous/distributed load (5 in figure 2) was modelled for half of the container weight of 110 kN. The other variant (B) involved modelling the beam with concentrated load (6 in figure 2) that was only applied to the ends of the container with a length of 5853 mm, i.e. where the container is attached to the frame. To each of these two points a load of 55 kN was applied, corresponding to a fourth of the weight of the entire container.

The beam model was assigned the properties of a linear-elastic material, i.e. S235 unalloyed steel having the following mechanical properties: Young's modulus $E = 210$ GPa, Poisson's ratio $\nu = 0.30$, yield stress $R_p = 235$ MPa, ultimate tensile strength $F_{tu} = 360$ MPa [10]. In all numerical calculations, maps of reduced stresses were created according to the Huber-Misses hypothesis. Some of these maps are presented later in this manuscript. Conclusions regarding the probable course of events were drawn by comparing the calculated stresses with the parameters of the material used in the beam structure.

2.1. Model for testing hypotheses 1 and 2 assuming that the bumper is resting on the road surface

The beam was modelled as a solid spatial element having the real cross-sectional dimensions and the same cross-section of the profile over the entire length (figure 2). Strength tests were carried out on a fragment of the beam located above the axles of the wheels of a semi-trailer 1. The beam support was modelled using elastic elements having a stiffness of 11.2 MNm^{-1} at points corresponding to the position of each of the three axles of the semi-trailer, which reflected the semi-trailer suspension and wheel stiffness [5]. At a distance of 565 mm from the front end, the beam was articulated in order to model support against a saddle 2. On the other end of the beam, at a distance corresponding to the position of the bumper relative to the structure, a reference point 3 was created and kinematically coupled with an end 4 of the support beam.

Every hypothesis was tested for two variants of load distribution on the semi-trailer beam. The first variant (A) involved uniform load distribution over the beam length between cargo securing points 5 (figure 2). The other variant (B) only took into account the distribution of load in container mounting points 6.

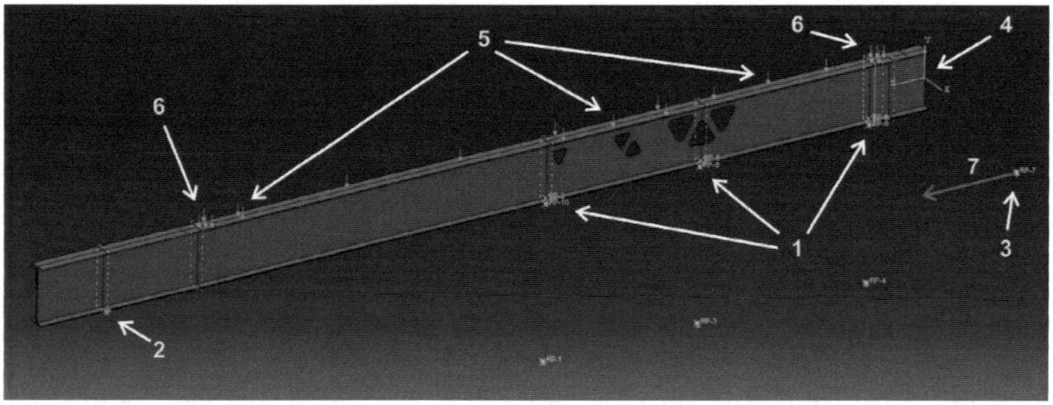

Figure 2. Model of longitudinal beam: 1 – spring support, 2 – articulated support on the saddle (fifth wheel), 3 – reference point on the bumper, 4 – rear end of the beam i, 5 – uniformly distributed load, 6 – concentrated load, 7 – force acting on the bumper.

The model was discretized using C3D10 tetragonal solid elements, which are 10-node finite elements with second-order shape function. The finite element mesh was additionally remeshed in the notches (where the highest stresses and strains were expected to occur) in order to locally increase computational accuracy.

2.2. Model for testing hypothesis 3 assuming that the bumper is subjected to point impact
The hypothesis assuming that the semi-trailer frame damage was caused by impact to the bumper was tested using the same geometric model as for hypotheses 1 and 2. However, the model had to be expanded to include force at reference point 3 (figure 2), which was to reflect the impact on the rear bumper. Only by analysing this force (its value being unknown) was it feasible to determine the effect of the impact to the rear bumper on the formation of stresses in the loaded semi-trailer frame.

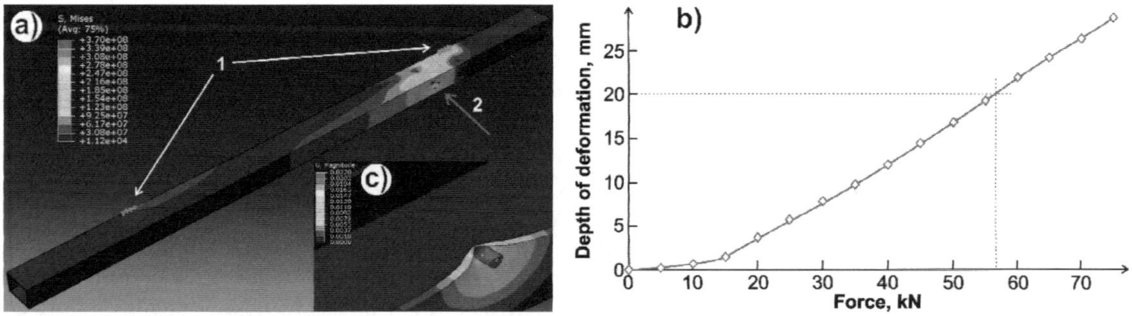

Figure 3. Impact force calculation: a) – rear bumper model and the distribution of stresses used for calculating rear impact force: 1 – fixed support, 2 – direction of indenter motion, b) – depth of deformation vs. force, c) bumper section deformed by the indenter.

Force (7 in figure 2) was determined by numerical analysis for a geometric model of the buffer shown in figure 3a. According to measurements of the real object, the bumper was a closed rectangular section with the dimensions of 120x80 mm and a wall thickness of 5 mm. It was attached to supports 1. It was assumed that the supports were constrained from moving in a direction 2.

The bumper model (figure 3a) was discretized using the same type of finite elements and the same material properties as those used for modelling the beam. The simulation consisted of establishing the relationship between the force and the depth of deformation caused by a non-deformable rounded indenter having 15 mm in diameter, pressing on the rear wall of the bumper in the location of the observed deformation 2. This relationship is illustrated in figure 3b. Based on the calculated deformation depth amounting to about 20 mm (figures 1f and 3b,c), the force acting on the bumper was estimated at 57 kN.

Having established that the force acting on the rear bumper is equal to 57 kN, the action of this force 7 was modelled for the entire beam, at the point marked as 3 (figure 2). This point was kinematically coupled with the beam end 4. The direction of this force was parallel to the longitudinal beam axis and its sense complied with that shown in the diagram.

3. Results and discussion

3.1. Numerical simulation of the case when the bumper is resting on the road surface

First of all, hypothesis 1 was tested, under which the bumper was resting on the road surface (3 in figure 2). The wheels of the semi-trailer got into the road defect and yet were in contact with the bottom of the cavity, and at the same time the semi-trailer was supported on the rear bumper. In the simulation of variant A, with the load uniformly distributed over the beam (5 in figure 2), the stresses did not exceed 225 MPa (figure 4). It was found that the highest stresses were located in the area with relief holes (figures 4b,c). The stress field of up to about 150 MPa extended from the upper flange of the longeron to the line marked by the upper edge of the triangular notches (1 in figures 4a,b). The greatest stress accumulation was observed at two points in the upper part of each of the triangular notches (2 in figure 4b). These points were located where the curvature of the notch changed, i.e. where the straight section of the notch ended and the arch-shaped region began. The simulated damage location is marked as 1 in figure 1b. Stresses calculated for the lower parts of the triangular notches are shown in figure 4c. It can be observed that the stresses exceeding 205 MPa only occur in the arch-shaped region, in the lower part of notch 3 located above the first axle of the semi-trailer wheels. A similar damage location obtained for the real object is marked as 1 in figure 1d.

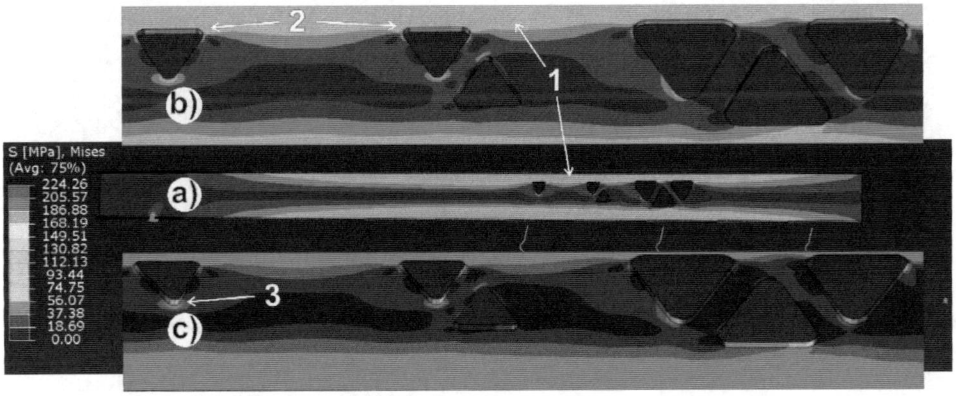

Figure 4. Map of stresses obtained by verification of hypothesis 1 (wheels are in contact with the road cavity bottom) for variant A (frame loaded over the entire container length): a) – general view, b) – view of the upper edge of the holes, c) – view of the bottom of the holes.

In the simulation of load variant B, it was assumed that the weight of the container is distributed in half only at two points (6 in figure 2) of the beam. The highest stresses amount to 170 MPa (figure 5) and also occur at points 1 where the curvature of the side of the triangular hole made in the web changes (increases). These points are located in the upper part of all triangular holes made in the beam web. Moreover, one can observe relatively high stresses at points 2, between the double hole. The stress field at the top of the beam, above the line joining the upper edges of the triangular holes, is quite uniform in terms of value, and the distribution of stresses is similar to that in figure 4b.

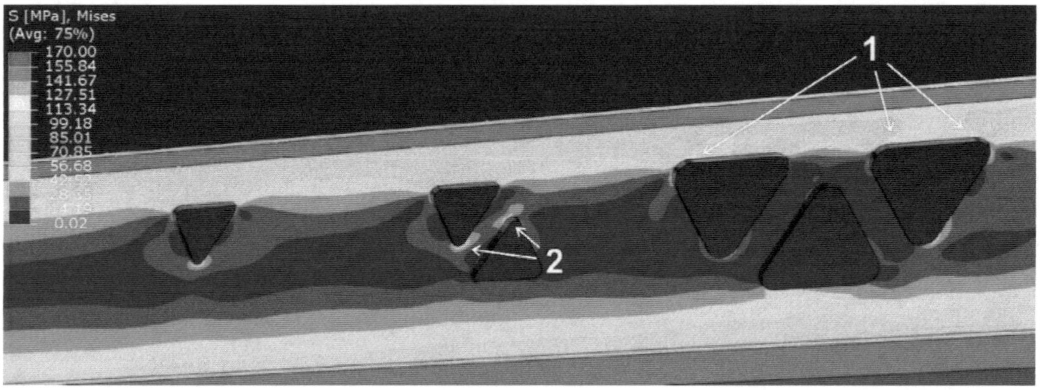

Figure 5. Map of stresses obtained by verification of hypothesis 1 (wheels are in contact with the road surface cavity bottom) for variant B (two-point load on the frame).

Next, simulations were performed to verify hypothesis 2, under which it was assumed that the semi-trailer was supported on the rear bumper while the wheels were not in contact with the bottom of the road surface cavity. For load variant A (uniform loading of frame), the maximum stresses are equal to 239 MPa (figure 6). They are located in the upper part of the beam, where the curvature of the triangular notches changes, and at the bottom of notch 1. Stresses of this magnitude could cause plastic deformation of the longeron material. In addition to that, stresses amounting to at least 150 MPa are observed in the beam regions 2 and 3 where the triangular notches are adjacent to each other. An equivalent of this location is marked as 1 in figure 1e. It can be observed that in terms of value and distribution, the simulated stresses are similar to those obtained in the simulation verifying hypothesis 1 for load model A (figure 4).

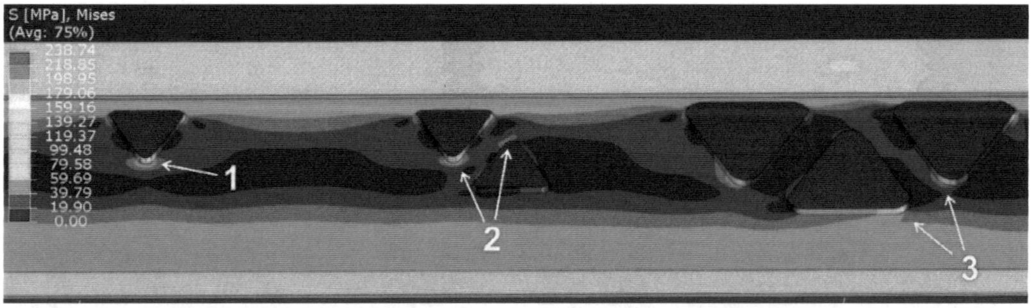

Figure 6. Map of stresses obtained by verification of hypothesis 2 (wheels are not in contact with the road surface defect bottom) for variant A (uniform load on the frame over the entire container length) .

For load variant B, the highest stresses induced in the longeron material due to the semi-trailer's resting on the rear bumper amount to only 170 MPa. The stress values and maps obtained for the two-point load model (6 in figure 2) are exactly the same as those obtained for the model presented in figure 5.

3.2. Simulation of the case when the bumper is subjected to a point impact
To verify hypothesis 3, under which the rear bumper was subjected to impact that could cause local deformation, it was assumed that at point 3 (figure 2) reflecting the real damage location, a force of 57 kN (figure 3) was applied. For variant A with a uniform loading of the longeron along the entire length of the 20-foot container, the maximum stresses were located in the lower part of the longeron, in the area where it was supported by the third axis. However, these stresses were low and amounted to 70 MPa.

Stresses obtained in this case are the highest for all tested models. For the case of point load in variant B, these stresses are equal to 241 MPa (figure 7). These values are sufficient for inducing plastic deformation of the material. It should be noted that the simulated stress distribution in the longeron is more symmetrical in the area with the cut-out holes. As a result of this symmetry, the stresses on the lower edge of the notches are higher, e.g. at the location marked as 2. Moreover, the maximum stresses concentrated on the upper edge of the triangular notches have different ranges (areas) and different maximum values than those observed at point 1. These relations are in total agreement with the damage shown in figure 1b. Also, the stresses between the vertices of the triangles denoted by 3 are significantly higher than those observed for other maps, which means that they could cause the damage observed in figure 1e.

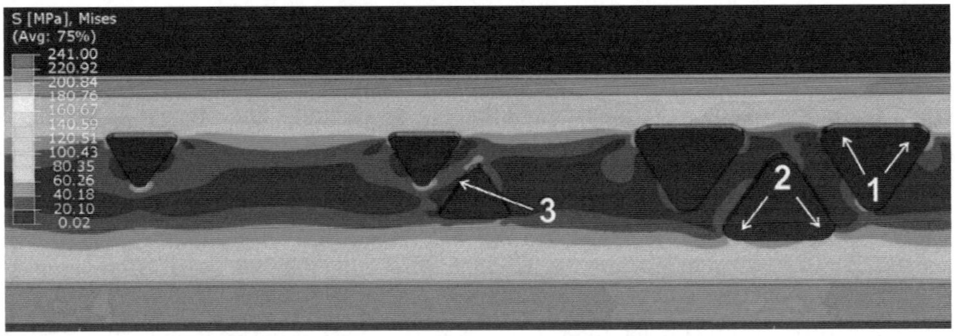

Figure 7. Map of stresses obtained by verification of hypothesis 3 (impact to rear bumper) for load variant B (point load on container edges).

3.3. Discussion

Table 1 gives the results of maximum stress location in the simulations of semi-trailer frame loading. For the three simulated variants, the approximate maximum stresses were calculated to range from 225 to 241 MPa. These stresses could cause plastic deformation observed in the damaged parts of the semi-trailer. The locations of the maximum stresses in the longeron observed for the three simulated variants were almost the same. They corresponded to the damage location marked in figures 1b,e. The simulation results did not unequivocally show any stress concentration that could cause the material discontinuity observed in figure 1c or the cracks marked as 1 in this figure. However, if we consider a sequence of two road events, such a significant damage of the frame, with the yield stress exceeding and material strengthening during the first event, seems possible as a result of the second event.

Table 1. Probability of beam damage versus maximum stress.

Hypothesis	Variant	Maximum stress, MPa	Probability of beam damage (Y)es/(N)o	Location of maximum stress in figure1
1	A	225	Y	b,e
	B	170	N	b
2	A	239	Y	b,e
	B	170	N	b
3	A	70	N	-
	B	241	Y	b,e

In the simulation, two points of stress concentration were determined on every horizontal edge of the triangular notches. In real conditions, only one damage of this type was induced (figure 1b). This discrepancy can be explained by the fact that the initiation of this damage type or other damage (in other triangular notches or between them) resulted in material unloading, and therefore no two cracks progressing from the same straight edge of the triangle were developed in any of the notches.

Due to the fact that the stresses simulated for longeron load variant A (for hypotheses 1 and 2) were higher than the yield stress of the material, it can be concluded that the correct model for simulating the performance of a fully operational semi-trailer is the one in which the loads are distributed uniformly over the entire length of the contact between the container and the semi-trailer frame.

In all tested load variants and all models used for verification of the damage hypotheses, the stress field was asymmetrical, which is marked in the stress maps (figures 4-7) as the large colourful area in the upper part of the beam. Although less dangerous compressive stresses are induced there under a normal load when the semi-trailer is moving on an even surface, under unusual loading conditions – such as those taken in the hypotheses tested in this study, these stresses may change their sign and, as a result, cause longeron damage.

Regardless of the load variant applied, the results of the strength tests show that in each of the three considered cases of structural performance, the highest reduced stresses are located in the triangular notches in the examined part of the longitudinal beam. Specifically, the highest reduced stresses are located in or near the rounded corners of the notches, which coincides with the location of cracks in the real structure.

In the analysis of the first two hypotheses of damage for the case of the semi-trailer's resting on the bumper, the load values reflect static impact, which means that the dynamic impact on point 3 in figure 2 was ignored in the simulation. According to the description of the event, the wheels fell into a cavity in the road surface, which made the semi-trailer hit the ground with the bumper's resting on the ground. By simplifying the model of this motion to free fall, it is possible to estimate the force F during the rear bumper's hitting on the ground. By comparing the exchange of potential energy of the semi-trailer when the bumper was at a height of $h = 300$ mm (figure 1f) into kinetic energy of vertical motion with a gravitational acceleration $g = 9.81\ \mathrm{ms}^{-2}$, it is possible to calculate the change in vertical velocity $\Delta v = \sqrt{2 \cdot g \cdot h} = 0.767\ \mathrm{ms}^{-1}$. From Newton's second law we obtain the equality $m \cdot \Delta v = F \cdot \Delta t$, which can be used to calculate the force. After substituting Δv and $\Delta t = 0.03$ s, which is the time of dynamic interaction between the bumper and the road, we determine the force to be equal to $F = 25.6 \cdot m$. This means that the dynamic interaction generated during the bumper's fall, which was ignored in the models, may be more than 2.5 times greater than the considered static force resulting from the mass m on the bumper if the longeron is supported by this bumper and the fifth wheel only. Taking into account the above dependence, the reduced stresses calculated for the first two hypotheses would be proportionally greater in the elastic deformation range (assuming linear deformability of the material), i.e. up to the yield stress.

4. Conclusions

Based on the results of this study, it was concluded that the damage of the longitudinal beam of the semi-trailer frame could be caused or conduced to by the semi-trailer's resting on the bumper. Ignored in modelling, the dynamic interaction generated during the bumper's fall may be more than 2.5 times greater than the static force resulting from the mass m on the bumper if the longeron was only supported by this bumper and the fifth wheel. One should reject the hypothesis that the damage of the longeron was caused by point impact to the rear bumper; however, this type of impact could cause deformation of the frame member. The correct model for simulating the loading of a semi-trailer in working order is that in which the load is distributed uniformly over the entire length of the contact between the container and the frame.

Acknowledgment

The project/research was financed in the framework of the project Lublin University of Technology-Regional Excellence Initiative, funded by the Polish Ministry of Science and Higher Education (contract no. 030/RID/2018/19).

References

[1] Q. F. Han, Q. B. Huang and S. D. Li, "Comparative Analysis of Static Strength of Traditional and Bionic Semi-trailer Frame," *Applied Mechanics and Materials*, vol. 470, pp. 320-324, 2013. doi:10.4028/www.scientific.net/AMM.470.320

[2] W. C. Zhang, X. B. Du, B. B. Ma, B. H. Pei, J. Chen and Z. X. Liu, "The Finite Element Analysis and Topology Optimization of Semi-trailer Frame,"*Advanced Materials Research*, vol. 424-425, pp. 90-93, 2012. doi:10.4028/www.scientific.net/AMR.424-425.90

[3] W. Cheng, C. Kaiwu, Q. Fukang and X. Jin, "Study on Fatigue Strength and Life of Freight Car Frame after Making Holes," *MATEC Web of Conferences*, vol. 175, pp. 1-4, 2018. doi:10.1051/matecconf/201817501035

[4] P. M. Zhang, W. Wang and W. C. Zhang, "Finite Element Modal Analysis of a Semi-trailer Frame," *Applied Mechanics and Materials*, vol. 685, pp. 199-203, 2014. doi:10.4028/www.scientific.net/AMM.685.199

[5] M. A. Abdelkareem, M. M. Makrahy, A. M. Abd-El-Tawwab, S. A. EL-Razaz, M. Kamal Ahmed Ali and M. M. Moheyeldein, "An analytical study of the performance indices of articulated truck semi-trailer during three different cases to improve the driver comfort," *Proceedings of the Institution of Mechanical Engineers, Part K: Journal of Multi-body Dynamics*, vol. 232, pp. 84-102, 2018. doi:10.1177/1464419317709895.

[6] R. Napierała, "Numerical-experimental static analysis of the effort of the crossbar-longeron in the semi-trailer frame," *AIP Conf. Proc.*, vol. 2078, pp. 1-6, 2018. doi:10.1063/1.5092101

[7] P. Działak and J. Karliński, "Comparative examination of the trailer frame in accordance with UIC 596-5," *Materials Today: Proceedings*, vol. 12, pp. 416–422, 2019.

[8] M. Kaczor and M. Januszka, "Optimization of semi-trailer design as a result of using the MTS road simulator," *Mechanik*, vol. 92, pp. 52-54, 2019. doi:10.17814/mechanik.2019.1.10

[9] A. Nieoczym, K. Drozd, A. Wójcik, "Geometric optimization of a beam detector for a WIM system", *Advances in Science and Technology Research Journal.*, vol. 12, pp. 233-241, 2018. doi:10.12913/22998624/97296

[10] EN 10025-2:2019, "Hot rolled products of structural steels - Part 2: Technical delivery conditions for non-alloy structural steels."

IOP Conf. Series: Materials Science and Engineering 1190 (2021) 012025 doi:10.1088/1757-899X/1190/1/012025

Additively Manufacturing an Indexable Insert

Jindřich Sýkora [1], Luboš Kroft [2]

[1] University of West Bohemia, i Družby 979/5, Plzeň, 31200, Czech Republic
[2] Tymákov 387, 332 01 Tymákov, Czech Republic

sykora@rti.zcu.cz

Abstract. Considering the constant advances in additive manufacturing, it makes sense to use this approach when creating a cutting tool. Currently, a lot of the tools used in machining are in the form of indexable inserts. This minimizes the amount of material used and ensures a swift change when a tool is worn out. Indexable inserts are currently made by pressing and sintering which requires a large investment in specialized mould tooling for each insert shape.
This paper investigates an additive approach to making indexable insert blanks. This manufacturing process does not require mould tooling and offers a large degree of freedom when it comes to making inserts of diverse shapes. In this work, the process of manufacturing CCMT inserts out of MS1 steel (X3NiCoMoTi 18-9-5) is investigated. MS1 is not an ideal material for a cutting tool, however, the findings can be translated to blanks additively manufactured out of tungsten carbide, cermets or ceramics.

1. Introduction

Additive manufacturing (AM) of metal parts is currently quite widely used in industrial applications. Using AM makes a lot of sense for parts with very complex shapes or parts that require complex internal cavities (e.g. injection moulds). The additive manufacturing process helps to minimize the material removal by machining as well as the need for complex fixtures. Moreover, it provides a solution for manufacturing otherwise impossible to make parts, for example internally cooled injection mould inserts [1], [2].

However, even with the rapid advances in additive manufacturing in recent years, conventional machining is still one of the most common methods used for part production. Therefore, the development of novel cutting tools is critical for the advancement of manufacturing as a whole. Arguably, one of the most important products in this industry is the indexable insert. It is the part of the tool that is responsible for the cutting action and is easily replaceable in the production environment. Tungsten carbide indexable inserts are the most common and they come in a variety of shapes and sizes. There are several ISO standardized types, but many more unconventional ones are used in milling, for example [3], [4].

The development of each of these unconventional inserts requires a mould in which the raw powder is pressed to create a blank of the desired shape. This is a very costly and time-demanding process that also limits geometrical freedom. In situations where the production volume of the inserts is not very large or when complex geometries or internal cavities are required, an additive approach to manufacturing could be more efficient. Prototyping a new indexable insert shape is one of the potential applications. After heat treatment, the blank is in many cases finished by grinding, even when using conventional manufacturing methods.

The idea of creating insert blanks additively has become increasingly relevant as new approaches to additively manufacturing tungsten carbide are being developed and the properties of the parts are close to those of conventionally manufactured tungsten carbide [5], [6].

Therefore, this paper investigates a method of producing indexable inserts by AM. Inserts of the CCMT shape are used here because they are very common and can be easily tested using existing tool holders. In this case, inserts are made out of hardened MS1 tool steel (X3NiCoMoTi 18-9-5), which is not an ideal material for cutting tools. [7] The findings can, however, be utilised in the process of additively manufacturing tools out of tungsten carbide, ceramics or even PCD to a large degree. Alternatively, a tool steel insert can be used as a carrier for a carbide, CBN or PCD cutting edge. [8]

2. Additive manufacturing of insert blanks

Firstly, additive manufacturing of indexable insert blanks requires a CAD model – in most cases, an STL format is needed. The design of the insert was inspired by the CCMT 120404 indexable insert sold by Iscar ČR s.r.o. [9]. However, the flank faces, rake face, and the bottom face of the insert were offset to allow for finish grinding. The screw hole of the insert was left at the nominal shape.

Secondly, the print data was prepared in the Materialise Magics software and the parts were connected to the build platform using block and cone support structures (see Figure 1). Then, using EOS Print, the parts were laid out on the building platform according to the recommendations of the machine manufacturer.

Lastly, the build process is started. Most often the disadvantage of additive manufacturing is the length of the build time and the cost of the raw material. In the case of the inserts, these disadvantages are close to insignificant. A cutting insert (CCMT 1204 in this case) is usually a reasonably small part, therefore, the build process is quite quick. The total build time when 3D printing 10 inserts is 106 minutes, resulting in less than 11 minutes per insert. The amount of raw powder needed for each insert is 9 g including the support structures.

Figure 1. Indexable insert with support structure – Prepared in Materialise Magics

3. Preparation of blanks for finishing

After AM, the inserts need to be post-processed. The first step is heat treatment of the entire build platform to reduce the stress induced by the build process and to harden the parts. Next, the parts are cut off the build platform and the support structures are removed.

The rake face and bottom face is then finished using a surface grinder. The cutting insert we used has a flank face angle of 7°. This needs to be taken into account while surface grinding. The amount of the material ground off these faces needs to be such that the diameter of the inscribed circle on the bottom face is 12.14 mm and the final height is 4.76 mm. This is important for the set up in the fixture (see Figure 4) used for finishing the bolt hole. The remains of the additive manufacturing process is a 0.044 mm raised rim that appears on the topmost surface of the part (see Figure 2). This needs to be taken into account when surface grinding to ensure the part's accuracy. The height of this feature was measured on an Alicona IFM G4 microscope.

Figure 2. Scan of the raised outline of the part taken on Alicona IFM G4

The surface grinding of the blanks was done in the following way. All the parts were secured on a magnetic chuck on the rake face and the remains of the support structures were ground flat. Secondly, all the parts were flipped and the lip on the edge of the blank was ground down. Thirdly the parts were flipped again and the bottom surface was ground to the final dimension. The thickness of the insert at this stage was 5.06 mm. Lastly, parts were flipped on this freshly ground bottom face and the rake face was ground bringing the insert to the final thickness of 4.76 mm.

The next step was finishing the bolt hole. The precision achieved by additive manufacturing is not sufficient for this bore, because in this case, it determines the positional accuracy of the insert in the tool holder. Therefore, the surface of this bore needs to be machined. For this procedure, a special tungsten carbide taper mill was used. It has a 44° taper which matches the taper of the head of the precision bolt used for securing the insert to the tool holder. A fixture is used to clamp the insert blanks (see Figure 4). The bore finishing process consists of finding the centre of the existing bore with a workpiece contact probe and milling out a tapered section (see Figure 3) on which the bolt head will rest.

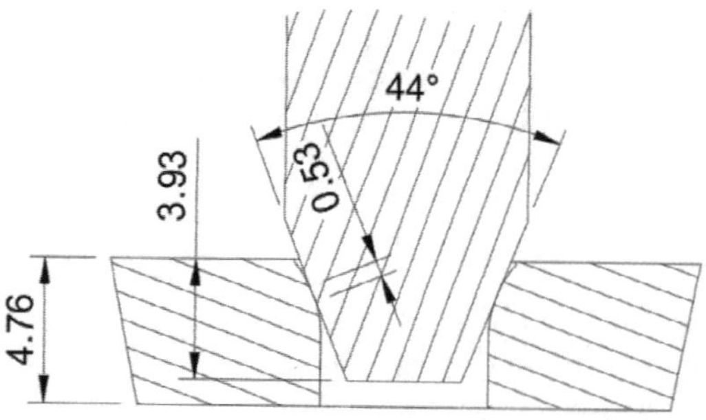

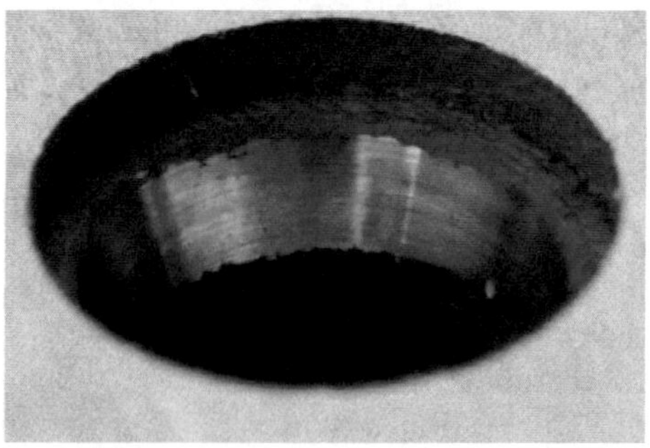

Figure 3. Bolt hole finishing

Figure 4. Fixture for clamping the insert blank for the bolt hole finishing operation

4. Grinding the inserts

The last step of the process is grinding the blank to its final dimensions and shape. For this final operation, which gives the insert its precise shape, a 5 axis Anca MX7 tool grinder is used. This grinder is designed to hold round bars with h6 precision. This meant that a clamping fixture for the inserts had to be designed and manufactured (see Figure 5). The fixture was produced to provide the highest concentricity of the M5 thread and the clamping cylinder. The insert is held in place by the precision bolt mentioned above. The tapered contact between the bolt and the insert together with the tolerances of the fixture determine the final precision and repeatability of the insert position. The tolerance of the insert position was smaller than 0.02 mm, measured on the Zoller Genius 3s.

Figure 5. Additively manufactured insert blank held in the grinding fixture

The 5 axis toolpath was created using NX CAM and machine movements were set up in such a way that the tool stays at the top of the part, ensuring sufficient delivery of the cutting fluid. The grinding parameters used for the inserts made out of MS1 steel are shown in Table 1.

This setup allows for the flank face as well as the rake face to be ground. This provides a large degree of freedom and various shapes of inserts can be created. The final shape of the insert is dictated by the manufacturing NC program, which can be changed quite easily. Therefore, it is a convenient finishing procedure for prototyping. Various corner radiuses, flank angles or rake geometries can be easily created.

Table 1. Cutting conditions of the 5 axis grinding process

Tool	SiC – DIA PRAHA 100x10x20 C49 80 K 9 V
Cutting speed	25 m/s
Feed rate	50 mm/min
RDOC	0.1 mm
ADOC	4.76 mm

5. Conclusions

The use of additive manufacturing could make the process of manufacturing prototype or low volume cutting inserts more efficient by reducing the need for mould tooling. Moreover, it makes the whole process much more flexible. This study presents a usable methodology for creating an indexable insert with the aid of additive manufacturing. It clearly shows the essentials of the whole process, starting with the design requirements and ending with a functioning product. Due to the current equipment limitations at our facility, only the process of manufacturing an insert made of MS1 steel was investigated. Although further research needs to be done, the findings should translate by a large degree to manufacturing inserts out of carbide, cermet or other suitable cutting materials. The steel inserts could also be equipped with a brazed PCD or CBN cutting edge.

To conclude, this paper shows a clear framework for making an indexable insert using additive manufacturing. However, more research needs to be done not only with more suitable cutting materials but also in utilizing other advantages of AM, such as manufacturing internal cavities

Acknowledgment(s)

This article was created under the project SGS-2019-008: Research and Development for Innovation in the Field of Manufacturing Technology - Machining Technology III.

References

[1] D. Herzog, V. Seyda, E. Wycisk, and C. Emmelmann, 'Additive manufacturing of metals', *Acta Mater.*, vol. 117, pp. 371–392, Sep. 2016, doi: 10.1016/j.actamat.2016.07.019.
[2] G. Mendible, J. Rulander, and S. Johnston, 'Comparative study of rapid and conventional tooling for plastics injection moulding', *Rapid Prototyp. J.*, vol. 23, pp. 344–352, Mar. 2017, doi: 10.1108/RPJ-01-2016-0013.
[3] K. Bobzin, 'High-performance coatings for cutting tools', *CIRP J. Manuf. Sci. Technol.*, vol. 18, pp. 1–9, Aug. 2017, doi: 10.1016/j.cirpj.2016.11.004.
[4] S. L. Marias, 'Round Tool Concepts: Indexable, Solid or Both', *Asia Pacific Metalworking Equipment News | Manufacturing | Automation | Quality Control*, 29-Jul-2019. [Online]. Available: https://www.equipment-news.com/round-tool-concepts-indexable-solid-or-both/. [Accessed: 09-Nov-2020]
[5] D. Bricín and A. Kriz, 'Processability of wc-co powder mixtures using slm additive technology', *MM Sci. J.*, vol. 2019, pp. 2939–2945, Jun. 2019, doi: 10.17973/MMSJ.2019_06_2018115.
[6] T. Josef, S. Tim, B. Timo, M. Markus, S. Gerhard, and S. Dirk, 'Laser sintering of tungsten carbide cutter shafts with integrated cooling channels', 2018, doi: 10.25341/D48G6C [Online]. Available: https://dr.ntu.edu.sg//handle/10356/88554. [Accessed: 09-Nov-2020]
[7] EOS GmbH, 'EOS MaragingSteel MS1'. 29-Jan-2020 [Online]. Available: https://www.eos.info/03_system-related-assets/material-related-contents/metal-materials-and-examples/metal-material-datasheet/werkzeugstahl_ms1_cx/ms1/ms-ms1-m280_m290_400w_material_data_sheet_05-14_en.pdf. [Accessed: 29-Jan-2020]
[8] G. Tadayyon, K. Seset, E. Parle, N. Murphy, and A. Ivankovic, 'Additive Manufacturing of Diamond Cutting Tools-Pushing the boundaries of 3D Printing', presented at the Irish Polymers and Materials Conference 2017, Dublin, Ireland, 2017, vol. 2017, doi: 10.13140/RG.2.2.18686.08009.
[9] ISCAR ČR s.r.o., 'CCMT 120404-SM'. [Online]. Available: https://www.iscar.com/eCatalog/item.aspx?cat=5566756&fnum=867&mapp=IS&app=20&GFSTYP=M&isoD=1. [Accessed: 09-Nov-2020]

IOP Conf. Series: Materials Science and Engineering 1190 (2021) 012026 doi:10.1088/1757-899X/1190/1/012026

Molecular Dynamics Investigation on the Effect of MWCNTs on Thermo-Mechanical Properties of Amorphous Cellulose Reinforced with Silicon Carbide Nanotubes

Ahmad Y. Al-Maharma[1], Franz Bamer[1], Bernd Markert[1]

[1]Institute of General Mechanics, RWTH Aachen University, Eilfschornsteinstraße 18, 52062, Aachen, Germany

almaharma@iam.rwth-aachen.de

Abstract. Cellulose is a biopolymer frequently implemented as fibers and membranes to fabricate high-performance and multi-functional composites. In this study, the molecular dynamics simulation is used to investigate the effect of CNTs on elastic modulus, ultimate strength, and toughness properties of silicon carbide nanotubes (SiCNTs) reinforced cellulose nanocomposite. Based on the simulation results, the temperature of the surrounding environment controls the stress transfer at the nanotube-cellulose interface. As the cellulose expands with increasing temperature creating more void regions at the filler-matrix interface, the load transfer from cellulose to nanotubes is reduced, leading to a noticeable reduction in tensile properties of cellulose composite. This behavior critically affects the tensile strength of cellulose composite reinforced with multi-walled CNTs (MWCNTs). The single-walled CNT with a high aspect ratio increases the cellulose composite's tensile strength and toughness, while the MWCNT introduces the most significant increase in cellulose stiffness. The difference in polarity between MWCNT and cellulose, along with the low aspect ratio of the external nanotube, would limit the capability of MWCNT in improving the tensile strength and toughness of cellulose. The single-walled SiCNTs are structurally unstable, and their stiffness degrades rapidly at high temperatures. Therefore, they are hybridized with CNTs to improve their stiffness and structural stability. SiCNT, which is hybridized with two CNTs, exhibits the best structural performance at 100, 300, and 500 K temperatures.

1. Introduction

Polymers synthesized from natural materials received a large portion of interest in recently published research due to the depletion of petroleum resources and the associated harmful impacts of the materials synthesized from these resources on climate change. The main objective of this study is to improve the tensile properties of amorphous cellulose under varying thermal environments. The stiffness, strength, and toughness are characterized at three different temperatures: 100, 300, and 500 K. To the best of the authors' knowledge, none of the previously published literature characterizes the mechanical properties of amorphous cellulose reinforced with either MWCNTs or hybrid multi-walled CNT/SiCNT in varying thermal environments. The available molecular dynamics (MD) studies in the published literature investigate the effect of single-walled CNTs [1] and double-walled CNTs [2] on cellulose properties. Improving cellulose properties through chemical treatments such as functionalization would compromise its properties such as strength and fracture toughness. Therefore, using mechanically high-performance nano-fillers is the optimum technique to preserve the original structure of cellulose while its properties are improved.

With increasing cellulose content, the mechanical properties of natural fibers are increased [3]. Due to attractive aspects of natural fibers of biodegradability, high specific strength, and moderate mechanical properties, plant fibers can be used in many applications of automotive and aerospace industries [4]. SiCNTs and CNTs can be used to reinforce cellulose to improve its thermal stability and material properties. Carbonous nano-fillers can increase cellulose's bulk, elastic, and shear moduli by 25.6%, 30%, and 32.9%, respectively. These improvements can be attributed to increased intermolecular hydrogen bonds and cohesive energy density [5].

SiCNTs have distinct properties of good chemical inertness, excellent electronic properties, and high thermal stability [6]. Moreover, cellulose nanocomposites reinforced with SiC have tensile properties and thermal stability higher than the neat cellulose polymer [7]. There is a need to hybridize SiCNTs with CNTs to improve their properties as the elastic modulus of SiCNT is evaluated to be half that of CNT with about 0.61 TPa [8]. The chemical modifications applied to the structure of cellulose and SiC nanofillers could compromise their material properties. For instance, applying functionalization based on amines like ($-NH_2$) and oxidization to modify the surface of SiC reinforcements could further improve their reinforcing effect up to a specific extent [9]. However, based on the results of MD simulation conducted on oxidized single-walled SiCNTs (SWSiCNT), the stiffness, strength, and toughness of the functionalized SiCNTs are decreased with increasing degrees of functionalization [10].

CNTs can improve cellulose's mechanical properties and hygro-thermal aging behavior since they have a high aspect ratio and good moisture resistance [11]. When these nanofillers reinforce polymers, they can significantly improve their properties so the resulting nanocomposites can be implemented in broad practical applications such as drug delivery systems and infiltrating membranes [12]. The ideal arrangements of carbon-carbon covalent bonds oriented along the axis of CNTs make them very strong with a remarkable strength-to-weight ratio. They exhibit very high tensile strength and Young's modulus values of 50-500 GPa and 1.50 TPa, respectively [13]. The combination of carbon and cellulose materials can improve the aspects of composite, such as tensile strength (23.1-277.5 MPa) and Young's modulus (1.14-16.6 GPa) [14]. The composite materials that combine CNTs and SiC show high stiffness values, superb thermal stability at extreme temperatures, and excellent heat resistance [15].

This research is organized as follows: first, details of the material configurations and their properties, along with their computational analyses, are described in Section 2. Then, simulation results on the mechanical properties of the cellulose composite are discussed in Section 3, along with a comparison to the neat cellulose polymer. Finally, conclusions are given in the last section.

2. Materials and test methods

MD simulation is used in this study to understand how the nano-fillers such as CNTs and SiCNTs can provide the structural reinforcing effect to cellulose polymer and comprehend the process of load transfer between cellulose and nanotubes at various temperatures. At the first stage of this research, composite models are designed with details discussed in Section 2.1. Then, the recent variant of all atoms force fields (OPLS 4) coupled with Desmond GPU solver described in Section 2.2 is used to characterize the tensile behavior of cellulose at three different temperatures of 100, 300, and 500 K.

2.1. Cellulose composite models

The cellulose matrix synthesized of linear D-beta-glucose polymer is first prepared according to the procedures discussed in [16]. Both MWCNT and multi-walled SiCNT (MWSiCNT) have three walls, and the first inner wall has a chiral index of (6,6). The inner separation space between the walls of MWCNT is determined to be 0.335 nm, while the distance among the walls of MWSiCNT is taken as

0.435 nm. MWSiCNT and MWCNT have armchair structures with lengths of 73.13 and 73.53 Angstrom and diameters of 21.62 and 27.41 Angstrom, respectively. The MWSiCNT is hybridized by replacing the first and second inner tubes with CNTs. To increase the specific mechanical properties of cellulose composites, the MWSiCNT, MWCNT, and their hybrid are arranged in rope-like structures to improve the load transfer between them and the cellulosic matrix. The SiCNT-rope is hybridized by substituting the two adjacent single-walled SiCNT with minor diameters with two SWCNTs at similar sizes. It is noteworthy that the geometry of the nanotube plays a critical role in affecting the properties of the final nanocomposite. The increasing aspect ratio of SiCNTs decreases the critical compressive forces of nanotubes. By fixing the side length of nanotubes, SiCNTs with smaller radii have more structural stability and less tendency toward buckling [17]. The material models are classified into two groups to understand the effect of aspect ratio and orientation of the nanotubes on the final properties of cellulose composites: (i) unidirectional (UD) CNTs and SiCNTs with periodic terminal atoms aligned along Y-axis; (ii) short length nano-fillers with hydrogen terminal atoms randomly positioned within the cellulosic polymer.

The essential elements used to generate the MD models of cellulose composites are shown in figure 1. Increasing the content of MWCNTs in cellulose composite reinforced with MWSiCNTs would make the material stiffer and increasing its density to the highest extent. Table 1 shows how the arrangements of MWSiCNT and MWCNT, in addition to their molecular weights, affect the overall density of cellulose composites. Hybridizing SiCNTs with CNTs has a trivial effect on the density of cellulose. It can be inferred from the table data that the rope-like structure of nanotubes creates a larger void space inside the cellulose composites relative to the traditional form of the multi-walled nanotubes (MWNTs) reducing the overall composite density.

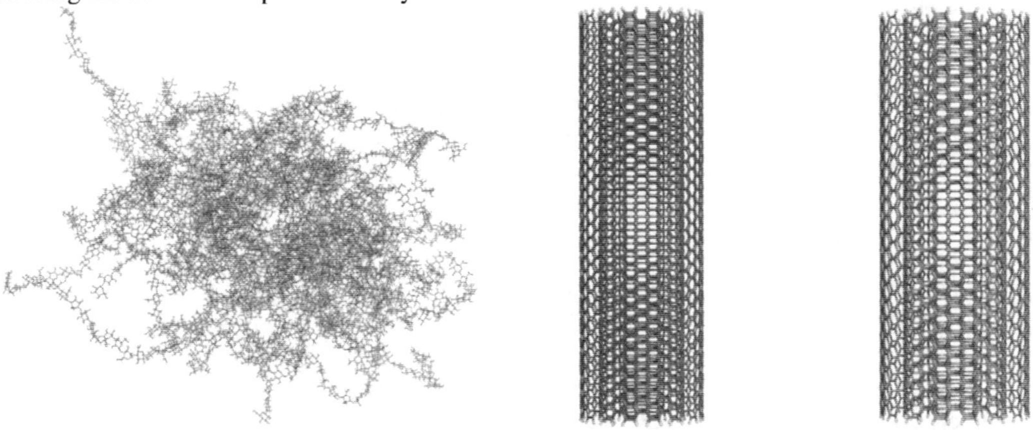

(a) (b) (c)

Figure 1. The fundamental elements used to construct the cellulose composite models: (a) amorphous cellulose, (b) MWCNT, and (c) MWSiCNT.

2.2. *The setup of MD simulation*

The OPLS force field defines the charges and positions of atoms, the optimum bond lengths, and dihedral angles distribution [18]. Energy minimization should be conducted before equilibrating the MD models using NPT and NVT ensembles to eliminate undesirable interactions, including clashes between atoms. The steepest descent (SD) and limited memory Broyden- Fletcher- Goldfarb- Shanno (L-BFGS) minimization methods are used to efficiently minimize large MD models' energy. The dihedral angles of the cellulose polymers are distributed in the unit cell so that minimum clashes among atoms can be guaranteed. The SD algorithm is superior to the conjugate gradient if the starting MD system is some way from the minimum energy [19]. NPT ensemble is then applied for 60 nanoseconds (ns) at 2 femtoseconds (fs) time step to relax the MD system and to increase its density based on Martyna Tobias

Klein barostat method [20]. To improve the distribution of cellulose polymers and to further increase the density of the MD models to the highest possible level, NVT ensemble is applied for 5 ns with 2 fs time step at T= 350 K. The latter ensemble is followed by another NVT ensemble for 3.36 ns with 1 fs time step at T= 500 K. The equilibration of MD models based on NVT ensemble is conducted based on Noose-Hoover thermostat as cellulose is flexible material [21]. The tensile properties are characterized under three different temperatures: 100 K, 300 K, and 500 K. NPT ensemble is used to expand the unit cell for 10 picoseconds (ps) with a time step of 2 fs.

Table 1. The effect of nanotubes' molecular weights and arrangements on the density of cellulose.

Material	Molecular weight (wt.%)			Average density (g/cm^3)	
	Cellulose	SiC	CNT	Traditional	Rope
Cellulose	100	0	0	1.41	-
Cellulose/UD-MWCNT	82.64	-	17.36	1.46	1.33
Cellulose/UD-MWSiCNT	78.25	21.75	-	1.42	1.30
Cellulose/UD-SiC-SiC-CNT	79.01	17.97	3.02	1.39	1.27
Cellulose/UD-SiC-CNT-CNT	80.46	10.84	8.70	1.38	1.26
Cellulose/short-MWCNT	82.36	-	17.64	1.48	1.35
Cellulose/short-MWSiCNT	77.84	22.16	-	1.45	1.38
Cellulose/short-SiC-SiC-CNT	78.63	18.31	3.06	1.44	1.39
Cellulose/short-SiC-CNT-CNT	80.10	11.06	8.84	1.44	1.41
Cellulose/ short- MWCNT-MWSiCNT	66.72	18.99	14.29	1.48	-

3. Results and discussions

The cellulose's structure contains voids that provide to its composite's high capability to attenuate the damaging effect of loads. Without nano-fillers, there is a need to apply more pressure and heat on neat cellulose to increase its stiffness at the expense of fracture toughness. This treatment would remove a large part of air pores making cellulose stiffer material. The stiffness values of multi-walled and rope-like arrangements of SiCNTs and CNTs are characterized in Section 3.1, whereas strength and fracture toughness properties for shot length nanotubes of SiCNTs and CNTs reinforced cellulose are studied in Section 3.2.

3.1. Aligned unidirectional nanotubes reinforced cellulose

The magnitude of improvement achieved on the stiffness of cellulose composites is controlled by the aspect ratio and the number of nanotubes placed in the direction of the load. As the values in figure 2 can tell, all arrangements of CNTs and SiCNTs increase the stiffness of cellulose, which is evaluated as 9.48 GPa agreeing well with the determined value in the literature of 9.37 GPa [16]. It can be noted from the same figure that cellulose composites reinforced with CNTs have higher stiffness relative to the ones reinforced only with SiCNTs. All nanotubes' arrangements can improve the cellulose stiffness as the MWSiCNT and MWCNT can increase the stiffness up to around 2 and 10 times of pure cellulose, respectively. The MWNTs are thick structures showing a higher resistance to the applied tensile load than single-walled nanotubes composing the rope-like structure, as can be concluded from the deformed cellulose composites shown in figure 3.

3.2. Short length nanotubes reinforced cellulose

The conventional method used to optimize the material properties of most bio-composites is simply the incorporation of MWCNTs into their structures. However, the addition of MWCNTs to MWSiCNT reinforced cellulose will displace more elastic cellulose from the matrix replacing it with stiff carbon atoms. Therefore, the stiffness of cellulose composite is significantly increased relative to neat cellulose, as demonstrated by the values listed in table 2. However, the strength of MWCNT-MWSiCNT reinforced cellulose is severely reduced even below that of pure cellulose at 100 K since the MWCNT has an incompatible interface with hydrophilic cellulose matrix and the CNT that has direct contact with cellulose has a lower aspect ratio relative to the internal tubes. This incompatibility in terms of polarity becomes more dominant and apparent at high temperatures, as shown in figure 4.

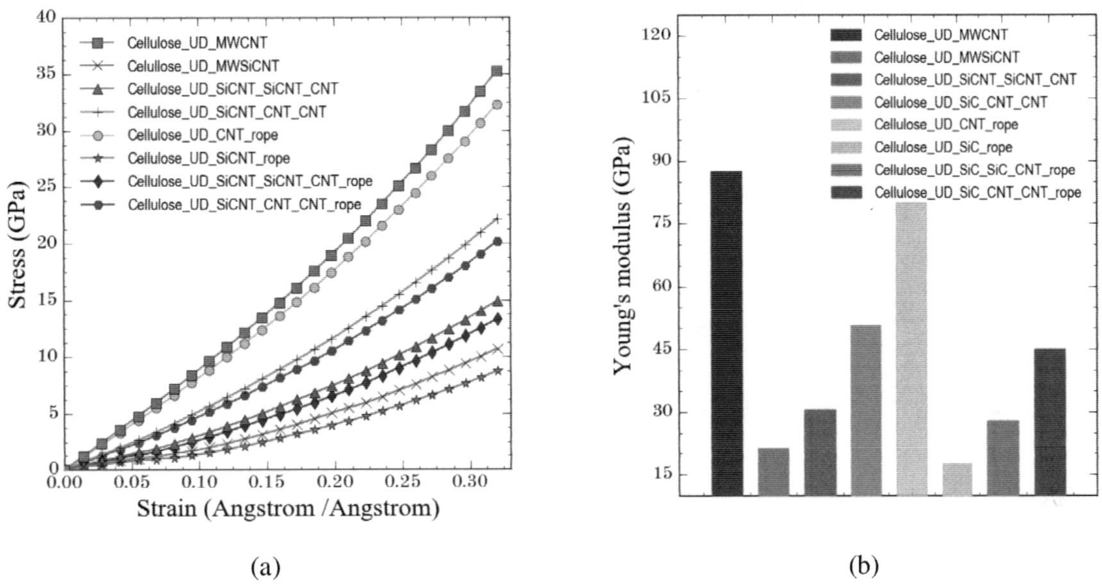

Figure 2. The effect of nanotubes arrangements on the stiffness of cellulose composites, which is characterized at T=100 K: (a) stress-strain relationships, (b) Young's modulus values.

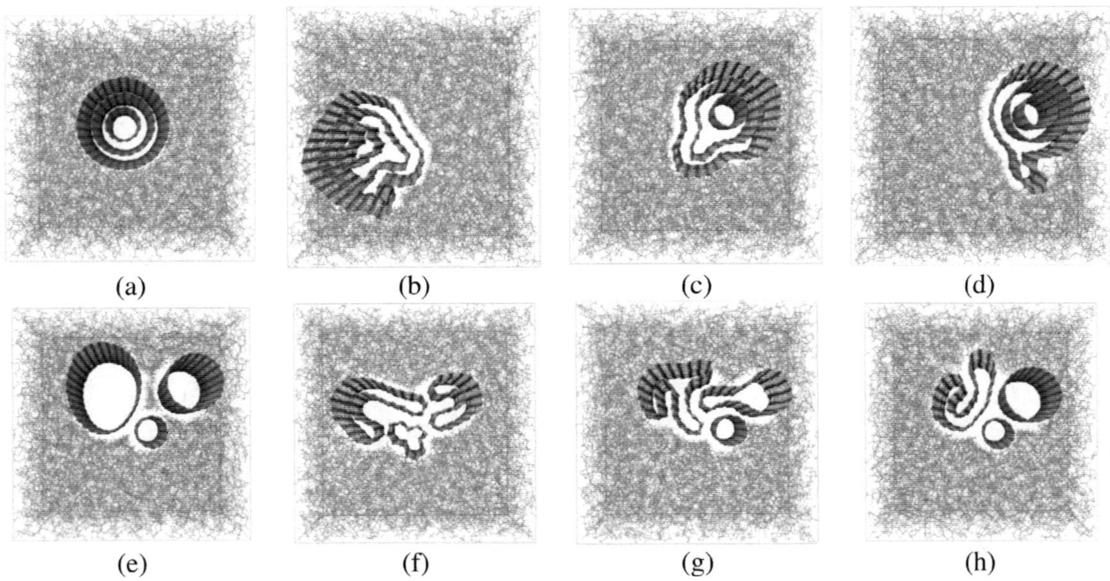

Figure 3. The cross-sections of deformed cellulose composites under tensile loading at T=100 K, which are reinforced with the following nanotubes: (a) UD-MWCNT, (b) UD-MWSiCNT, (c) UD-SiC-SiC-CNT, (d) UD-SiC-CNT-CNT, (e) UD-CNTs-rope, (f) UD-SiCNTs-rope, (g) UD-SiC-SiC-CNT-rope, and (h) UD-SiC-CNT-CNT-rope.

Table 2. The effect of temperature on tensile properties of cellulose/SiC-CNT hybrid composites.

Material	T= 100 K			T= 300 K			T= 500 K		
	E (GPa)	*S* (MPa)	*G* (MJ/m^3)	*E* (GPa)	*S* (MPa)	*G* (MJ/m^3)	*E* (GPa)	*S* (MPa)	*G* (MJ/m^3)
Cellulose	9.48	643.73	165.98	6.74	420.89	108.76	4.86	302.62	64.28
Cellulose/ MWCNT	12.72	671.5	187	10	533.63	138.86	8.31	347.98	82.3
Cellulose/ MWSiCNT	12.03	735.9	178.54	9.83	493.28	121.19	5.07	298.22	78.28
Cellulose/ MWCNT-MWSiCNT	12.21	592.89	172.03	10.22	451.2	121.68	5.81	312.89	74.14
Cellulose/ SiC-SiC-CNT	10.09	720.83	178.18	8.62	545.96	134.04	4.87	327.57	80.14
Cellulose/ SiC-CNT-CNT	11.25	703.41	182.82	9.70	526.89	136.89	7.21	357.97	92.06
Cellulose/ CNTs-rope	11.60	710.69	197.62	8.66	499.54	132.66	8.29	360.82	90.35
Cellulose/SiCs-rope	10.62	500.08	144.91	7.90	328.72	100.95	6.15	253.24	61.91
Cellulose/ SiC-SiC-CNT-rope	11.23	665.2	161.47	9.43	397.56	111.07	5.19	312.22	76.14
Cellulose/ SiC-CNT-CNT-rope	12.44	700.88	191.91	11.84	602.24	144.5	8.41	396.22	86.04

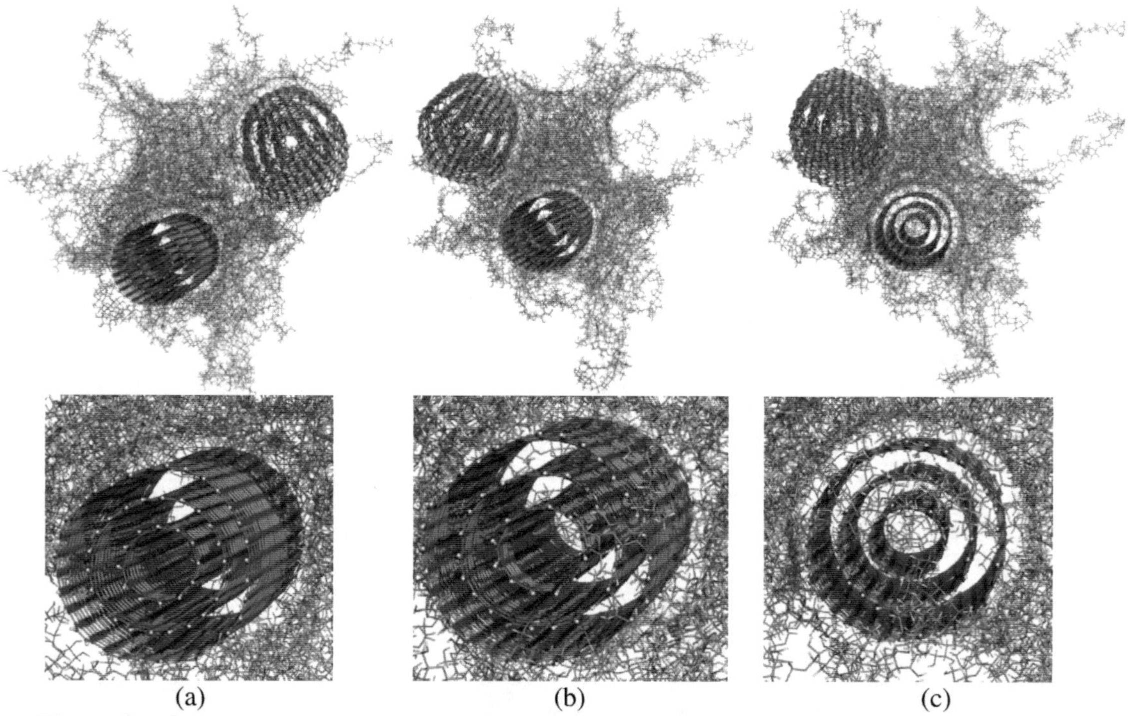

Figure 4. The generating of interfacial voids between MWCNT and cellulose with increasing temperatures: (a) 100 K, (b) 300 K, and (c) 500 K.

The rope-like structures shown in figure 5 prove that the single-walled SiCNTs are structurally unstable. Through hybridizing SiCNTs with CNTs, we will ensure that the hybrid nanotubes will have good interlocking with cellulose (needed for strength), more void spaces are generated as the CNTs with small diameters will replace larger internal tubes of SiCNTs (needed for toughness), and the high aspect ratio stiff CNTs placed in the tube core will improve the tube capability to resist loads (needed for stiffness). The hybrid nanotube of SiC-SiC-CNT has a thick external wall with the capability to transfer load from cellulose. It functions perfectly to improve cellulose's strength and toughness at low to moderate temperatures (100-300 K), but its stiffness degrades rapidly at high temperatures (T= 500 K). The hybrid nanotube of SiC-CNT-CNT has a thin external wall with good interlocking with cellulose at low temperatures (T= 100 K), but it becomes structurally unstable with increasing temperatures (T= 300 and 500 K). Nevertheless, this configuration has a stable stiffness at all temperatures due to its stiff core.

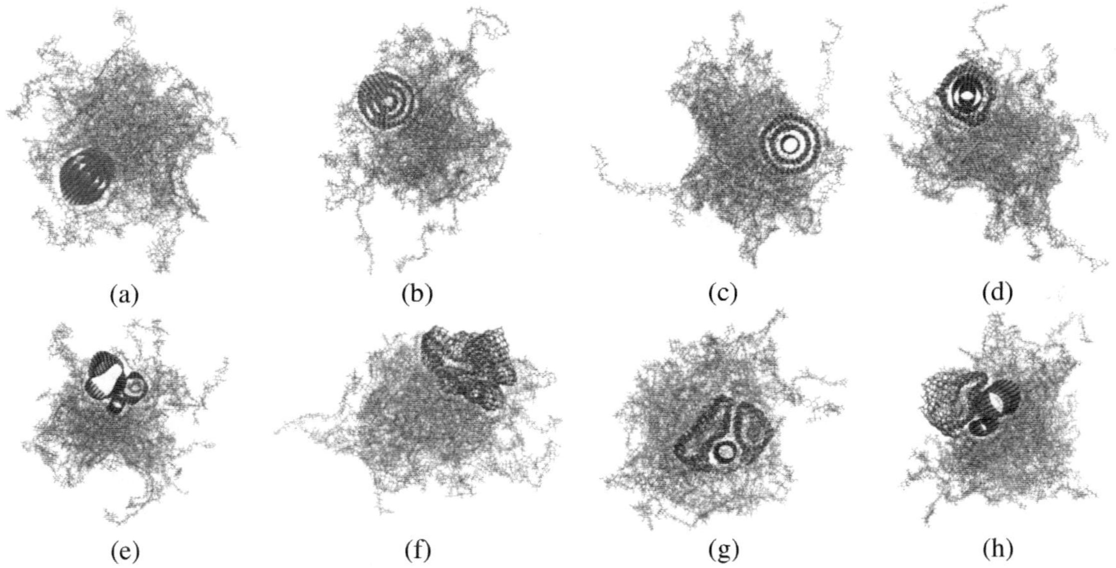

Figure 5. The deformed cellulose composites reinforced with short nanotubes at T=100 K: (a) MWCNT, (b) MWSiCNT, (c) SiCNT-SiCNT-CNT, (d) SiCNT-CNT-CNT, (e) CNTs-rope, (f) SiCNTs-rope, (g) SiC-SiC-CNT-rope, and (h) SiCNT-CNT-CNT-rope.

The stress-strain curves characterized at low and high temperatures of cellulose composites reinforced with CNTs and SiCNTs are shown in figure 6. It is confirmed from the results shown in figure 6.a that CNTs play a critical role in improving the strength and fracture toughness of cellulose reinforced with rope-like configurations of nanotubes. The highest strength peaks are noticed at cellulose reinforced with CNT-rope since all CNTs, especially those with high aspect ratio, have direct contact with cellulose matrix at T=100 K. At room temperature of T=300 K, the only configuration that shows remarkable poor performance is SiCNT-rope reinforced cellulose as shown in figure 6.b since the low aspect ratio, thin, and single-walled SiCNT is structurally unstable with increasing temperatures. The rope-like configuration, modified with two high aspect ratio CNTs, exhibits the highest tensile strength. Figure 6.c shows that the tensile properties of SWSiCNT are reduced significantly at 500 K, and these tubes play a damaging role in the cellulose structure. It can be concluded from the previous discussion that the hybridization of SiCNTs with CNTs would improve the values of stiffness and toughness at all temperature ranges. The stiffness and toughness values of cellulose reinforced with only CNTs (CNT-rope & MWCNT) are stable at all temperatures (100, 300, and 500 K) and do not experience any severe degradation. Furthermore, the SiCNTs have effective interlocking with cellulose matrix at low and high temperatures, as shown in figure 6.d increasing the tensile strength of cellulose composite.

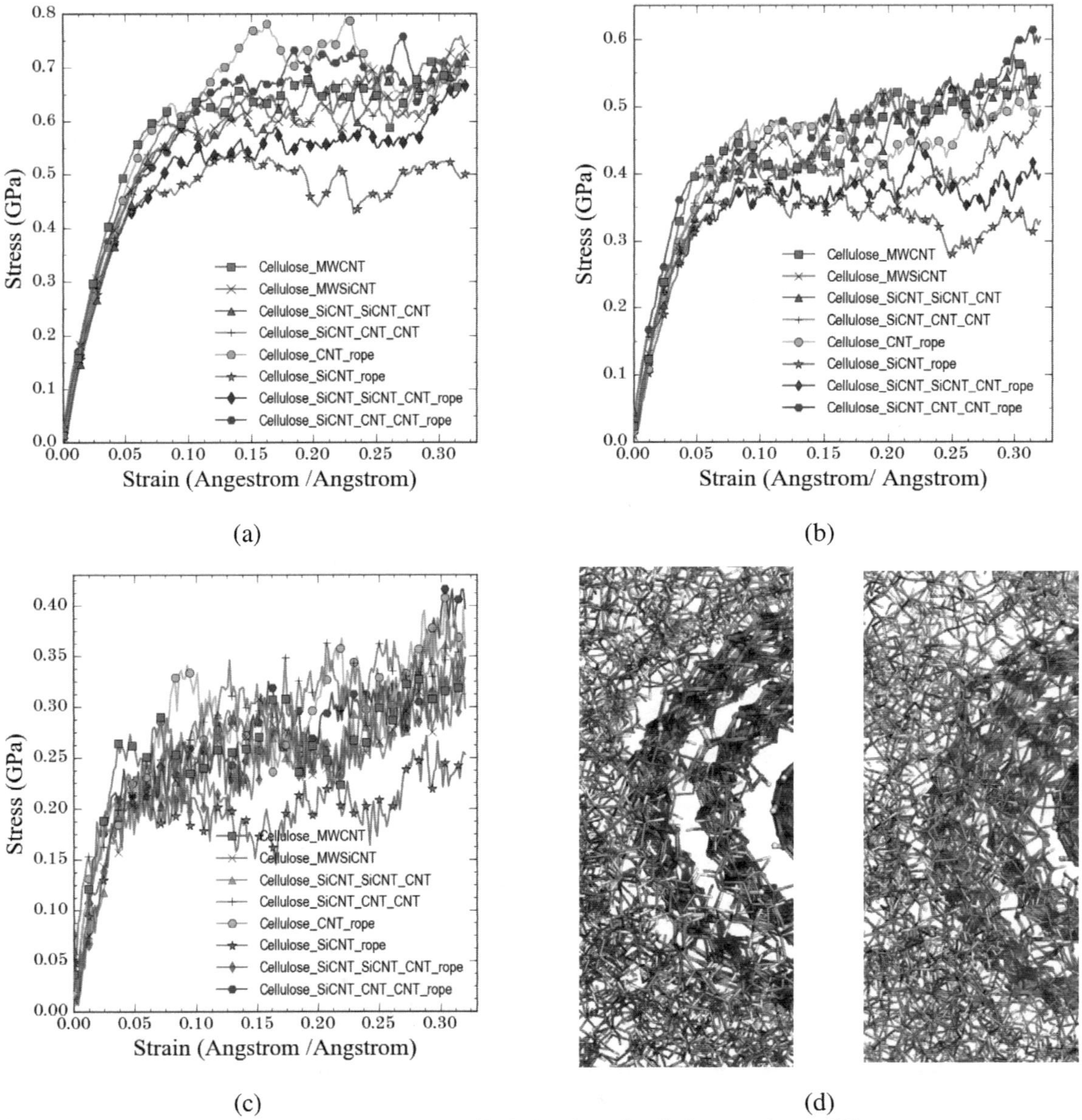

Figure 6. The effect of SiCNT/CNTs on tensile behavior of cellulose at three different temperatures: (a) 100 K, (b) 300 K, and (c) 500 K, and (d) effective interlocking of thick SiCNT wall with cellulose at 100 K (left) to 500 K (right).

4. Conclusions

CNTs are high mechanical performance nanofillers with competitive stiffness and strength values. However, these nanofillers have poor compatibility with biomaterials like cellulose due to the difference in polarity. SiCNTs have an effective interlocking with hosting cellulose matrix. However, their stiffness is degraded fast with increasing temperatures. Additionally, the SWSiCNT is structurally unstable and cannot be used as standalone reinforcements in varying thermal environments. Incorporating MWCNT directly into MWSiCNT/ cellulose composite without hybridization would increase the stiffness of the composite at the expense of strength and toughness properties. Therefore, CNTs are used to hybridize SiCNTs to improve cellulose composites' stiffness, strength, and toughness properties. Two models of hybrid MWNTs are considered. The first model has a thick external wall of double-walled SiCNTs, while the other has SWSiCNT. The hybrid MWNT of SiCNT-SiCNT-CNT shows superb strength and fracture toughness values with a moderate increase in stiffness at 100 and 300 K. This can be justified to the large void space created inside the core of MWNT after substituting SiCNT with high aspect ratio CNT and good interlocking of the thick external wall of SiCNT with cellulose. At 500 K, the stiffness of SiCNT-SiCNT-CNT is degraded, and the composite shows the stiffness of neat cellulose. Therefore, additional SiCNT is replaced with CNT to improve the stiffness, strength, and toughness at high temperatures.

The results of the MD simulation confirm that at a low temperature of 100 K, the cellulose polymers are adequately compressed to the nanotubes leading to effective load transfer from cellulose to fillers. To improve the tensile properties of cellulose composites at higher temperatures of 300 and 500 K, chemical functionalization is needed at nanotube- matrix interface, but care must be taken during this process to avoid compromising the structural aspects of cellulose.

Acknowledgment
The first author would like to acknowledge the financial support provided by the German Academic Exchange Service (DAAD) to accomplish this research.

References

[1] M. Shishehbor and M.R. Pouranian, "Tuning the Mechanical and Adhesion Properties of Carbon Nanotubes Using Aligned Cellulose Wrap (Cellulose Nanotube): A Molecular Dynamics Study," *Nanomaterials*, vol. 10(1), pp.154, 2020.

[2] Y. Liu, Y. Wang, Y. Nie, C. Wang, X. Ji, L. Zhou, F. Pan, and S. Zhang, "Preparation of MWCNTs-graphene-cellulose fiber with ionic liquids," *ACS Sustain. Chem. Eng.*, vol. 7(24), pp. 20013-21, 2019.

[3] M. M. Ahmed, H. N. Dhakal, Z. Y. Zhang, A. Barouni, and R. Zahari, "Enhancement of impact toughness and damage behaviour of natural fibre reinforced composites and their hybrids through novel improvement techniques: A critical review," *Compos. Struct.*, vol. 259, pp.113496, 2021.

[4] L. Thomas and M. Ramachandra, "Advanced materials for wind turbine blade-A Review," *Mater. Today. Proc.*, vol. 5(1), pp. 2635-40, 2018.

[5] D. Du, C. Tang, Y. Tang, L. Yang, and J. Hao, "Molecular simulation on the mechanical and thermal properties of carbon nanowire modified cellulose insulating paper". *Compos. Struct.*, vol. 261, pp. 113283, 2021.

[6] V.C. Tony, C.H. Voon, C.C. Lee, B.Y. Lim, M.M. Arshad, S.C. Gopinath, K.L. Foo, A.R. Ruslinda, U. Hashim, and M.N. Nashaain. "Novel synthesis of silicon carbide nanotubes by microwave heating of blended silicon dioxide and multi-walled carbon nanotubes: The effect of the heating temperature," *Ceram. Int.*, vol. 42(15), pp. 17642-9, 2016.

[7] SK. Kisku, S. Dash, and SK. Swain, "Dispersion of SiC nanoparticles in cellulose for study of tensile, thermal and oxygen barrier properties," *Carbohydr. Polym.*, vol. 99, pp. 306-10, 2014.

[8] J. Hurst, "Boron Nitride Nanotubes, Silicon Carbide Nanotubes, and Carbon Nanotubes—A Comparison of Properties and Applications," *InNanotube Superfiber Materials*, William Andrew Publishing, pp. 267-287, 2014.

[9] J. Bose, J. Dasgupta, U. Adhikari, and J. Sikder, "Tuning permeation characteristics of cellulose acetate membrane embedded with raw and amine-functionalized silicon carbide nanoparticle for oil-water separation," *J. Water Process. Eng.*, vol. 41, pp. 102019, 2021.

[10] M. Eghbalian, R. Ansari, and S. Rouhi, "Mechanical properties of oxygen-functionalized silicon carbide nanotubes: A molecular dynamics study," *Physica B Condens.* Matter, vol. 610, pp. 412939, 2021.

[11] V. Gautam, K.P. Singh, and V. L. Yadav, "Preparation and characterization of green-nano-

composite material based on polyaniline, multi-walled carbon nano tubes and carboxymethyl cellulose: for electrochemical sensor applications". *Carbohydr. Polym.*, Vol. 189, pp. 218-28, 2018.

[12] Y. Khairnar, D. Hansora, C. Hazra, D. Kundu, S. Tayde, S. Tonde, J. Naik, and A. Chatterjee, "Cellulose bionanocomposites for sustainable planet and people: A global snapshot of preparation, properties, and applications," *Carbohydr. Polym. Technol. Appl.*, vol. 2, pp. 100065, 2021.

[13] BV Basheer, J.J. George, S. Siengchin, and J. Parameswaranpillai, "Polymer grafted carbon nanotubes—Synthesis, properties, and applications: A review," *Nano-Struct. Nano-Objects*, vol. 22, pp. 100429, 2020.

[14] YD. Dong, H. Zhang, G.J. Zhong, G. Yao, and B. Lai, "Cellulose/carbon composites and their applications in water treatment–a review," *Chem. Eng. J.*, vol. 405, pp. 126980, 2021.

[15] K. Aly, M. Lubna, and P.D. Bradford, "Low density, three-dimensionally interconnected carbon nanotube/silicon carbide nanocomposites for thermal protection applications," *J. Eur. Ceram. Soc.*, vol. 41(1), pp. 233-43, 2021.

[16] X. Zhang, M.A. Tschopp, M.F. Horstemeyer, S.Q. Shi, and J. Cao, "Mechanical properties of amorphous cellulose using molecular dynamics simulations with a reactive force field," *Int. J. Model. Identif. Control*, vol. 18(3), pp. 211-7, 2013.

[17] R. Ansari, S. Rouhi, M. Aryayi, and M. Mirnezhad, "On the buckling behavior of single-walled silicon carbide nanotubes," *Sci. Iran.*, vol. 19(6), pp. 1984-90, 2012.

[18] E. Harder, W. Damm, J. Maple, C. Wu, M. Reboul, J.Y. Xiang, L. Wang, D. Lupyan, M.K. Dahlgren, J.L. Knight, and J.W. Kaus, "OPLS3: a force field providing broad coverage of drug-like small molecules and proteins," *J. Chem. Theory Comput.* , vol. 12(1), pp. 281-96, 2016.

[19] B. Gautam, "Energy Minimization. InHomology Molecular Modeling-Perspectives and Applications, *IntechOpen*, 2020.

[20] S.M. Rogge, L. Vanduyfhuys, A. Ghysels, M. Waroquier, T. Verstraelen, G. Maurin, and V. Van Speybroeck, "A comparison of barostats for the mechanical characterization of metal–organic frameworks," *J. Chem. Theory Comput.*, vol. 11(12), pp. 5583-97, 2015.

[21] V. Rühle, "Berendsen and nose-hoover thermostats," *Am. J. Phys*, 2007.

IOP Conf. Series: Materials Science and Engineering 1190 (2021) 012027 doi:10.1088/1757-899X/1190/1/012027

Investigation of the Influence of the Airflow Around a Grinding Wheel on the Coolant Supply

Roman Stabauer [1], **Wilfried Saxler**[1]

[1] University of Applied Sciences RFH-Cologne, Institute of Manufacturing and Tooling Technology (iWFT), Vogelsanger Str. 295, Cologne 50825, Germany

roman.stabauer@rfh-koeln.de

Abstract. The fast rotation of the grinding wheel creates an airflow near its surface, which causes a deflection of the coolant jet. To obtain an efficient and reliable grinding process, it is necessary to consider the airflow when dimensioning the coolant supply. For this purpose, the airflow around rotating grinding wheels was analyzed and characterized. In this research the airflow velocities around the grinding wheel were measured using a Prandtl tube, high-speed records were used to analyze the interactions between coolant supply and the airflow and positions for an effective air deflection were examined. It is found that the influence of the airflow especially depends on the cutting speed and the porosity of the grinding wheel, the airflow is not constant along the width of the grinding wheel and without air deflection the coolant jet requires an exit velocity of at least 50% of the grinding wheel peripheral speed.

1. Introduction

The material removal with a large variety of cutting edges and high cutting speeds is characteristic for grinding. As a result, it is possible to achieve tight tolerances for form and dimension as well as high rates in material removal because of the development of synthetic, extremely hard cutting materials. Compared to machining with geometrically defined cutting edges higher specific energies are needed. An adequate lubrication and cooling are especially necessary when grinding with high material removal rates to avoid undesired structural properties. The precise use of cooling lubricant could impact positively on the tolerances for form and dimension, the surface quality and the boundary zone. The fast rotation of the grinding wheel creates an airflow near the surface, which causes a deflection of the coolant jet. In the context of high-speed grinding, consideration of airflow effects around the grinding wheel become more important. In addition, the increasing demand for sustainability in manufacturing technology is shifting the focus to the efficient use of cooling lubricants in line with economic and ecological requirements. [1; 2]

The airflow around the grinding wheel is a layer of air created directly on the wheel surface by boundary layer friction and adhesion when the wheel rotates. [2] The creation of the airflow is essentially influenced by three factors: the cutting speed, the surface topography (grit size and dressing conditions) and the porosity of the grinding wheel. The air is absorbed through the pores in the grinding wheel and is circulated by rotation. The friction between the air layer and the grinding wheel surface builds up airflow and the centrifugal force accelerates the air continuously outward from the grinding wheel. This creates an airflow which, due to its radial alignment to the grinding wheel axis, counteracts the cooling lubricant supplied.

There are different studies conducted on the airflow around the grinding wheel. For example, *Alenius* and *Johansson* performed measurements to determine air movements and the transport of grinding particles [3]. *Okubo et al.* developed a technology to reduce the required amount of coolant supply whereby air supplied above the grinding point from the side of the wheel effectively intercepts the airflow around the grinding wheel [4]. *Klocke*, *Beck* and *Tawakoli* prove the existence of the airflow, analyzed the profile image and provide statements on the jet pressure required to pass the airflow [5–7].

In in this paper, the airflow around rotating grinding wheels is analyzed and characterized. For this purpose, different methods for analyzing the airflow were tested. The main objective is to give the user an understanding of the influence of airflow on the grinding process and coolant supply as well as to provide information on how to reduce the impact of the airflow. For this research, the flow velocities of air movement around the grinding wheel were measured using a Prandtl tube, high-speed records were used to analyze the interactions between coolant supply and the airflow and positions for an effective air shielding were examined.

2. Experiment Methods

The experiments were conducted on a surface and profile grinding machine ABA ECOLINE 1006 CNC with improved spindle for cutting speeds up to 80 m/s. For the research of the influence of airflow on the coolant supply different types of grinding wheels were used. One open-pored aluminum oxide grinding wheel (400 x 30 x 127 mm) which is vitrified bonded and has a grit size of 60. The second grinding wheel is also vitrified bonded but sintered aluminum oxide (400 x 30 x 127 mm) with a grit size of 90. These two wheels are reinforced for the use up to 80 m/s. A synthetic resin bonded diamond grinding wheel (400 x 20 x 127 mm) with a grit size of 126 was also examined. A needle nozzle with a total cross-section of 15.7 mm^2 was used for the coolant supply with emulsion (5%).

For measuring the velocity of the airflow a Prandtl tube was used. This tube has two ports, one for measuring the dynamic and one for the static pressure. The relationship between the measured dynamic pressure and flow velocity can be described by Bernoulli's principle (1).

$$p_{dyn} = \frac{1}{2} * \rho * v^2 \quad (1)$$

A stainless steel tube with an elliptical head was used. The inner diameter of the tube is 0.3 mm throughout. The outer diameter is 2.3 mm at the tip and 4 mm at the shaft. In combination with a pressure gauge, the differential pressure can be measured. When measuring the flow velocity with a Prandtl tube, it should be noted that it only measures the airflow in one direction. For video recording a high-speed-camera Keyence VW-9000 was used.

3. Results and discussions

The research carried out on the airflow can be divided into three parts. First of all the airflow around the grinding wheel is characterized by measuring the airflow velocity in radial and axial direction of the grinding wheel using a Prandtl tube. Afterwards, the functional relationships of the coolant supply and the airflow are analysed using a high-speed camera. Furthermore, investigations regarding airflow deflection are presented.

3.1. Characterization of the airflow around the grinding wheel

The first investigations were aimed at characterizing the airflow around the grinding wheel and visualizing its profile. For this purpose, the flow velocities in the radial and axial direction of the grinding wheel were measured using a Prandtl tube. First, the flow velocity of the airflow was investigated as a function of the radial distance to the grinding wheel. Here, the measuring point of the Prandtl tube was positioned exactly in line with the rotational axis below the grinding wheel and the flow velocities were measured tangentially at different distances (**Figure 1**). The Prandtl tube could be positioned close to the wheel surface due to its small diameter of 1 mm. The peripheral speed of the grinding wheel v_s was varied between 20 and 60 m/s. These speeds are above the common machining speeds for the present conventional grinding wheel.

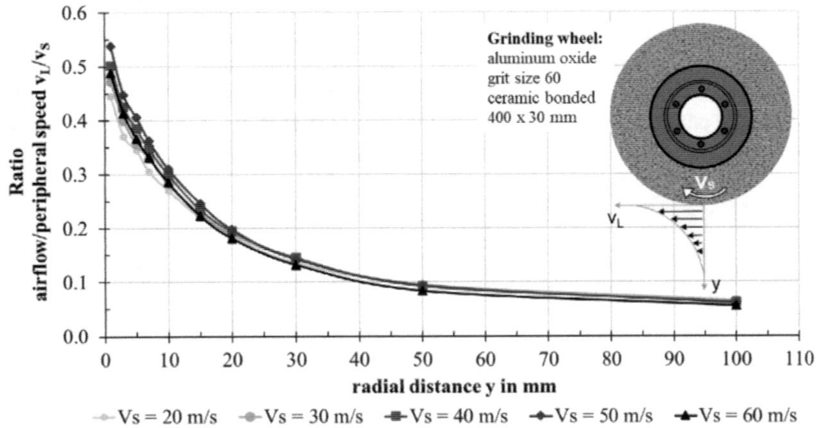

Figure 1. Ratio of the airflow velocity v_L to the peripheral speed v_S of the grinding wheel with different radial distances to the surface

The velocity of the airflow shows an exponential decrease of the flow velocity with increasing radial distance to the surface of the grinding wheel. It should be noted that the airflow measured 1 mm from the grinding wheel surface only applies about half the speed of the peripheral speed. This can be observed at all grinding wheel peripheral speeds between 20 and 60 m/s.

The same test setup was additionally carried out with different types of grinding wheels. In addition to the aluminum oxide, a finer sintered aluminum oxide and a diamond grinding wheel were used, as well as a plane steel wheel. The results with a peripheral speed v_s of 35 m/s are shown in Figure 2.

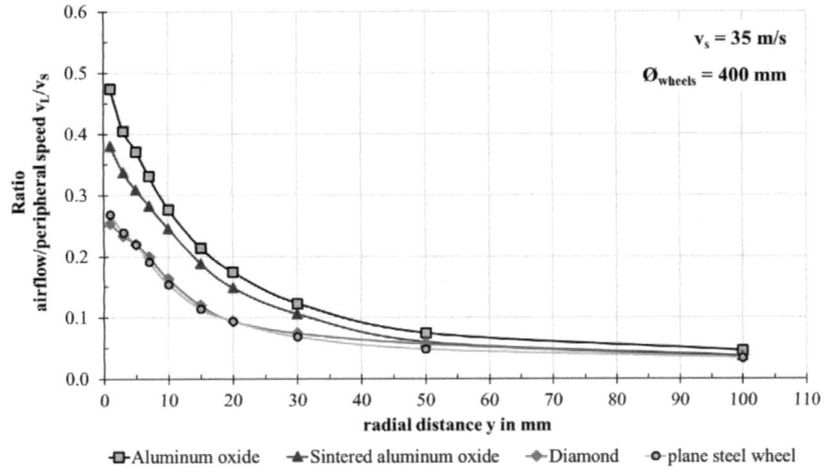

Figure 2. Ratio of the airflow velocity v_L to the peripheral speed v_S of different types of grinding wheels with different radial distances to the surface

The exponentially decreasing course of the airflow velocity with increasing radial distance is found in similar form in all types of grinding wheels. It is noticeable that the influence of the airflow decreases with a finer grit size or a smoother wheel surface. A rough and open-pored grinding wheel can carry more air with it and accelerate it more. A diamond grinding wheel with a closed resin bonding has no porosity and can be compared to a plane steel wheel in terms of airflow. Almost no difference can be observed here, although the surface roughness differs significantly. The porosity has therefore a bigger influence on the airflow than the topography of the grinding wheel surface.

The plot of the flow velocity of different grinding wheels close to the grinding wheel surface with variation of the wheels peripheral speed shows an almost linear course (Figure 3). Depending on the topography of the grinding wheel, however, the linear curves have different gradients (μ_L).

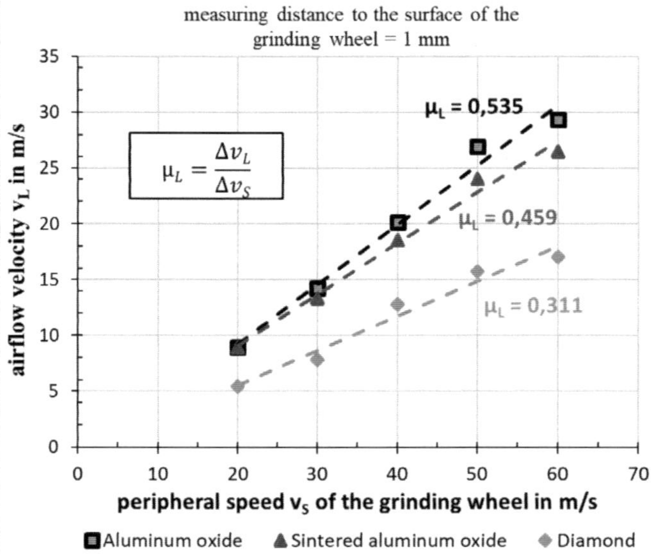

Figure 3. Influence of grinding wheel topography on the formation of the airflow at different peripheral speeds

The graph of the open-pore and rougher aluminum oxide grinding wheel (grit size 60) has a steeper gradient than the graph of the more dense and finer sintered aluminum oxide grinding wheel (grit size 90). Compared to this, the graph of the closed diamond grinding wheel with a very fine surface topography has a much flatter slope. By defining a gradient μ_L for various types of grinding wheels, it is possible to describe the formation of the air cushion at different wheel peripheral speeds.

In addition to characterizing the airflow in the radial direction, the behavior of the airflow in the axial direction, i.e. across the wheel width, was also investigated. The result of the research of the open-pored aluminum oxide grinding wheel is shown below as an example (**Figure 4**).

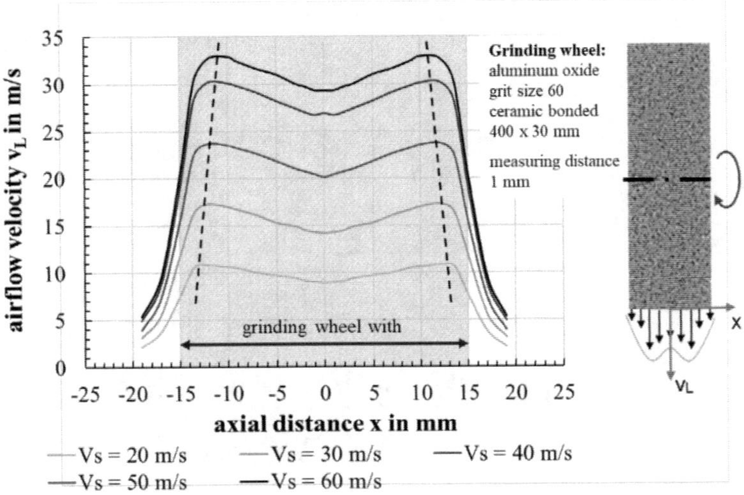

Figure 4. Measurement of the airflow velocity v_L with different distances in axial direction

The width of the grinding wheel used is 30 mm, so that the measuring points +15 and -15 mm represent the edges of the wheel. Starting from the center axis of the grinding wheel, the airflow has a symmetrical image. The maxima are not, as possibly expected, exactly in the center of the wheel, but at a distance of approx. 10 to 13 mm from the center axis. The greater the peripheral speed of the grinding wheel, the further the highest flow velocity of the air cushion "moves" to the center axis. The air on the sides of the grinding wheel is accelerated centrifugally due to the rotation. In addition, there is the airflow at the periphery of the grinding wheel. These combine in the boundary area to form this characteristic velocity profile.

As an interim conclusion on the characterization of the airflow, it can be noted that different grinding wheels have a similar profile of the airflow. However, there are strong differences in the characteristics of the airflow depending on the structure of the wheel surface and the porosity. It should also be noted that the airflow varies along the width of the wheel. Furthermore, the shape of the airflow strongly depends on the peripheral speed of the grinding wheel.

3.2. Analysis of the functional relationships between the coolant supply and the airflow

To analyze the functional relationships between the coolant supply and the airflow around the grinding wheel, test series were carried out with a high-speed camera. Here, for example, it was observed how the coolant jet, which hits the grinding wheel tangentially, behaves when there is an increase of the peripheral speed of the grinding wheel or the exit speed of the coolant jet.

There are image sections from the recorded videos in Figure 5. In the tests on image sections 1-3, the peripheral speed of the grinding wheel was increased from 25 to 35 m/s, and the coolant jet exit speed was kept constant at 15 m/s. It can be seen that at 35 m/s the coolant jet is deflected by the airflow so strong that it no longer hits the wheel surface tangentially and thus the coolant cannot be carried into the contact zone. At low speeds of the grinding wheel, the coolant jet is less affected and can hit the wheel tangentially. In image sections 4 and 5, at a peripheral speed of 35 m/s, the coolant jet exit speed is increased to 30 m/s and it can be clearly seen that the jet is no longer significantly influenced by the airflow, hits the wheel tangentially and is carried along by it. Such a high exit velocity is useful for breaking through the airflow. However, if this also increases the flow rate too much, the grinding process is no longer efficient and the contact zone can be heavily flooded. Aquaplaning effects are possible in this case. The adhesion of the cooling lubricant to the grinding wheel surface also decreases if the coolant jet velocities are too high.

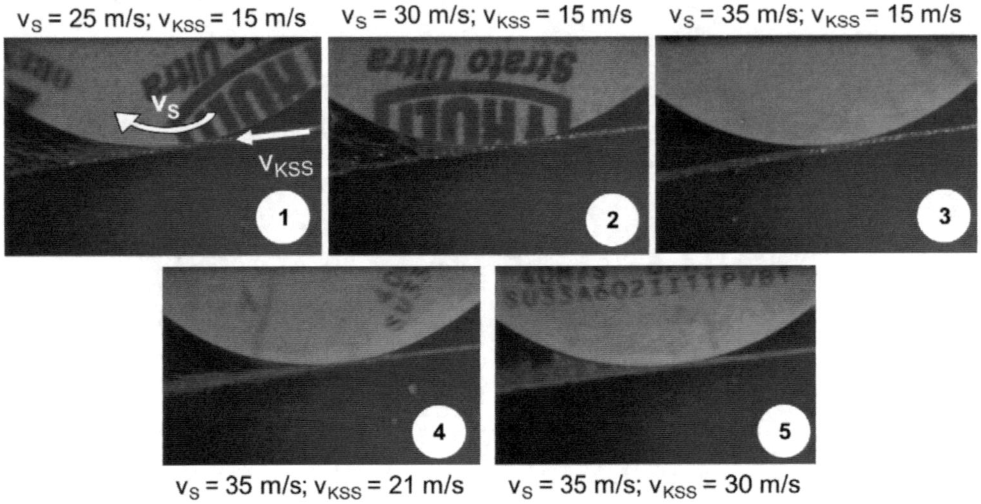

Figure 5. High-speed records to analyze the functional relationships between the coolant supply and the airflow

3.3. Investigations regarding the position for airflow deflection

The purpose of this study is to determine the position at which the airflow should be deflected and how quickly the airflow rebuilds after deflection.

First, the airflow velocity was measured in the 6 o'clock position at different positions of the air deflector plate (air shield), so that the effectiveness of the air shield position could be determined. The effectiveness of the airflow deflection describes how many percent of the airflow velocity is deflected in contrast to a setup without air shield. The air shield was placed at five different positions around the grinding wheel as close as possible to the wheel surface (Figure 6 right). With the grinding wheel rotating at peripheral speeds of 25 - 35 m/s, the effectiveness of the airflow deflection was measured and displayed without the influence of a workpiece (Figure 6 left).

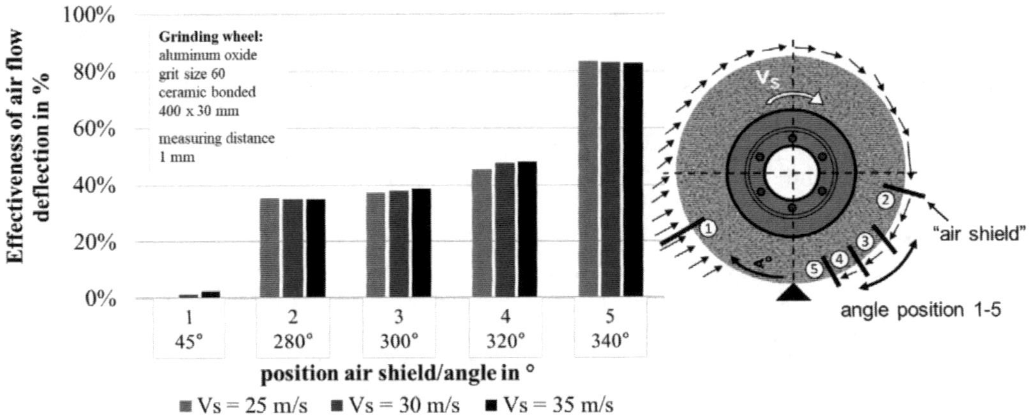

Figure 6. Effect of different positions for airflow deflection

As a result, it can be noted that the closer the deflector plate is to the measuring point or the contact zone, the more effective the air deflection. However, a useful airflow deflection can already take place at a more distant point (angular position 280°) to the contact zone at the presented velocities. If the air shield is located at this point, the airflow can already be reduced by approx. 40%. Of course, this

IOP Conf. Series: Materials Science and Engineering 1190 (2021) 012027 doi:10.1088/1757-899X/1190/1/012027

depends on the grinding wheel topography. It is also noticeable that the peripheral speed of the grinding wheel does not have a great influence on the effectiveness of the airflow deflection.

By using the high-speed camera, it was additionally possible to investigate the effect of the airflow deflection with coolant supply (Figure 7).

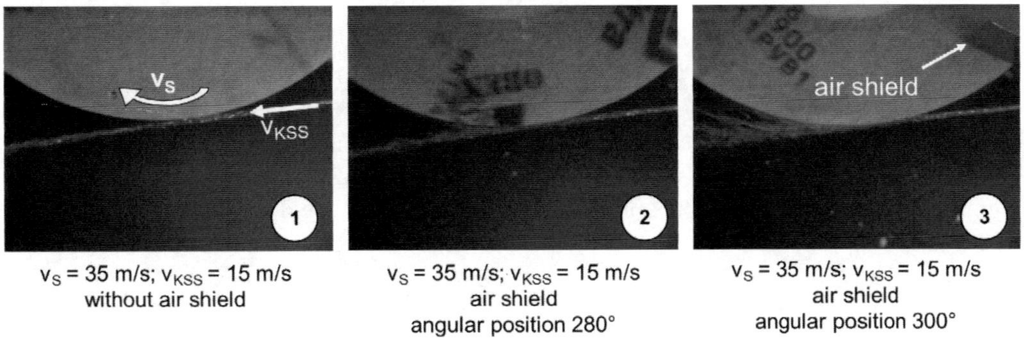

Figure 7. High-speed records for the investigation of the coolant jet influence with airflow deflection

In the left image section 1, the coolant jet is shown without the air shield at a peripheral speed of 35 m/s. The jet is deflected by airflow so much that it cannot hit the grinding wheel tangentially and misses it. In the image section 2, an air shield was placed in angular position 2 (Figure 6). The jet is deflected less and hits the grinding wheel but can't adhere to the wheel effectively. The air shield in image section 3 was attached in angular position 3 (Figure 6). Here, the jet can hit the wheel tangentially and is pulled along by the wheel. The airflow is deflected very effectively and the coolant jet is no longer strongly influenced.

In addition to the position of the shield on the grinding wheel, the position of the air shield radially to the wheel surface was investigated. This is very important for the user, as it represents the extent to which it is necessary to readjust the air shield in the event of wheel wear. The air shield was placed in angular position 4 (Figure 6) and moved away from the grinding wheel surface at different wheel peripheral speeds (Figure 8).

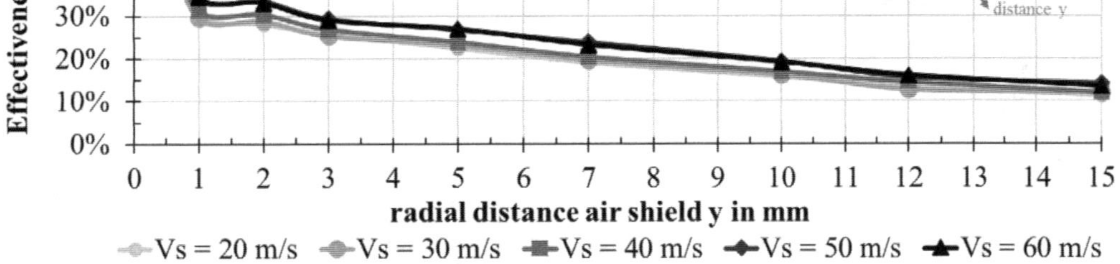

Figure 8. Effect of different radial distances of the air shield for airflow deflection

When looking at the measurement results, it is noticeable that a deflector plate causes a significant airflow deflection at a radial distance of up to 1 mm from the grinding wheel surface. Here, the airflow can be deflected by more than 50%. For a more effective deflection, the air shield must be positioned closer to the contact zone. At a distance of 3 mm and more, only a slight effect due to the air shield can be seen. The different influences of the air shield on the airflow at the various wheel peripheral speeds are almost the same in the present investigations.

4. Conclusions

The production process of grinding is particularly characterized by its high geometrical accuracy as well as high surface qualities when machining hard materials. High cutting speeds are often used for this purpose, which lead to the formation of an airflow around the grinding wheel. This airflow prevents the cooling lubricant from reaching the contact zone, therefore there is a risk of thermal damage in the boundary zone. If the airflow is not considered when dimensioning the coolant supply, process quality and reliability will be significantly affected.

The results of the research shown here allow a characterization of the airflow. It should be noted that the characteristics of the airflow depend both on the surface roughness of the grinding wheel, but especially on the cutting speed and the porosity. In addition, the flow velocity of the airflow is not constant over the width of the grinding wheel. This must be considered when dimensioning the coolant supply, because cooling lubricant should be supplied at least at the speed of the airflow. With the presented measurement setup the flow velocities of the airflow around the grinding wheel can be measured precisely. Near the surface, the airflow velocities are only approx. 50% of the peripheral speed of the grinding wheel. Closed and fine grinding wheels even have significant lower airflow velocities. Although the airflow should always be deflected as close as possible to the contact zone, larger distances can also be sufficient for the process. If an air shield is used to deflect the airflow, a readjustment in the event of grinding wheel wear is necessary after a distance of approx. 1 mm from the grinding wheel surface.

In the future, further investigations will be carried out to determine the influence of the grinding wheel topography and the grinding wheel type on the characteristics of the airflow. The aim is to provide a model for different types of grinding wheels that describes the characteristics of the airflow, so that the coolant supply can be optimized and efficiently dimensioned. The maximum capacity of the grinding wheel to absorb cooling lubricant also has an influence on this. In addition, the airflow is to be simulated with flow simulations and thus allow the dimension of the coolant supply without experimental investigations. Alternative "air shields" will also be developed and investigated, which don't require readjustment in the event of wheel wear.

Acknowledgment(s)

Parts of the shown investigations were carried out within a ZIM cooperation project, funded by the German Federal Ministry for Economic Affairs and Energy based on a decision of the German Bundestag. We would like to thank the company Grindaix GmbH for their cooperation and support in the collaborative research project "GrIntCool".

References

[1] R. Vits, "Technologische Aspekte der Kühlschmierung beim Schleifen", Dr.-Ing. Diss., Aachen, 1985.
[2] M. Meister, *Vademecum des Schleifens*. München: Hanser, pp. 384–400, 2011.
[3] S. Alenius und J. Johansson, „Airflows and Particle Distribution Around a Rotating Grinding Wheel", *Aerosol Sci. Technol.*, Bd. 25, Nr. 2, pp. 121–133, 1996.
[4] S. OKUBO, Y. MIURA, T. KASUGA, H. MORITA, T. YOSHIMI, und S. MURAKAMI, „S1302-1-3 Development of Minimized Coolant Supply Technology in Cylindrical

Grinding", *Proc. JSME Annu. Meet.*, Bd. 2010.4, pp. 255–256, 2010.

[5] F. Klocke, *Fertigungsverfahren 2: Zerspanung mit geometrisch unbestimmter Schneide*, 5th Edition Springer Vieweg, pp. 130-135, 2017.

[6] T. Beck, "Kühlschmierstoffeinsatz beim Schleifen mit CBN". Dr.-Ing. Diss., Aachen: Shaker, pp. 50-77, 2002.

[7] T. Tawakoli, Seminar Moderne Schleiftechnologie, und Fachhochschule Furtwangen, Hrsg., *Moderne Schleiftechnologie: neue Entwicklungen und zukünftige Trends aus der Praxis und Forschung*. Villingen-Schwenningen, 1998.

Improving the Accuracy of Deep Drawn Fiber-Metal Laminate Parts by Preliminary Surface Treatment

Dietrich Voswinkel [1], Hüseyin Sapli [2], Dennis Kloidt 1, Thomas Heggemann[2], Werner Homberg [2], Olexandr Grydin [1], Mirko Schaper [1]

[1] Lehrstuhl für Werkstoffkunde (Chair of Materials Science), Paderborn University, Warburger Straße 100, 33098 Paderborn, Germany
[2] Umformende und Spanende Fertigungstechnik (Chair of Forming and Machining Technology), Paderborn University, Warburger Straße 100, 33098 Paderborn, Germany

voswinkel@lwk.upb.de

Abstract. In the present study, fiber-metal laminates (FMLs) were formed by deep drawing after the metallic component was pretreated by laser structuring, sandblasting or anodizing in order to reduce the height of process-related wrinkles. Such surface texturing is commonly performed to improve the bond strength between the adhesively joined components of the FML by selectively increasing the surface area. This results in improved stability of the finished FML part due to increased resistance to delamination. First, the adhesion properties induced by each process were quantified using shear tensile tests. The shear strength was increased for all processes compared to the unstructured surface. The surface structures were characterized by roughness measurements and scanning electron micrographs. The surface properties adjusted by the different processes also contributed to the part accuracy of the deep-drawn FMLs. During forming, the matrix material contained in the unidirectional fiber composite plastic tends to build up against the forming direction, resulting in wrinkling between the flange and the cup shell. The increased adhesive strength of the metallic component impedes the flow of the matrix material and decreases the height of the resulting wrinkles. The height of the wrinkles was determined using a coordinate measuring machine on the formed FML cups.

1. Introduction

The climate crisis is forcing automobile, aircraft and ship builders to make a major change in the use of materials. Lightweight materials are necessary to reduce energy consumption and thus minimize CO_2 emissions. The use of fiber-metal laminates (FML) is one promising way to meet the stringent requirements, as they combine advantageous properties such as the ductility of steel and the high stiffness of carbon fiber reinforced plastics (CFRP) into high-performance hybrid materials. [1, 2]

The most common way of joining FML components together is adhesive joining. This technique ensures a two-dimensional force transmission as well as the galvanic isolation of the components to avoid contact corrosion. [3, 4] The surfaces should always be sufficiently pretreated beforehand. Prerequisite is a grease-free surface, which can be achieved by simple cleaning, e.g., with acetone. In addition, the surface can be enlarged or roughened to achieve a mechanical bond between the cured adhesive and the substrate. [5] Improved adhesion between fiber composite plastic and metal significantly enhances the strength properties of an FML component [6]. Various methods are available for this purpose. A very well-known process is sandblasting, in which compressed air is used to blast the abrasive at high speed onto the surface, thus deforming it. The resulting undercuts in the surface structure create spaces into which the adhesive can flow and cure during the joining process. This produces a mechanical bond that ensures good adhesion properties. [7] Another method for surface pretreatment of metals is laser structuring with a pulsed laser source, which has both a cleaning effect and can enlarge the surface microscopically and nanoscopically. While the microscopic surface expansion originates from the craters that form during the resolidification of the material, the nanoscopic surface expansion results from particles that are ejected from the molten pool and then land back on the surface and solidify. [8, 9] Another process is anodizing, an electrolytic pretreatment to create nanoporous oxide layers with excellent wetting properties, making it a very good basis for adhesive joining processes. [10, 11]

It is possible to further process fiber-metal laminates by forming operations such as deep drawing. This opens up a wide range of potential applications, including the production of highly rigid car body components. [4, 12, 13] During forming, the matrix material contained in the unidirectional fiber composite tends to build up against the forming direction, resulting in wrinkles between the flange and the cup shell. [12] The increased bond strength between the metallic component and the fiber composite plastic impedes the flow of the matrix material and could reduce the height of the resulting wrinkles. This would lead to improved manufacturing accuracy of FML components, another positive effect in addition to the good strength properties. The aim of this paper is to investigate the influence of surface structuring of zinc coated steel sheets by means of sandblasting, laser structuring and anodizing on the adhesion properties of the fiber-metal laminate and the resulting manufacturing properties such as the accuracy of deep drawn FML cups. The process chain is presented below.

2. Experimental setup

2.1. Materials

The metallic component subjected to the various surface pretreatments was CR290Y490T-DP-EG [14], a zinc coated dual-phase steel. The material thickness was 0.55 mm. This material is mainly used in the automotive industry due to its excellent lightweight properties. The second component of the FML was a 0.3 mm thick unidirectional carbon fiber reinforced polymer (CFRP) prepreg called SIGRAPREG C U230-0-/NF-E320 [15]. Furthermore, the film adhesive HexBond™ 677 [16], which was specially developed for joining metals with fiber composites, was chosen to produce the shear tensile specimens. No adhesive was used in the combined forming process as the components were intrinsically joined.

2.2. Surface pretreatment

The steel sheets, the metallic component in the FML, were pretreated with three different processes. First, the steel sheets were cut into two different shapes. While 20 sheet metal strips measuring 100 x 22 mm^2 were cut on a plate shear for the shear tensile tests, 40 circular sheet metal blanks with a diameter of 180 mm were laser cut for the deep drawing tests. The sheets were then pretreated with the appropriate processes, the sheet strips being textured on both sides and the circular blanks on one side.

The sandblasting was carried out with a compressed air unit made by MHG of the SMG 160 S/D type and a corundum with a grain size of F220 as the abrasive. A blasting pressure of 3.5 bar was employed with an angle of 70 - 80° to the surface. The laser structuring was performed with a Q-switched pulsed YVO4 laser (MD-X1520C) and the following parameters: pulse frequency f = 25 kHz, scan speed v = 1500 mm/s, output power Pa = 25 W, line spacing L = 0.08 mm. The direction of the laser patterning on the circular sheet metal blanks is shown in Figure 1. In contrast to laser-structured surfaces, sandblasted and anodized surfaces have direction-independent properties, which is why a schematic representation is not necessary.

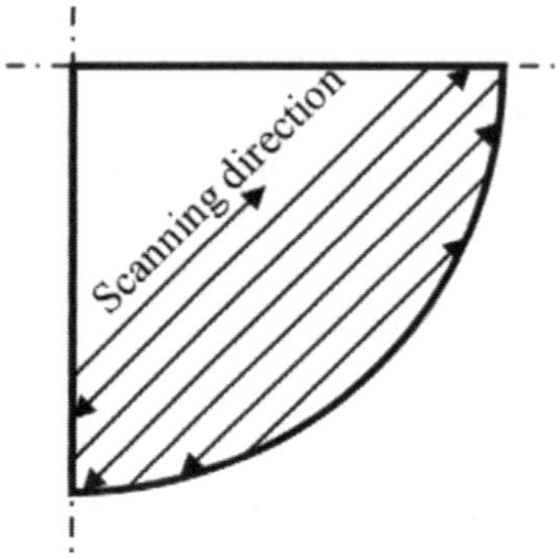

Figure 1: Scanning direction of the laser on the circular sheet metal blanks

The anodizing was carried out with a self-developed system for anodizing zinc coated steel sheets up to a size of 640 x 380 mm², which is shown in Figure 2. The tank of the equipment was filled with an electrolyte in which the sheet to be anodized was immersed before the current source was switched on for a defined time and current. A water solution of potassium hydroxide (KOH) was used as the electrolyte for a substance quantity of 1 Mol. The current density was 2.5 A/dm² and the anodizing time 1 min.

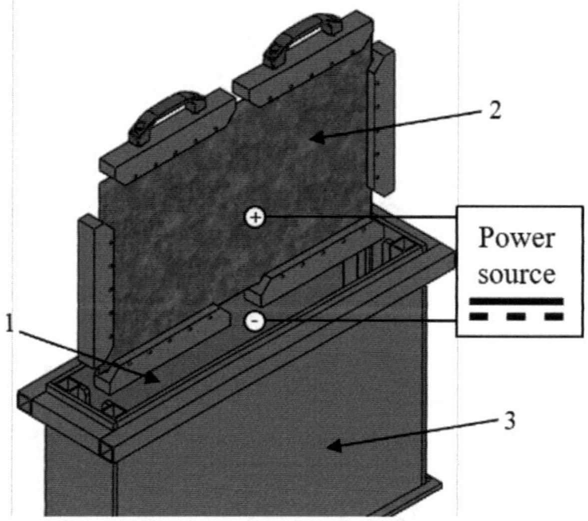

Figure 2: Schematic diagram of the anodizing unit for the galvanized steel sheets 1: lead cathode, 2: steel sheet, 3: tank for electrolyte

2.3. Surface characterization

The surface roughness values Ra and Rz of the differently pretreated surfaces were determined with the HOMMEL-ETAMIC T8000 tactile roughness tester, each surface being measured five times. The arithmetic mean was then calculated from the results. Subsequently, the generated surface morphology was examined in a Zeiss Ultra Plus scanning electron microscope (SEM). The images were taken with a secondary electron detector (SE2) at an accelerating voltage of 5 kV and a working distance of 3.6 - 3.7 mm.

2.4. Lap shear tests

For the shear tensile tests, first the double overlapped shear tensile specimens were fabricated. The already cut out sheet metal strips were each cured with two CFRP strips with the adhesive in a suitable fixture at 150°C and a joining pressure of 350 kPa for 4 min. These were the ideal adhesive curing conditions, as specified by the manufacturer. The overlap length of the joining zone was 6.5 mm. The arrangement of the components shown in Figure 3 reduces bending during the shear tensile tests.

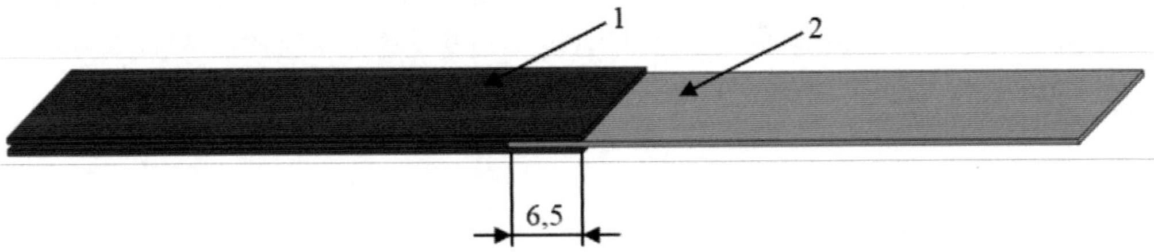

Figure 3: Double overlapped shear tensile specimen 1: CFRP strips, 2: steel strip

Subsequently, the shear tensile tests were performed on the MTS Criterion® electromechanical universal testing machine at a test speed of 1 mm/min. The shear strength was obtained by measuring the maximum force at failure of the bonded joint and dividing it by the bond area. Five shear tensile tests were performed for each surface condition, i.e. anodized, sandblasted, laser textured and as-received.

2.5. Deep drawing of FML

As can be seen in Figure 4, the active die elements for the combined forming and curing process consisted of a die, a blankholder and a punch, into which heating cartridges were inserted. The punch is cylindrical and has a diameter of d_p = 99 mm and an edge radius of r_p = 10 mm. The cylindrical die has an inner diameter of d_D = 105 mm and a drawing edge radius of r_D =10 mm. The drawing depth used in the tests was h = 37 mm. In addition, temperature-stable fluoropolymer foils (s_0 = 12 μm) were applied, which were inserted between the die elements and the semi-finished product. This film served to prevent contamination of the die elements by escaping epoxy resin and to minimize friction between the tool surface and the FML blank.

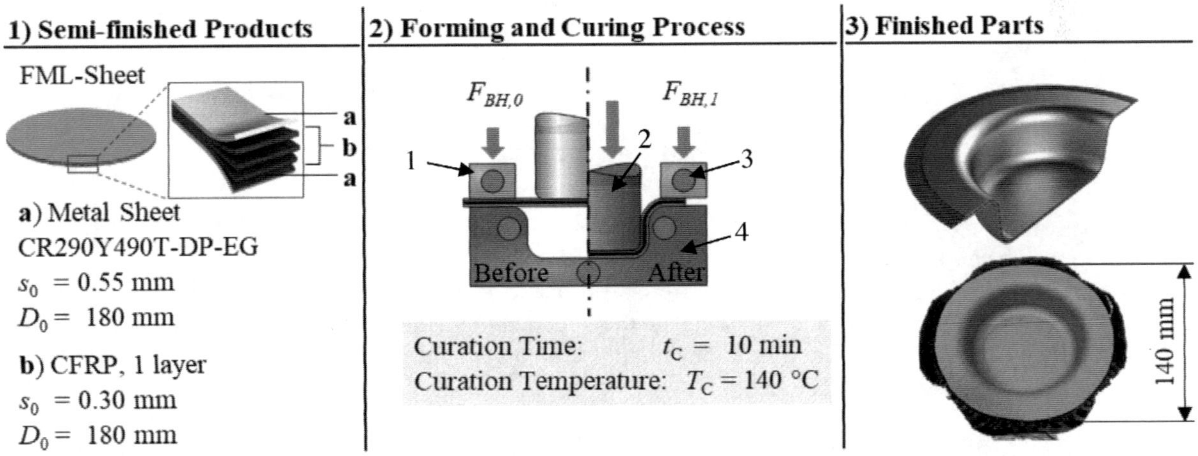

Figure 4: Schematic image of the semi-finished product and the combined forming and curing process, 1: blank holder, 2: punch, 3: heating cartridges, 4: die

The die elements were placed in a high-precision column guide frame to keep the blank holder and die ring parallel. Then, the entire assembly was placed in a hydraulic press. After the active die elements were tempered to the curing temperature of T_C = 140 °C, the FML blanks were centered within the die assembly and the thermoforming process was started. The blankholder applied normal stresses to the FML blank, thus suppressing wrinkling of the flange area caused by tangential compressive stresses during deep drawing of the blank. The blankholder force $F_{BH,1}$ was 260 kN, the force F_P exerted by the punch was 500 kN. After the forming process, the deep-drawn FML blank remained in the closed mold for a curing time of t_C = 10 min. These curing conditions for the CFRP prepreg were based on previous studies as they result in optimum curing of the epoxy resin of the matrix material and thus maximum strength of the prepregs.

2.6. Measurement of the wrinkle height

The height h_{wr} of the wrinkles formed during deep drawing above the cup height z_{md} (see Figure 5) was measured with the Zeiss Contura G2 7/7/6 RDS coordinate measuring machine.

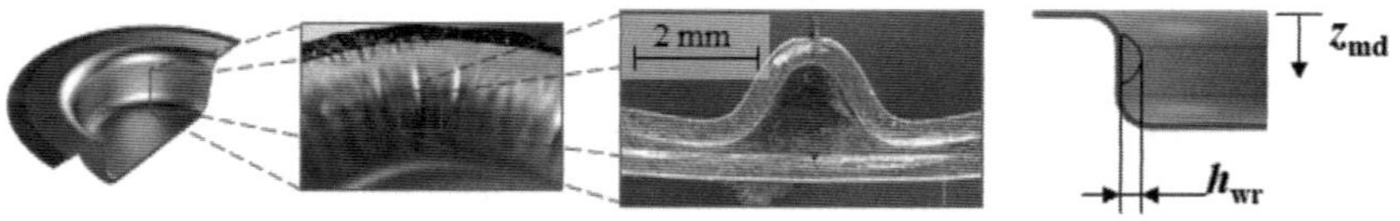

Figure 5: Wrinkle height z_{md} and measuring depth h_{wr}

The measurement is carried out over six measuring depths z_{md}, in which the probe records 2000 measuring points on a circular path in each case. The measurement process is illustrated in Figure 6, the measuring height z_{md} refers to the flange underside of the cup.

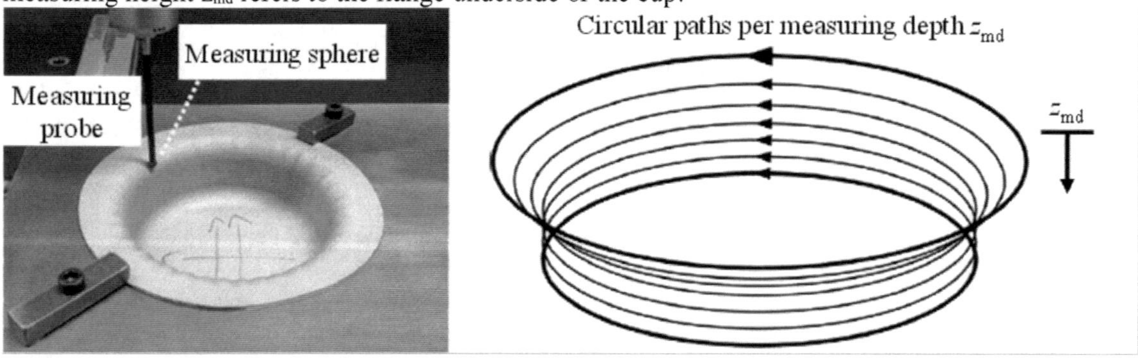

Figure 6: Measurement of the wrinkle height

3. Results

3.1. Surface morphology

Figure 7 shows the SEM images of the different surface textures, each with two different magnifications. When comparing the anodized surface (Figure 7 b_1) with the initial state (Figure 7 a_1), only a slight difference can be seen, as the bright elevations visible in the initial state were slightly reduced by the anodizing process. At higher magnification, however, there is no longer any similarity between the surfaces. The zinc flakes shown in Figure 7 a_2 have completely transformed into porous zinc oxide. Furthermore, the sandblasted surface (Figure 7 c_1 and c_2) indicates a very strong plastic deformation, as deep indentations and undercuts are visible. Figure 7 d_1 shows the zinc surface melted and resolidified by laser structuring. Higher magnification demonstrates that particles have settled in the craters.

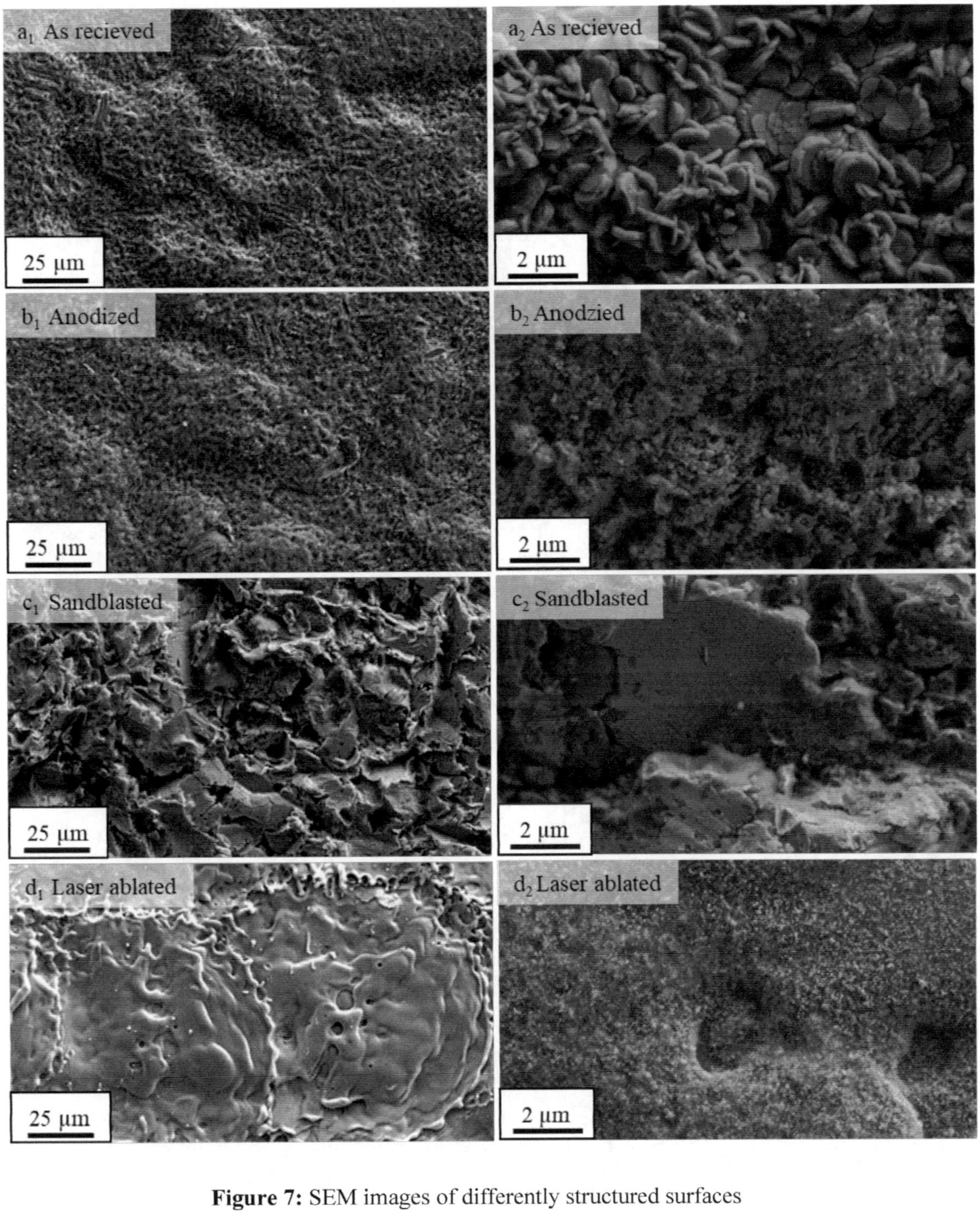

Figure 7: SEM images of differently structured surfaces

3.2. Surface roughness

Figure 8 shows the surface roughness Ra and Rz as a function of the surface texturing. It can be seen that the highest roughness is achieved by sandblasting, with Ra increasing by approx. 30%, and Rz by 128% compared to the as recieved state. In contrast, anodizing reduced both values by 37% and 20%, respectively. Laser structuring did not significantly change the Ra value, the Rz value increased by 18%.

IOP Conf. Series: Materials Science and Engineering 1190 (2021) 012028 doi:10.1088/1757-899X/1190/1/012028

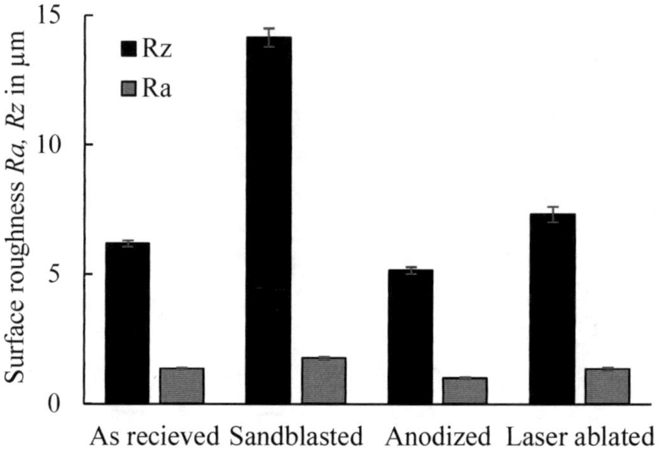

Figure 8: Surface roughness referred to differently structured surfaces

3.3. Shear strength

The shear strength was improved by all three surface pretreatments compared to the as-received condition, as shown in Figure 9. Sandblasting increased the shear strength by 20%, anodizing by 16% and laser structuring by 14%. The standard deviation for the as recieved shear tensile tests is the highest compared to the textured metal surface shear tensile tests. Sandblasting produced the lowest standard deviation of results, followed by anodizing and laser structuring.

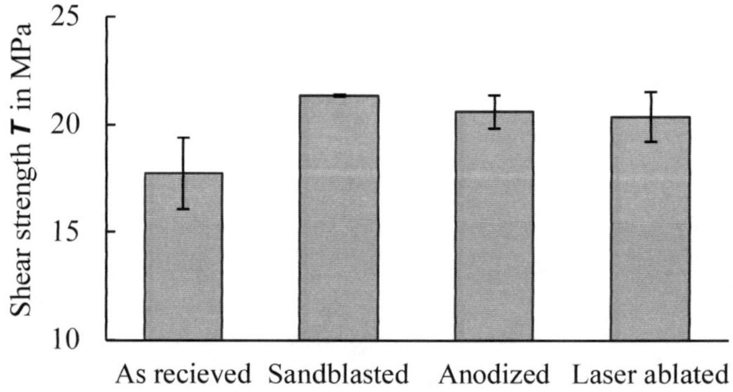

Figure 9: Shear strength referred to differently structured surfaces

3.4. Wrinkle height

In terms of wrinkle height, the surface textures had different effects, which can be seen in Figure 10. An anodized surface, for instance, had a negative effect on the flow behavior of the epoxy resin. While wrinkles with a height of $h_{wr} = 0.52$ mm can be observed at a measuring depth of $z_{md} = 2$ mm, they decrease steadily to $h_{wr} = 0.16$ mm at $z_{md} = 17$ mm, increase again and then decrease to a value of $h_{wr} = 0.12$ mm at $z_{md} = 27$ mm. In contrast, treatment of the sheet blanks by sandblasting and laser structuring resulted in lower wrinkle formation. Laser structuring yielded wrinkles with a height of approx. $h_{wr} = 0.40$ mm at a measuring depth of $z_{md} = 2$ mm, which decrease to a value of $h_{wr} = 0.09$ mm at $z_{md} = 17$ mm. The further course is almost analogous to the reference values. Sandblasting had the greatest positive influence of the three methods used. On these test specimens, wrinkles with a height of approx. $h_{wr} = 0.41$ mm were detected $z_{md} = 2$ mm, which, however, decrease with the largest drop to a value of h_{wr} = 0.04 mm at z_{md} = 17 mm. Although this drop also occurs in the sandblasted specimens, it is much shallower, as at a measuring depth of z_{md} = 22 mm, the wrinkles had a height of h_{wr} = 0.08 mm.

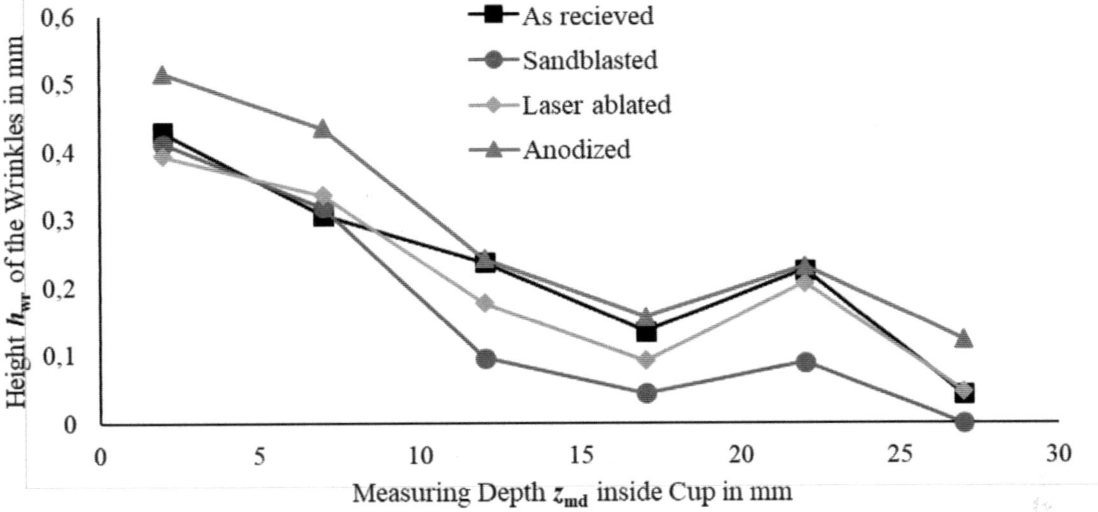

Figure 10: Height of the wrinkles h_{wr} above the measurement depth z_{md}

4. Discussion

After anodizing, the surface roughness decreased as the topography of the starting material was slightly smoothed. The nanoporosity induced by the formation of porous zinc oxide could not be measured by measurements with the tactile roughness tester. However, the adhesion properties in the shear tensile test were improved compared to the as-received condition. This is probably due to the additional mechanical interlock between the steel substrate and the adhesive, as the adhesive flows into the pores during joining and cures there, which was already observed in an earlier study by Engelkemeier et al. [10]. Furthermore, sandblasting and laser structuring also led to a significant improvement of the adhesive properties. While the highly deformed sandblasted surface is ideal for creating a mechanical bond between the adhesive and the substrate, laser structuring results in surface enlargement at both the macroscopic and nanoscopic levels. A combination of the two methods has a positive effect on the adhesive properties as well. Although the results of the shear tensile tests with double or single overlapped shear tensile specimens should be taken with caution, since they do not represent a realistic load case of FML components, they are suitable for a qualitative statement on whether the surface pretreatment adds value compared to the untreated surface. [17]

Despite the improved adhesion properties of the anodized zinc coated steel sheets, wrinkling during deep drawing of the FML parts could not be reduced. In the range from z_{md} = 2 mm to z_{md} = 7 mm it even increased in comparison with the as-received condition.

However, it may be worthwhile to look for correlations between the surface roughness values and the wrinkle height, as sandblasting yielded both the roughest surface and the best wrinkle height, and laser structuring slightly increased the roughness value while decreasing the wrinkle height. It appears that macroscopic surface enlargement, as occurs with sandblasting and to some degree with laser patterning, is beneficial in preventing wrinkling due to accumulation of the matrix material. In contrast, nanoscopic surface enlargement, as in anodizing and to some extent in laser structuring, is ineffective for improving the accuracy of deep-drawn FML parts. Furthermore, it should be considered that especially with laser ablation, the parameterization is extremely complex and at the same time has a great influence on the resulting surface properties. Thus, it has not been conclusively determined that laser structuring could not achieve similar results to sandblasting. For this purpose, further parameter studies should be carried out with regard to laser structuring.

5. Conclusions

In this study, the influence of the surface pretreatments sandblasting, anodizing and laser structuring on the adhesion properties between components in FML parts was investigated. Subsequently, the components were formed to test whether the surface pretreatment also has a positive effect on the manufacturing accuracy.

- Surface roughness is greatly increased by sandblasting and slightly increased by laser structuring. It is slightly reduced by anodizing.
- The three processes significantly improve the adhesion properties. With the parameters used in this study, the best results were achieved by sandblasting.
- The development of the wrinkle height is analogous to the surface roughness. The lowest wrinkle heights were measured after sandblasting, while laser structuring only slightly reduced the wrinkle height and anodizing slightly increased it.
- A direct correlation between surface roughness, i.e., macroscopic surface morphology, and wrinkle obstruction is possible and should be further investigated.

Acknowledgment(s)

The authors gratefully acknowledge the financial support from the European Regional Development Fund (ERDF) and the state of North Rhine-Westphalia (NRW) within the project HyOpt.

References

[1] Kavitha K, Vijayan R and Sathishkumar T 2020 Fibre-metal laminates: A review of reinforcement and formability characteristics *Materials Today: Proceedings* **22** 601–5

[2] Sinmazçelik T, Avcu E, Bora M Ö and Çoban O 2011 A review: Fibre metal laminates, background, bonding types and applied test methods *Materials & Design* **32** 3671–85

[3] Ramakrishnan K R, Sarlin E, Kanerva M and Hokka M 2021 Experimental study of adhesively bonded natural fibre composite – steel hybrid laminates *Composites Part C: Open Access* **5** 100157

[4] Zinn C, Bobbert M, Dammann C, Wang Z, Tröster T, Mahnken R, Meschut G and Schaper M 2018 Shear strength and failure behaviour of laser nano-structured and conventionally pre-treated interfaces in intrinsically manufactured CFRP-steel hybrids *Composites Part B: Engineering* **151** 173–85

[5] Rotella G, Alfano M, Schiefer T and Jansen I 2015 Enhancement of static strength and long term durability of steel/epoxy joints through a fiber laser surface pre-treatment *International Journal of Adhesion and Adhesives* **63** 87–95

[6] Voswinkel D, Kloidt D, Grydin O and Schaper M 2021 Time efficient laser modification of steel surfaces for advanced bonding in hybrid materials *Prod. Eng. Res. Devel.* **15** 263–70

[7] Bonpain B and Stommel M 2018 Influence of surface roughness on the shear strength of direct injection molded plastic-aluminum hybrid-parts *International Journal of Adhesion and Adhesives* **82** 290–8

[8] Min J, Wan H, Carlson B E, Lin J and Sun C 2020 Application of laser ablation in adhesive bonding of metallic materials: A review *Optics & Laser Technology* **128** 106188

[9] Wan H, Min J, Lin J, Carlson B E, Maddela S and Sun C 2021 Effect of laser spot overlap ratio on surface characteristics and adhesive bonding strength of an Al alloy processed by nanosecond pulsed laser *Journal of Manufacturing Processes* **62** 555–65

[10] Engelkemeier K, Mücke C, Hoyer K P and Schaper M 2019 Anodizing of electrolytically galvanized steel surfaces for improved interface properties in fiber metal laminates *Adv Compos Hybrid Mater* **2** 189–99

IOP Conf. Series: Materials Science and Engineering 1190 (2021) 012028 doi:10.1088/1757-899X/1190/1/012028

[11] Engelkemeier K, Sun A, Voswinkel D, Grydin O, Schaper M and Bremser W 2021 Zinc Anodizing: Structural Diversity of Anodic Zinc Oxide Controlled by the Type of Electrolyte *ChemElectroChem* **62** 265

[12] Heggemann T, Homberg W and Sapli H 2020 Combined Curing and Forming of Fiber Metal Laminates *Procedia Manufacturing* **47** 36–42

[13] 2018_Camberg-Engelkemeier_LightweightDesign_Top-down design of tailored fiber-metal laminates

[14] thyssenkrupp steel *DP-W® und DP-K®Produktinformation für Dualphasen-Stähle* https://www.thyssenkrupp-steel.com/media/content_1/publikationen/produktinformationen/dp_w_dp_k/thyssenkrupp_dp-w_dp-k_produktinformation_steel_de.pdf

[15] sgl carbon *SIGRAPREG® Pre-Impre-gnated Materials* https://www.sglcarbon.com/en/markets-solutions/material/sigrapreg-pre-impregnated-materials/

[16] HEXCEL *HexBond™ 677Modified epoxy film adhesive* https://www.hexcel.com/user_area/content_media/raw/HexBond_677_DataSheet.pdf

IOP Conf. Series: Materials Science and Engineering 1190 (2021) 012029 doi:10.1088/1757-899X/1190/1/012029

Evaluating Coordinated Cooperative Control of Three Active Car Systems Using Fuzzy-Logic

Vase Jordanoska [1], Darko Danev [1], Vasko Changoski [1]

[1] University "Ss. Cyril and Methodius", Faculty of Mechanical Engineering, st. Rugjer Boshkovikj no. 18, Skopje, Republic of Macedonia

vase.jordanoska@mf.edu.mk

Abstract. Vehicle systems automation is an inevitable process as a result of the growing safety requirements and the trend of increasing car operation autonomy by reducing driver's influence. The paper covers the coordinated control operation of three active systems to assess the improvement of vehicle response. Direct yaw control through selective wheel braking (DYC), active front steering control (AFS) and active suspension normal force control (NFC) are included. The control method is fuzzy logic. The modelling is done in Matlab/Simulink, where 14-DOF nonlinear full vehicle model and 3-DOF reference model were introduced. Simulations were done for three manoeuvres: cornering event, single lane change and fishhook manoeuvre, at a speed of 130 km/h and dry road surface with adhesion coefficient of 0.9 was assumed. Coordination was done with adjusting scaling factors for each control. Results show improvements in the transient response behaviour of the vehicle compared with four vehicles: vehicle without any active system, vehicle with coordinated DYC and AFS, vehicle with coordinated DYC and NFC and vehicle with coordinated AFS and NFC. The biggest impact can be seen in results for sideslip angle and roll angle, coordinated cooperative control here gives lowest values in the transient period during the manoeuvre. Although it gives better results for the other variables as well. The goal of the coordinated cooperative control is fulfilled by improving the stability and handling of the vehicle, which helps the driver to maintain the feeling of manoeuvrability and the sense for the direction of movement of the vehicle. The system for direct yaw control through selective wheel braking has major impact but the coordination with the other two systems lowers the braking force per wheel. The system for active front steering control has lower impact but at lower speeds helps for maintaining better direction of vehicle movement. Considering these three active systems act simultaneously during high speeds and critical manoeuvres next step would be to define sequence of activation depending on the currently measured vehicle behaviour variables.

1. Introduction

Safety is still the main purpose in the research area of motor vehicles, but also automotive technology is going into direction of making vehicles autonomous. Steps between include automated vehicle's systems. Automation of vehicle systems lowers the human (driver) influence in order to increase safety and stability of the vehicle, considering the human effect a main purpose for majority of car accidents [1]. Vehicle's system automation starts with main vehicle systems which enable and support its drivability: steering, braking, suspension, but also automation of powertrain subsystems. Integrated control system of the front steering angle compensation and the braking force distribution with the application of model-matching control technique based on optimal control theory is proposed in [2].

Authors also worked on improved behavior under severe driving conditions with active control of either front or rear angle combined with yaw moment in [3]. Integrated vehicle chassis control is proposed in [4] to coordinate steering, braking/traction and active stabilizer via loop structure. The main-loop calculates the stabilizing force using sliding mode control, while the servo-loop optimally distributes force to each tire, using SQP approach. Three different cases of combining lateral force and direct yaw moment control have been investigated in [5]. A simulation of a closed-loop driver–vehicle system during quick lane change with braking is used to show significant influence of the combined control on vehicle stability and responsiveness. Coordinated control of AFS and DYC is presented in [6] based on optimal guaranteed cost theory. A rollover prevention controller for vehicles with a high center of gravity is proposed in [7]. Active suspension with lateral acceleration as a disturbance was designed and controller gains were obtained through the LQ SOF method. And ESP was designed to maintain the maneuverability of the controlled vehicle. A two-layer hierarchical control architecture is proposed in [8] for integrated control of ASS and ESP where the upper-layer is designed to coordinate the interactions between the ASS and the ESP. Coordination of active front steering and rear braking in a driver-assist system for vehicle yaw control is investigated in [9]. The coordination is achieved through a suitable gain scheduled Linear Parameter Varying controller. A combined AFS and DYC controller is presented in [10] with good robustness against in-vehicle network-induced time-varying delays for 4 wheel-independent-drive electric vehicles. An H∞-based LQR tracking controller is introduced and adopted in the control synthesis. A review study on integrated active steering and braking control for vehicle yaw stability system is conducted in [11]. Gain-scheduled vehicle handling stability control via integration of active front steering and suspension systems is presented in [12]. Model and controller for an active front-wheel steering (AFS) system is proposed in [13] which are designed to be investigated in integrated dynamics control systems. Article [14] introduces the β phase plane method to determine the stability state of the vehicle and electronic stability program fuzzy controller to improve the stability of vehicle driving on a low adhesion surface at high speed. An advanced control system is designed in [15] which integrate three fuzzy logic controllers including Direct Yaw-moment Control (DYC), Active Roll-moment Control (ARC) and Active Front Steering (AFS) to enhance vehicle cornering and overturning stability. In this paper step forward is made from [15] adding one more active system in order to improve vehicle lateral dynamics and rollover stability. This paper uses fuzzy-logic control for coordinated cooperative control of three active systems. Developed control algorithm includes electronic stability control system, active front steering and front active suspension system.

2. Vehicle Dynamic Model

The mathematical model used for representation of the vehicle is based on a 14-DOF full vehicle model, developed by [16] and used by [17] (Figure 1). The 14-DOF nonlinear model is implemented in Matlab/Simulink to describe the dynamic behavior of the vehicle. Vehicle mass is taken as a lumped mass at the center of gravity and has six degrees of freedom, including linear and angular movements along X, Y and Z axis. It represents the sprung mass of the vehicle. There are another four lumped masses of wheels i.e. unsprung mass, each one having two degrees of freedom: vertical travel and wheel rotation. The 14-DOF vehicle model is sufficient to study the dynamic behavior of the vehicle in longitudinal, lateral, and vertical direction. Assumptions include that vehicle body is being modelled as rigid, steer angles of the outer and inner wheel are same, tires have contact with the ground all the time and the aerodynamic force is neglected.

A spring and a damper at each corner connect sprung mass to four unsprung masses. Thus, the vertical motion of sprung mass is given by:

$$m_{pm} \cdot \ddot{z}_{pm} = F_{pr\_pl} + F_{am\_pl} + F_{pr\_pd} + F_{am\_pd} + F_{pr\_zl} + F_{am\_zl} + F_{pr\_zd} + F_{am\_zd} \quad (1)$$

Equation of sprung mass pitch motion:

$$I_y \cdot \dot{\omega}_y = \left(F_{pr\_zl} + F_{am\_zl} + F_{pr\_zd} + F_{am\_zd}\right) \cdot l_z - \left(F_{pr\_pl} + F_{am\_pl} + F_{pr\_pd} + F_{am\_pd}\right) \cdot l_p \quad (2)$$

Equation of sprung mass roll motion:

$$I_x \cdot \dot{\omega}_x = \left(F_{pr\_pl} + F_{am\_pl} + F_{pr\_zl} + F_{am\_zl}\right) \cdot \frac{b}{2} - \left(F_{pr\_pd} + F_{am\_pd} + F_{pr\_zd} + F_{am\_zd}\right) \cdot \frac{b}{2} \quad (3)$$

Equation of vertical motion of each wheel:

$$m_{nm\_ij} \cdot \ddot{z}_{nm\_ij} = F_{t\_ij} - F_{pr\_ij} - F_{am\_ij} \quad (4)$$

where index ij stands for the wheel position (i - front, rear; j - left, right).

The handling model has longitudinal, lateral and yaw motion of the sprung mass and wheel spin motion at each of the four wheels. The equation of longitudinal motion according to Newton's second law is given by:

$$m \cdot a_x = F_{xpl} \cos\theta_p - F_{ypl} \sin\theta_p + F_{xpd} \cos\theta_p - F_{ypd} \sin\theta_p + F_{xzl} + F_{xzd} \quad (5)$$

Analogously to the (5), lateral motion is given by:

$$m \cdot a_y = F_{ypl} \cos\theta_p + F_{xpl} \sin\theta_p + F_{ypd} \cos\theta_p + F_{xpd} \sin\theta_p + F_{yzl} + F_{yzd} \quad (6)$$

The sum of yaw moment is given with:

$$I_z \cdot \dot{\omega}_z = -\frac{b}{2} F_{xpl} \cos\theta_p + \frac{b}{2} F_{xpd} \cos\theta_p - \frac{b}{2} F_{xzl} + \frac{b}{2} F_{xzd} + \frac{b}{2} F_{ypl} \sin\theta_p - \frac{b}{2} F_{ypd} \sin\theta_p + l_f F_{xpl} \sin\theta_p + l_f F_{ypl} \cos\theta_p + l_f F_{xpd} \sin\theta_p + l_f F_{ypd} \cos\theta_p - l_z F_{yzl} - l_z F_{yzd} + M_{zpl} + M_{zpd} + M_{zzl} + M_{zzd} \quad (7)$$

The equation of motion for each wheel rotation is:

$$I_t \cdot \dot{\omega}_{ij} = T_{v_{ij}} - T_{k_{ij}} - F_{xij} \cdot r_t \quad (8)$$

Geometry and mass parameters used in the vehicle model correspond to vehicle with higher centre of gravity, such as sport utility vehicles or crossover vehicles.

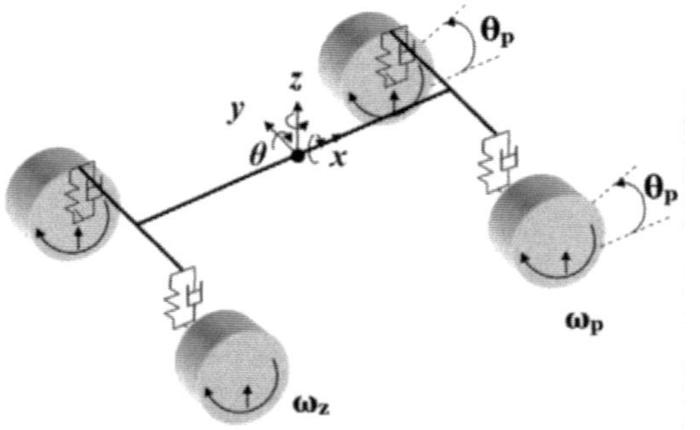

Figure 1. 14 DOF vehicle model

2.1. Tire Model

The longitudinal and lateral tire forces are modelled using Pacejka's magic formula tire model. Pacejka's model is implemented in Matlab/Simulink to obtain the nonlinear behavior of tires [18, 19, 20].

2.2. Tire Model

A reference linear model with three degrees of freedom is adopted in order to obtain the desired values in relation to which the control algorithm makes calculations for corrective actions. The three degrees of freedom are lateral acceleration, yaw velocity and roll angle. The reference model calculates the desired: yaw velocity (ω_{z_pos}), side-slip angle (β_{c_pos}) and roll angle (ϕ_{pos}) according to the change in steering angle and vehicle speed, given by 9 to 11 [21].

$$\omega_{z\_pos} = \frac{v_x}{l+\left(\frac{m\cdot(l_z\cdot K_{\delta z}-l_p\cdot K_{\delta p})}{2\cdot K_{\delta p}\cdot K_{\delta z}\cdot l}\right)\cdot v_x^2}\cdot\theta_p \tag{9}$$

$$\beta_{c\_pos} = \frac{l_z-\left(\frac{l_p\cdot m}{2\cdot K_{\delta z}\cdot l}\right)\cdot v_x^2}{l+\left(\frac{m\cdot(l_z\cdot K_{\delta z}-l_p\cdot K_{\delta p})}{2\cdot K_{\delta p}\cdot K_{\delta z}\cdot l}\right)\cdot v_x^2}\cdot\theta_p \tag{10}$$

$$\phi_{pos} = 0 \tag{11}$$

3. Control Design

Fuzzy-logic is a method of nonlinear control and is effective in manipulating the uncertainty and nonlinearity associated with complex system control. According to [22] fuzzy-logic is a proven effective tool for control of vehicle dynamics. By adjusting the membership functions and linguistic rules through testing a real vehicle, best combination of control sets can be obtained. It has been concluded that Gaussian and triangular membership functions give approximately equally well results and generally better than other types of membership functions.

3.1. Modeling of Electronic Stability Control

Control actions are performed by calculating yaw moment T_{yaw} and roll moment T_ϕ according to (12) and (13), where F_{kpi} и F_{kzi} denote the braking force of the front and rear wheel, respectively.

$$T_{yaw} = F_{kpi}\cdot\frac{b_p}{2} + F_{kzi}\cdot\frac{b_z}{2} \tag{12}$$

$$T_\phi = T_{\phi ff} + T_{\phi fb} \tag{13}$$

The principle of brake distribution is to brake the front and rear wheel on the same side of the vehicle. According to the steering wheel angle and the difference between the actual and the reference yaw velocity, the direction of the compensation moment can be determined. Thus can be determined which wheels should be braked. The principle for defining which wheels need to be braked is given in Table 1 [14].

Table 1. Principe of choosing braking wheels

Reference yaw velocity ω_{z_ref}	Yaw velocity deviation $(\omega_z - \omega_{z\_ref})$	Characteristic of steering	Braked wheels
$\omega_{z\_ref}>0$ (left turn)	$(\omega_z - \omega_{z\_ref})\geq 0$	oversteer	front and rear right side
	$(\omega_z - \omega_{z\_ref})<0$	understeer	front and rear left side
$\omega_{z\_ref}<0$ (right turn)	$(\omega_z - \omega_{z\_ref})>0$	understeer	front and rear right side
	$(\omega_z - \omega_{z\_ref})\leq 0$	oversteer	front and rear left side

To avoid frequent activation of the electronic stability control system, β-$\dot{\beta}$ phase plane method is used [14]. Fuzzy logic control calculates the compensation yaw moment T_{yaw} based on the yaw velocity error ω_z and the sideslip angle β_c. Five membership functions are assigned to each control input, two trapezoidal and three triangular. Direct yaw moment is obtained with a scaling coefficient, with defined eleven membership functions, of which two trapezoidal and nine triangular.

Rollover stability control regime has a feedforward and feedback control. Feedforward control inputs are steering wheel angle θ_{ut} and vehicle longitudinal speed v_x. Feedback control inputs are roll angle error $e(\phi)$ and roll rate error $\dot{e}(\phi)$. The corrective active roll moment T_ϕ is the sum of both outputs: feedforward and feedback moments. Rollover control uses only Gaussian membership functions.

3.2. Modeling of Active Front Steering Control

In active front steering system the actual turning angle θ_p is a difference between the input angle by the steering wheel θ_{ut} and the corrective angle generated by the fuzzy-controller θ_{corr}. The steering actions are performed by accurate calculations of the correct turning angle of the front wheels as follows:

$$\theta_{corr} = \theta_{ut} - \theta_p \tag{14}$$

Fuzzy-logic control estimates the corrective angle based on yaw velocity error $e(\omega_z)$ and the steering wheel angle generated by the driver θ_{ut}. Scaling factor is also used. The number of membership functions and linguistic rules for both inputs and for the output are defined the same manner like in direct yaw moment control [15].

3.3. C. Modeling of Suspension Normal Force Control

Studies by [22], as well as [23] have shown the potential benefits of adapting the normal force through active suspension system control during turning maneuvers. Figure 2 shows a schematic representation of a single wheel suspension system where the force generated by the active and passive components can be determined by:

$$F_{sus} = c(z_{np} - z_p) + F_a + k(\dot{z}_{np} - \dot{z}_p) \tag{15}$$

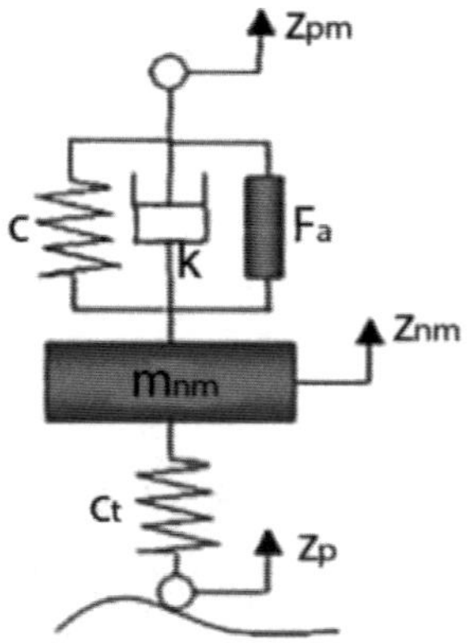

Figure 2. Quarter car model

In 15 F_a is the controlling force in the system. Positive force causes acceleration of the body up and negative vertical acceleration of the wheel. By including the input acceleration and deceleration values, it has been observed that the application of activation force on the front angle, opposite to the turn direction, gives the best results, in terms of the lateral force impact during turning. It also gives the lowest roll movement, according to the combinations examined [22]. Therefore, an active system only on the front axle of the vehicle is chosen to model.

First controller considers yaw velocity error $e(\omega_z)$ and its derivative $\dot{e}(\omega_z)$. The input values have the opposite sign for left and right wheel. Both inputs have assigned per five Gaussian membership functions. Inputs for the second controller are steering wheel angle and longitudinal speed of the vehicle. In addition, Gaussian membership functions are assigned, five for the speed and seven for the angle. The system output is the normal control force, which is a sum from both controllers.

4. Coordinated Cooperative Control of Car Systems

Primary coordinated control was made for pairs of systems and their impact was analysed. Then coordinated cooperative control of all three active systems was comprised. It is based on coordinated control of electronic stability control (ESC) and active suspension normal force control (ASNFC) and then the active front steering control (AFSC) was added. Coordination was made by adaptation of the used scaling coefficients. In that manner every system contributes with a certain share in dynamic behaviour corrective actions. The structure of the coordinated control of those three systems: ESC, ASNFC and AFSC, is given on Figure 3 [24].

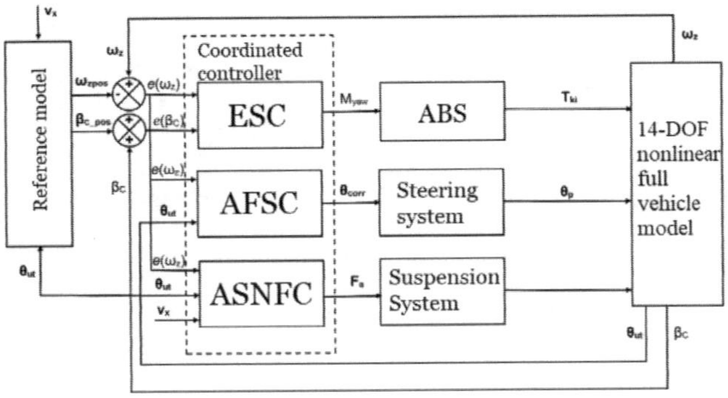

Figure 3. Structure of active control system

For the needs of the research three maneuvers were used in simulations to analyze vehicle's response in critical conditions, which belong in the category of open-loop test methods. First two are turning in a curvature and single lane-change according to ISO 7401 and the third one is fishhook maneuver developed by NHTSA.

5. Simulation Results and Discussions

All manoeuvres are performed at a speed of 130 km/h as it is limit speed on national highways. It is assumed a dry road surface with adhesion coefficient ϕ=0.9. This paper only includes results for step input and sinusoidal input of one period.

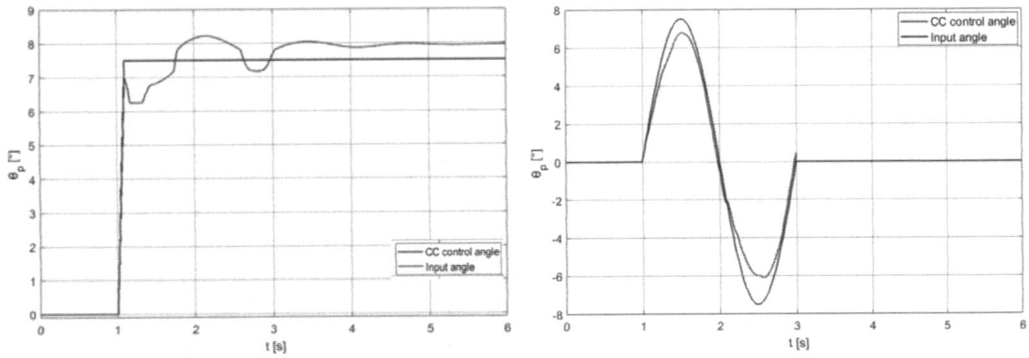

Figure 4. Wheels input angle (left - step steer, right - sine input)

Figure 4 shows the input angle on the wheels transferred through the steering wheel by the driver for the vehicle without active control (black line) and for the vehicle with coordinated cooperative control (brown line). Vehicles' dynamic response is given by yaw velocity (Figure 5), sideslip angle (Figure 6) and roll angle (Figure 7). Also, trajectory of vehicles is given on Figure 8.

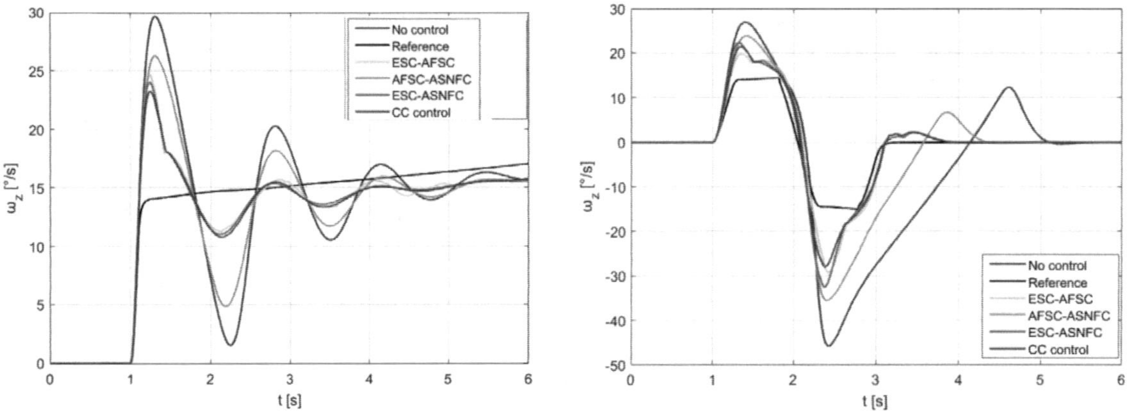

Figure 5. Yaw velocity (left - step steer, right - sine input)

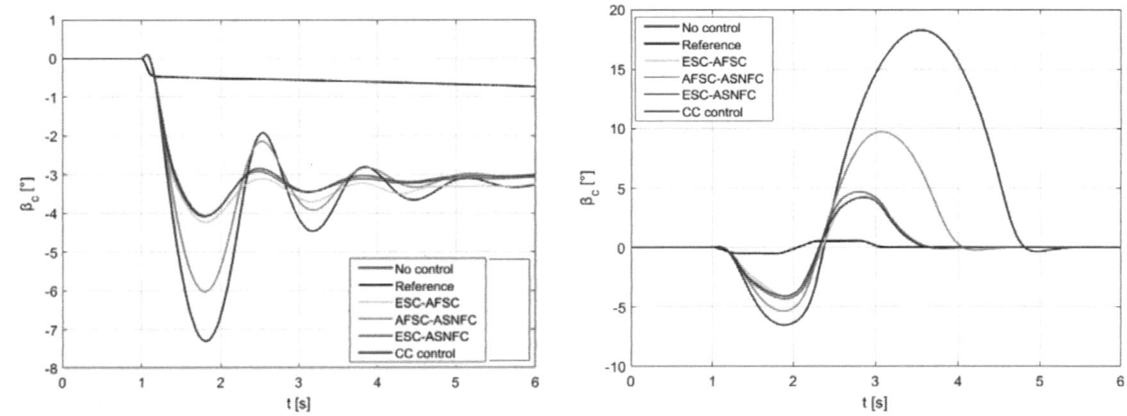

Figure 6. Sideslip angle (left - step steer, right - sine input)

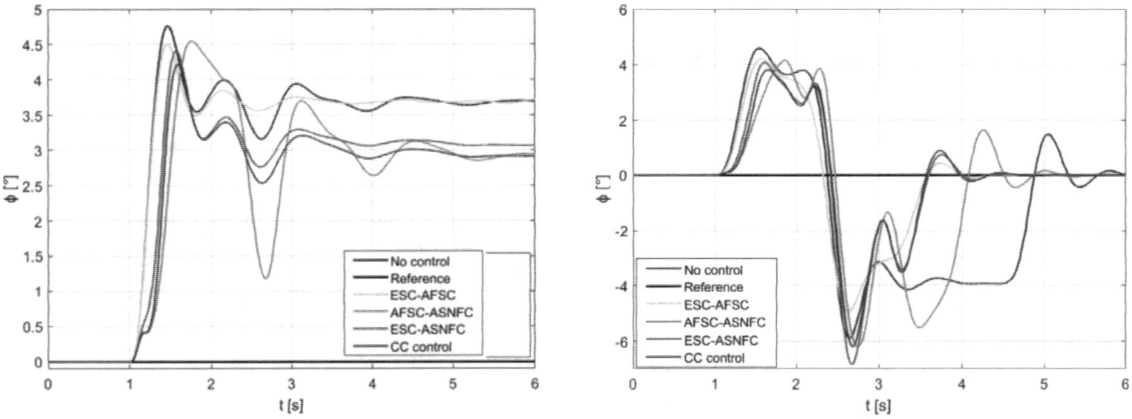

Figure 7. Roll angle (left - step steer, right - sine input)

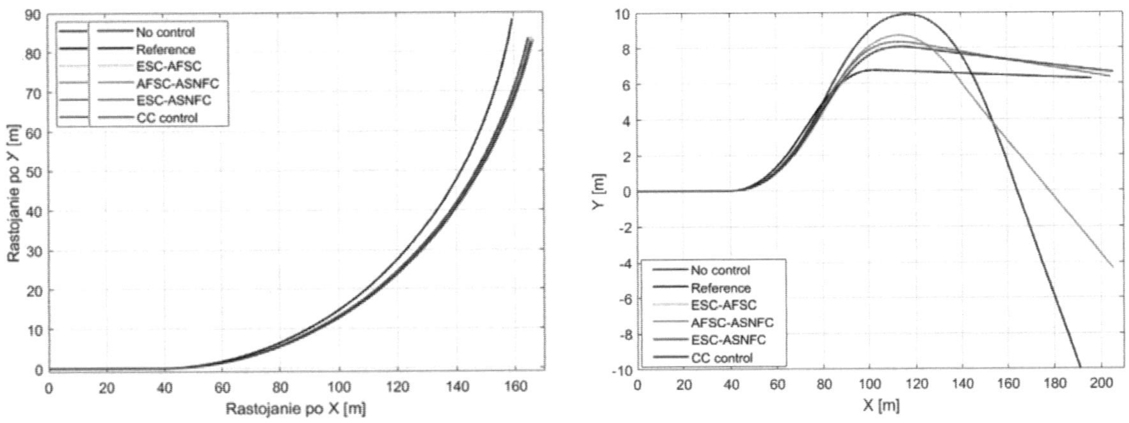

Figure 8. Trajectory (left - step steer, right - sine input)

Figures give comparison of vehicle without active control systems, vehicles with coordinated control of pair of active systems and vehicle with coordinated control of all three systems.

Improvement of vehicle dynamic response is visible for all variables considered: yaw velocity, sideslip angle and roll angle. It leads to the conclusion that stability and thus safety is increased. Improvements are seen not just in lowering of variable transient magnitude but also in the time response.

Coordinated control of ESC, AFSC and ASNFC gives slightly better results than coordinated control of ESC and AFSC or of ESC and ASNFC. It points out the effectiveness of the inclusion of ESC into vehicle dynamics control.

The main benefit is seen in the values of the side slip angle and the roll angle, where the coordinated control of three active systems provides superior results, i.e. gives lowest values in the transient response. Coordinated control requires lower braking forces on wheels compared to single ESC system and also lower normal control forces in the suspension system compared to single ASNFC system. Since ESC works through individual wheel braking, coordinated action of the three active systems affects the transverse forces of wheels, which means it does not lead to a large reduction in the longitudinal velocity of the vehicle. AFSC has less impact at high speeds but its corrective actions provide better understanding of the direction of motion for the driver.

Further improvement of the defined control system can be done through defining sequence of activation depending on the currently measured vehicle behaviour variables.

References

[1] Road traffic injuries. (2020, February 7). Retrieved from https://www.who.int/news-room/fact-sheets/detail/road-traffic-injuries.

[2] M. Nagai, M. Shino, and F. Gao, "Study on integrated control of active front steer angle and direct yaw moment," *JSAE review*, vol. 23(3), pp. 309-315, 2002.

[3] M. Shino, P. Raksincharoensak, and M. Nagai, "Vehicle handling and stability control by integrated control of direct yaw moment and active steering," *Proceedings of AVEC*, vol. 23, 2002.

[4] D. Li, S. Du, and F. Yu, "Integrated vehicle chassis control based on direct yaw moment, active steering and active stabiliser," *Vehicle System Dynamics*, vol. 46(S1), pp. 341-351, 2008.

[5] O. Mokhiamar, and M. Abe, "Active wheel steering and yaw moment control combination to maximize stability as well as vehicle responsiveness during quick lane change for active

vehicle handling safety," *Proceedings of the Institution of Mechanical Engineers, Part D: Journal of Automobile Engineering*, vol. 216(2), pp. 115-124, 2002.

[6] X. Yang, Z. Wang, and W. Peng, "Coordinated control of AFS and DYC for vehicle handling and stability based on optimal guaranteed cost theory," *Vehicle System Dynamics*, vol. 47(1), pp. 57-79, 2009.

[7] S. Yim, Y. Park, and K. Yi, "Design of active suspension and electronic stability program for rollover prevention," *International journal of automotive technology*, vol. 11(2), pp. 147-153, 2010.

[8] H. Xiao, W. Chen, H. Zhou, and J.W. Zu, "Integrated control of active suspension system and electronic stability programme using hierarchical control strategy: theory and experiment," *Vehicle System Dynamics*, vol. 49(1-2), pp. 381-397, 2011.

[9] M. Doumiati, O. Sename, L. Dugard, J.J. Martinez-Molina, P. Gaspar, and Z. Szabo, "Integrated vehicle dynamics control via coordination of active front steering and rear braking," *European Journal of Control*, vol. 19(2), pp. 121-143, 2013.

[10] Z. Shuai, H. Zhang, J. Wang, J. Li and M. Ouyang, "Combined AFS and DYC control of four-wheel-independent-drive electric vehicles over CAN network with time-varying delays," *IEEE Transactions on Vehicular Technology*, vol. 63(2), pp. 591-602, 2013.

[11] M.K. Aripina, Y.M. Samb, A.D. Kumeresanb, M.F. Ismailc, and P. Kemaod, "A review on integrated active steering and braking control for vehicle yaw stability system," *Journal Technology*, vol. 71(2), pp. 105-111, 2014.

[12] X. Jin, G. Yin, C. Bian, J. Chen, P. Li, and N. Chen, "Gain-scheduled vehicle handling stability control via integration of active front steering and suspension systems," *Journal of Dynamic Systems, Measurement, and Control*, vol. 138(1), pp. 014501 1-12, 2016.

[13] J. Song, "Active front wheel steering model and controller for integrated dynamics control systems," *International Journal of Automotive Technology*, vol. 17(2), pp. 265-272, 2016.

[14] L. Jin, X. Xie, C. Shen, F. Wang, F. Wang, S. Ji, X. Guan, and J. Xu, "Study on electronic stability program control strategy based on the fuzzy logical and genetic optimization method," *Advances in Mechanical Engineering*, vol. 9(5), pp. 1-13, 2017.

[15] A. Elhefnawy, H. Ragheb, and S. Hegazy, "Design of an integrated yaw-roll moment and active front steering controller using fuzzy logic control," *SAE International Journal of Vehicle Dynamics, Stability, and NVH*, vol. 1(2017-01-1569), pp. 270-282, 2017.

[16] T. Shim, and D. Toomey, "Investigation of active steering/wheel torque control at the rollover limit maneuver," *SAE Transactions*, (2004-01-2097), pp. 1133-1140, 2004.

[17] J.D. Setiawan, M. Safarudin, and A. Singh, "Modeling, simulation and validation of 14 DOF full vehicle model," *InInternational Conference on Instrumentation, Communication, Information Technology, and Biomedical Engineering 2009*, pp. 1-6, 2009.

[18] E. Bakker, H.B. Pacejka, and L. Lidner, L., "A new tire model with an application in vehicle dynamics studies," *SAE transactions*, pp. 101-113, 1989.

[19] H.B. Pacejka, and E. Bakker, "The magic formula tyre model," *Vehicle system dynamics*, vol. 21(S1), pp. 1-8, 1992.

[20] H.B. Pacejka, and I.J.M. Besselink, "Magic formula tyre model with transient properties," *Vehicle system dynamics*, vol. 27(S1), pp. 234-249, 1997.

[21] R. Rajamani, "Vehicle dynamics and control," *Springer Science & Business Media*, 2011.

[22] C. Margolis, and T. Shim, "Integrated control of suspension and front steering to enhance vehicle handling," *Proceedings of the Institution of Mechanical Engineers, Part D: Journal of Automobile Engineering*, vol. 221(4), pp. 377-391, 2007.

[23] M. Blank, and D. Margolis, "The effect of normal force variation on the lateral dynamics of automobiles," *SAE Technical Paper*, (No. 960484), 1996.

[24] V. Jordanoska, "Coordinated cooperative control of active automotive systems for improved vehicles' stability and safety," April 2020 [Doctoral dissertation, University "Ss. Cyril and Methodius" Faculty of Mechanical Engineering - Skopje].

Tribo-Mechanical Behavior of Geopolymer Composites with Wasted Flax Fibers

Patrycja Bazan [1], Barbara Kozub [1], Kinga Korniejenko [1], Rihards Gailitis [2], Andina Sprince [2]

[1] Faculty of Material Engineering and Physics, Cracow University of Technology, Jana Pawła II 37, 31-864 Cracow, Poland
[2] Faculty of Civil Engineering, Riga Technical University, Kipsalas 6A/B, LV-1048 Riga, Latvia

patrycja.bazan@pk.edu.pl

Abstract. The paper presents the research results on the mechanical and tribological properties of geopolymers reinforced with natural fibers. As a matrix, conventional fly ash and dry construction sand in proportion 1:1 were used. Materials were activated with Na activator. Wasted flax tow in an amount of 0.25-1% by weight was introduced as a reinforcement. The test results showed that with the increase in the fiber content, the bending strength increases, while the compressive properties are comparable to the matrix material. Based on the obtained research results, the optimal content of flax tow was determined at the level of 0.5% by weight. In addition, the abrasion resistance of geopolymer composites was determined using the so-called Boehme shield. This test consists of measuring the change in sample height. The sample is placed on a disc sprinkled with abrasive powder, then pressed and put into rotation. The abrasion results showed that the addition of fibers in the form of waste pulps does not reduce the abrasion resistance, which should be considered as a positive effect. Research also revealed that wasted flax fibers could be successfully used as a reinforcing material in geopolymeric materials as an alternative to the utilization of waste fibers, and additionally may also have a positive effect on the mechanical and tribological properties of geopolymers materials.

1. Introduction

The term "geopolymers" refers to inorganic polymers, which are amorphous aluminosilicate materials synthesized in a strongly alkaline environment at a temperature up to 100° C. The structure of geopolymers is very similar to natural stones, which is why they are often called "artificial stones". Geopolymers consist of long chains - copolymers of aluminum and silicon oxides, as well as metallic cations and bound water that stabilize them. In addition to well-defined polymer chains, the material usually contains various mixed phases, such as silicon oxide, unreacted aluminosilicate substrate, and sometimes crystallized zeolite-type aluminosilicates [1-2].

Geopolymer materials are of great interest to scientists, and many studies have already been published on this topic. In the analysis of the available literature, the most widely conducted works were on the use of fly ash, metakaolin, rice husk ash, or slag for the production of geopolymer or alkali-activated materials. Published studies indicate that geopolymer materials show excellent mechanical strength and durability, resistance to a corrosive environment, and very high temperatures, makes them materials that show great potential in construction applications [3-12].

However, geopolymeric materials have disadvantages, one of them is a low fracture toughness and thus a reduction in strength properties. The solution to this problem is, for example, the introduction of reinforcing fibers, which can significantly reduce the propagation of cracks [13], hence the research on geopolymer composites reinforced with short fibers [14]. The commonly used fibers in geopolymer matrices include organic, inorganic, and natural fibers. Polyethylene, polypropylene, and aramid fibers are used as organic fibers, steel and carbon fibers are among the inorganic fibers. The advantage of organic and inorganic fibers is their high mechanical strength and repeatability compared to natural fibers [15-17]. However, natural fibers are getting more popular; for example, applied natural fibers are flax, sisal, banana, coconut, cotton, raffia, jute, and basalt fibers. These fibers are mainly introduced to increase the resistance to tensile and compressive stresses [2].

Due to the increasing environmental burden, research was also undertaken on the use of waste materials as reinforcement in geopolymer composites. Examples of such fibers are rockwool waste or waste cotton stalk materials. Research has shown that non-recyclable waste can be a valuable raw material for producing geopolymeric materials [18-19].

This paper presents the basic tests of the strength and wear properties of geopolymer composites reinforced with waste flax fibers to indicate the potential of using waste fibers as an alternative to the standard thermal utilization process. This research aimed to emphasize the principle that should be followed when making decisions regarding appropriate methods and technologies of waste disposal. It is. the principle of sustainable development helps in proper waste management, and such management aims to consider economic and environmental aspects while using waste in the widest possible use.

2. Materials and Methods

2.1. Materials

The geopolymer matrix was made of fly ash from the Skawina CHP Plant (Skawina, Poland) and a fine dry saturated construction sand at a ratio of 1:1. The fly ash applied in this study is composed mainly of silica and alumina and includes less than 4% calcium oxide [2]. The precise proportion of each phase in the fly ash is presented in Table 1. Based on an analysis of the diffraction patterns, it can be concluded that the conventional fly ash was characterized by a high quartz content, mullite, hematite, and alite. These are the most important components that react in the geopolymerization process [21]. In the microscopic photo (Figure 1), globular fly ash particles up to 20 μm in size (point no. 1), unburned carbon (point no.2), as well as particles containing high contents of aluminum and silicon oxide, which was confirmed in the EDS analysis can be seen. Spectrometric analysis showed that the fly ash contains silicon, aluminum, and calcium compounds and low iron content, which positively affects the geopolymerization process.

Table 1. XRD analysis of fly ash

Sample ID	Identified phases	Chemical formula
Fly ash	Quartz	SiO_2
	Mullite	$Al_6Si_2O_{13}$
	Hematite	Fe_2O_3
	Alite	Ca3SiO5

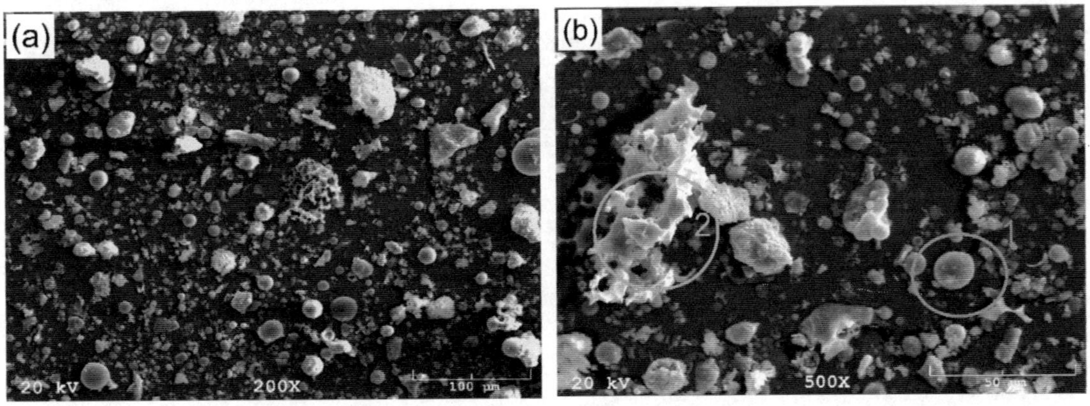

Figure 1. SEM image of applied fly ash: (a) magnification 200 x, (b) magnification 500 x.

Flax tow (Hempiflax Group Sp. Z o.o., Cracow) was used to reinforce geopolymer composites. The reinforcement was added in the amount of 0.25% (sample designated as 0.25W), 0.5% (0.5W), 0.75% (0.75W) or 1% (1W) by weight of dry components (Table 2). For comparison, a reference sample based on matrix material without any additives was made (samples designated as BM).

A 10 molar (10 M) solution of sodium hydroxide and sodium water glass R-145 (with a molar modulus of 2.5 and a density of about 1.45 g/cm^3) were combined as an alkaline activator in the ratio of 1:2. The alkaline solution was prepared by pouring the aqueous sodium silicate solution into flakes of technical sodium hydroxide dissolved in water. The solution was mixed and allowed to reach equilibrium concentration and ambient temperature. The fly ash, construction sand, flax tow, and alkali solution were mixed to form a homogeneous paste for about 15 minutes in a slow-speed mixing machine to prepare the geopolymer masses.

In the next step, the prepared masses were poured into molds, which were vibrated on a vibrating table to remove air bubbles. The tightly wrapped foil molds were heated at 75°C in a laboratory dryer for 24 hours. The samples were unmolded and stored under laboratory conditions (temperature about 20 ° C, relative humidity about 50%). After 28 days, mechanical tests were done.

Table 2. Designation of the manufactured composites.

Index	Mixture Proportion (% by Weight)			NaOH Solution	Density (g/cm^3)
	Fly Ash	Sand	Flax tow		
BM	50	50	-	10 M sodium hydroxide solution + water glass (1200 mL in total)	1.83 ± 0.03
0.25W	49.87	49.87	0.25		1.81±0.06
0.5W	49.75	49.75	0.5		1.79±0.05
0.75W	49.62	49.62	0.75		1.71±0.08
1W	49.5	49.5	1		1.65±0.1

2.2. Methods

2.2.1. Density

Before performing the strength tests, the density of the samples was determined using a geometric method. The density of each geopolymer composition was determined as the average of the four measurements. The samples were dimensioned with a laboratory caliper with a measurement accuracy of 0.01 mm. The weight of the samples was measured with a laboratory precision analytical scale RADWAG PS 200/2000.R2 (maximum load: 200/2000 g; reading accuracy: 0.001 / 0.01 g).

2.2.2. X-ray diffraction (XRD) and microscopic analysis

X-ray diffraction (XRD) was carried out with an X'Pert Pro MPD diffractometer (PANalytical, Malvern, UK) with CuKα radiation at 30 mA and 40 kV. The 2θ angle was varied from 20 ° to 53 ° with a step of 0.04 ° and an accumulation time of 7 s for each step.

Microscopic observations and energy dispersive spectroscopy (EDS) were conducted using a scanning electron microscope Joel JSM-820 with the EDS (IXRF, Inc., Austin, TX, USA) X-ray detector.

2.2.3. Strength tests

Compressive strength tests were performed according to EN 12390-3 standard ("Testing of hardened concrete. Compressive strength of specimens"). The tests were performed on a Matest 3000 kN universal testing machine (Matest, Treviolo, Italy) with a speed of 0.05 MPa/s. The dimensions of the specimens are according to EN 12390-3 (approximately) 50 mm × 50 mm × 50 mm.

Flexural strength tests were also performed according to the EN 12390-5 standard ("Hardened concrete tests. Flexural strength of test specimens"). Tests were also performed on a Matest 3000 kN universal testing machine with a speed of 0.05 MPa/s. Prismatic specimens with dimensions approximately 50 mm × 50 mm × 200 mm were made according to EN 12390-5.

2.2.4. Abrasive wear test

Abrasion tests were carried out using the Boehme shield method according to the PN-EN 13892-3: 2005 standard. For the test, samples with dimensions of 71 mm x 71 mm x 71 mm were prepared, the initial dimensions were measured with a caliper, and then the sample was placed in the holder of the device, and pressure equal to 294N was applied. Corundum powder was used as the abrasive, four grinding cycles were carried out for 22 turns each. After each cycle, the sample was rotated by 90° to eliminate abrasion unevenness.

3. Results and discussions

Table 2 shows the density results of the produced composites. Along with the increase in the share of fibers, the density of the composite decreases, which is related to the lower density of the flax fibers, the density of which is 1.37 g / cm3 [20]. Figure 2 shows a histogram of particle size distribution and cumulative particle size distribution curve for the fly ash used in the test. Figure 3 presents a histogram of the particle size distribution for the construction sand. The sieve analysis results showed that fly ash consisted of particles below 200 microns in the amount of about 80%, including particles below 100 micrometers about 30%, and particles above 200 microns constituted about 12%. Diaz et al. studied factors impacting the suitability of fly ash as a source material for geopolymers, including the effect of fly ash particle size on geopolymer properties. The study showed that particle size distribution is one of the most relevant physical properties that affect the reactivity of fly ash and the subsequent geopolymer product. Since much of the reaction proceeds at the particle-liquid boundary, the finer the particles, the larger the surface area and the higher the reactivity of the fly ash [21]. The addition of sand to geopolymeric materials is justified to increase the strength. Kuenzel et al. proved that the addition of sand increases the viscosity of the paste. The mechanical strength increases significantly when the sand content is above 25% by volume. In addition, the sand reduces the formation of cracks due to significantly lower linear shrinkage. It is related to the formation of a supporting network with a predetermined void volume. The higher the sand content, the higher the compressive properties [22].

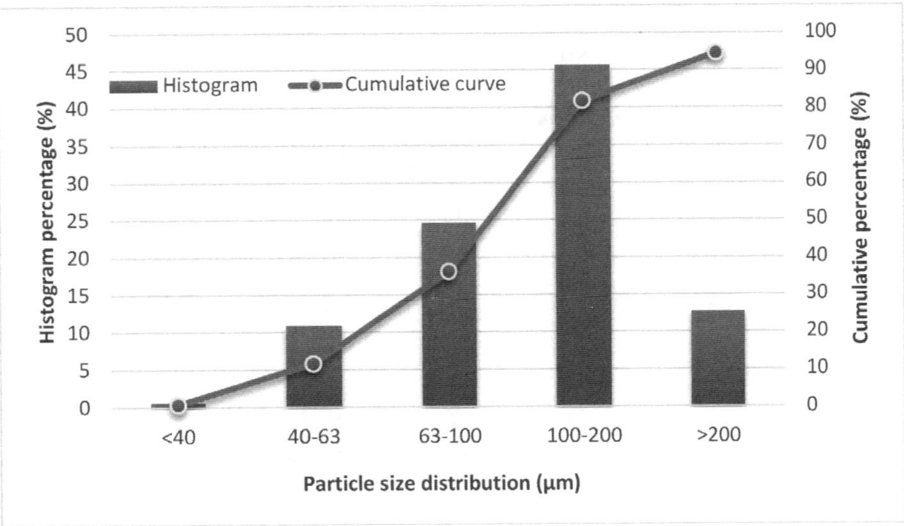

Figure 2. Histogram of particle size distribution and cumulative particle size distribution curve for the fly ash used in the test

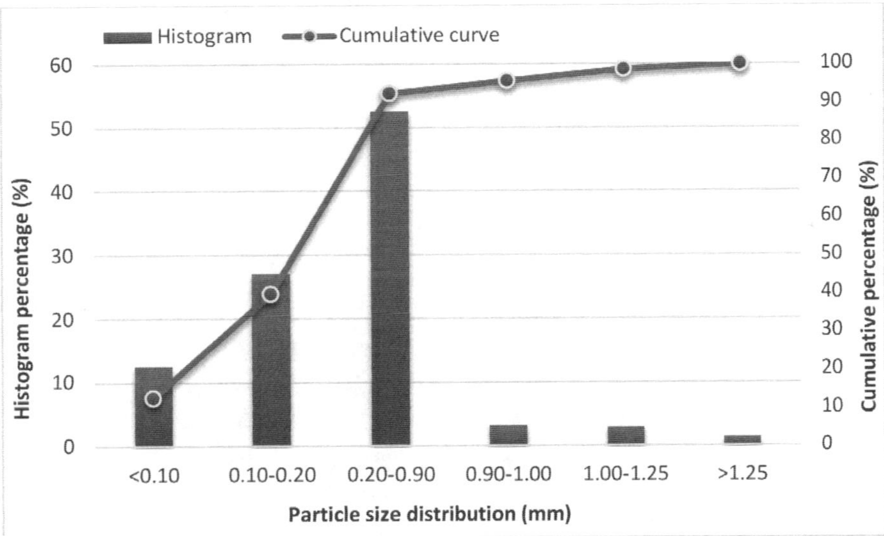

Figure 3. Histogram of particle size distribution and cumulative particle size distribution curve for the construction sand used in the test

Much attention has been devoted to the mechanical and physical properties of fiber-reinforced composites, which are greatly influenced by the type, quantity, and morphology of the reinforcing fibers and the interfacial bonding efficiency between the fibers and the matrix of the composite. Variables such as fiber content, orientation, aspect ratio, and interfacial strength are essential to the final balance of composite strength properties. The addition of short fibers to geopolymeric materials leads to composites that show significant improvements in mechanical properties such as bending and compression strength. The next stage of the research was to determine the strength properties during the flexural and compression test. Both the bending and the compressive strength increased with increased waste fiber content, although the improvement was more noticeable in the bending strength. Figure 4 presents a comparison of the compressive strength of testes materials. The sample contained 0.5% of waste fibers was characterized by the highest compressive strength (46.3 MPa). The further increase in the content of the fibers caused a slight decrease in the compressive strength, which may be related to the deterioration of the workability and the problem of even distribution of the fibers in the matrix [17, 23-24]. In the case of the bending strength presented in figure 5, as the fiber content increased, the bending stress resistance also increased. However, the noticeable effect was achieved only when the fiber content exceeded 0.75% by weight.

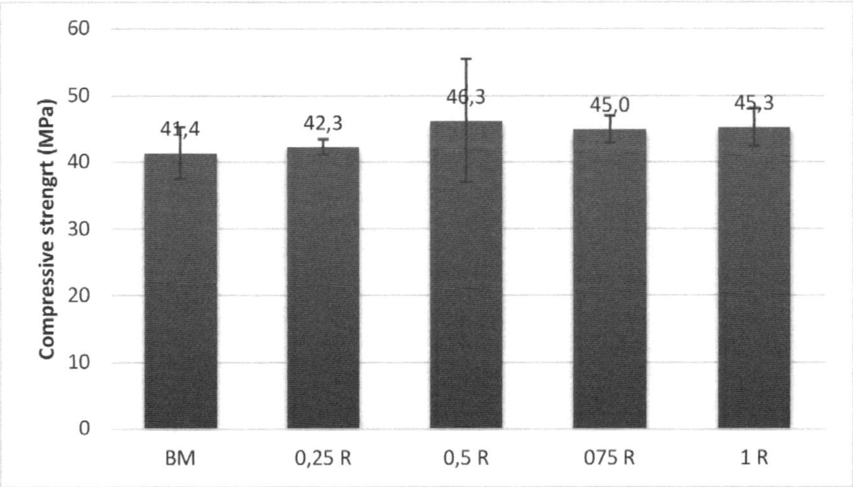

Figure 4. Comparison of compressive strength of tested materials

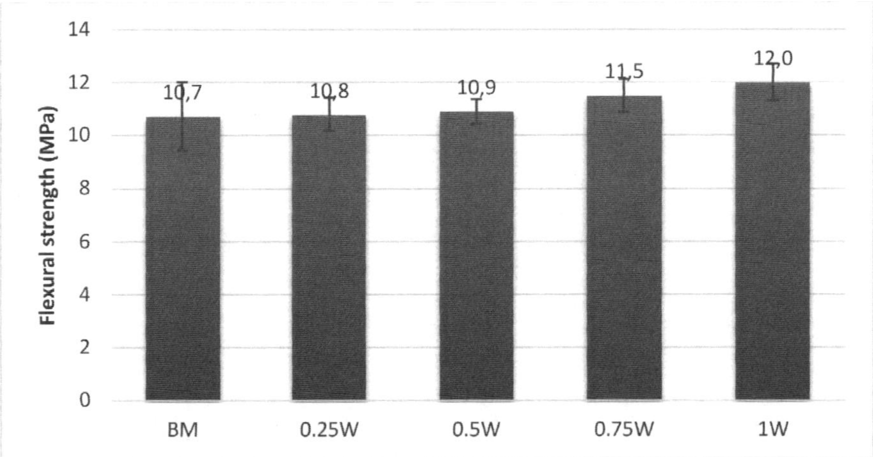

Figure 5. Comparison of flexural strength of tested materials

The incorporation of fibers into geopolymer materials can also affect the wear rate. Fibers can act in many ways, ranging from reducing the degree of wear, limiting the propagation of cracks, to increasing the degree of wear, because if the fibers are pulled out of the matrix and additionally remain in the friction zone, they can act as an additional abrasive [25]. Table 3 presents a summary of the results obtained from the abrasion test for the manufactured composites. The results showed a similar resistance to wear, which is a favorable effect due to the type of applied fiber like tow.

Table 3. Abrasive test results

Index	Parameter		
	Average height before the test (mm)	Average height after the examination (mm)	Average height loss (mm)
BM	71 .2±0.5	70 .4± 0.4	0 .9±0.02
0.25W	71 .2±0.7	70 .4±0.2	0 .8±0.03
0.5W	71 .4±0.3	70 .6±.0.5	0 .8±0.01
0.75W	70 .9±0.1	70 .2±0.2	0 .7±0.02
1W	71 .5±0.2	70 .6±0.3	0 .9±0.01

The form of the fibers, such as pulp, makes it difficult to evenly distribute the fibers in the matrix of the composite and may contribute to a deterioration of the strength properties as well as the wear resistance because the fibers pulled out from the matrix are drawn as bundles of fibers which may result in greater loss of material. However, in this case, such behavior was not observed, which suggests a fairly good connection between the composite components. It may be related to quite good penetration of geopolymeric material between the fibers, which results in good bonding of the matrix material with the reinforcement.

4. Conclusions

The paper presents the possibility of using waste flax pulp. Strength tests revealed that the introduction of fibers in this form positively affects the mechanical properties. Introduction 1% by weight of fibers increased compressive and flexural strength above 10%. An interesting result was also noticed performing the abrasion resistance tests because no increase in the wear of such materials was seen, which is a positive behavior. These studies indicate the potential of using waste fibers as a process of their utilization. Additionally, this type of fibers may correctly fulfill the role of reinforcement in geopolymer composites. In summary, waste fibers show great potential and can be successfully used as reinforcement of geopolymeric materials

Acknowledgment(s)

The presented works were carried out as part of the project: "Geopolymer foams with low thermal conductivity produced based on industrial waste as an innovative material for the circular economy", which is financed by the Polish National Center for Research and Development under the LIDER X program. Grant No.: LIDER/31/0168/L-10/18/NCBR/2019.

References

[1] J. Davidovits, "Inorganic polymeric new materials," *Journal of Thermal Analysis*, vol. 37, iss.8, pp.1633-1656, 1991

[2] P. Bazan, B. Kozub, M. Łach, K. Korniejenko, "Evaluation of Hybrid Melamine and Steel Fiber Reinforced Geopolymers Composites," *Materials*, vol. 13, pp. 5548, 2020

[3] M. Sitarz, i. Hager, M. Choińska, "Evolution of mechanical properties with time of fly-ash-based geopolymer mortars under the effect of granulated ground blast furnace slag addition," *Energies*, vol.13, pp.1135, 2020

[4] Z. Zhang, J.L. Provis, A. Reid, H. Wang, "Geopolymer foam concrete: An emerging material for sustainable construction," *Constr Build Mater*. Vol. 56, pp. 113-127, 2014,

[5] L.N. Assi, S.A. Ghahari, E. Deaver, D. Leaphart, P. Ziehl, "Improvement of the early and final compressive strength of fly ash-based geopolymer concrete at ambient conditions," *Constr Build Mater*, vol. 123, pp.806-813, 2016

[6] T. Luukkonen, M. Sarkkinen, K. Kemppainen, J. Rämö, U. Lassi, "Metakaolin geopolymer characterization and application for ammonium removal from model solutions and landfill leach-ate," *Appl Clay Sci*. vol. 119, pp. 266-276, 2016

[7] H. Wang, H. Li, F. Yan, "Synthesis and mechanical properties of metakaolinite-based geopolymer," *Colloids Surfaces A – Physicochem Eng Asp*. Vol. 268, pp. 1-6, 2005
[8] P. Duxon, S.W. Mallicoat, G.C. Lukey, W.M. Kriven, J.S. Van Deventer, "The effect of alkali and Si/Al ratio on the development of mechanical properties of metakaolin-based geopolymers," *Colloids Surfaces A – hysicochemical Eng Asp*. Vol. 292, pp. 8-20, 2007
[9] K. Kaur, J. Singh, M. Kaur, "Compressive strength of rice husk ash based geopolymer: The effect of alkaline activator," *Constr Build Mater*. Vol. 169, pp. 188-192, 2018
[10] S. Detphan, P. Chindaprasirt, "Preparation of fly ash and rice husk ash geopolymer," *Int J Miner Metall Mater*. Vol 16, iss. 6, pp. 720-720, 2009
[11] T.W. Cheng, J.P. Chiu, "Fire-resistant geopolymer produce by granulated blast furnace slag," *Miner Eng*. Vol. 16, pp. 205-210, 2003
[12] H. Nguyen, V. Carvelli, E. Adesanya, P. Kinnunen, M. Illikainen, "High performance cementitious composite from alkali-activated ladle slag reinforced with polypropylene fibers," *Cem Concr Compos*. Vol. 90, pp. 150-160, 2018.
[13] P. Sukontasukkul, P. Pongsopha, P. Chindaprasirt, S. Songpiriyakij, "Flexural performance and toughness of hybrid steel and polypropylene fibre reinforced geopolymer," *Constr. Build. Mater*. Vol. 161, pp. 37-44, 2018, 161,
[14] C. Sotelo-Pina, E.N. Aguilera-Gonzalez, A. Martinez-Luevanos, "Geopolymers: Past, Present and Future of Low Carbon Footprint Eco-materials. In *Handbook of Ecomaterials*; L.M.T. Martinez, O.V. Kharissowa, B. Kharisov, B., Eds.; Springer: Cham, Switzerland, 2019; pp. 2765–2785.
[15] K. Korniejenko, E. Frączek, E. Pytlak, M. Adamski, "Mechanical properties of geopolymer composites reinforced with natural fibers," *Procedia Eng*. Vol. 151, pp. 388-393, 2016
[16] N. Ranjbar, M. Zhang, "Fiber-Reinforced geopolymer composites: A review," *Cem. Concr. Compos*. Vol. 107, pp. 103498, 2020
[17] K. Korniejenko, W.-T. Lin, H. Šimonová, "Mechanical Properties of Short Polymer Fiber-Reinforced Geopolymer Composites," *J. Compos. Sci*. vol. 4, pp. 128-150 2020
[18] P. Kinnunen, J. Yliniemi, B. Talling, M Illikainen, M. "Rockwool waste in fly ash geopolymer composites," *Journal of Material Cycles and Waste Management*, vol. 19 iss.3, pp. 1220-1227, 2017
[19] B. Zhou, L. Wang, G. Ma, X. Zhao, and X. Zhao, X. "Preparation and properties of bio-geopolymer composites with waste cotton stalk materials," *Journal of Cleaner Production*, vol. *245*, pp. 118842, 2020
[20] M. Le Gall, P Davies, N. Martin, C. Baley, "Recommended flax fibre density values for composite property predictions," *Industrial crops and products*, vol. *114*, pp. 52-58, 2018
[21] E.I. Diaz, E.N. Allouche, S. Eklund, "Factors affecting the suitability of fly ash as source material for geopolymers," *Fuel*, vol. *89, iss*. 5, pp. 992-996 2010
[22] C. Kuenzel, L. Li, L. Vandeperre, A.R. Boccaccini, C.R. Cheeseman, "Influence of sand on the mechanical properties of metakaolin geopolymers," *Construction and Building Materials*, vol. *66*, pp. 442-446, 2014
[23] D.P. Dias, C. Thaumaturgo, "Fracture toughness of geopolymeric concretes reinforced with basalt fibers," *Cement and concrete composites*, vol. *27, iss*. 1, pp. 49-54, 2005.
[24] F.U.A. Shaikh, "Tensile and flexural behaviour of recycled polyethylene terephthalate (PET) fibre reinforced geopolymer composites," *Construction and Building Materials*, vol. *245*, pp. 118438, 2020
[25] A. Abdelbary, A. *Wear of polymers and composites*. Woodhead Publishing.2014

IOP Conf. Series: Materials Science and Engineering 1190 (2021) 012031 doi:10.1088/1757-899X/1190/1/012031

Effect of Residual Stress on Cracking of Hot Die-Forged Brass Fittings

Lenka Kunčická [1,2], Marek Benč [1], Violetta Andreyachshenko [3]

[1] Institute of Physics of Materials, Czech Academy of Sciences, 61662 Brno, Czech Republic
[2] Faculty of Materials Science and Technology, VSB-Technical University of Ostrava, 17. listopadu 2172/15, 708 00 Ostrava, Czech Republic
[3] Department Metallurgy and Energy, Ekibastuz Technical and Engineering Institute named after the Academician K. Satpayev, Ekibastuz, Kazakhstan

kuncicka@ipm.cz

Abstract. Lead brasses, i.e. copper-based alloys with additions of zinc and lead, are widely used for production of specific components in the industry and commerce. Brass components are usually produced under hot conditions, i.e. via hot die forging (with the advantageous applications of induction heating), the semi-products for which are typically rods extruded from cast billets. Finally, the components are machined to their exact shapes. Nevertheless, such a complicated production process is prone to introduce defects in the structures of the final components. The herein presented paper deals with investigations of a brass component (pre-shaped gas valve, i.e. fitting) manufactured via hot die forging. The fitting exhibited cracking in a specific location after forging and the presented investigation attempts to characterize the factors, which could have possibly induced the occurring cracking, from the viewpoint of stress distribution. The experimental investigation focused on characterization of the structure in the vicinity of the cracked area with the particular focus on the occurrence of residual stress via scanning electron microscopy. This analysis was supported with the results of stress-strain distribution during forging acquired via numerical simulation of the die forging processing step (with implemented induction heating prior forging, in accordance with the real industrial conditions) performed using the finite element method (FEM). According to the acquired results, the cracking occurred in a location featuring local inhomogeneity of stress. The inhomogeneous distribution of stress in the particular location most probably developed as the result of the geometry of the forging die, which also introduced instabilities in the material plastic flow during forging.

1. Introduction

Regardless the exact composition and geometry of the final product, commercial purity (CP) copper, as well as alloys, compounds, and composites based on copper are widely used in the industry and commerce, primarily due to their superior properties (e.g. electric and thermal conductivities, non-magnetism, and corrosion resistance) [1,2]. Probably the most characteristic property of CP Cu is the high electric conductivity; given by this feature, CP Cu is widely used for conductors [3]. Innovative Cu-based composite materials are also typically fabricated with the aim to maintain high electric conductivity of the final composite, however, together with enhancing specific mechanical properties, such as strength and hardness [4–6]. On the other hand, the primary aim of the production of numerous Cu-based alloys is to fabricate constructional materials with favourable properties suitable for various applications (such as for screws, bolts, nuts, valves, etc.) [7,8]. Cu-based alloys with probably the widest range of possible applications are brasses, which are compounds based on copper, zinc, and other selected elements. The selection and overall content of the alloying elements is the major factor influencing the phase composition of the final brass, and consequently its specific properties [1].

Lead brasses are created by combining Cu and Zn with the additions of Pb. Among the typical properties of Cu-based alloys is also their good machinability, which facilitates the production process of the final products, as most of the components are subjected to final machining. As regards the lead brasses, the most advantageous machining behaviour was documented for alloys containing less than 3 wt. % of Pb; García et al. [9] reported that the solubility of Pb in Cu is very low by the effect of which Pb forms fine precipitated particles in the structures, and Pantazopoulos and Vazdirvanidis [10] documented that the content of lead in the brass influences formation of chips during machining, since Pb forms discrete island of precipitated particles at the α/β interfaces. Despite the fact that Xiao et al. [11] reported the positive effects of heat treatment performed at temperatures between 400 °C and 600 °C followed by subsequent quenching on modification of structures of biphasic brasses, based on the heretofore performed research, the Pb content in lead brasses should be kept under 3 wt. % to ensure facile finishing of the produced components. Lead brasses are used to produce medical gas valves. The production process of these components is quite complicated and consists of multiple technological steps: the procedure begins with casting, which is typically followed by extrusion performed to more or less homogenize the structure and properties of the cast billet, hot forging the primary aim of which is to transform the extruded semi-product into the pre-finished required shape and dimensions, and final machining performed to finish the shape and surface of the final brass gas valve. Each technological step is designed with the intention to enhance the performance of the final product. On the other hand, each step can result in the introduction of defects, such as segregations, cavities, pores, cracks, surface defects, etc., which can eventually decrease the lifetime of the gas valve.

Researchers have been dealing with the possibility to eliminate the risk of formation of defects during hot processing of brasses. For example, annealing [12], and optimization of the production process [13] were implemented to modify the structures and grains' morphologies of brasses in order to positively influence their final properties. Generally, the heretofore performed research has documented that the phase composition and grain size are the most important factors affecting the final performance and mechanical and utility properties of brass products. Based upon these findings, the application of severe plastic deformation (SPD) methods, based on imposing high amounts of shear strain into metallic materials in order to impart substantial grain refinement and consequently enhance their final performance and properties [14–18], has a great potential for production of high-quality brass products. This hypothesis has recently been proven by Naser Radhi et al. [19], who documented the positive effects of processing of a brass via the Equal Channel Angular Pressing (ECAP) SPD method on its final mechanical properties. Nevertheless, the presence of residual stress within the processed structure, or stress inhomogeneity during deformation processing, are of non-negligible importance for the lifetime and durability of the final products, too, as both these factors can contribute to possible nucleation and development of cracks [20–22].

According to the available literature, no study documenting the stress-strain relations within a brass gas valve (fitting) produced under real industrial conditions with the occurrence of defects (cracking) has been published. The aim of the presented study is thus to numerically predict the distribution of the effective strain and stress within the forged fitting, with a special focus on the area the cracking in which occurred. The stress-strain parameters are supplemented with evaluation of other important influencing factors, such as the material flow and distribution of temperature. The predicted results are supported by experimental data from microscopic analyses and microhardness measurements.

2. Materials and Methods

The presented study was performed to provide numerical and experimental characterization of selected factors within a lead brass (CuZn40Pb2) fitting, i.e. pre-shaped gas valve, produced by GCE company (Gas Control Equipment, Chotěboř, Czech Republic), which exhibited cracking during hot die forging (see Figure 1a for the axial longitudinal cut through the fitting the cracked area in which was highlighted). The fitting was manufactured from a cast and subsequently extruded semi-product (rod), which was subsequently subjected to induction heating to the processing temperature of 680 °C, followed by die forging. The chemical composition of the brass the fitting from which was produced, acquired via inductively coupled plasma atomic emission spectroscope, i.e. ICP-AES, was the following (in wt. %): Cu 58.10, Zn 39.6, Pb 1.75, Fe 0.27, Sn 0.20, Ni 0.07, Al 0.004, Si 0.003.

2.1. Numerical analyses

The first step performed to investigate the parameters of interest within the forged fitting during processing was assembling the FEM numerical simulation of die forging with the help of Forge NxT® commercial software. To be able to compare the predicted results with the experimentally acquired ones, the geometrical dimensions, as well as mechanical properties of the FEM assembly, were selected to be identical to the experimental conditions (see Figure 1b for the numerical model of the brass fitting). The dies were defined to be rigid, while the brass billet was defined to be a deformable body meshed with tetrahedral elements with activated re-meshing (intensive plastic deformation was expected to occur). According to the real processing conditions, the friction was defined via using a graphite lubricant, and the defined tools' speed was 3 mm/s. The selected boundary conditions for the elevated processing temperature of 680 °C were the thermal conductivity of 120 W/(mK), thermal expansion coefficient of 2.2×10^{-5} (K^{-1}), specific heat of 380 J/kgK, Young's modulus of 30 GPa, Poisson's ratio of 0.35, emissivity of 0.7, and density of 8440 kg/m^3. The simulation was defined via the elastic-plastic model with implemented Newton–Raphson convergent algorithm and the mechanical properties of the used brass were acquired experimentally using SETARAM device (servo-hydraulic torsion plastometer), and subsequently imported to the material flow stress database of the Forge NxT software. The deformation behaviour of the brass semi-product was characterized with the help of the Haensel-Spittel equation (e.g. [23–25]), the values of the regression coefficients for which were A = 8039.59 MPa, m_1, m_2, m_3, and m_4 were -0.00835, -0.00099, 0.1693, and -0.00924, respectively, and m_5, m_6, m_7, and m_8 were 0.

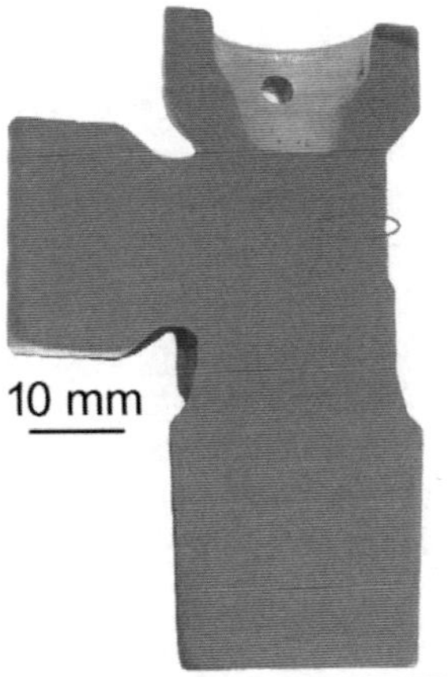

(a)

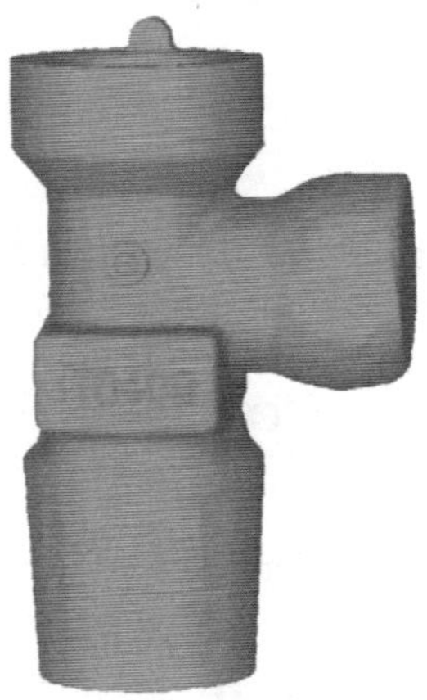
(b)

Figure 1: Axial longitudinal cut through die-forged brass fitting with highlighted cracked area (a); numerical model of fitting (b).

2.2. Experimental analyses
The mechanical properties were evaluated via Vickers microhardness measurements. These were performed in two significant locations (two perpendicular lines), in a line along the cracked surface, and in a line from the crack towards the internal volume of the die-forged fitting (for depiction of the measured locations see Figure 2). The loading time for each indent was 10 s, and the load for each was 200 gf. Finally, graphs from both the sets of measurements were drawn.

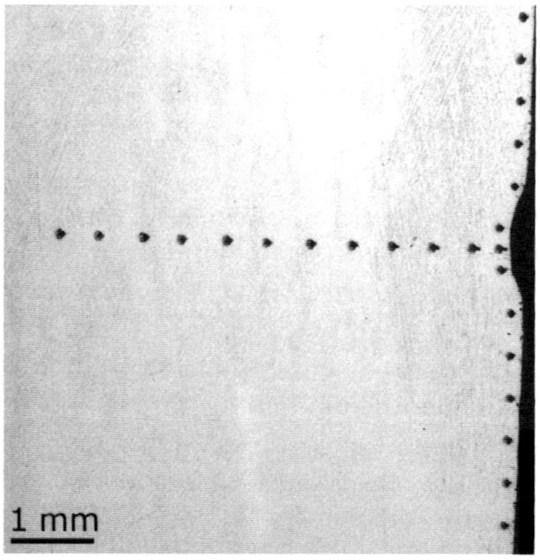

Figure 2: Locations of microhardness measurements.

Subsequent experimental analyses were performed on a sample acquired from a longitudinal axial cut through the fitting (see Figure 1a) via optical microscopy using Olympus DSX100 digital optical microscope, and scanning electron microscopy, Tescan Lyra 3 XMU FEG/SEMxFIB device equipped with Ultim Max EDS detector and Symmetry EBSD detector for which was used. The optical microscopy was used to observe the material flow and compare the real flow of grains with the precited material flow vectors. The the EDS observation was performed to detect possible presence of precipitates and/or oxides.

3. Results and discussions
3.1. Material flow
The first predicted parameter was the material flow, since inhomogeneities in material flow are among the possible factors which could have imparted the generation of cracks. This parameter was assessed via depicting the material flow vectors. Figure 3a shows the vectors of material flow on the axial longitudinal cut through the simulated fitting at the end of forging. As can be seen in the figure, the volume of the material had a strong tendency to flow towards the upper neck of the fitting (i.e. in the horizontal direction in Figure 3a), which is in accordance with the tendency of the material to flow in the line of least resistance. The simulation of material flow thus revealed that this parameter featured significant inhomogeneities, as presupposed. The area the cracking in which subsequently occurred (depicted with arrow in Figure 3a) featured evident aggravation of material flow, as there are almost no flow vectors evident in this location.

The predicted material flow was experimentally confirmed by optical microscopy analysis (Figure 3b), which also revealed the tendency of the material volume of the brass rod to flow towards the neck of the forged component. The cracked area evidently corresponds to the area the predicted material flow in which was aggravated (the area of crack is in Figure 3b depicted by an arrow).

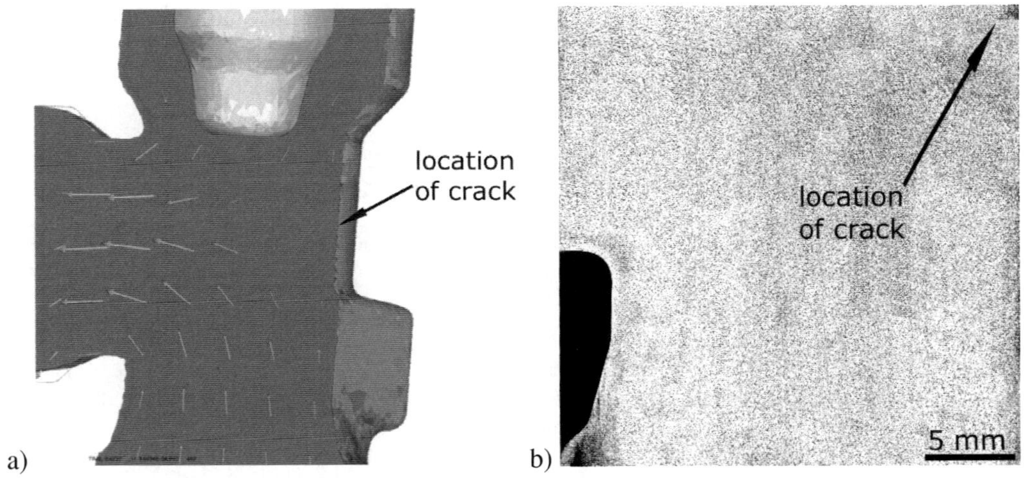

Figure 3: Predicted vectors of material flow on axial longitudinal cut through the fitting at the end of forging (a); optical microscopy image depicting flow of grains (b).

3.2. Temperature distribution

The inhomogeneous material flow goes hand in hand with changes in the distribution of temperature throughout the forged component. Before forging, the brass rod was homogeneously pre-heated via induction to the processing temperature of 680 °C. Nevertheless, the temperature and its distribution during forging was influenced by mutual effects of several factors. The first important factor is that the pre-heated brass rod was inserted into a die pre-heated to a lower temperature, which instantaneously imparted heat transfer from the hotter brass to the cooler die. The second important factor affecting the temperature is friction between the brass and the die. Friction is generally known to affect the processed material by influencing its (surface) temperature [27]. Nevertheless, as the material flow of the brass was inhomogeneous, the temperature of the forged product was inhomogeneously affected by friction, too, and thus local decreases in temperature occurred.

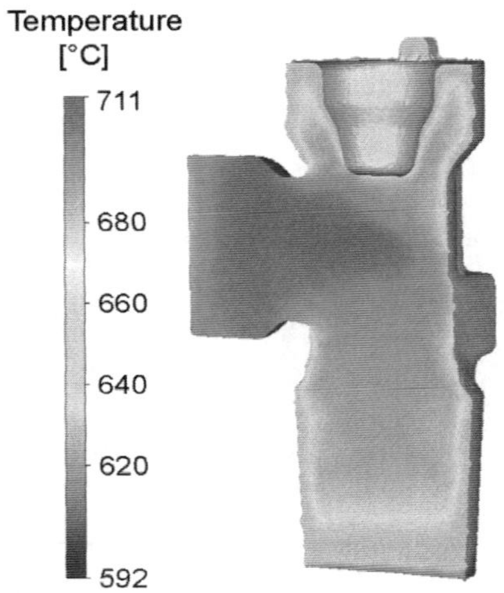

Figure 4: Predicted temperature distribution on axial longitudinal cut through the fitting at the end of forging.

This is documented by Figure 4, the temperature distribution throughout the forged brass product in which is depicted. The figure shows that the temperature was the highest in the neck area of the fitting, i.e. in the area the material flow in which was the most intense. On the other hand, the temperature was the lowest in the peripheral areas of the conical part of the fitting, including the area in which the crack occurred, which exhibited temperature drop to approx. 650 °C.

3.3. Stress-strain parameters
The inhomogeneous temperature distribution, together with the inhomogeneous material flow, non-negligibly affects the stress-strain relations within the forged fitting.

Figure 5a shows a detail of the area of crack with predicted stress distribution during forging. As can be seen, this particular location featured evident stress inhomogeneity, areas exhibiting tensile and compressive stresses in which fade into one another. Such stress distribution was most probably imparted by the mutual effect of the geometry of the die, which introduced the aggravated material flow in this location, resulting in the observed local temperature drop. These factors most probably imparted increased friction, and consequently the development of residual stress, which could not further be relaxed. The inhomogeneous stress state was a significant contributor to the cracking.

The predicted effective imposed strain along the axial longitudinal cut through the fitting is depicted in Figure 5b. The figure shows that the effective strain was the highest in the shaped areas of the forged fitting (depicted by red colour). Nevertheless, the majority of the conical part of the fitting, together with certain locations in its shaped part, including the area the material flow in which was aggravated, featured lower effective strain values (compared to the shaped areas). As can be seen, the area of the crack is located in the region featuring effective imposed strain inhomogeneity; the blue colour depicting the effective strain of around 1 changes rapidly into the green colour, i.e. the effective strain increases rapidly to the values of around (and more than) 2 in this location.

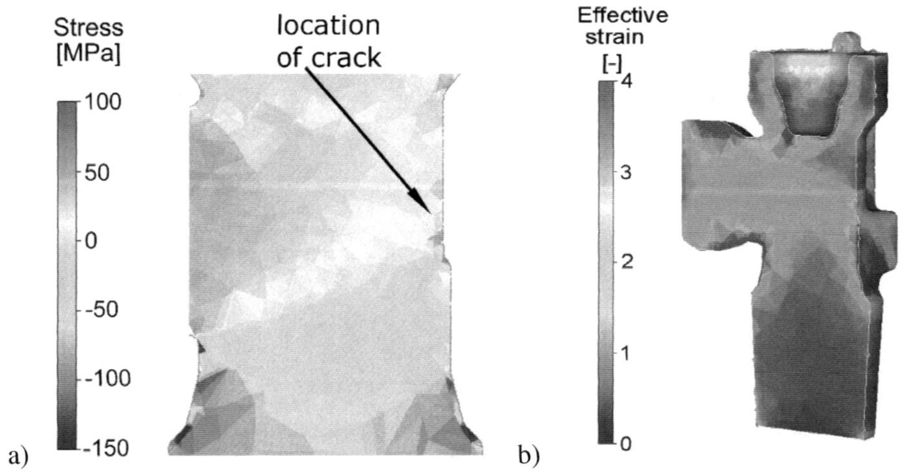

Figure 5: Predicted stress distribution along cracked area, detail from axial longitudinal cut through the fitting (a); predicted effective imposed strain along axial longitudinal cut through the fitting (b).

3.4. Structure analysis and Microhardness
The predicted parameters were subsequently verified experimentally via structure analyses and microhardness measurements.

Figure 6a depicts a SEM-SE image of the cracked area; given by the aggravated material flow and decreased (sub)surface temperature in this location, the material evidently relaxed the developed residual stress via "stemmed" cracking. The following analyses were focused on mapping of the chemical composition in the cracked area in order to evaluate possible presence of oxides. Figure 6b shows the maps of the substantial elements resulting from the EDS analyses. The crack most probably originated during forging since the stems of the crack evidently contained oxygen. The cracked area also exhibited depletion of copper, however, higher contents of zincs. The analyses thus point to the presence of zinc oxides in the crack.

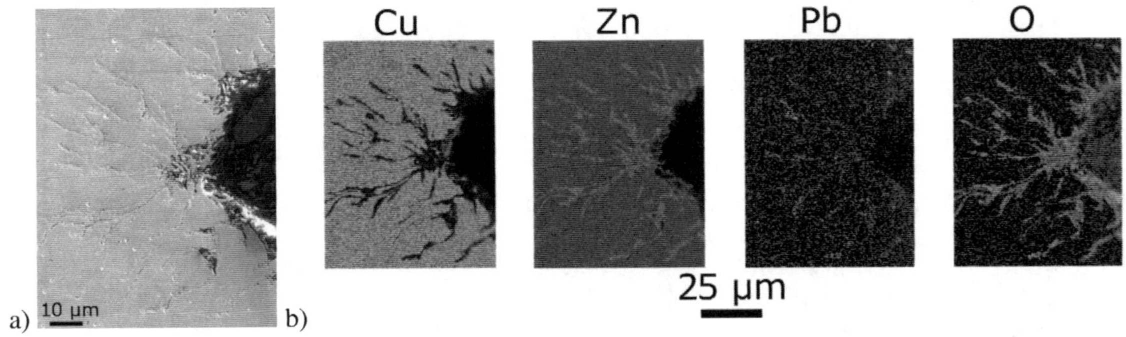

Figure 6: SEM-SE image of cracked area (a); SEM-EDS maps of elements present in cracked area (b).

Last but not least, Vickers microhardness analyses were performed. Figure 7a shows the distribution of microhardness along the cracked surface of the forged fitting, while Figure 7b shows the profile of microhardness in the direction perpendicular to the cracked surface, i.e. towards the internal volume of the fitting (see Figure 2 for the depiction of the measured locations).

The surface of the forged product evidently featured the highest measured HV values, although sudden drops of microhardness can be observed in the exact locations of the "stems" of the crack. These finding corresponds to the structure observations, i.e. to the presence of hard and brittle oxides. The location of the most prominent crack (point at the distance of 5.5 mm in Figure 7a) was then the starting point for the second (perpendicular) line of measurement. Figure 7b depicts that the HV values decreased already in 1 mm distance from the surface of the forged component and remained more or less constant towards its internal volume; the average microhardness value in the internal volume of the forged fitting was 102.3 HV.

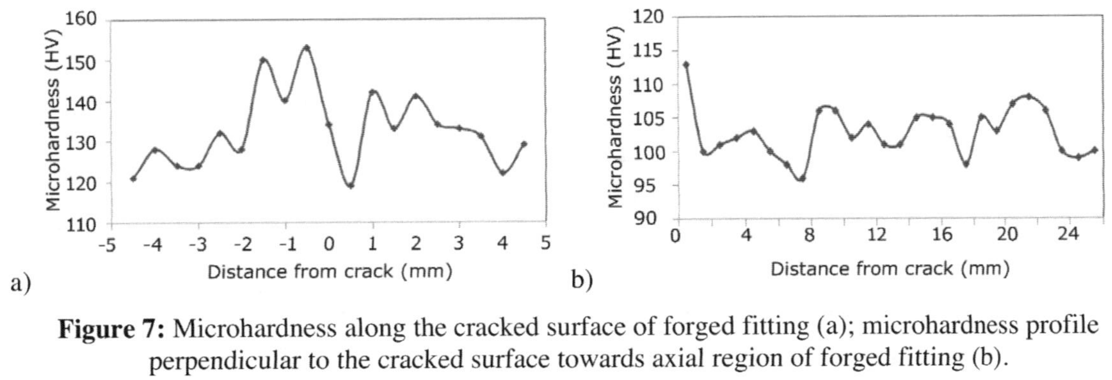

Figure 7: Microhardness along the cracked surface of forged fitting (a); microhardness profile perpendicular to the cracked surface towards axial region of forged fitting (b).

4. Conclusions

The aim of the herein presented study was to numerically and experimentally analyse the stress-strain relations within a brass fitting exhibiting cracking during hot die forging, including selected important influencing factors. The investigations revealed that the critical location the cracking in which occurred exhibited significant aggravation of the material flow, which resulted in accelerated heat transfer from the heated brass billet to the cooler die and imparted temperature drop in this location.

These factors were the most important phenomena behind the development of inhomogeneous stress state in the critical location and subsequent occurrence of "stemmed" cracking. Supplementary experimental analyses documented that the crack contained hard and brittle zinc oxides, which also resulted in local increase in (sub)surface Vickers microhardness values.

Acknowledgment(s)
This research was supported by the 19-15479S project of the Grant Agency of the Czech Republic.

References

[1] A. Russell, K.L. Lee, "Structure-Property Relations in Nonferrous Metals," 1st ed., *John Wiley & Sons, Inc.*, New Jersey, 2005.
[2] M. Kreuzeder, M.-D. Abad, M.-M. Primorac, P. Hosemann, V. Maier, D. Kiener, "Fabrication and thermo-mechanical behavior of ultra-fine porous copper," *J. Mater. Sci.* vol. 50, pp. 634–643, 2014.
[3] X.H. Chen, L. Lu, K. Lu, "Electrical resistivity of ultrafine-grained copper with nanoscale growth twins," *J. Appl. Phys.* vol. 102, p. 083708, 2007.
[4] L. Kunčická, R. Kocich, K. Dvořák, A. Macháčková, "Rotary swaged laminated Cu-Al composites: Effect of structure on residual stress and mechanical and electric properties," *Mater. Sci. Eng. A* vol. 742, pp. 743-750, 2019.
[5] T. Sapanathan, S. Khoddam, S.H. Zahiri, A. Zarei-Hanzaki, "Strength changes and bonded interface investigations in a spiral extruded aluminum/copper composite," *Mater. Des.* vol. 57, pp. 306–314, 2014.
[6] L. Kunčická, R. Kocich, "Deformation behaviour of Cu-Al clad composites produced by rotary swaging," *IOP Conf. Ser. Mater. Sci. Eng.* 369, p. 012029, 2018.
[7] I. Hamidah, A. Solehudin, A. Hamdani, L. Hasanah, K. Khairurrijal, T. Kurniawan, R. Mamat, R. Maryanti, A.B.D. Nandiyanto, B. Hammouti, "Corrosion of copper alloys in KOH, NaOH, NaCl, and HCl electrolyte solutions and its impact to the mechanical properties," *Alexandria Eng. J.* vol. 60, pp. 2235–2243, 2021.
[8] Y. Geng, Y. Ban, B. Wang, X. Li, K. Song, Y. Zhang, Y. Jia, B. Tian, Y. Liu, A.A. Volinsky, "A review of microstructure and texture evolution with nanoscale precipitates for copper alloys," *J. Mater. Res. Technol.* vol. 9, pp. 11918–11934, 2020. https://doi.org/10.1016/j.jmrt.2020.08.055.
[9] P. García, S. Rivera, M. Palacios, J. Belzunce, "Comparative study of the parameters influencing the machinability of leaded brasses," *Eng. Fail. Anal.* vol. 17, pp. 771–776, 2010.
[10] G. Pantazopoulos, A. Vazdirvanidis, "Characterization of the microstructural aspects of machinable alpha-beta phase brass" - *2013 - Wiley Analytical Science*, 2006.
[11] Y.H. Xiao, C. Guo, X.Y. Guo, "Constitutive modeling of hot deformation behavior of H62 brass," *Mater. Sci. Eng. A.* vol. 528, pp. 6510–6518, 2011.
[12] J. Matsumoto, H. Anada, M. Furui, "The effect of grain size and amount of β phase on the properties of back-torsion working in 60/40 brass," *Adv. Mater. Res.*, pp. 661-666, 2007.
[13] G. Pantazopoulos, "Leaded brass rods C 38500 for automatic machining operations: A technical report," *J. Mater. Eng. Perform.* vol. 11, pp. 402–407, 2002.
[14] A.B. Naizabekov, V.A. Andreyachshenko, R. Kocich, "Study of deformation behavior, structure and mechanical properties of the AlSiMnFe alloy during ECAP-PBP," *Micron* vol. 44, pp. 210–217, 2013.
[15] R. Kocich, A. Macháčková, L. Kunčická, "Twist channel multi-angular pressing (TCMAP) as a new SPD process: Numerical and experimental study," *Mater. Sci. Eng. A.* vol. 612, pp. 445–455, 2014.
[16] L. Kunčická, R. Kocich, P. Král, M. Pohludka, M. Marek, "Effect of strain path on severely deformed aluminium," *Mater. Lett.* vol. 180, 280–283, 2016.

[17] U.M. Iqbal, S. Muralidharan, "Optimization of die design parameters and experimental validation on twist channel angular pressing process of AA6061-T6 aluminium alloy," *Mater. Res. Express*. vol. 6, p. 0865f2, 2019.

[18] L. Kunčická, R. Kocich, V. Ryukhtin, J.C.T. Cullen, N.P. Lavery, "Study of structure of naturally aged aluminium after twist channel angular pressing," *Mater. Charact.* vol. 152, pp. 94–100, 2019.

[19] H. Naser Radhi, M. Talib Mohammed, A.M.H. Aljassani, "Influence of ECAP processing on mechanical and wear properties of brass alloy," *Mater. Today Proc.* vol. 44, pp. 2399–2402, 2021.

[20] L. Kunčická, R. Kocich, C. Hervoches, A. Macháčková, "Study of structure and residual stresses in cold rotary swaged tungsten heavy alloy," *Mater. Sci. Eng. A*. vol. 704, pp. 25–31, 2017.

[21] N. van den Berg, H. Xin, M. Veljkovic, "Effects of residual stresses on fatigue crack propagation of an orthotropic steel bridge deck," *Mater. Des.* vol. 198, p. 109294, 2021.

[22] L. Kunčická, R. Kocich, P. Strunz, A. Macháčková, "Texture and residual stress within rotary swaged Cu/Al clad composites," *Mater. Lett.* vol. 230, pp. 88–91, 2018.

[23] R. Kocich, "Effects of twist channel angular pressing on structure and properties of bimetallic Al/Cu clad composites," *Mater. Des.* vol. 196, p. 109255, 2020.

[24] R. Kocich, L. Kunčická, A. Macháčková, "Twist Channel Multi-Angular Pressing (TCMAP) as a method for increasing the efficiency of SPD," *IOP Conf. Ser. Mater. Sci. Eng.* vol. 63, p. 012006, 2014.

[25] L. Kunčická, A. Macháčková, N.P. Lavery, R. Kocich, J.C.T. Cullen, L.M. Hlaváč, "Effect of thermomechanical processing via rotary swaging on properties and residual stress within tungsten heavy alloy," *Int. J. Refract. Met. Hard Mater.* vol. 87, pp. 1–15, 2020.

[26] B. Beausir, J.J. Fundenberger, "Analysis Tools for Electron and X-ray diffraction," *ATEX - software*, www.atex-software.eu, (2017).

[27] R. Kocich, M. Kursa, A. Macháčková, "FEA of Plastic Flow in AZ63 Alloy during ECAP Process," *Acta Phys. Pol. A*. vol. 122, pp. 581–587, 2012.

IOP Conf. Series: Materials Science and Engineering 1190 (2021) 012032 doi:10.1088/1757-899X/1190/1/012032

Correlation of Texture and Residual Stress in Twist Channel Angular Pressed Laminated Al/Cu Composites

Radim Kocich [1], Ludmila Krátká [1], Petr Král [2]

[1] Faculty of Materials Science and Technology, VSB-Technical University of Ostrava, 17. listopadu 2172/15, 708 00 Ostrava, Czech Republic
[2] Institute of Physics of Materials, Czech Academy of Sciences, 61662 Brno, Czech Republic

radim.kocich@vsb.cz

Abstract. Composites are generally materials consisting of a minimum of two phases or components, which are selected for their advantageous properties the combination of which provides the final composite material with better specific properties than those of (the original) single-phase materials and alloys. One of the composite types is clad composites, or laminated composites, consisting of layers of individual (metallic) materials. The presented study deals with laminated Al/Cu composite billets, which were processed via the severe plastic deformation (SPD) method of twist channel angular pressing (TCAP). The primary advantage of the SPD methods in general is their ability to introduce significant shear strain, resulting in substantial refinement of the grains of the processed materials, which consequently enhances their properties. The study is specifically focused on assessing the effects of single and double TCAP pass on the specific lattice rotations induced via the severe shear strain within the reinforcing Cu wires, which typically manifest in the changes in residual stress distribution and modifications in texture. The results revealed that after the first TCAP pass, the imposed strain was not homogeneous across the cross-section of the TCAP-ed composite sample as the individual Cu wires featured differences in the distribution of residual stress, as well as in the observed ideal texture orientations. Nevertheless, the second TCAP pass introduced substantial homogenization of the imposed shear strain, which manifested in a more homogeneous distribution of residual stress across the individual wires compared to the first pass. Also, the distribution of ideal texture orientations was more homogenized after the second pass, as all the Cu fibres exhibited the dominance of ideal shear *A* fibre texture.

1. Introduction

Composite materials typically consist of more than two different components separated by discrete interfaces [1,2]. The combination of metals featuring different advantageous properties brings about the possibility to create innovative materials with enhanced properties. Composites usually consist of a matrix and other reinforcing phases, and are typically characterized according to the reinforcing phase (e.g. discontinuous composites with reinforcing particles or short fibres) [3]. However, regardless the type of the reinforcing element, the grain size of the matrix, and size and distribution of the reinforcing elements are the primary factors influencing the final mechanical and utility properties of the composite [3]. By this reason, application of specific processing methods, such as the techniques of severe plastic deformation (SPD), for example equal channel angular pressing [4,5], twist channel (multi) angular pressing [6–8], twist extrusion [9], high pressure torsion [10], high pressure sliding

[11], accumulative roll bonding [12], etc., can be advantageous, among for conventional alloys, also for composite materials, from the viewpoint of mutually achieving high quality of bonding, and ultra-fine-grained (UFG) structure.

Clad composites, also denoted as layered materials or hybrid materials, are a specific type of composite materials consisting of (the minimum of) two layers of different metals bond at discrete interfaces [13]. The preparation methods of various types of these materials (e.g. Ti/Cu [14], Al/Mg [15], Al/steel [16], Mg/steel [17], etc.) have been studied by researchers worldwide. They typically combine rolling and other technologies, such as explosive bonding [18], or apply modifications of the rolling technology, such as asymmetrical rolling [19], which was applied to fabricate Al/Cu clad composites [20].

Composites combining both the Al and Cu are specific, since both the component metals exhibit superior electric conductivity. Optimized processing providing composites combining both the Al and Cu results in the production of materials with relatively light weight, and lower price compared to copper. Al/Cu clad composites have prospective application e.g. in the automotive, aerospace, and electrotechnics [21,22]. Due to the light weight and advantageous mechanical properties, they are expected to replace steel in specific applications in aircrafts and vehicles in the near future; researchers have documented satisfactory durability of the composites during dynamic loading at strain rates up to 100 s^{-1} [23]. Such high shear strain rates support refinement of grains, and impart friction at the interfaces due to differences in the plastic flows of the metals, which eventually supports bonding of the cladded metals via atomic diffusion. However, the internal heat generated by the mentioned structural phenomena increases the actual processing temperature, which can impart formation of intermetallic phases at the interfaces.

Since the conventional forming methods, such as rolling, forging, extrusion, etc., have limitations, unconventional forming methods, such as the mentioned SPD methods, has become popular also to manufacture clad composites. They are beneficial primarily since they can impart (significant) grain refinement, introduce uniform distribution of secondary (precipitated) particles, and homogenize the structures. Moreover, the applied severe shear strain supports breakage of possibly present oxide layers at the interfaces of the component metals, and diffusion via generating lattice defects [24]. Several works documented significant enhancement of the mechanical and utility properties of clad composites prepared via SPD methods. The methods with probably the highest potential is high pressure torsion (HPT) [25]. By applying very high pressure, HPT enables to impose substantially high strains without introducing defects to the processed samples. Moreover, the high shear strain supports redistribution of precipitated (secondary) particles. As regards the mostly applied and researched equal channel angular pressing (ECAP) method, the shear strain imposed into the composite during a single pass ECAP is not as high as e.g. during HPT [26]. Nevertheless, the amount of the imposed strain can be increased by implementing modifications of the process, e.g. by modifying the dies, or by implementing (partial) back pressure [27].

The twist channel angular pressing (TCAP) method is based on the ECAP method, however, it combines a bend deformation zone with a twist deformation zone (i.e. main deformation zone, MDZ, and twist deformation zone, TDZ) in a single channel. This particular die design enables to introduce severe shear strain into the processed (composite) specimen during a single pass [28]. The method has been successfully applied before to provoke the formation of uniform UFG structures within single-phase materials. The efficiency of a single pass TCAP was reported to be higher than the efficiency of double pass ECAP, as regards the structure refinement and its homogenization [29].

The herein presented study is primarily focused on investigating the effects of single and double pass TCAP on the orientations of grains within the reinforcing Cu wires inserted in the Al matrix of the newly designed Al/Cu clad composite. The texture analyses are supported with the distribution of residual stress depicted via internal grains misorientations, the combination of which provides a complex view on the deformation behaviour of the composite components during TCAP.

2. Materials and Methods

The primary aim of the work was to evaluate the character of grains within the reinforcing Cu wires of a self-designed Al/Cu clad composite after TCAP processing. The particular focus was on performing the correlation between the observed grains orientations and the presence of internal grains misorientations, i.e. residual stress, within the structures. The original component metals were commercially pure (CP) Cu of 99.97% purity (composition (in wt. %): Cu plus 0.0073 Ni, 0.0060 Sn, 0.0030 Zn, 0.0030 Fe, 0.0023 Si), and CP Al of 99.97% purity (composition of (in wt. %): Al plus 0.125Fe, 0.023Cu, 0.021Zn, 0.016Mn, 0.016Mg, 0.015Ti, 0.10Si). The designed clad composite was composed of an Al sheath and five inserted Cu wires, see Figure 1a for schematic depiction of the cross-section of the designed composite.

The study deals with single and double pass TCAP. Schematic depiction of the TCAP die can be found elsewhere (e.g. [30]). The billets were at first processed via a single pass, and subsequently subjected to a double pass TCAP via the deformation route *C* (the billet was inserted to the second pass rotated by 180° - see ref. [31] for the details of the individual TCAP deformation routes and strain paths). Figure 1b shows the composite billets processed via single and double TCAP passes. Before assembling the billets, the component metals were heat treated for 30 min at 500 °C in an electric furnace to relax possible residual stress originating from previous processing, and homogenize the structures. The original billets were designed with square cross-sections and their dimensions were 12 mm × 12 mm × 130 mm (length). Room-temperature TCAP processing was performed using a hydraulic press with the extrusion rate of 5 mm s^{-1}, MoS_2 was used as lubricant.

a) 

b) 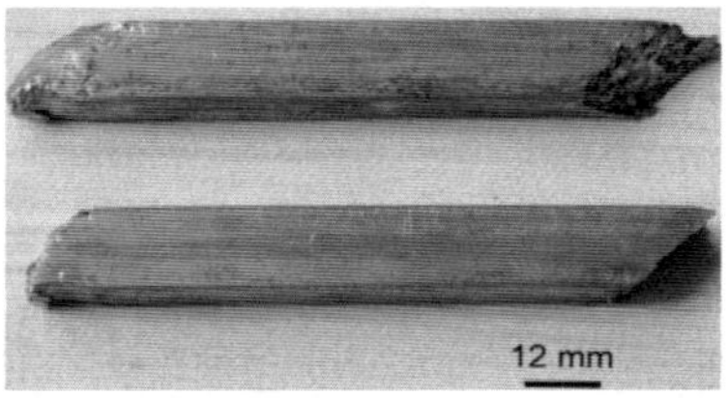

Figure 1: Schematic depiction of designed composite billet (a); experimentally processed composite billets (b).

The analyses of texture and grains misorientations within the clad composite billets processed via TCAP were performed using scanning electron microscopy (SEM), by the electron backscatter diffraction (EBSD) method in particular. The used microscope was Tescan Lyra 3 XMU FEG/SEMxFIB system equipped with Symmetry EBSD detector. The composite samples were prepared by manual grinding and subsequently polished via a combination of manual and electrolytic polishing. Scanning by EBSD was performed with the accelerating voltage of 20 kV, and steps of scanning of 50 nm. The subsequent evaluation of the scanned data was performed using AZtecCrystal 1.1 software by Oxford Instruments, and ATEX software [32].

3. Results and discussions

3.1. Texture characteristics

The effects of both the twist and main deformation zones on the developments of textures, i.e. grains orientations, within the reinforcing Cu wires of the extruded Al/Cu cladded billets were evaluated via stereographic projection, i.e. pole figures (PFs). The original Cu wires subjected to the initial heat treatment featured more or less random grains orientations with the tendency to form recrystallization texture of the ideal cube orientation (not shown here). However, after the first TCAP pass, the texture orientations acquired for the individual reinforcing Cu wires exhibited differences.

Figures 2a to 2e depict the {001}, {011}, and {111} pole figures for the Cu wires located in the upper left corner, upper right corner, axis, bottom left corner, and bottom right corner of the composite billet processed via single TCAP pass, respectively. The figures document that the highest texture intensity of almost six times random was observed for the axial Cu wire (Figure 2c). This wire exhibited a strong tendency to form ideal *A* fibre shear texture orientation ({111}||shear plane according to Beyerlein and Tóth [33]). The analyses further showed that the peripheral Cu wires exhibited lower maximum texture intensities, and also different tendencies to form the ideal orientations. However, the couples of the Cu wires located in the diagonals across the cross-section of the composite processed via single TCAP pass exhibited the tendencies to mirror. The Cu wire located in the upper left corner of the composite billet exhibited the preferential tendency to form the ideal *Ab* shear texture orientation (Figure 2a). This orientation belongs to the ideal *A* texture fibre and is defined by the Euler angles of $\phi 1=180°$, $\phi=35.26°$, and $\phi 2=45°$. Similar tendency was observed also for the Cu wire located in the bottom right corner of the composite (Figure 2e), although for the upper left Cu wire, the ideal *Ab* orientation was slightly shifted towards the ideal *Bb* orientation (defined by the Euler angles of $\phi 1=180$, $\phi°=54.74°$, and $\phi 2=45°$). The Cu wire located in the bottom left corner of the composite (Figure 2d) then exhibited the tendency to form the ideal *C* fibre shear texture orientation (defined by the Euler angles of $\phi 1=90$, $\phi°=45°$, and $\phi 2=0°$), similar to the Cu wire located in the upper right composite corner (Figure 2b). However, the ideal *C* fibre texture orientations were for both the wires shifted by approximately 10°. The observed slight shifts from the ideal texture fibre orientations, i.e. shift of the ideal *Ab* shear texture orientation towards the ideal *Bb* shear texture orientation for the upper left corner Cu wire, and shift of the ideal *C* shear texture orientation by approx. 10° for the bottom left corner Cu wire, can most probably be attributed to the effect of the twist deformation zone.

Figures 3a to 3e depict the {001}, {011}, and {111} PFs for the Cu wires located in the upper left corner, upper right corner, axis, bottom left corner, and bottom right corner of the composite processed via double TCAP pass, respectively. Figure 3c depicts that the highest ideal shear texture intensity was again observed for the axial Cu wire, however, the ideal *A* fibre orientation was for this Cu wire shifted towards the ideal *A2* orientation (defined by the Euler angles of $\phi 1=144.7°$, $\phi°=45°$, and $\phi 2=0°$) [34]. Strong tendencies to form the ideal *A2* texture orientation was observed also for the Cu wires in the upper left and upper right corners of the billet processed via double TCAP (Figures 3a and 3b), and the upper right Cu wire (Figure 3b) also featured the presence of ideal *A1* texture orientation (defined by the Euler angles of $\phi 1=35.3°$, $\phi°=45°$, and $\phi 2=0°$). As the Cu wires exhibited the tendency to mirror the characteristic ideal texture orientations across the cross-sectional diagonals also after the second TCAP pass, the ideal shear texture orientations observed for the Cu wire in the bottom right corner of the composite were comparable to the ideal orientations observed for the upper left Cu wire, and the ideal shear texture orientations observed for the Cu wire in the bottom left corner of the composite were comparable to the ideal orientations observed for the upper right Cu wire. Nevertheless, the maximum observed texture intensities were lower for the bottom couple of Cu wires than for the upper one (contrary to the texture intensities observed after the first TCAP pass, where the maximum texture intensity values were higher for the bottom couple of the Cu wires than for the upper one). These phenomena are imparted by the character of the TCAP die and by the effect of three independent strain paths affecting the composite billet during passing through the die [29]. Also, TCAP processing imparts severe grain refinement. Nevertheless, the introduced grain refinement features slight inhomogeneity across the billet cross-section, according to the imposed effective strain in the particular location [35]. Decrease in the ideal texture intensity is thus most probably related also to the grain refinement and consequent structure softening occurring in the particular location. The mentioned phenomena are also non-negligibly related to the development of residual stress.

IOP Conf. Series: Materials Science and Engineering 1190 (2021) 012032 doi:10.1088/1757-899X/1190/1/012032

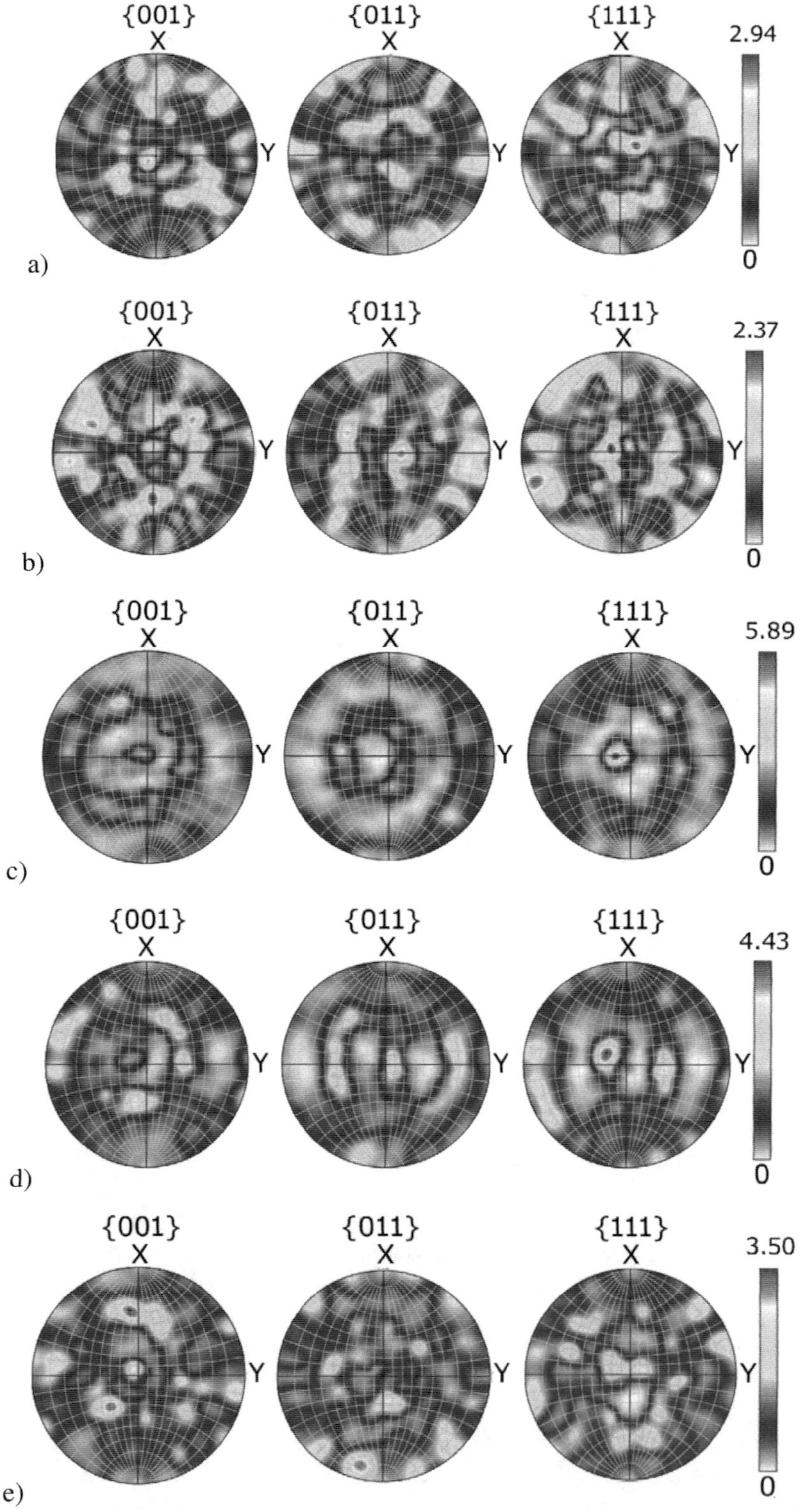

Figure 2: Pole figures for individual Cu wires within composite billet extruded via single pass TCAP: upper left (a); upper right (b); axial (c); bottom left (d); bottom right (e).

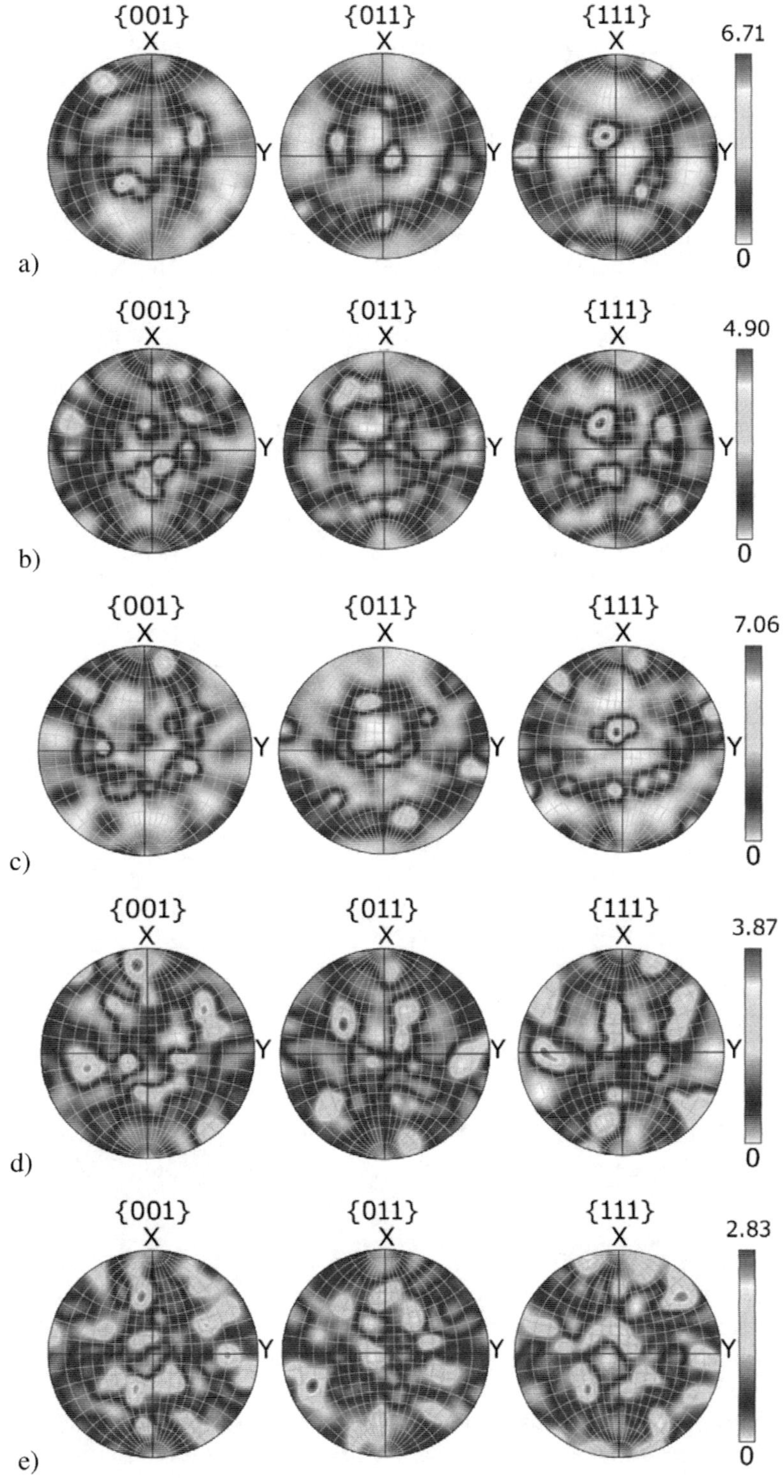

Figure 3: Pole figures for individual Cu wires within composite billet extruded via double pass TCAP: upper left (a); upper right (b); axial (c); bottom left (d); bottom right (e).

3.2. Residual stress

Figures 4a and 4b document the internal grains misorientations (depicted in rainbow colour scheme from 0° to 15°) pointing to the presence of residual stress in the reinforcing Cu wires within the composite billets processed via the single and double TCAP pass, respectively. The figures show that the presence and distribution of the residual stress within the axial Cu wire was comparable after the first and second TCAP pass. On the other hand, for the peripheral Cu wires, the presence of residual stress was more significant after the second pass, as the scans for the wires of this composite billet featured greater densities of the areas exhibiting higher internal grains misorientations. However, the high misorientations, i.e. residual stress, were more or less homogeneously distributed throughout the structures, which points to the lack of significant stress peaks within the Cu wires' structures (especially for the bottom couple of the Cu wires, which corresponds to the lower maximum texture intensities observed for these wires). Mutual comparison of the presence and distribution of the residual stress within the structures of the peripheral Cu wires and the axial Cu wire shows that the presence of residual stress was more pronounced in the peripheral Cu wires after both the single and double TCAP passes. The results of residual stress analyses are in accordance with the analyses of texture – as mentioned, less significant presence of the residual stress corresponded to the relatively lower intensities of ideal shear texture orientations for the bottom couple of the Cu wires after the second TCAP pass. This was true also for the upper couple of the Cu wires after the first TCAP pass.

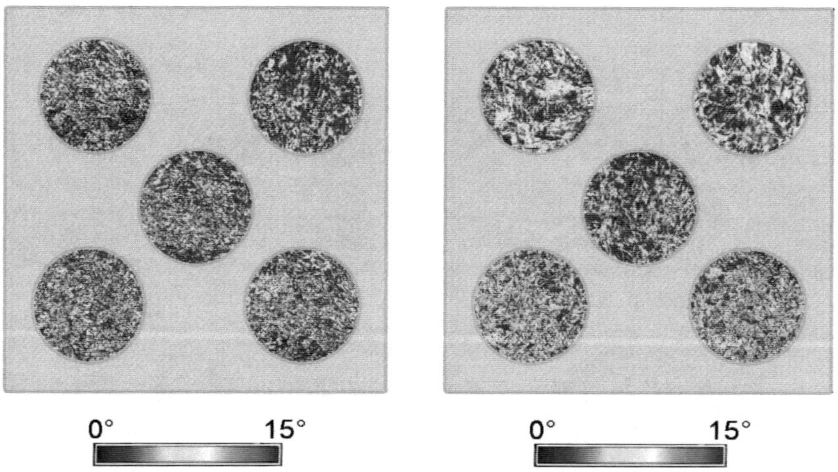

a) b)

Figure 4: Residual stress depicted via grains misorientations for individual wires within composite billets processed via: single pass TCAP (a); double pass TCAP (b).

4. Conclusions

The aim of this study was to perform experimental investigations of self-designed laminated Al/Cu composites processed via the twist channel angular pressing (TCAP) method. In particular, the study focused on evaluating the effects of single and double TCAP pass on modifications in texture and residual stress within the reinforcing Cu wires. The results showed that the first TCAP pass introduced satisfactory imposed strain to impart formation of ideal shear texture fibre orientations, however, the strain was not homogeneous across the cross-section of the processed composite as the individual Cu wires featured differences in their ideal orientations. Nevertheless, processing via double TCAP pass (deformation route *C*) imparted substantial homogenization of the imposed shear strain, as all of the reinforcing Cu fibres exhibited the dominant presence of ideal *A* fibre shear texture. The results of the analyses of residual stress corresponded to the results of texture analyses; the Cu fibres exhibiting higher texture intensities also exhibited greater presence of internal grains misorientations pointing to the presence of residual stress.

Acknowledgment(s)
This research was supported by the 19-15479S project of the Grant Agency of the Czech Republic.

References

[1] F.L. Matthews, R.D. Rawlings, "Composite materials: engineering and science", *Woodhead Publishing*, pp. 470, 1999.

[2] A.R. Studart, R.M. Erb, R. Libanori, "Bioinspired Hierarchical Composites," *Springer International Publishing, Cham*, pp. 287–318, 2015.

[3] L. Kunčická, A. Macháčková, R. Petrmichl, Z. Klečková, M. Marek, "Optimizing Induction Heating of WNiCo Billets Processed via Intensive Plastic Deformation," *Appl. Sci.* vol. 10, p. 8125, 2004.

[4] P. Lukáč, R. Kocich, M. Greger, O. Padalka, Z. Szaraz, "Microstructure of AZ31 and AZ61 Mg alloys prepared by rolling and ECAP," *Kov. Mater.* vol. 45, pp. 115-120, 2007.

[5] A.I. Alateyah, M.M.Z. Ahmed, Y. Zedan, H.A. El-Hafez, M.O. Alawad, W.H. El-Garaihy, "Experimental and Numerical Investigation of the ECAP Processed Copper: Microstructural Evolution, Crystallographic Texture and Hardness Homogeneity," *Metals* vol. 11, p. 607, 2021.

[6] L. Kunčická, R. Kocich, V. Ryukhtin, J.C.T. Cullen, N.P. Lavery, "Study of structure of naturally aged aluminium after twist channel angular pressing," *Mater. Charact.* vol. 152, pp. 94-100, 2019.

[7] R. Kocich, A. Macháčková, L. Kunčická, "Twist channel multi-angular pressing (TCMAP) as a new SPD process: Numerical and experimental study," *Mater. Sci. Eng. A* vol. 612, pp. 445–455, 2014.

[8] R. Kocich, L. Kunčická, A. Macháčková, "Twist Channel Multi-Angular Pressing (TCMAP) as a method for increasing the efficiency of SPD," *IOP Conf. Ser. Mater. Sci. Eng.* vol. 63, p. 012006, 2014.

[9] X. Wang, M. He, Z. Zhu, K. Xue, P. Li, "Influence of twist extrusion process on consolidation of pure aluminum powder in tubes by equal channel angular pressing and torsion," *Trans. Nonferrous Met. Soc. China.* vol. 25, pp. 2122–2129, 2015.

[10] H. Kim, H. Ha, J. Lee, S. Son, H.S. Kim, H. Sung, J.B. Seol, J.G. Kim, "Outstanding mechanical properties of ultrafine-grained Al7075 alloys by high-pressure torsion," *Mater. Sci. Eng. A* vol. 810, p. 141020, 2021.

[11] Y. Takizawa, T. Kajita, P. Kral, T. Masuda, K. Watanabe, M. Yumoto, Y. Otagiri, V. Sklenicka, Z. Horita, "Superplasticity of Inconel 718 after processing by high-pressure sliding (HPS)," *Mater. Sci. Eng. A* vol. 682, pp. 603–612, 2017.

[12] R. Kocich, A. Macháčková, F. Fojtík, "Comparison of strain and stress conditions in conventional and ARB rolling processes," *Int. J. Mech. Sci.* vol. 64, pp. 54–61, 2012.

[13] C.W. Wu, R.Q. Hsu, "Extrusion analysis and workability prediction of three-layer composite hexagonal clad rods, " *J. Mater. Process. Technol.* vol. 174, pp. 312–317, 2006.

[14] M. Hosseini, H. Danesh Manesh, "Bond strength optimization of Ti/Cu/Ti clad composites produced by roll-bonding, " *Mater. Des.* vol. 81, pp. 122–132, 2015.

[15] L. Guo, J. Wang, X. Yun, Z. Chen, "Fabrication of aluminum–magnesium clad composites by continuous extrusion, " *Mater. Sci. Eng. A* vol. 802, p. 140670, 2021.

[16] M. Movahedi, A.H. Kokabi, S.M. Seyed Reihani, "Investigation on the bond strength of Al-1100/St-12 roll bonded sheets, optimization and characterization," *Mater. Des.* vol. 32, pp. 3143–3149, 2011.

[17] I.K. Kim, S.I. Hong, "Effect of component layer thickness on the bending behaviors of roll-bonded tri-layered Mg/Al/STS clad composites," *Mater. Des.* vol. 49, pp. 935–944, 2013.

[18] S.N. Polyanskii, V.S. Kolnogorov, "Cladded Steel for the Oil and Gas Industries," *Chem. Pet. Eng.* vol. 38, pp. 703–707, 2002.

[19] X. Li, G. Zu, M. Ding, Y. Mu, P. Wang, "Interfacial microstructure and mechanical properties of Cu/Al clad sheet fabricated by asymmetrical roll bonding and annealing," *Mater. Sci. Eng. A*

vol. 529, pp. 485–491, 2011.

[20] X. Li, G. Zu, P. Wang, "Effect of strain rate on tensile performance of Al/Cu/Al laminated composites produced by asymmetrical roll bonding," *Mater. Sci. Eng. A* vol. 575, pp. 61–64, 2013.

[21] P.R. Matli, U. Fareeha, R.A. Shakoor, A.M.A. Mohamed, "A comparative study of structural and mechanical properties of Al–Cu composites prepared by vacuum and microwave sintering techniques," *J. Mater. Res. Technol.* vol. 7, pp. 165–172, 2018.

[22] J. Zhang, B. Wang, G. Chen, R. Wang, C. Miao, Z. Zheng, W. Tang, "Formation and growth of Cu–Al IMCs and their effect on electrical property of electroplated Cu/Al laminar composites," *Trans. Nonferrous Met. Soc. China.* vol. 26, pp. 3283–3291, 2016.

[23] Y.I. Matvienko, S.S. Polishchuk, A.D. Rud, O.Y. Popov, S.A. Demchenkov, O.M. Fesenko," Effect of graphite additives on microstructure and mechanical properties of Al–Cu composites prepared by mechanical alloying and sintering," *Mater. Chem. Phys.* vol. 254, p. 123437, 2020.

[24] O. Bouaziz, Y. Bréchet, J.D. Embury, "Heterogeneous and Architectured Materials: A Possible Strategy for Design of Structural Materials," *Adv. Eng. Mater.* vol. 10, pp. 24–36, 2008.

[25] L. Kunčická, T.C. Lowe, C.F. Davis, R. Kocich, M. Pohludka, "Synthesis of an Al/Al2O3 composite by severe plastic deformation," *Mater. Sci. Eng. A* vol. 646, pp. 234–241, 2015.

[26] P. Král, J. Staněk, L. Kunčická, F. Seitl, L. Petrich, V. Schmidt, V. Beneš, V. Sklenička, "Microstructure changes in HPT-processed copper occurring at room temperature," *Mater. Charact.* vol. 151, pp. 602–611, 2019.

[27] L. Kunčická, R. Kocich, J. Drápala, V.A. Andreyachshenko, " FEM simulations and comparison of the ecap and ECAP-PBP influence on Ti6Al4V alloy's deformation behaviour," *Met. 2013 22nd Int. Met. Mater. Conf.*, pp. 391–396, 2013.

[28] R. Kocich, M. Greger, M. Kursa, I. Szurman, A. Macháčková, "Twist channel angular pressing (TCAP) as a method for increasing the efficiency of SPD, "*Mater. Sci. Eng. A* vol. 527, pp. 6386–6392, 2010.

[29] L. Kunčická, R. Kocich, P. Král, M. Pohludka, M. Marek, "Effect of strain path on severely deformed aluminium, "*Mater. Lett.* vol.180, pp. 280–283, 2016.

[30] R. Kocich, L. Kunčická, P. Král, A. Macháčková, "Sub-structure and mechanical properties of twist channel angular pressed aluminium," *Mater. Charact.* vol. 119, pp. 75–83, 2016.

[31] R. Kocich, J. Fiala, I. Szurman, A. Macháčková, M. Mihola, "Twist-channel angular pressing: effect of the strain path on grain refinement and mechanical properties of copper," *J. Mater. Sci.* vol. 46, pp. 7865–7876, 2011.

[32] B. Beausir, J.J. Fundenberger, "Analysis Tools for Electron and X-ray diffraction," *ATEX - software*, www.atex-software.eu, 2017.

[33] I.J. Beyerlein, L.S. Tóth, "Texture evolution in equal-channel angular extrusion," *Prog. Mater. Sci.* vol. 54, pp. 427–510, 2009.

[34] L.S. Toth, P. Gilormini, J.J. Jonas, "Effect of rate sensitivity on the stability of torsion textures", *Acta Metall.* vol 36, pp. 3077-3091, 1988.

[35] R. Kocich, L. Kunčická, M. Mihola, K. Skotnicová, "Numerical and experimental analysis of twist channel angular pressing (TCAP) as a SPD process," *Mater. Sci. Eng. A* vol. 563, pp. 86–94, 2013.

Influence of Rotary Swaging on Creep Behaviour of P92 Steel

Petr Kral [1], Jiri Dvorak [1], Lenka Kuncicka [1], Radim Kocich [1], Marie Kvapilova [1], Marie Svobodova [2], Vaclav Sklenicka [1]

[1] Institute of Physics of Materials, Czech Academy of Sciences, Zizkova 22, 616 62 Brno, Czech Republic
[2] UJP PRAHA a.s., 156 10 Praha-Zbraslav, Czech Republic

pkral@ipm.cz

Abstract. This work investigates the influence of rotary swaging on creep behaviour and microstructure heterogeneity of P92 martensitic steel. Its as-received coarse-grained state was processed by rotary swaging at room temperature. The bar of 30 mm in diameter was reduced at room temperature down to 15 mm by rotary swaging. The value of equivalent strain imposed during rotary swaging was about 1.4. Microstructures were investigated by scanning electron microscope Tescan Lyra 3 equipped with NordlysNano EBSD detector operating at an accelerating voltage of 20 kV with specimen tilted at 70°. It was found that the processing by rotary swaging led to the formation of heterogeneous microstructure containing a mixture of fine nearly equiaxed grains and large grains, which were significantly, elongated parallel to the swaging direction. Constant load tensile creep tests in tension and compression were conducted at 923 K under different applied stresses. In order to evaluate the influence of microstructural heterogeneity on the creep behaviour, the creep tests were performed both in longitudinal and also transverse directions to the rotary swaging. The results of tensile creep tests revealed that specimens manufactured in the direction parallel to the rotary swaging exhibited significantly higher creep life and strain to fracture in comparison with the results measured in the transverse direction. The creep results showed that creep tests performed at medium stresses exhibit a pronounced minimum creep rate. The microstructure results suggest that creep behaviour depends on the microstructure coarsening during creep testing and the orientation of the grain boundaries with respect to the applied stress.

1. Introduction

The rotary swaging (RS) is one of the severe plastic deformation (SPD) methods [1-3] that enables to impose relatively large strain into material. Thus, the standard coarse-grained microstructure is refined by subsequent transformation of subgrain- into grain- boundaries [4,5]. Usually the microstructure processed by SPD at room temperature is homogeneous and completely ultrafine-grained (UFG) when the equivalent strain higher than 10 is imposed in the material. However, the strain imposed during rotary swaging is not so high compared to other SPD methods [6], such as high-pressure torsion (HPT) or equal-channel angular pressing (ECAP).

It was observed that SPD-processed microstructures with the imposed strain about one are heterogeneous and contain predominantly low-angle grain boundaries (LAGBs). These materials may exhibit higher creep life than its coarse-grained undeformed state [7,8]. However, the creep properties of SPD-processed materials are usually tested only in one section, usually parallel to the direction of deformation [7]. In other sections, the creep behaviour may be different, especially at low imposed strains when the microstructure is heterogeneous. In the present time there are no works investigating creep behavior of RS-processed materials in different sections. The aim of this work is comparison of creep behavior of RS-processed P92 steel in the different sections with respect to the direction of rotary swaging.

2. Experimental material and procedures

The experimental material used in present work was advanced tungsten modified 9%Cr P92 steel. The chemical composition is shown in Table 1. The as-received coarse-grained state was normalized at 1323 K/60 min/air and tempered at 1013 K/140 min/air. This state is referred to herein as the received state.

Table 1. Chemical composition (wt.%) of the investigated P92 pipe.

Element	C	Cr	Mo	W	Si	Mn	V	Nb	P	N	Al	S
wt.%	0.11	8.58	0.33	1.67	0.37	0.48	0.23	0.06	0.013	0.037	0.017	0.005

The von Mises equivalent strain imposed by rotary swaging [9,10] was estimated by $\varepsilon_{eq} = \ln(D_0/D_n)^2$, where D_0 is the initial diameter ~ 30 mm and D_n is the final diameter ~ 15 mm after application of RS at room temperature. The imposed equivalent strain was about 1.4.

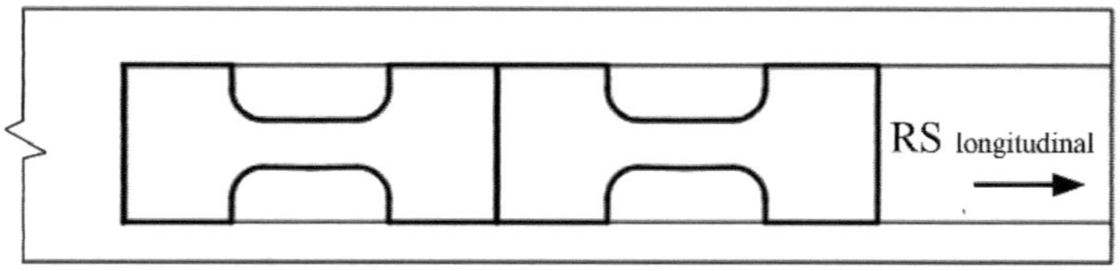

a)

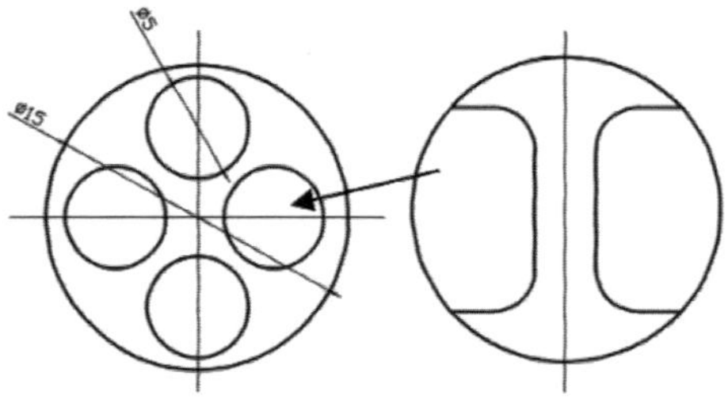

b)

Figure 1. Scheme of arrangement of tensile specimens on the RS-processed rod; a) longitudinal and b) transverse direction

The tensile (Figure 1) and compression specimens were manufactured in longitudinal and transverse sections with respect to the rotary swaging direction. Constant load tensile creep tests were conducted at 923 K in protective argon atmosphere. For tensile creep tests were used flat specimens with the gauge length of 10 mm and the cross section of 3 x 1 mm^2 cut in the longitudinal section. The flat tensile specimens cut from transverse section had the gauge length of 4 mm and the cross section of 3 x 1 mm^2. The tensile creep test at constant load were run up to the final fracture. Uniaxial compression tests at constant stress were conducted on cylindrical specimens with initial height of 10 mm and diameter of 5 mm. The compression creep tests were interrupted at creep strain about 0.2 – 0.25. Microstructure investigations were performed using a scanning electron microscope (SEM, Tescan Lyra 3 equipped with NordlysNano EBSD detector operating at accelerating voltage of 20 kV with specimen tilted at 70°). EBSD was used to determine the misorientations θ between neighbouring grains. Misorientation θ = 15° was taken to distinguish HAGBs from LAGBs.

3. Results and discussions

3.1. Creep behaviour

Figure 2 shows the comparison of tensile creep test of specimens manufactured in the longitudinal and transverse directions. The results demonstrate that transverse creep specimens exhibit faster minimum creep rate $\dot{\varepsilon}_{min}$, shorter creep life and lower ductility in comparison with longitudinal creep specimens. The longitudinal creep specimens also exhibit increase of fracture strain ε_f with a decrease of applied stress from 150 MPa to 80 MPa. However, the opposite results can be found in the creep of transversal tensile specimens.

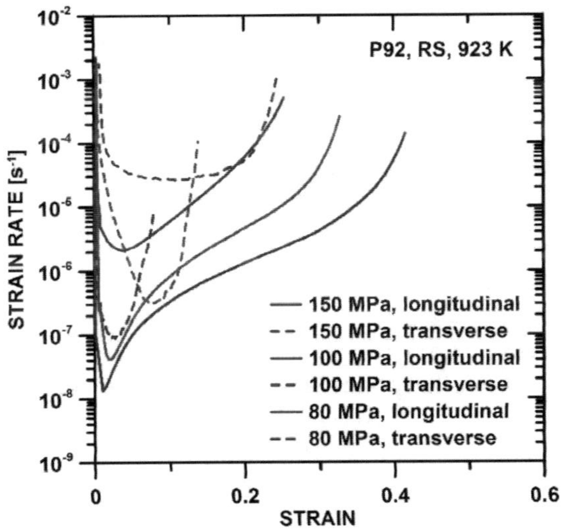

a)

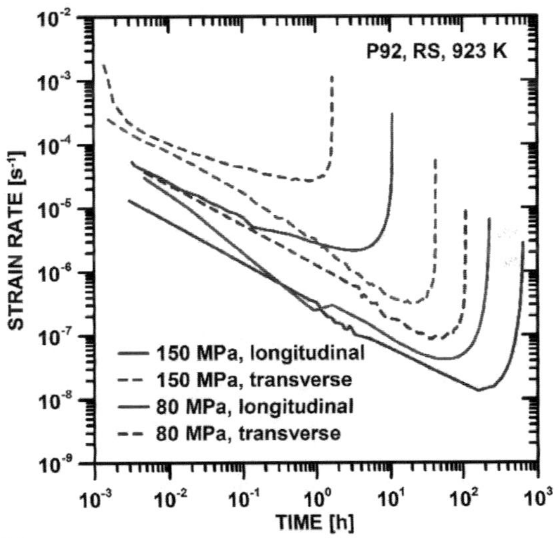

b)

Figure 2. Tensile creep curves for longitudinal and transverse RS-processed specimens; a) strain rate vs. strain; b) strain rate vs. time.

One can see that specimen parallel to RD tested at 80 MPa exhibits a pronounced softening after reaching of minimum creep rate. The softening slows down in the strain region of about 0.1 – 0.3. In this region the largest creep strain occurs.

The pronounced softening after reaching of $\dot{\varepsilon}_{min}$ is also seen in the strain rate – strain dependence (Figure 2) for transverse specimen tested at 80 MPa. However, after the pronounced softening, there is no slowing down of the creep rate but fracture processes occur.

The differences in creep behaviour of tensile specimens manufactured in longitudinal and transverse sections can be caused by microstructure heterogeneity that leads to faster fracture processes in transverse specimens.

Figure 3 shows the compression creep tests at constant stress of longitudinal and transverse specimens. The compression tests were interrupted at creep strain about 0.2 – 0.25.

Although the compression creep tests are not influenced by fracture, the minimum creep rates $\dot{\varepsilon}_{min}$ of longitudinal specimens are about order of magnitude slower compared to transverse compression specimens.

In Figure 3 is also shown the stress change test of longitudinal specimen. The stress was increased from 100 to 150 MPa at creep strain about 0.05. The creep results demonstrate that $\dot{\varepsilon}_{min}$ of longitudinal specimen after stress increase from 100 to 150 MPa at strain of about 0.05 is faster than $\dot{\varepsilon}_{min}$ of longitudinal specimen tested only at 150 MPa. The faster $\dot{\varepsilon}_{min}$ after strain about 0.05 can be explained by microstructure changes occurring during creep testing, such as microstructure coarsening [6,8].

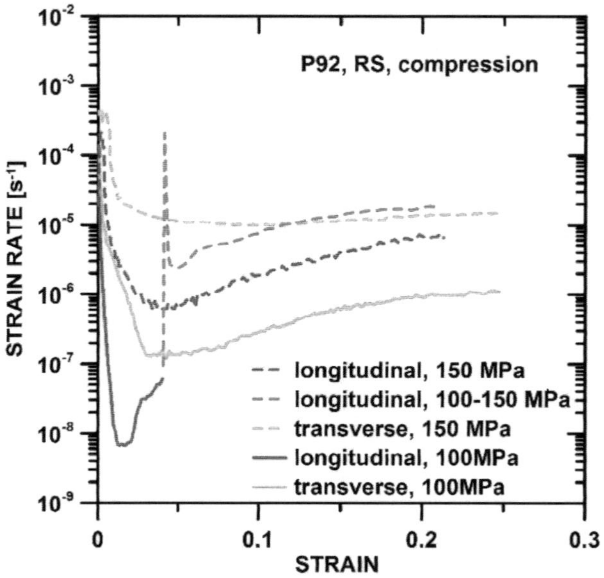

Figure 3. Strain rate vs. strain curves for longitudinal and transverse specimens tested in compression at 923 K.

3.2. Microstructure after creep

Figure 4 shows the microstructure processed by RS and after subsequent creep testing at 100 MPa in the longitudinal and transverse sections. In Figure 4a are shown the directions of applied stress in the case of longitudinal (σ_L) and transverse (σ_T) specimens. One can see that the creep testing led to the significant coarsening of microstructure. The tensile longitudinal specimen reached fracture strain about 0.34 (Figure 2a). However, the microstructure observed in transverse section after RS processing and after subsequent creep showed no significant grain coarsening during creep. The tensile specimen tested at 100 MPa reached fracture strain only about 0.15. For this reason, the higher fracture strain in longitudinal specimen compared to transverse specimen can be caused by significant coarsening of microstructure during creep testing.

Figure 4d shows the cumulative frequency of boundaries measured in longitudinal and transverse section of P92 steel processed by RS and after subsequent creep. One can see that creep testing led to the decrease of LAGBs in the microstructure. The decrease of LAGBs, formed by RS processing at room temperature, during creep testing leads to the predominance of HAGBs in the microstructure. It is seen that decrease of LAGBs after creep is larger in the coarsened microstructure of longitudinal section than in microstructure of transverse section. For this reason, the contribution of grain boundary mediated processes to the total creep strain is higher [11-13] in the longitudinal compared to the transverse specimen.

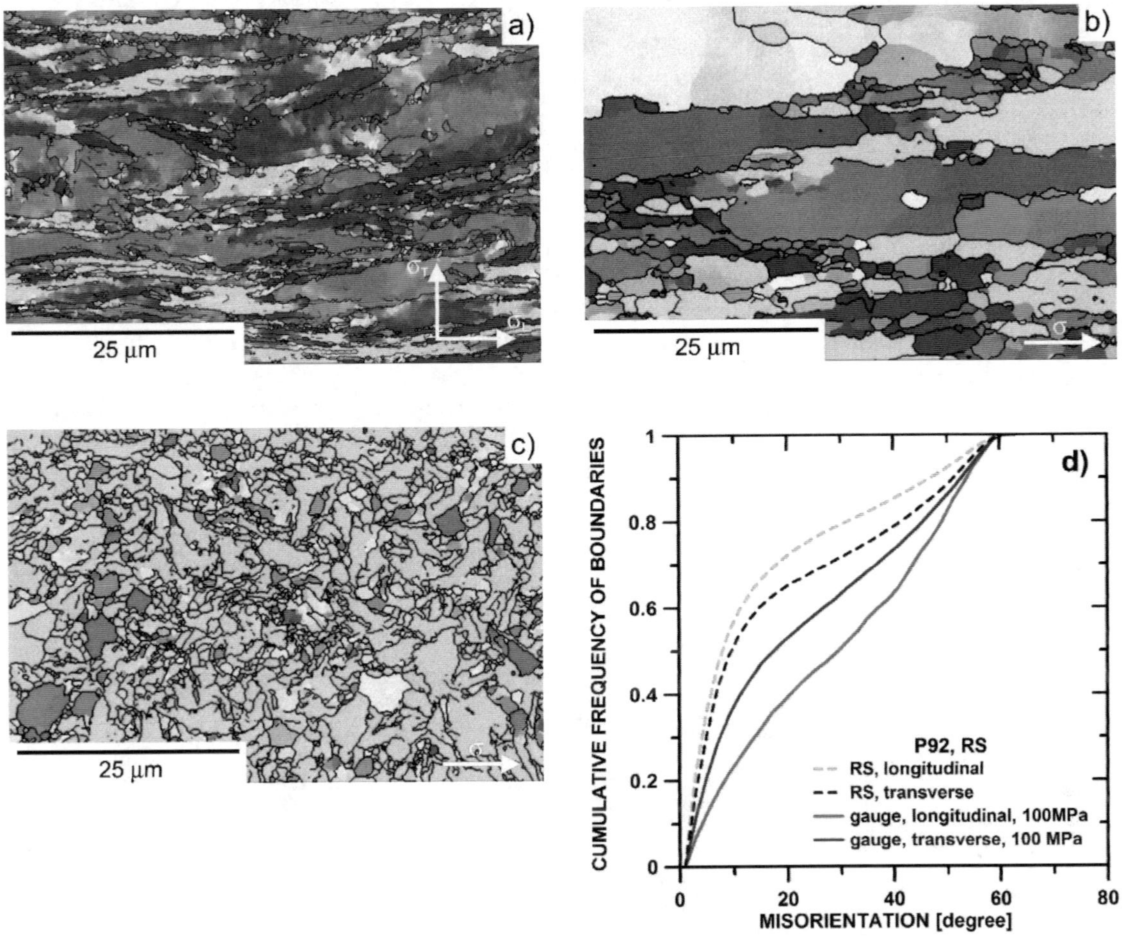

Figure 4. Microstructure of P92 steel; a) processed by RS (longitudinal section), b) after subsequent creep at 100 MPa (longitudinal section), c) after subsequent creep at 100 MPa (transverse section) and d) cumulative frequency of boundaries measured in longitudinal and transverse section.

Figure 5 shows the cavities near fracture surface in longitudinal RS-processed specimen tested at 150 MPa. It is seen that cavities are predominantly formed along boundaries parallel to the stress axis. Although the cavities often nucleate on boundaries which are transverse to the stress axis. In the specimen manufactured in the longitudinal section cavities are distributed along boundaries parallel to the stress axis. It can be suggested that the nucleation of cavities is associated with the high local stress concentration at large carbides distributed along grain boundary [11,14].

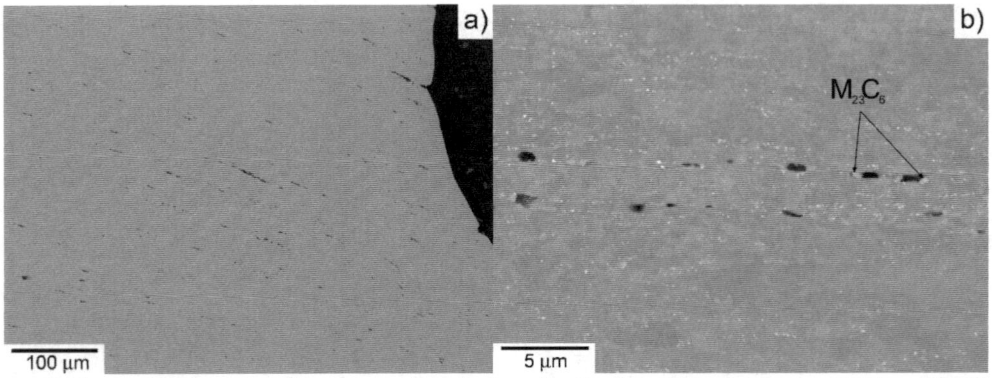

Figure 5. Cavitation near the fracture surface in longitudinal tensile specimen tested at 923 K and 150 MPa; a) long cavities approximately parallel to the stress axis, b) formation of cavities along boundaries. Stress axis is horizontal.

Figure 6 shows the cavities near fracture surface in the transverse RS-processed specimen tested at 150 MPa. The results demonstrate that near fracture surface can be seen many small cavities. These small cavities are formed at grain boundary triple points (Figure 6b) and some of them look like wedge-type cracks. One can see that cavities propagate and interlinked together along transverse grain boundaries. The previous studies suggested that the cavities at triple points can be caused by grain boundary sliding insufficiently accommodated by high local plastic deformation [15-17]. However, the fracture strain of transverse specimen significantly decreases with decreasing value of applied stress.

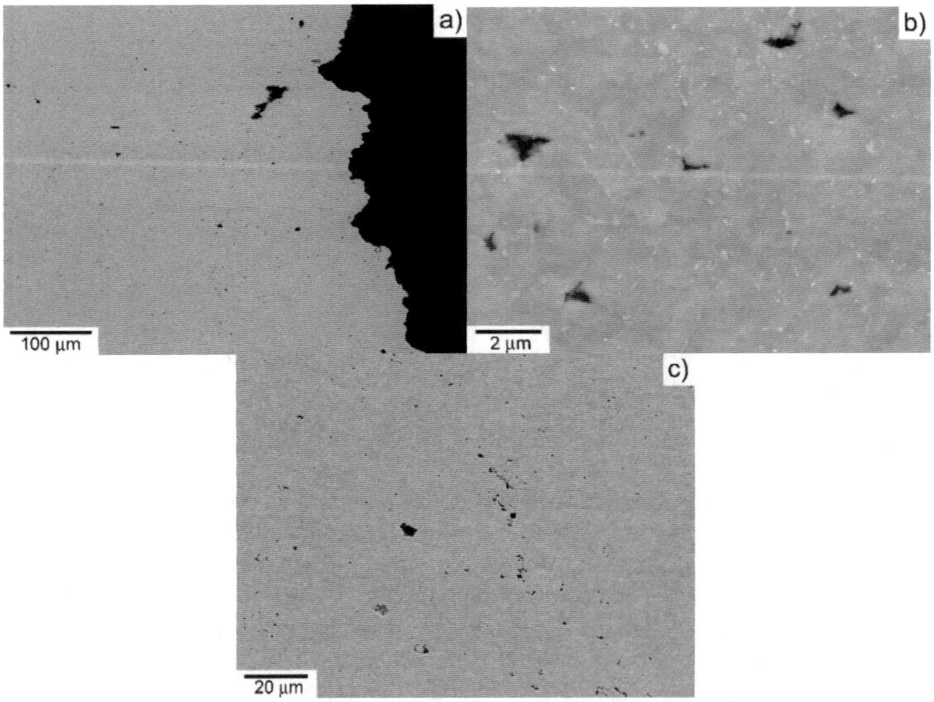

Figure 6. Cavitation in transverse tensile specimen tested at 923 K and 150 MPa; a) cavities near the fracture surface, b) formation of wedge cracks at triple points, c) interlinkage of small cavities along boundaries approximately perpendicular to the stress axis. Stress axis is horizontal.

It can be caused by the fact that boundaries of elongated grains formed during RS-processing (Figure 4a) are, in the case of transverse specimen, predominantly perpendicular to the stress axis. Many works [11,14] reported that cavities nucleate and propagate particularly along boundaries transverse to the stress axis. For this reason the formation and grow of cavities can be easier in the transverse specimen which led to the premature fracture at lower creep strain compared to longitudinal specimen.

4. Conclusions

Creep tensile and compression tests were conducted on an advanced creep-resistant 9% Cr P92 steel to evaluate the creep behaviour of RS-processed steel in longitudinal and transverse section. The creep tests were performed at 923 K and with an applied stress range from 80 to 150 MPa.

The main results are as follows:

1. The RS-processed P92 steel exhibits shorter creep life and lower ductility in transverse section compared to the longitudinal creep specimens.

2. The higher creep strain in longitudinal tensile specimens led to the significant coarsening of microstructure.

3. The formation of cavities and resulting ductility of RS-processed P92 steel depends on the orientation of the grain boundaries with respect to the applied stress.

Acknowledgment(s)

The authors acknowledge financial support from the Czech Science Foundation (grant No. 19-18725S). The work was supported in part by a grant-in-aid from MEXT, Japan, for scientific research (A) (No. J19H00830).

References

[1] R.. Kocich, L. Kunčická, C.F. Davis, T.C. Lowe, I. Szurman, A. Macháčková, Deformation behavior of multilayered Al–Cu clad composite during cold-swaging, *Mater. Des.*, vol. 90, pp. 379–388, 2016.

[2] L. Kunčická, R. Kocich, P. Strunz, A. Macháčková, Texture and residual stress within rotary swaged Cu/Al clad composites. *Mater. Lett.*, vol. 230, pp. 88-91, 2018.

[3] R. Kocich, L. Kunčická, P. Král, P. Strunz, Characterization of innovative rotaryswaged Cu-Al clad composite wire conductors, *Mater. Des.*, vol. 160, pp. 828–835, 2018.

[4] R.Z. Valiev, R.K. Islamgaliev, I.V. Alexandrov, Bulk nanostructured materials from severe plastic materials, *Prog. Mater. Sci.*, vol. 45, pp. 103-189, 2000.

[5] O. Renk, R. Pippan, Saturation of Grain Refinement during Severe Plastic Deformation of Single Phase Materials: Reconsiderations, Current Status and Open Questions, *Mater. Trans.*, vol. 60, pp. 1270 – 1282, 2019.

[6] P. Král, J. Dvořák,V. Sklenička,T. Masuda,Y. Tang, Z. Horita, L. Kunčická, K. Kuchařová, M. Kvapilová, M. Svobodová, Effect of severe plastic deformation on creep behaviour and microstructure changes of P92 at 923K, *Kovove Mater.*, vol. 59, pp. 141–148, 2021.

[7] P. Kral, J. Dvorak, V. Sklenicka, T.G. Langdon, The Characteristics of creep in metallic materials processed by severe plastic deformation, *Mater. Trans.*, vol. 60, pp. 1506-1517, 2019.

[8] P. Kral, J. Dvorak, W. Blum, V. Sklenicka, Z. Horita, Y. Takizawa, Y. Tang, L. Kunčická, R. Kocich, M. Kvapilova, M. Svobodova, The effect of predeformation on creep strength of 9% Cr steel, *Materials*, vol. 13, 5330, 2020.

[9] L. Kunčická, R. Kocich, J. Dvořák, A. Macháčková, Rotary swaged laminated Cu-Al

composites: Effect of structure on residual stress and mechanical and electric properties, *Mater. Sci. Eng.*, vol. 742, pp. 743-750. 2019.

[10] A. Macháčková, L. Krátká, R. Petrmichl, L. Kunčická, R. Kocich, Affecting structure characteristics of rotary swaged tungsten heavy alloy via variable deformation temperature. *Materials*, vol. 12, 4200, 2019.

[11] M.E. Kassner, Fundamentals of Creep in Metals and Alloys. Second edition, Elsevier, Amsterdam, 2009.

[12] W. Blum, X.H. Zeng, A simple dislocation model of deformation resistance of ultrafine-grained materials explaining Hall–Petch strengthening and enhanced strain rate sensitivity, *Acta Mater.*, vol. 57, pp. 1966–1974, 2009.

[13] P. Král, J. Dvořák, A. Jäger, M. Kvapilová, Z. Horita, V. Sklenička, Creep properties of aluminium processed by ECAP, *Kovove Mater.*, vol. 54, pp. 441–451, 2016.

[14] J. Čadek, Creep in Metallic Materials, Elsevier Science Publishers, Amsterdam, The Netherlands 1988.

[15] Y. Takigawa, J. V. Aguirre, E. M. Taleff, K. Higashi, Cavitation during grain-boundary-sliding deformation in an AZ61magnesium alloy, *Mater. Sci. Eng. A*, vol. 497, pp. 139–146, 2008.

[16] R. B. Figueiredo, S. Terzi, T. G. Langdon, Using X-ray microtomography to evaluate cavity formation in a superplastic magnesium alloy processed by equal-channelangular pressing, *Acta Mater.*, vol. 58, pp. 5737–5748, 2010.

[17] Y. Takizawa, T. Kajita, P. Kral, T. Masuda, K. Watanabe, M. Yumoto, Y. Otagiri, V. Sklenicka, Z. Horita, Superplasticity of Inconel 718 after processing by high-pressure sliding(HPS), *Mater. Sci. Eng. A*, vol. 682, pp. 603-612, 2017.

IOP Conf. Series: Materials Science and Engineering 1190 (2021) 012034 doi:10.1088/1757-899X/1190/1/012034

Design and Numerical Characterization of a Human Ankle Prosthesis

Copilusi Cristian[1], Geonea Ionut[1], Margine Alexandru[1], Constantin Andra[1]

[1]University of Craiova, Faculty of Mechanics 107, Craiova, Dolj, Romania

cristian.copilusi@edu.ucv.ro

Abstract. This paper addresses to a finite element analysis in dynamic conditions of a human ankle joint prosthesis based on a cam mechanism design. The proposed ankle joint prosthesis mechanism is a parameterized one and was designed in order to fulfil main human locomotion tasks namely walking during daily activities. A proper 3D CAD model has been developed and the proposed cam mechanism ankle prosthesis was modelled and simulated into MSC Visual Nastran environment. Results of simulations demonstrate the engineering feasibility of the design and validate the proposed parameterized prototype.

1. Introduction

Today there are many persons with locomotion deficiencies due to several dis-eases or accidents, from where these lost some human leg segments. For these persons different research centers develop human lower limb prosthesis which can improve or increase the human locomotion ability during daily living activities and moreover.

Significant existing prostheses are presented in [1, 2, 3, 4, 5] with different de-signs, especially dedicated to human ankle joint. In addition, a main common drawback can be considered the ankle joint angular limitation from where is no equivalent mechanism of this human joint, only a dumper. Kinematic analysis methods and techniques can be created to model the human natural walking, like in [1, 7, 8]. They can be also used as input points for dynamic analyses and to design human lower limb prosthesis equivalent to hip, knee and ankle joints. A similar model with the one proposed through this research was designed in [11, 12], and it was validated for an amputee namely a human male subject. Thus, our challenge is to adapt this to a human female subject, with different anthropometric data. In this case, we keep the same functionality and structure principle. This will be represented by a cam mechanism that will be evaluated from dynamic viewpoints, with finite element method for fulfilling human ankle joint mobility during walking. The proposed ankle joint prosthesis was designed with a low-cost concept and an equivalent DOFs to a natural human ankle joint. This design was characterized with results from simulations in a dynamic mode, of ankle prosthesis basic functionality.

2. Human walking experimental analysis

Nowadays there are different equipments and specific software which can be used to understand and to analyze human motions through walking experimental tests for biomechanics characterization. For this we use a motion analysis equipment called CONTEMPLAS and experimental analysis were carried out at Craiova University [9].

IOP Conf. Series: Materials Science and Engineering 1190 (2021) 012034 doi:10.1088/1757-899X/1190/1/012034

Our experimental research will be focused on the human ankle joint motion in order to acquire this in a polynomial form for using it in virtual simulations as input data for the parameterized ankle joint prosthesis. The experimental analysis principle is to place reflective markers on the desired joints, and with the aid of two cameras we track trajectories and angles in real time. A proper experimental analysis was performed in case of a female with 1.68m height, 65kg, age 28 by using a reflective marker on each human joint centre.

A snapshot during experimental analysis is provided in figure 1 and in figure 2 is shown the obtained result namely human ankle joint motion law developed by the analyzed female. A complete gait was performed in 1.4 seconds.

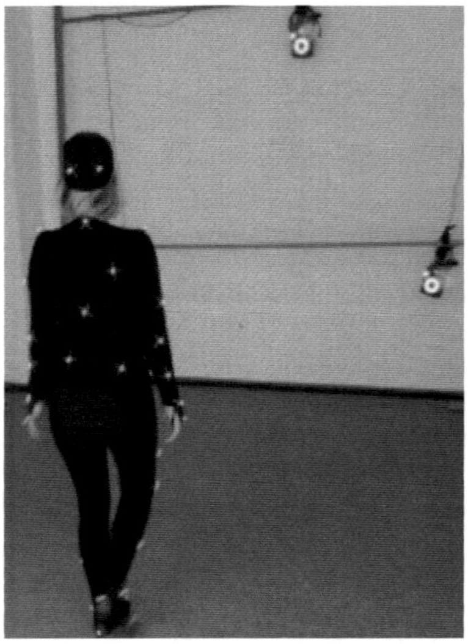

Figure 1. A snapshot from experimental analysis

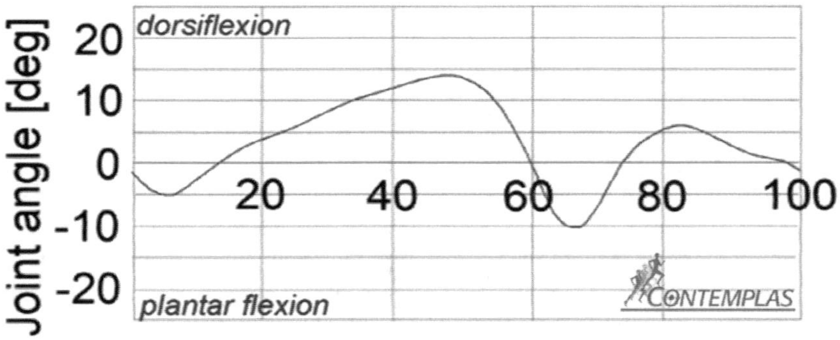

Figure 2. Ankle joint angle [degrees] vs. Gait [%]

By having in sight the data literature reported in [10], it was obtained an ankle angular amplitude of 23degrees during walking and this is a reasonable value situated between 20 to 30 degrees. The polynomial form of this motion law it has an accuracy of 95% from the represented curve reported in figure 2 and has the following form:

$$\alpha(t) = -2.39812t^6 - 1.77398012t^5 - 0.0004t^4 - \\ -0,0006t^3 - 0,0389t^2 + 0,9908t - 7,3084 \tag{1}$$

3. Ankle prosthesis cam mechanism concept

The proposed ankle prosthesis has on its structure a cam mechanism which will actuate a shock absorber during gait, and also this will facilitate a proper support for the shank component. A structural scheme of the proposed mechanism is shown in figure 3.

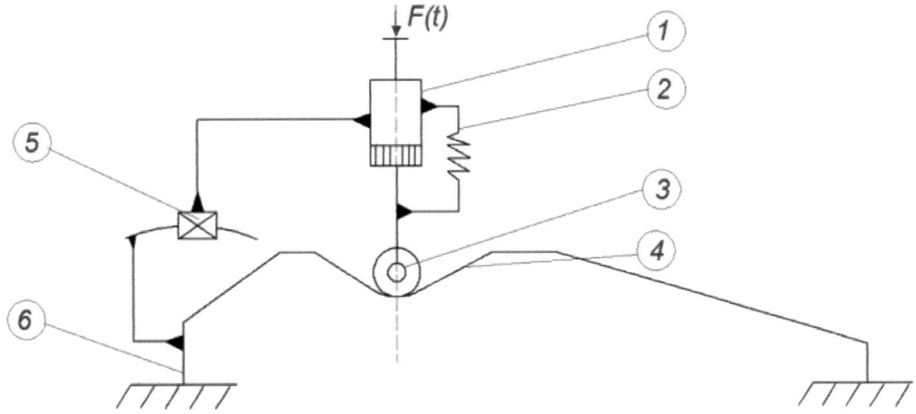

Figure 3. Ankle joint prosthesis structural scheme

The cam mechanism will command and control the shock absorber and from figure 3 it can be identified the following components: 1- shock absorber connect-ed to tibia element; 2 – spring; 3 – cam follower connected with the shock absorb-er rod, 4 – the cam integrated on the human artificial foot; 5 – translational joint with a specific profile which respect human ankle motion law; 6 – artificial foot. This is a planar mechanism with a single DOF and details regarding of this mech-anism design can be remarked in [11, 12].

4. Ankle prosthesis cam mechanism design

Based on the cam mechanism model and by having in sight the anthropometrical data of the chosen human female subject from the experimental analysis, a parameterized ankle prosthesis was designed with the aid of CATIA 3D modelling software. This is shown in figure 4.

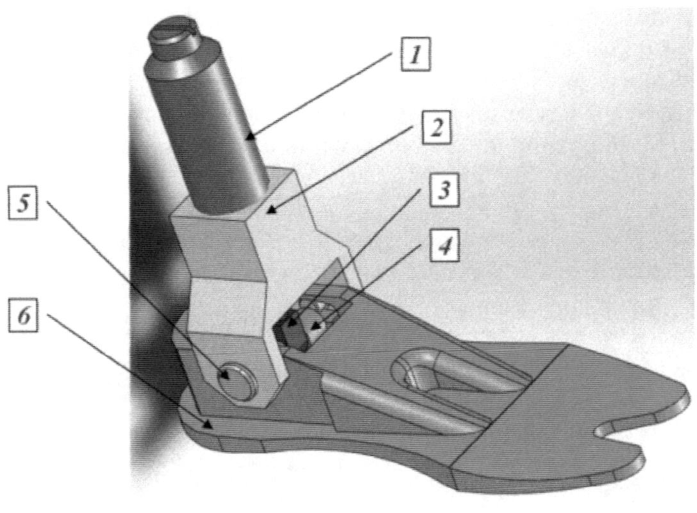

Figure 4. Identifying the components of the designed ankle joint prosthesis virtual model

From figure 4 it can be remarked the following components: 1 – tibia component – which in this case is represented by the shock absorber (it has the dumper adjusting posibility and it will attach to the artifical shank) ; 2 – shock absorber support; 3 – shock absorber rod; 4 – roller connected to the shock's rod and it will slide on the cam profile described by the dumping displacement; 5 – bolt which will slide on a imposed profile by the human ankle joint motion law; 6 – Greissinger foot. A crossection of the proposed cam mechanism human ankle joint prosthesis is shown in figure 5.

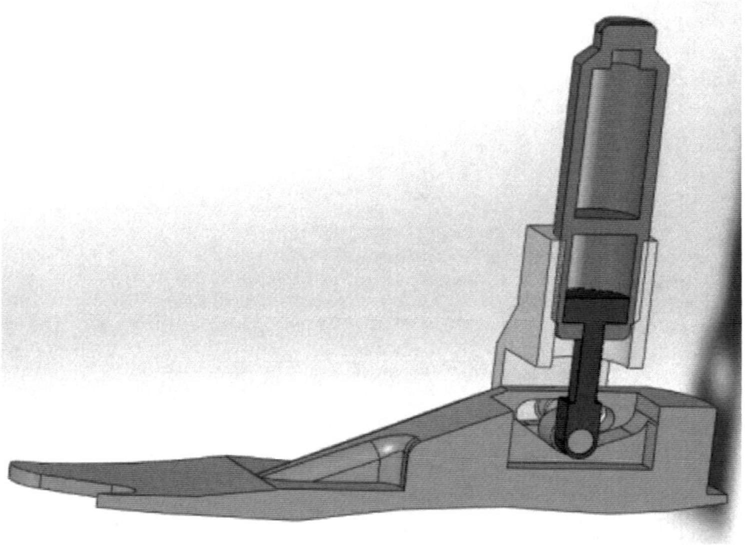

Figure 5. A detailed view of the ankle joint prosthesis virtual model with cam mechanism

5. Ankle prosthesis dynamic analysis

By having in sight the designed virtual model of the proposed ankle joint prosthesis mechanism, this was imported in MSC Visual Nastran environment.

For each component it was defined proper material characteristics, namely for rubber, steel and aluminium alloy. The entire model was meshed with tetrahedral finite elements and it was count a number of 44058 finite elements. As regarding the loads and initial conditions the foot component was fixed in space and at the shock absorber head it was applied a force variable in time obtained for a single gait from [1, 3]. By having in sight the literature overview, this force should be higher than initial force generated by the proper mass of the analyzed human female.

Thus in our case, by computing the mathematical models reported in [1, 3], this will have a values multiplied by 2, which was recorded in critical situations like stairs climbing or climbing on a angular surface.

The force variation F(t) is presented in diagram from figure 6 and in figure 7 are shown the applied loads on the parameterized virtual model. Contact forces and friction forces were considered active ones during virtual simulations, according with the values presented in table 1. These values were applied in case of contact between cam profile and cam follower. For the sliding action in case of bolt no 5, friction will be neglected, due to the add of a bearing elements in case of the real prototype.

As regarding the time setup for virtual simulations in dynamic conditions, this was the same during experimental analysis evaluation for performing a single gait, namely 1.4 seconds.

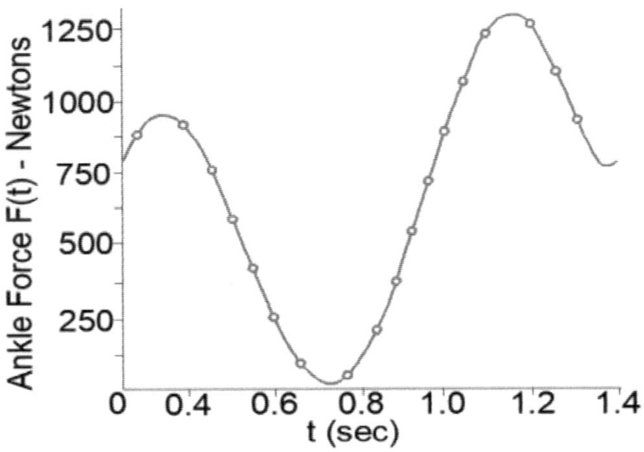

Figure 6. Force variation *F(t)* applied on shock absorbers head during time

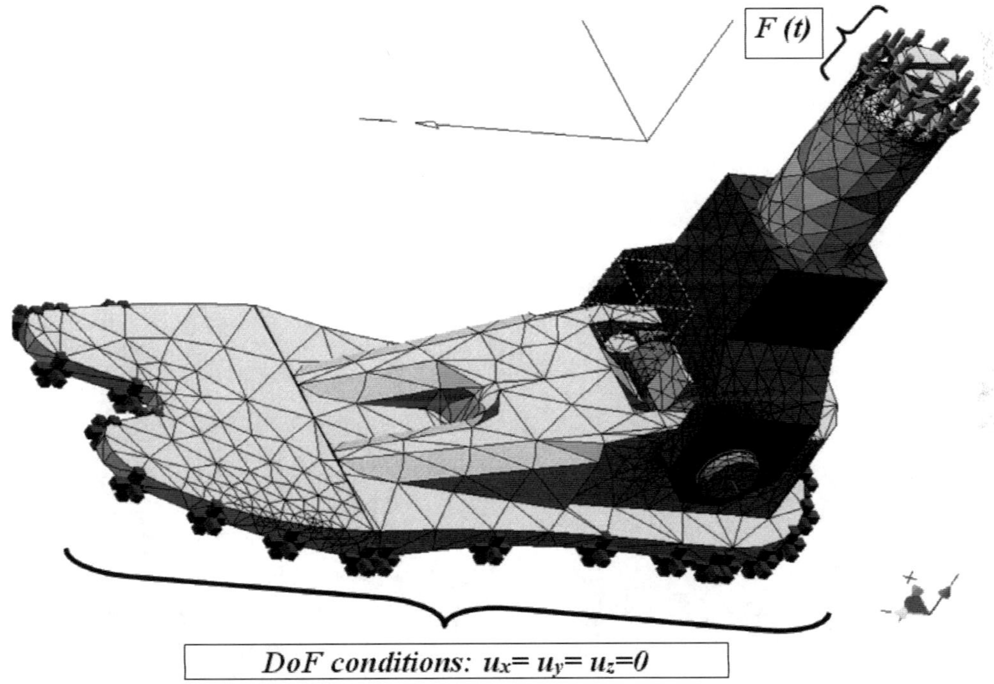

Figure 7. Initial conditions and loadings applied on the ankle prosthesis

Table 1. Input parameters for dynamic simulation.

Parameter	Type/Value	*Units*	Parameter	Type/Value	*Units*
Contact μs	0,65	-	**Contact μd**	0,45	-
Penetration depth	0,075	mm	**Force exponent**	1,65	-
Friction force	Coulomb	N	**Damping**	34	N·sec/mm

According with the figure 6, the ankle joint motion law was applied as a polynomial form at the level of component no. 5 by having in sight the equation 1. The mentioned load was applied as a polynomial form, onto shock absorber body and this will have a sliding motion imposed by the link no 5 profile from figure 3. The force equation has the following form:

$$F(t) = -1154.22967t^6 + 4238.6521t^5 - 9554.2369t^4 +$$
$$+615.33456t^3 - 1654.2367t^2 + 183.7746t + 0.5587325 \quad (2)$$

Some snapshots acquired during simulations with the dynamic response represented through stress, are shown in figures 8 and 9. Thus, the dynamic response was characterized by stress, deformations and displacements depending on time through diagrams reported in figures 10, 11 and 12.

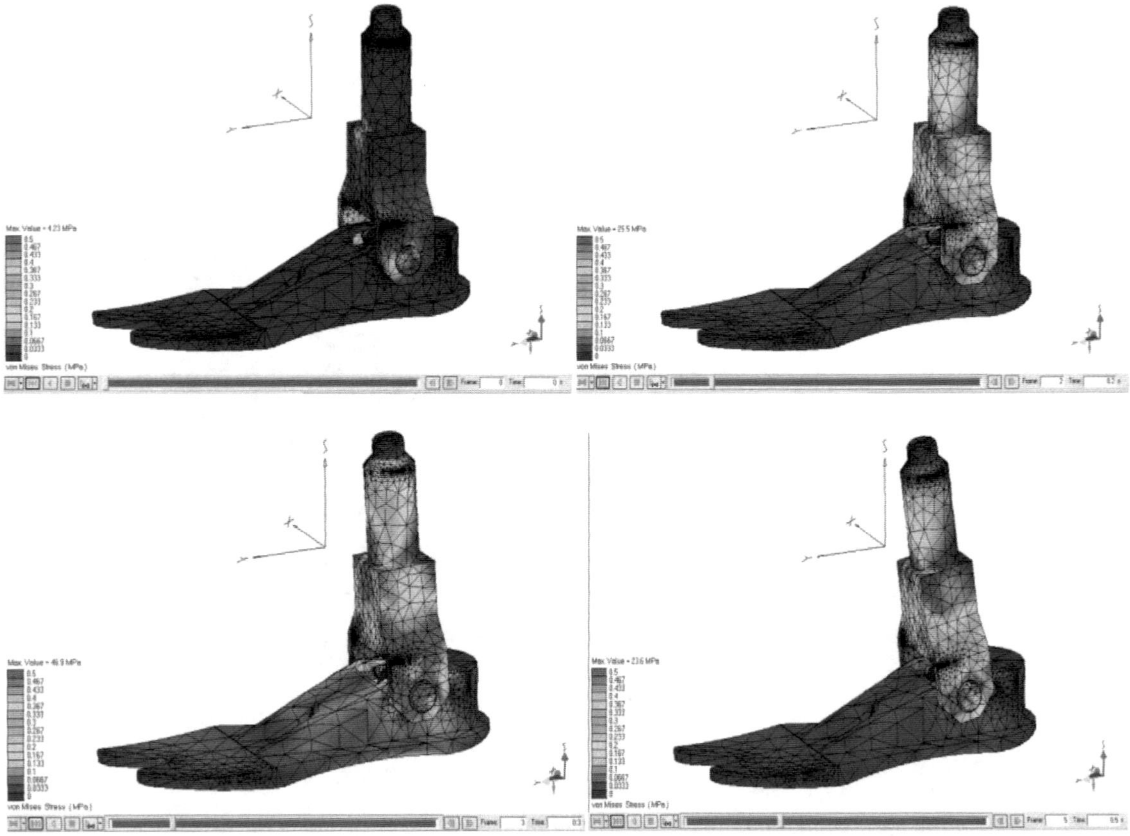

Figure 8. Ankle joint prosthesis von Misses stress vs. time interval 0 to 0.5 [sec]

IOP Conf. Series: Materials Science and Engineering 1190 (2021) 012034 doi:10.1088/1757-899X/1190/1/012034

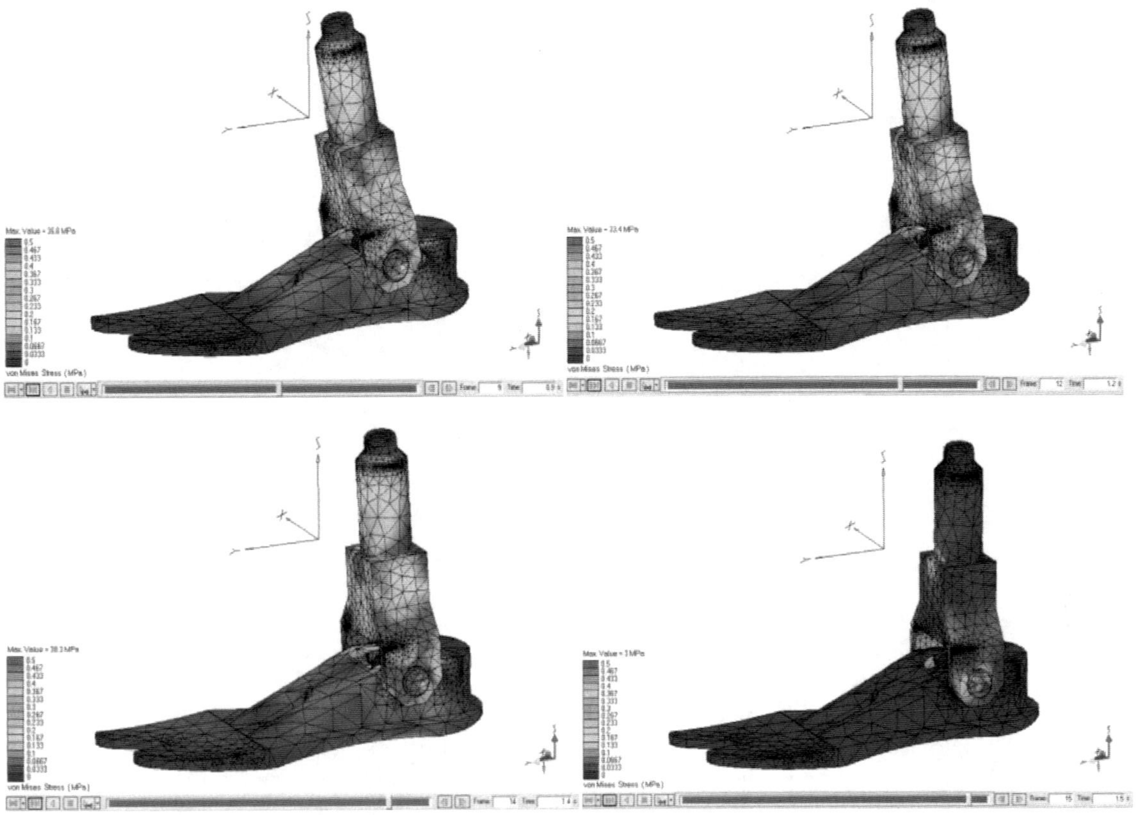

Figure 9. Ankle joint prosthesis von Misses stress vs. time interval 0.6 to 1.4 [sec]

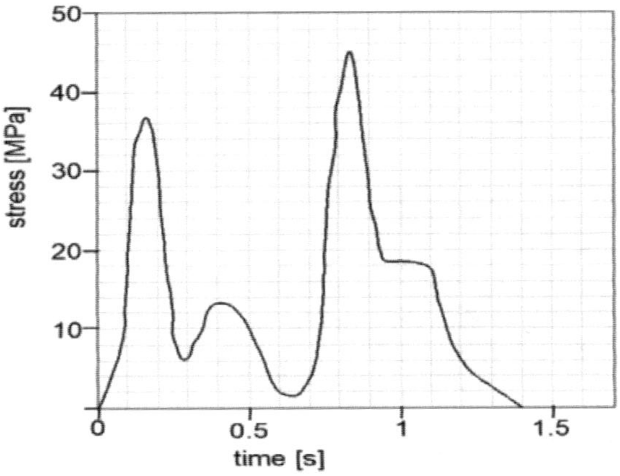

Figure 10. Von Misses stress variation vs. time [sec]

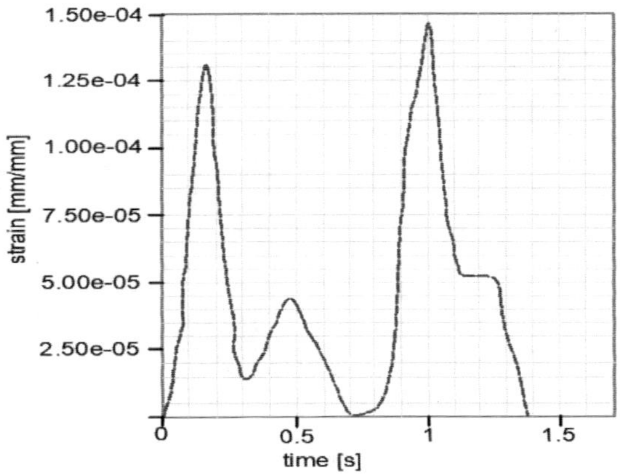

Figure 11. Von Misses strain variation vs. time [sec]

By having in sight, the obtained results reported in figures 10, 11 and 12, it can be remarked that the paths from von Misses stress and strain have the same shape. The maximum stress value was recorded with a value of 46.9Mpa, as it can be seen in figure 10 and this was situated at the contact between cam follower and cam profile. Other high values were recorded in the area of component no.5 and the shock absorber support.

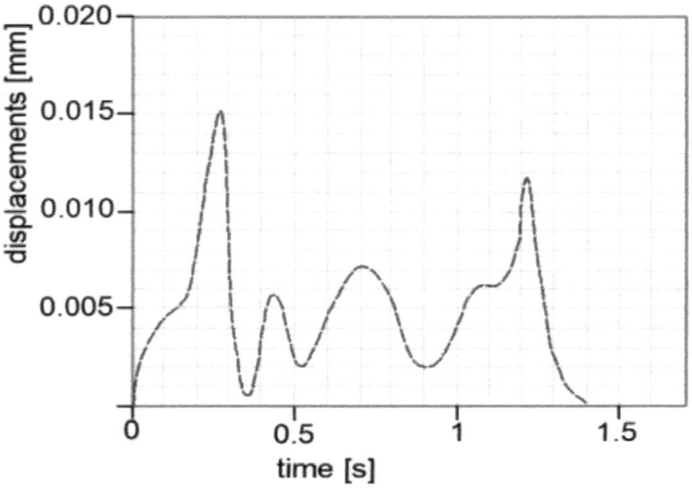

Figure 12. Total displacements variation during time [sec]

Similar values were recorded for total mechanical strain as a value of 0.01438%, which means that all the rigid components were have small deformations, as it can be remarked in figure 11.

In case of total displacements, there were recorded high values in the area of shock absorber segment and his support. The obtained value was 0.015123 millimeters according with the graph from figure 12. Thus, this dynamic behavior was related to friction forces between cam follower and cam profile, but also due to the bolt component no 5 during sliding motion on the foot component path.

The obtained results encourage to elaborate this prototype and to test it in real daily walking activities of a female amputee.

6. Ankle prosthesis prototype

According with the obtained results, which make this study feasible, the proposed ankle joint prosthesis has a good behaviour and a prototype was fabricated. This is presented in figures 13, 14 and 15.

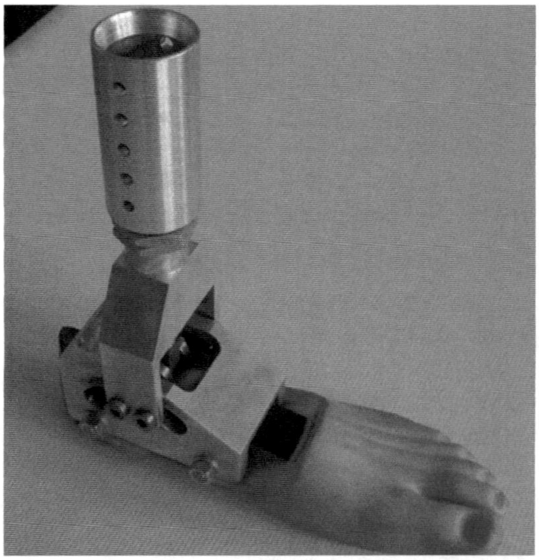

Figure 13. Ankle joint prosthesis prototype – lateral view

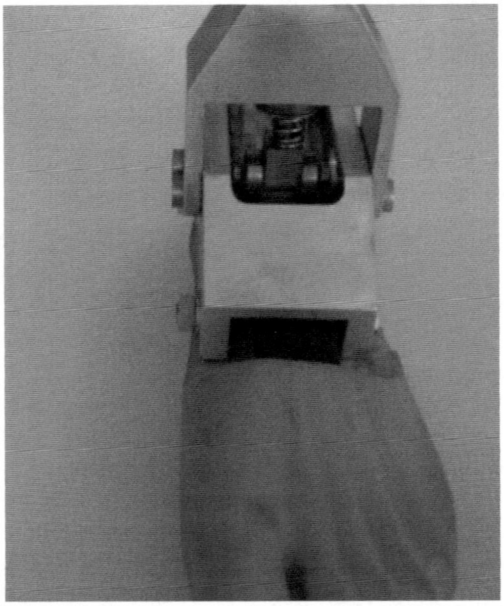

Figure 14. Ankle joint prosthesis prototype – front view

Figure 15. Ankle joint prosthesis prototype assembled with a complete knee and ankle prosthesis

Only a part of components was fabricated from zero point, other were purchased, namely the shock absorber from FESTO Pneumatics and the foot was Greissinger Plus type provide from Otto-bock. In order to reduce friction, the designed prototype was equipped with ball bearings at the level of cam follower component and also at the level of component no 5 according with figure 4.

7. Conclusions

A new prototype of a human ankle joint prosthesis for human locomotion rehabilitation is proposed and equipped with a cam mechanism.

Simulation results have outlined suitable performance for suitable user-oriented operation in walking during daily living activities, although its design may require additional components in future developments.

The proposed design is fairly simple wearable and lightweight with adjustable structure and its operation is fulfilled with a shock absorber integrated in tibia segment

References

[1] Copilusi C., "Research regarding applicable mechanical systems in medicine". *Phd thesis*. 2009.
[2] Zhu, J., Wang, Q., and Wang, L., "On the Design of a Powered Transtibial Prosthesis with Stiffness Adaptable Ankle and Toe Joints", *IEEE Trans. Ind., 61(9)*. pp. 4797–4807. 2014.
[3] Dickinson AS, Steer JW,Worsley PR., "Finite element analysis of the amputated lower limb: a systematic review and recommendations", *Med Eng Phys*. 43:1–18. 2017.
[4] Brånemark R, Berlin Ö, Hagberg K, Bergh P, Gunterberg B, Rydevik B., "A novel osseointegrated percutaneous prosthetic system for the treatment of patients with transfemoral amputation: a prospective study of 51 patients", *Bone Joint J*. 96:106–13. 2014.
[5] Shepherd M.K., Rouse E.J., "Design and validation of a Torque-Controllable knee exoskeleton for sit-to-stand assistance", *IEEE/ASME Trans Mechatron*. 22(4):1695–1704. 2017.
[6] Windrich M, Grimmer M, Christ O, Rinderknecht S, Beckerle P., "Active lower limb prosthetics: a systematic review of design issues and solutions", *Biomed Eng Online*. 15:5–19. 2016.
[7] Gao, F., Liu, Y. N., and Liao, W. H., "Design of Powered Ankle-Foot Prosthesis With Nonlinear Parallel Spring Mechanism", *Journal of Mechanical Design ASME*. Vol. 140. pp. 055001/1-055001/8. 2018.
[8] Radcliffe C.W., "Four-bar linkage prosthetic knee mechanisms: kinematics, alignment and prescription criteria", *Prosthet Orthot Int*. 18(3):159–173. 1994.
[9] CONTEMPLAS Equipment User Manual. Accessed in 2016.
[10] Williams M., "Biomechanics of human motion", *W.B. Saunders Co. Philadelphia and London*, 1996.
[11] Copilusi, C., Dumitru, N., Rusu, L., and Marin, M., "Cam Mechanism Kinematic Analysis Used in a Human Ankle Prosthesis Structure", *World Congress on Engineering (WCE), London*, June 30–July 2, pp. 1316–1320. 2010.
[12] Copilusi, C., Dumitru, N., Margine A., and Geonea I., "Dynamic Analysis of a Human Ankle Joint Prosthesis", *Annual Session of a Scientific Papers "IMT ORADEA 2019". IOP Conf. Series: Materials Science and Eningeering* 568. 2019.

IOP Conf. Series: Materials Science and Engineering 1190 (2021) 012035 doi:10.1088/1757-899X/1190/1/012035

On the Appropriate Preparation and Realization of Production Process Quality Attribute Measurement

Eliška Cézová [1]

[1] Czech Technical University in Prague, Faculty of Mechanical Engineering, Department of Designing and Machine Components, Czech Republic

eliska.cezova@fs.cvut.cz

Abstract. In this paper we are dealing with the methodology of design, preparation and implementation of quality features measurement of the production process. The aim is to develop a methodology for measuring parts in the structure. This article describes the procedure of the sample process management methodology.

1. Introduction

The manufacturing plant produces products that are expected to be of required quality and the producer should be able to determine and maintain the desired price of the product. In order to achieve the required or higher product quality, it is necessary to have a well-organized and technically mastered production process. Furthermore, it is necessary to eliminate unnecessary downtime in the production process, e.g. to avoid wasting time in moving the workpiece on the production line, etc. In some cases it might be appropriate to replace human labor with machines to increase the production efficiency, reduce costs and eliminate errors eventually introduced by human factor. However, the quality of the final product is usually determined in a fundamental way by the pre-production stages, especially by the design. At this stage it is essential to take into account the environmental conditions, resources and the capabilities of the contractor (manufacturer). Already when creating the technical documentations and technical drawings, a suitably designed measuring basis for quality verification must already be taken into account.

2. Description of pre-production stages

Before starting the production process, it must always be clear what will be the final product and who are customers that will be buying it. It should be a-priori known when (how often) the product will be innovated, to keep the customer satisfied with the product. The producer must be sure that the product will be still in demand. Therefore, it is necessary to set in advance the production objectives and the procedures leading achieve and verify them (process monitoring). First, let's take a look on the prerequisites for a good quality and efficient production process:

ISO documentation: The benefit of the established and effective ISO documentation according to the standard ČSN EN ISO 9001: 2016 (Quality Management System) lies in the formalization of the company's internal processes, which results in the use of approved and verified procedures.

Properly maintained metrological order: All measuring devices must be in accordance with legal metrology, i.e. a controlled process according to the Act on Metrology No. 505/1990, Coll. as amended. The aim of this disposal is to minimize influence of the measuring device, environment, used measurement methods, operator and used software equipment. The measurement process defined

IOP Conf. Series: Materials Science and Engineering 1190 (2021) 012035 doi:10.1088/1757-899X/1190/1/012035

in this case is in some cases subject of analysis of the measurement system (accuracy, repeatability, linearity, ...), which provides us with information on whether the measurement system meets the intended purpose, is objective, or whether our measurements will be accepted by the customer. The measurement process is also influenced by professionally qualified staff and the environment.

Customer requirements: The customer enters the design process both by defining required product properties (parameters, tolerances,...), but also requirements for verifying quality of the production process (indicators of competence, performance, management of control diagrams, etc.) .

Recording of measurement data: Before data collection begins, it should be specified which data will be monitored, recorded and how frequently. Moreover it is necessary to decide whether to check every piece or every *x*-th piece. [1]

Scheme of the procedure before the analysis of the production process is described in Figure 1. According to [4] it is possible to set other possible approaches of the methodology of complex design of the control diagram. Here we follow the division of data into continuous and discrete.

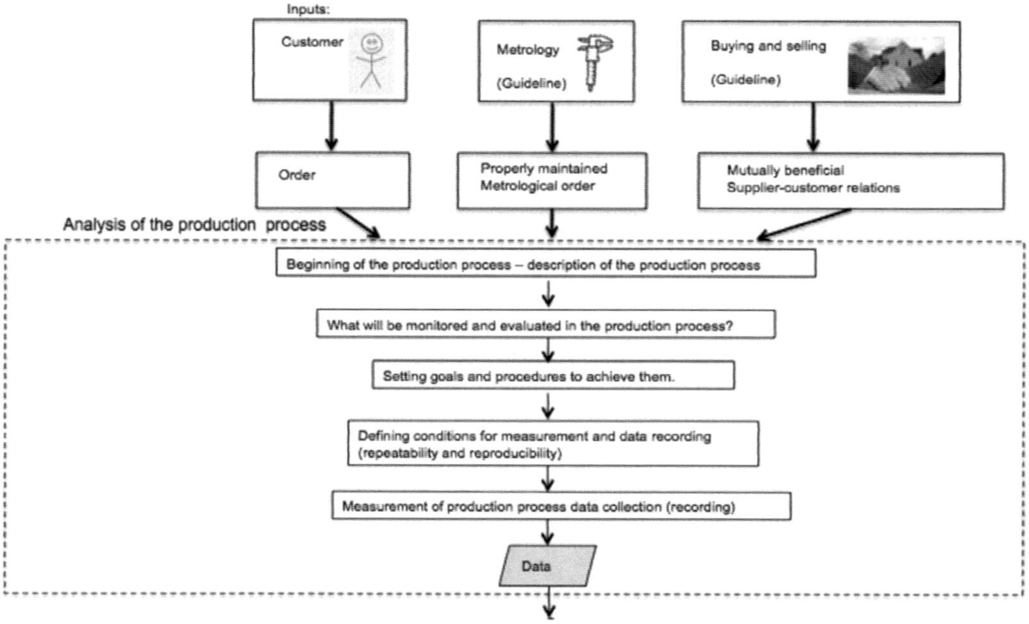

Figure 1. Scheme of the procedure before the analysis of the production process

3. Process analysis

There is an increasing need to use new and faster technologies to measure and control parts in the mechanical engineering. Calibrated gauges are used for higher production quality and more accurate production. Inspection jigs are used in mass production for the inspection of manufactured parts. Inspection jigs are single-purpose devices without evaluation electronics, which enable cheap, fast and simple inspection of manufactured components with minimal demands on the operator. The jig operator verifies for example that the inserted inspected part has shape and dimensions that conform to specifications. The inspection elements are fixed on the base plate of the jig, so that they correspond to the model of the inspected component within the specified tolerances. The devices should meet basic important parameters, such as the handling given device, quick and easy inspection of parts and assemblies, reliability and accuracy of the measurements. When designing the product, 3D construction programs are used, guaranteeing high productivity and clarity. The use of products reduces scrap and increases accuracy in the production process, further increases labor productivity and reduces the price of the product.

Serial production has an effect on the design of fixtures. In the production of a large number of manufactured pieces, a better preparation is required for a small number of pieces, a simpler and cheaper preparation. If workpieces of similar size and shape are produced at a given point in time, they can be combined into one jig. In terms of technological feasibility, it is necessary to study the production documentation, including production processes. These are, for example, the quality of the machined surfaces, the prescribed dimensions, the size of the machine tool, the number of operations, sequences and operations, etc. The design of the jig must not allow the reverse insertion of the object. Positioning and clamping of the workpiece and its release and removal from the jig must be possible in a short time. All sharp edges must be rounded due to possible injury to the operator. When using delimiting mandrels with the same line diameters, it is necessary to distinguish them (by color, mark). In practice, the POKA-YOKE method is used.

When analyzing the production process, we use basic statistical methods, which include, among others:

1) Calculation of the sampling characteristics. These characteristics (sampling mean, range, standard deviation, ...) are used for the analysis of measured values (data). It is important to realize that the resulting value of the selection characteristic is its point estimate and depends on the method of data collection, unlike the process parameters, which we estimate through the selection characteristics.

2) Histogram and box plot belong among the most important graphical tools for the basic process analysis. It provides us with the information about the shape of the probability distribution, process settings, variability, etc. .

3) Control chart is a graphical representation of a (typically sequential) test with the value of a controlled variable or its parameter. The scope for continuing process monitoring is the band defined by the LCL (lower control limit) and UCL (upper control limit), while the scope for stopping the process is the areas outside these limits. Using appropriately selected control diagrams, we can obtain information (signal) that a detectable cause of variability has started to act in the process, which needs to be identified, eliminated and such measures taken so that it cannot be repeated. We usually monitor the selected process parameters at specified time intervals, when we perform so-called inspections. Inspection consists of one or more measurements. From these measurements we calculate the sampling characteristics, most often the sampling mean and the sampling standard deviation. These statistical characteristics provide us with information about the changes of parameters of the studied process quality feature as shown in Figure 2.

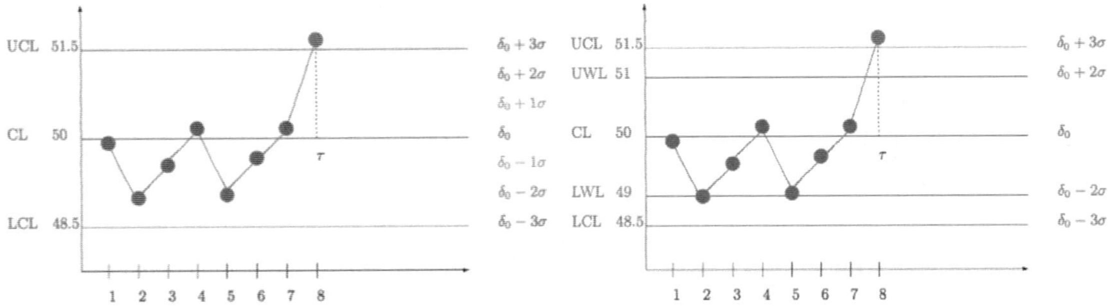

Figure 2. Shewhart control chart (left CL = 50 and σ = 0,5) and right diagram extended by warning limits.

The choice of the control diagram depends on various factors among them the most important are:
- sensitivity of the control diagram;

- process changes;
- occurrence of correlated data;
- number of quality characteristics monitored simultaneously;
- correlation between monitored quality characteristics.

The best known type of control diagram is the Shewhart control diagram, as shown in Figure 2. In practice, this type alone is not sufficient, so many other types of control diagrams have been invented. Control diagrams can be divided into one-dimensional and multidimensional, where each type is best suited for a different process. Control diagrams can be divided into groups according to various aspects:

- according to the nature of the monitored quantity to:

  • control measurement diagram for quantitative (measurable) data obtained by measuring and recording numerical values of the characteristics for each unit in the considered group during inspection

  • a control chart by comparison for qualitative data plotting the number of discrepancies between the observed units and the standard.

- according to the number of monitored variables to:

  • one-dimensional, always following only one characteristic. This includes, for example, the classical Shewhart plot for the mean or variance

  • multidimensional, monitoring several characteristics simultaneously in a single graph. When displaying them, the so-called Hotelling distances are often used.

- according to the number of inspections needed to calculate the last value:

  • classical, using data found only during the last inspection (always drawing only the last measured value of the monitored characteristic)

  • using *k* previous inspections (measurements) to calculate the actual evaluation value. This includes control charts of moving totals, either unweighted (MA diagram) or weighted (EWMA diagram)

  • using the entire measurement history in a given cycle to calculate the last characteristic. This includes so-called CUSUM diagrams or zone diagrams.

- according to the frequency of inspections

  • classic with a constant time between individual inspections,

  • adaptive, in which the frequency of inspections changes based on the last value.

Before analyzing the process itself, it is necessary to answer the following questions:
- which quantity / quantities need to be monitored?
- Which properties of the production process are influenced by this quantity and vice versa, how are they influenced by this process?
- What exactly does it mean from an operational point of view that the "process is under control"?
- What are the risks when the process is "out of control"?
- Can the process continue during the identification of the identifiable cause (signal verification)? What is the risk?
- Can the process continue during maintenance operations? What is the risk?
- What are the cost items (losses) and their quantification?

Only after answering these questions can we continue to design a regulatory diagram, which should be implemented in the following steps:

1) Analysis of the production process. In the first step, we will perform an analysis of the production process, the result of which will be the determination of a controlled variable, the behavior of which we will monitor. We should focus on the influences that affect it and how it affects this process. From an operational point of view, we should determine when the process is under control and when out of control. Whether it is necessary to stop the process in case of finding a detectable cause or not, and whether the production process continues during maintenance or not.
2) Stochastic analysis. In the second step, we analyze the process in terms of its stochastic properties, i.e. we determine the probability distribution of the controlled variable and its time to failure. We find out the probability distribution of the repair or maintenance time. Using the (auto) correlation function in time, we analyze the dependencies of controlled variables.
3) Selection of control diagram. The third step is the most important point of the whole design, where we select the most suitable type of control diagram, which signals the detectable cause in time and minimizes the occurrence of false signals. Individual control diagrams have different sensitivities to different types of deviations (changes). For example, the Shewhart control diagram is able to very effectively capture a large shift (sudden changes) in the monitored process level. On the other hand, CUSUM is much more suitable when the process change is gradual or there has been a small shift. They can be used in various conditions, which also depend on the technical equipment and the complexity of the calculation of the individual control diagrams. We choose the control diagram according to the nature of the controlled variable, if we monitor more characteristics at the same time, we choose according to the number of controlled variables. We take into account the dependence structure of the observed (stochastic) process. When choosing a control diagram, we emphasize not only its ease of use in operation, but we also take into account economic aspects.
4) Selection of the optimal maintenance strategy. In the fourth step, we select the optimal maintenance strategy. Preventive maintenance can significantly reduce production costs and increase its quality. To do this, it is necessary to determine the optimal intervals and ranges of scheduled maintenance. From the point of view of the economic-statistical design of the control diagram, we will continue to consider complete maintenance, which is performed either at predetermined regular intervals or always after signaling by the control diagram.
5) Economic-statistical optimization. The economic-statistical design takes into account both economic parameters and statistical properties of the control diagram. The statistical properties of the control diagram include the ARL value. This serves primarily to optimally set the parameters of the control diagram, by which we reduce the costs not only of statistical control, but also of the loss of quality of the controlled process.
Optimized parameters include the sampling range, the length of the interval between inspections in hours, and the width of the control limits, which are usually reported in multiples of the standard deviation of the observed steady state process in which the process is under statistical control.
From the point of view of the economic design of the control diagram, we assume that there are costs for measurement (inspection), which can be divided into variable and fixed costs. At an unknown moment ν, a detectable cause occurs that moves the process out of statistical control. If the process is in a state under statistical control, it means that a false signal has occurred, we only count on the cost of finding a false signal. If the signal is justified, we count on the cost of finding and removing the detectable cause.
6) Rules for implementation. In the last step of the control diagram design, we focus on the correct way of its application, where we set the rules of use and ensure their compliance in an appropriate manner. We will introduce rules for the use of control diagrams in the organizational rules of the given production company, by which we will determine the conditions of measurement and sampling, we will assign the powers and responsibilities of individual workers for the correct measurement and evaluation of samples. We will reserve space for the analysis of samples and ensure timely intervention in the detection of an authorized signal.
Average Run Length (ARL) is the average run length of a process, often just the average number of inspections that take place before signaling that a process has reached a statistically unmanageable

state. This quantity depends on the probabilistic behavior of the process. The ARL should be maximal if the process is in a statistically controlled state.

Figures 3 and Figure 5 schematically show the procedure to be followed in the case of data with a normal distribution. [2], [3], [4].

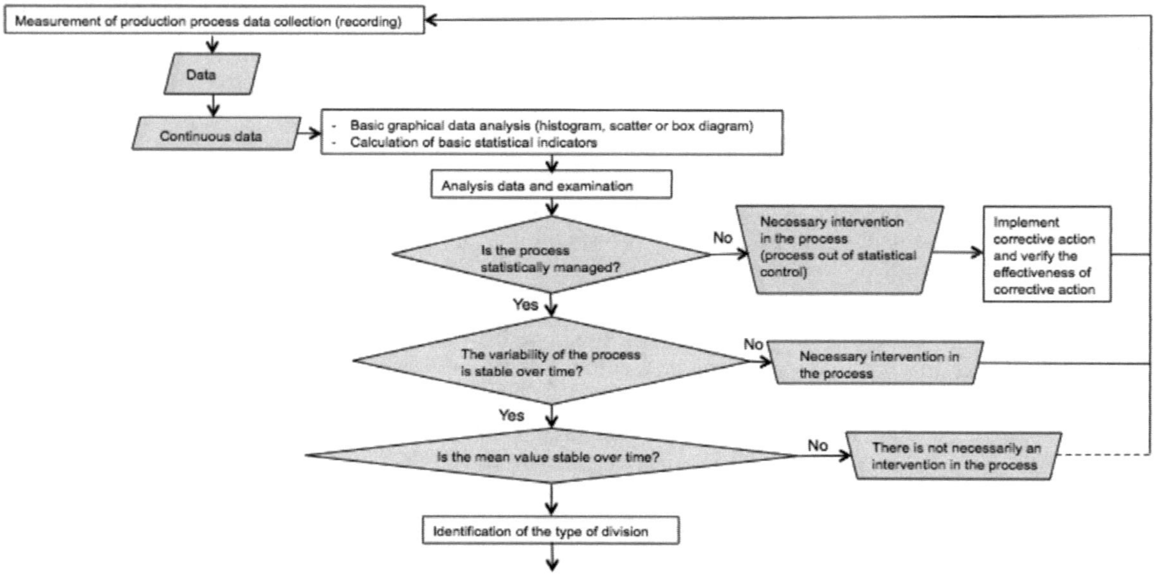

Figure 3. Scheme for continuous data the analysis of the production process

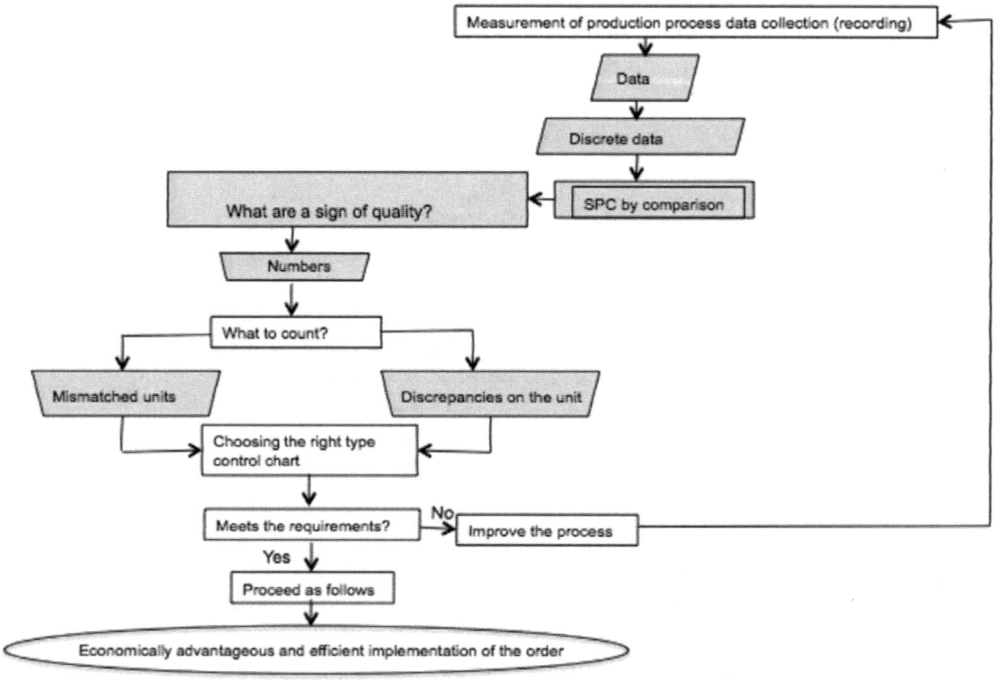

Figure 4. Scheme for discrete data the analysis of the production process

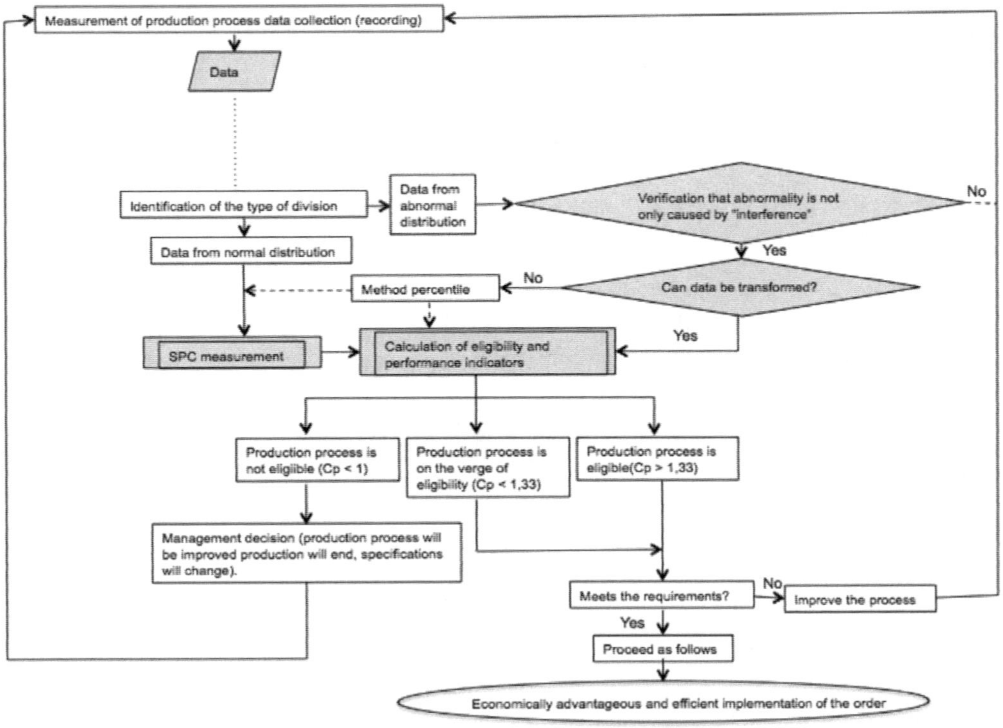

Figure 5. Scheme continuation for continuous data the analysis of the production process (with C_p)

4. Eligibility and performance indicators process

A number of statistical software tools is used to evaluate the competence and performance of the production process. Unfortunately, they are not always used completely correctly, it is always necessary to verify the assumptions so that their conclusions and interpretation are not misleading. In the case of a normal division of the quality mark, the formulas below are used. If there is no normal distribution of the quality characteristic, a different formula will be used to calculate the eligibility and performance indicators. It can be, for example, the Weibull distribution, the log-normal distribution, etc. If the values of the monitored quality trait are not normally distributed, the Johnson transform can be used so that the newly transformed data is then distributed normally N (0; 1).

Figure 5 shows at which stage of the production process the calculation of competence and performance indicators is performed.

- The C_p competency indicator does not takes into account the issue of process centering, but characterizes only what we are able to achieve.
- The C_{pk} competency indicator, on the other hand, takes into account degree of centering of the process achieved, and characterizes what we have actually achieved.
- The performance indicators P_p and P_{pk} are based on variability over a longer period of time, assessed by the total standard deviation for this period.

The parameters C_p and P_p do not consider the position parameter. Thus, the process may have a high C_p and P_p value while producing a high percentage of mismatched pieces.

If the process is centered, the proportion of non-conforming pieces decreases, which is achieved by the indicators C_{pk} and P_{pk}. The introduction of these indicators is based on the assumption of a normal distribution, where μ is the location parameter, i. e. the mean value of the normal distribution.

$$C_p = \frac{USL - LSL}{6\sigma}, P_p = \frac{USL - LSL}{6s_{TOT}}; C_p \geq P_p$$

$$C_{pk} = min\left(\frac{USL - \mu}{3\sigma}; \frac{\mu - LSL}{3\sigma}\right); P_{pk} = min\left(\frac{USL - \mu}{3s_{TOT}}; \frac{\mu - LSL}{3s_{TOT}}\right); C_{pk} \geq P_{pk}$$

If the value of $C_p < 1$, the process is not eligible, $C_p = 1$ the process is close to eligibility and $C_p \geq$ 1,33 the process is eligible. [5]

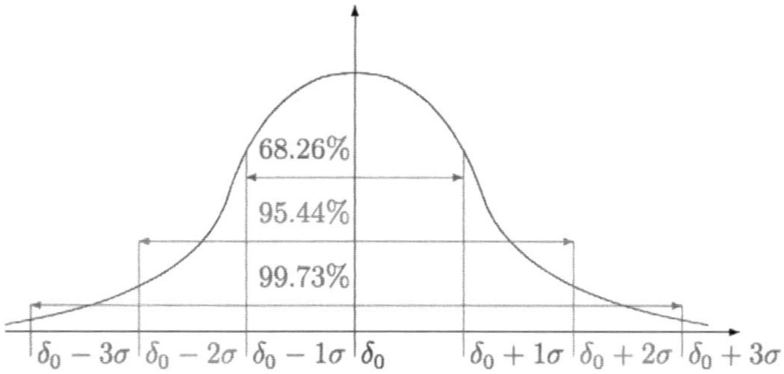

Figure 6. Probability density for Shewhart's control chart with normal distribution $X \sim N(\delta_0, \sigma^2)$

Assuming normality we have:
- In the interval < δ_0 - σ; δ_0 + σ> lies 68.26% of all observations, outside this interval lies 2 * 15.87%, i.e. 31.74%.
- In the interval < δ_0 - 2σ; δ_0 + 2σ> lies 95.44% of all observations, outside this interval lies 2 * 2.28%, i.e. 4.56%.
- In the interval < δ_0 - 3σ; δ_0 + 3σ> lies 99.73% of all observations, outside this interval lies 2 * 0.135%, i.e. 0.27% (2,700 ppm).
- In the interval < δ_0 - 4σ; δ_0 + 4σ> lies 99.994% of all observations, outside this interval lies 2 * 0.003%, i.e. 0.006% (60 ppm).
- In the interval < δ_0 - 5σ; δ_0 + 5σ> lies 99.99994% of all observations, outside this interval lies 2 * 0.00003%, i.e. 0.00006% (0.6 ppm).
- In the interval < δ_0 - 6σ; δ_0 + 6σ> lies 99.999999999% of all observations, outside this interval lies 2 * 0.000000001%, i.e. 0.000000002% (0.002 ppm).

Mismatch rate - ppm (parts-per-million) expresses the number of mismatched pieces in a million series. The number 10-6 ppm then expresses the probability of occurrence of a mismatched piece. [5]

5. Conclusions

Each measurement in design case specific. Whether it is a classic measurement of lengths or the measurement of geometric tolerances, e.g. throwing, etc., it is always necessary to verify the assumptions for the use of statistical methods to avoid erroneous conclusions. As mentioned earlier, it is necessary to verify the assumptions of normality for the use, whether of control diagrams or for the use of competence and performance indicators. In many cases, in practice there is no theoretically pure solution, often because of inability to meet certain assumptions. The practitioner, however, must always find a solution, must decide in a specific case and give the best possible answer to the question asked. He has to intuitively choose the optimal, practically possible solution from all that are offered to him.

IOP Conf. Series: Materials Science and Engineering 1190 (2021) 012035 doi:10.1088/1757-899X/1190/1/012035

Acknowledgment(s)
This work was partly funded by the Czech Ministry of Education under the Institutional support for the development of research organization No.:RVO12000 for Faculty of Mechanical Engineering of the Czech Technical University in Prague.

References

[1] Cézová E., "Metrology in application", Request 06, CQR, ISBN 978-80-01-03709-6, pp. 43 - 52,
[2] Cézová E., "Economic-statistic design control chart", Request 08, CQR VUT Brno, ISBN 978-80-214-3774-6, 2008
[3] Cézová E.; Dohnal G., "Economical and statistical optimization of control chart", Robust 2012, ISSN 1210-8022, pp. 141 – 151, 2012
[4] Král J., "Methodology for a complex design of control chart", Robust 2012, ISSN 1210-8022, pp. 182 – 193, 2012
[5]https://www.csq.cz/fileadmin/user_upload/Spolkova_cinnost/Odborne_skupiny/Statisticke_metody/sborniky/2005/06_-_4_-_Zpusobilost.pdf

AUTHOR INDEX